Bronisław Malinowski

Coral Gardens and their Magic

A Study of the Methods of Tilling the Soil and of Agricultural Rites in the Trobriand Islands

With 3 Maps, 116 Illustrations and 24 Figures

VOLUMEN ONE
The Description of Gardening

SEVERUS

Malinowki, Bronoslaw: Coral Gardens and their Magic. A Study of the Methods of Tilling the Soil and of Agricultural Rites in the Trobriand Islands. With 3 Maps, 116 Illustrations and 24 Figures.
Volume One: The Description of Gardening

Hamburg, SEVERUS Verlag 2013
Nachdruck der Originalausgabe von 1935

ISBN: 978-3-86347-646-5
Druck: SEVERUS Verlag, Hamburg, 2013
Umschlagbild: Coral garden. © Jenny Huang

Der SEVERUS Verlag ist ein Imprint der Diplomica Verlag GmbH.

Bibliografische Information der Deutschen Nationalbibliothek:
Die Deutsche Nationalbibliothek verzeichnet diese Publikation in der Deutschen Nationalbibliografie; detaillierte bibliografische Daten sind im Internet über http://dnb.d-nb.de abrufbar.

CORAL GARDENS
AND THEIR MAGIC

PLATE 1

DISPLAY OF FOOD IN A VILLAGE

"Distributions of food are one of the most usual and characteristic ceremonial acts in the Trobriands. They attain their most elaborate and quantitatively biggest form at mortuary commemorative feasts" (Part I, Sec. 7)

CORAL GARDENS
AND
THEIR MAGIC

A Study of the Methods of
Tilling the Soil and of Agricultural
Rites in the Trobriand Islands

by

BRONISŁAW
MALINOWSKI

With 3 Maps, 116 Illustrations
and 24 Figures

VOLUME ONE

THE DESCRIPTION
OF GARDENING

TO MY WIFE

PREFACE

ONCE again I have to make my appearance as a chronicler and spokesman of the Trobrianders, the Melanesian community so small and lowly as to appear almost negligible—a few thousand "savages", practically naked, scattered over a small flat archipelago of dead coral—and yet for many reasons so important to the student of primitive humanity. The Trobrianders are typical representatives of the South Sea natives in general. They have been remarkably tenacious of their own culture and of their old traditions, and we can study in them the ways and manners of Oceania as it flourished for ages, unknown and untouched by Europeans.

In this book we are going to meet the essential Trobriander. Whatever he might appear to others, to himself he is first and foremost a gardener. His passion for his soil is that of a real peasant. He experiences a mysterious joy in delving into the earth, in turning it up, planting the seed, watching the plant grow, mature, and yield the desired harvest. If you want to know him, you must meet him in his yam gardens, among his palm groves or on his taro fields. You must see him digging his black or brown soil among the white outcrops of dead coral and building the fence, which surrounds his garden with a "magical wall" of prismatic structures and triangular supports. You must follow him when, in the cool of the day, he watches the seed rise and develop within the precincts of the "magical wall", which at first gleams like gold among the green of the new growth and then shows bronzed or grey under the rich garlands of yam foliage.

The side of tribal life described in these volumes is perhaps less sensational than the sailing and trading and witchcraft known to the readers of the *Argonauts of the Western Pacific*; it may have less direct appeal to our curiosity than the customs of courtship and marriage treated in *The Sexual Life of Savages*. But it is at least as important for our knowledge of the Trobrianders, of Oceanic civilisations and, I venture to say, even of human nature in general.

For, on the one hand, nothing bears so directly on man's economic nature as the study of primitive forms of tilling the soil. The manner in which so-called savages produce their primary sustenance, store it and handle it, the way in which they surround it with magical and religious beliefs, open problems of the relation between man and environment of some importance to economic philosophy. On the other hand, agriculture and its consequences enter very deeply into the social organisation of our South Sea community—and of any community for that matter; they form the foundation of political power and of domestic arrangements; they are the mainstay of the obligations of kinship and of the law of marriage. Thus in many ways the perusal of the present book may add to our knowledge of primitive economic organisation, political order, and domestic life.

One aspect of Trobriand gardening is very prominent and may raise problems of wider implication: I mean the relation between purely economic, rationally founded and technically effective work on the one hand, and magic on the other. The organising function of magic and of belief is, from the theoretical point of view, perhaps the most important contribution which our knowledge of Trobriand agriculture will allow us to make to the study of man. It comes out more clearly there than in the relation of magic to work found in the *kula* or in the conduct of native courtship, although these cast valuable light on the rôle of magic in human affairs.

No human beings, at whatever stage of culture, completely eliminate spiritual preoccupations from their economic concerns. Whether we pray for our daily bread or for adequate rainfall; whether the Divine Kings of Africa exercise their powers for fertility and moisture; whether the Trobrianders and other Melanesians or Polynesians offer prayers or perform magic for success in fishing, in trade or in sailing—the relation between supernatural means of controlling the course of events and the rational technique is one of the most important subjects for the Sociologist. But as regards the Trobrianders, there is no other aspect of tribal activity as fully and as naturally controlled by magic as the tilling of the soil. And incidentally, in no other aspect of my studies was I so fortunate in collecting information, in translating magical spells and commenting on them, as in the material presented in this book.

And this I owe mainly to some of my native informants, in whom I was exceptionally fortunate. The Anthropologist takes full credit for some of his discoveries, but for the real toil, as well as for the degree of intelligence in the approach, he can only take part credit. To a large extent my informants are responsible for the correct interpretation and perspective, for the sincerity and relevance of what is contained in these volumes, as also in my other ethnographic writings. In discussing garden magic I was specially favoured by having in Bagido'u, one of the leading garden magicians of the area, an excellent collaborator. He was a naturally gifted man, second to none in the knowledge of tribal lore, and since he was ill and not very active physically, he had plenty of time to place at my disposal. He died soon after I left the district. When I look back at my work with Bagido'u, I often feel that he must have had some dim recognition that he was the last repository of a vanishing world. He was so keen to give me the right understanding of his magic; he spent so much time and real care on seeing that I had the correct wording of the spells and had grasped their meaning—that in a way I think he knew that he was leaving them to posterity. It is difficult to reconstruct the attitude of the natives in such matters, to do full justice to their insight and yet not to fall into exaggeration or false sentiment. The fact remains, however, that I personally, and those who are interested in the ethnography of the Trobrianders, owe a good deal to him as also to some of those other informants who will be met with in the course of the following chapters.

In one respect the present book differs fundamentally from all I have published previously: that is, in the full treatment given to the language of agriculture. For the first time I am able here fully to document my ethnographic contribution from the linguistic point of view. This is not due to the absence from my field notes of the same, or of a reasonably comparable quantity of texts, commentaries, sayings and terminologies to validate the statements which I have made in *Argonauts of the Western Pacific* or in *The Sexual Life of Savages*; in my booklet on *Myth* or in *Crime and Custom*. The reason is, that earlier in my career there would have been no chance of publishing as full a linguistic documentation as has become possible now, when the interest in the Trobrianders and in more detailed ethnographic accounts has on the whole increased. I trust that the theoretical parts of this book: the Introductions to the Linguistic Supplement and to the Magical Formulae (Parts IV and VI), will add to this interest and to the understanding that such full documentation is necessary; and that they will justify the methods here adopted.

In connection with this I would like to add that, when I first presented the manuscript of my earliest ethnographic account—*Argonauts of the Western Pacific*—to the publishers, it was rejected by some half-dozen of the biggest firms. It required the insight of the late Mr. Swan Stallybrass—whose bibliographic work as well as his publishing activities have assured him a place in the history of British books—to recognise the importance of an adequate documentation; he accepted my first book three days after it had been offered to him, without any cuts or restrictions The acceptance of the present book, with its full linguistic documentation, is also a new departure, and I wish to express my indebtedness to Mr. Stanley Unwin for his vision and initiative in this publishing venture.

As in most ethnographic books, so here, too, I have a long list of acknowledgments to make. How much I owe those who helped me in my field-work, I stated fully in the acknowledgments to my first book: Mr. Robert Mond, who financed my expedition; Professor Seligman who directed me to the Papuan field and helped me throughout in more ways than one; the Governments of Papua and of the Commonwealth of Australia; but above all my late friend Billy Hancock, and my friend Raphael Brudo who, as I am writing these lines, is on his way back to the Trobriands.

In the working out of my material I received considerable help from research assistants provided by the Rockefeller Foundation; notably Dr. J. Obrebski and Miss Agnes Drew, who assisted me in eliciting the material from my field notes; Mrs. R. C. Mathers, Miss Girsavicius and Miss Margaret Read who helped me to throw this material into shape. The technological parts of the book were scanned by my friend Dr. Peter Buck of Honolulu, while some linguistic chapters were read by Mr. Edwin Smith, and I am indebted to both of them for valuable suggestions. Also the manuscript in its semi-final form was read at one of my seminars, a chapter by each member, criticised and discussed, and from this I, and I hope the book too, have derived considerable profit. Some of my pupils have worked right through the book with me. I wish here to mention more specifically Drs. Wagner, Nadel and M. Fortes, Miss Hilda Beemer and Mr. Godfrey Wilson, Dr. Sjoerd Hofstra and Dr. Keesing, and Miss Marjory Lawrence. Since both for my research assistants and the majority of my research pupils I am indebted to the possibilities opened by the Rockefeller Foundation, I should like to say how much my work has been fostered by this Institution, personified to me by Director Edmund E. Day and Vice-President Gunn, by Mr. Kittredge and Dr. Van Sickle. This is but a feeble acknowledgment of what I owe to them personally and scientifically, and to the Institution for which they are working.

I have dedicated this work to my wife because I believe it is the best I have produced or am ever likely to produce. Her suggestions and critical advice in this, as in other aspects of my research, have been the most valuable and effective inspiration to me.

B. M.

April 1934

Department of Anthropology
University of London
London School of Economics

CONTENTS OF VOLUME ONE

PART ONE

INTRODUCTION: TRIBAL ECONOMICS AND SOCIAL ORGANISATION OF THE TROBRIANDERS

Page 3

PART TWO

GARDENS AND THEIR MAGIC ON A CORAL ATOLL

CHAPTER		PAGE
I.	General Account of Gardening	1
II.	The Gardens of Omarakana: Early Work and Inaugurative Magic	84
III.	The Gardens of Omarakana: Preparing the Soil and Planting the Seed	110
IV.	The Gardens of Omarakana: The Magic of Growth	137
V.	Harvest	159
VI.	The Customary Law of Harvest Gifts	188
VII.	The Work and Magic of Prosperity	218
VIII.	Structure and Construction of the Bwayma	240
IX.	A Comparative Glance at Trobriand Gardening	273
X.	The Cultivation of Taro, Palms and Bananas	290
XI.	The Method of Field-Work and the Invisible Facts of Native Law and Economics	317
XII.	Land Tenure	341

PART THREE

DOCUMENTS AND APPENDICES

Page 385

LIST OF PLATES

PLATE	FACING PAGE
1. Display of Food in a Village	*frontispiece*
2. The Chief's Polygamous Family	8
3. Sea-going Canoe under Sail	9
4. Fishing Canoe on the Lagoon	8
5. Catching Fish in the Air	16
6. Scene in a Lagoon Village	16
7. The Beach of Teyava	17
8. Fishing Canoes in the Creek of Tukwa'ukwa	24
9. Men practising a Dance	24
10. Village Street during the Season of *Milamala*	25
11. A Small Mortuary Distribution of Tubers	32
12. Smoking of Fish for a Festive Eating	33
13. Specialist at Work	33
14. A Ceremonial Food Offering in the Exchange of Tubers for Fish	40
15. Arrival of a Fishing Fleet	41
16. The Display of Agricultural Produce during the *Milamala*	41
17. The Culminating Achievement of Gardening	56
18. On the Road Crossing the Jungle	57
19. On the Road through a Harvested Garden	57
20. View over a New Garden	64
21. A Yam Garden in full Development	65
22. A Taro Garden	65
23. The Magical Corner	80
24. A Family Group in the Garden	80
25. The Men of the Gardening Team Planting Taytu	81
26. Aesthetics in Gardening	88
27. The Paramount Chief supervising his Harvest Tribute	89
28. Bagido'u in Front of his Storehouse	88
29. Bagido'u at the First Burning	96
30. Site of an Early Garden	96
31. The Centre of Ceremonial Activities	97
32. An Exuberant Patch of Vegetation	97
33. View of a Garden after the Ritual Burning	112

PLATE FACING PAGE

34. The Burning of the Garden 113
35. The Second Burning 113
36. A Family Group Clearing their Garden 120
37. A Stony Garden in the South 121
38. A Plot subdivided into Squares 121
39. Stile in a Garden Fence 128
40. *Kamkokola* 128
41. Big and Miniature *Kamkokola* 128
42. A *Karivisi* 129
43. Communal Planting 144
44. Man in the Act of Tilling 144
45. Training of Vines 145
46. The Taytu Vine 145
47. Corner of fully grown Garden 152
48. The Show Portion of the Harvest 152
49. Display of First Fruits: Taro Harvested after the Inaugural Rite 153
50. Display of First Fruits: Yams Harvested after the Inaugural Rite 153
51. Awaiting the Magician for Ceremonial Harvesting of Taytu . 160
52. Inaugural Harvest Ceremony for Taytu in Teyava 160
53. Shaving the Taytu Tubers 161
54. Very Small Harvest Arbour 161
55. Medium Sized Harvest Arbour 168
56. A Large Harvest Arbour 168
57. Details of Large Arbour 169
58. Harvesting Party on the Way to the Garden 169
59. Harvest Party on the Way to the Village 176
60. Large Party of Carriers entering the Village 176
61. Harvest Party Resting 177
62. Construction of Small Heap in Front of the new Owner's Storehouse 177
63. Construction of Large Heap 184
64. Very Large Heap in Construction 185
65. Large Decorated Heap 185
66. *Liku* in Process of Construction 192
67. Close-up of *Liku* seen on Plate 66 192

PLATE FACING PAGE

68. The *Buritila'ulo* Challenge 193
69. Lateral View of the Crate Shown on Plate 68 194
70. Competing Villages Displaying their Largest Yams 193
71. Filling the *Liku* in Wakayse 195
72. The Main Storehouse of the Trobriands 195
73. Storehouses in Omarakana during a Temporary Inundation 224
74. Empty Stores Ready for Filling 224
75. The Ceremonial Filling of *Bwayma* in Yalumugwa 225
76. The Filling of the Stores in Progress 228
77. Details of Filling a Storehouse 229
78. Where the Road from the Sea Strikes the Outskirts of Omarakana 229
79. The Main *Bwayma* of Kasana'i Filled 230
80. A Ramshackle Storehouse 230
81. Storehouses which have been almost Emptied 231
82. Filled Storehouses 232
83. The Chief's Yam-House and Shelter in Omarakana 233
84. Large Covered Platform in Bwoytalu 233
85. Platform in Front of Show *Bwayma* 240
86. People Resting on *Bwayma* Platforms 240
87. The Main Storehouse of the Trobriands, Empty 241
88. The Chief's *Bwayma* just before Rebuilding 248
89. The Framework of a Roof 248
90. Construction of a Small *Bwayma* 249
91. Three Small *Bwayma*, one in Process of Construction . . . 249
92. Technique of Thatching a Roof 256
93. Yam-House built on Wooden Piles 256
94. Typical Village Street 257
95. Two *Bwayma* with Spacious Front Platforms 257
96. Houses and *Bwayma* in Oburaku 258
97. A Village Street showing two types of Storehouse 258
98. Sitting Platform and Storehouse in Bwoytalu 259
99. Toy *Bwayma* with Owner Leaning against it 259
100. Bwaydeda the Garden Magician of Obowada 280
101. Nasibowa'i with the Ceremonial Axe on his Shoulder . . . 280
102. Chanting the Spell over the *Kamkokola* 281
103. People Gathering at the Stile for the *Kamkokola* Ceremony . 281

PLATE FACING PAGE

104. The Materials for Magical Rite and Religious Ceremony . 285
105. The Food for the Spirits Displayed 284
106. The Recital of the Spell 285
107. Men with Magical Leaves 288
108. Rite on the *Kamkokola* 288
109. Navavile, the Garden Magician of Oburaku 289
110. Two Garden Magicians of Sinaketa on their Rounds . . 296
111. Motago'i at a Magical Corner of a Taro Garden 297
112. Glimpse of a New Taro Garden 297
113. The Planting of Taro 304
114. Taro Garden on Stony Soil 304
115. A Magical Rite in the Taro Garden 305
116. A Banana Patch on the *Momola* 305

LIST OF MAPS AND FIGURES

FIGURE PAGE

1. Map of the Trobriand Islands 2
2. Plan of Omarakana 25
3. Chart of Time-reckoning 50
4. Diagram of a Field (*Kwabila*) 89
5. The Two Main Standard Plots (*Leywota*) 100
6. Growth of Taytu: the Sprouting of the Old Tuber 140
7. Growth of Taytu: the New Vine 141
8. Diagram of Garden Arbours 174
9. *Kaytubutabu* Pole 303
10. Taboo Marks on Palm 305
11. Plan of the *Baku* of the Village of Yalumugwa 384
12. Harvest Display in Omarakana . . . . . . *facing page* 392
13. Map of Omarakana Garden Lands . . . . . *facing page* 430
14. Chart of Magic and Work 436
15. Taro Plant . . . . (Volume II) 105

DIAGRAMS

I–XII. The Structure of the Storehouse 263–272

ANALYTICAL TABLE
OF CONTENTS

Part One

Introduction: Tribal Economics and Social Organisation of the Trobrianders

Sec. 1. THE SETTING AND SCENERY OF NEW GUINEA GARDENS (*Pp.* 3–6). First impressions of New Guinea scenery.—Few traces of man and his works on the mainland.—Different picture in the Trobriands.—Approaching the Trobriands.—Ethnographic data about the inhabitants.—Sailing from the Amphletts.

Sec. 2. THE HABITAT AND PURSUITS OF THE TROBRIANDERS (*Pp.* 6–7). The cultural affinities of these natives.—The topography of their archipelago.—An environmentalist makes successful guesses.—The districts of the main island.

Sec. 3. FIRST IMPRESSIONS OF TROBRIAND GARDENS (*Pp.* 7–12). Crops, agriculture, trade and fishing.—The paramount importance of gardening.—Fishing, hunting and collecting in economic perspective.—First visit to the island at harvest-time.—Description of village and variety of garden scenery.—The chief at work with his wives.—Communal and individual labour.—Simplicity and unobtrusiveness of magical rites.—The importance of gardens and of the garden magician.

Sec. 4. ECONOMIC DISTRICTS: THE TILLERS AND CRAFTSMEN (*Pp.* 12–16). Specialisation in fishing and agriculture.—The fishing villages of the northern coast.—Difficulty of giving numerical data on fishing.—The "land-lubbers" of Tilataula. Expert stone-polishers at work.—Industries further west.—The wood-carvers of Bwoytalu and Ba'u.

Sec. 5. ECONOMIC DISTRICTS: THE FISHERMEN (*Pp.* 16–20). The basket-makers of Luya.—Kavataria and its fishermen.—Fishing in the coral outcrops.—An ancestral myth of fishing magic.—Monopoly in this method of fishing.—Seine fishing in Teyava and Osaysuya.—Native astronomers.—Fishing for spondylus shell.—Decorative armshells.—Preference for agriculture and fishing, in spite of inducements offered by pearling.

Sec. 6. WHAT INDUSTRIAL SPECIALISATION LOOKS LIKE IN MELANESIA (*Pp.* 20–23). Industrial specialisation not monopoly.—

Agriculture everywhere predominant.—Natives at work in slack agricultural seasons.—Exclusive access to certain materials.—How an industrial community looks.—Expectations of finding large-scale industry disappointed.—A few old women make pots.—Continuity of the tradition in handicraft.—Estimate of output.

Sec. 7. THE VILLAGE AND WHAT HAPPENS IN IT (*Pp.* 23–30). Approaching a village.—Village clusters.—The rings of dwellings and stores.—A trip to Omarakana, through village groves and untouched jungle.—The central place of the capital.—Seasonal calendar in the village reflecting the dominance of gardening.—Ceremonial food displays.—Dancing seasons and competitive enterprises.—Mortuary feasts and redistribution of food.—Fish and pork as welcome relish.—Eating in small groups.—The sociological principles of food distributions.—The commissariat of ceremonies, expeditions, and industrial production.—Harvesting the new crops.—The round of the seasons.

Sec. 8. WHAT HAPPENS IN THE HOUSE & TO THE HOUSEHOLD (*Pp.* 30–32). The family and its home.—The interior of the house and its background.—The domestic larder.—Starting for the day's work. —The evening meal.—Different types of food.—Staple food, relish and dainty.—Taboo on cooking near storehouses.—The importance of gardening.—The sociology of food exchange.

Sec. 9. THE CONSTITUTION OF TROBRIAND SOCIETY (*Pp.* 33–40). The family and the clan.—The position of the chief.—Physical elevation and dignity of rank.—The position of women.—The four ancestral clans.—Two units: household and village community.—Patrilocal marriage.—Childhood and adolescence within the family.—Severance from home at maturity.—The unity of the family still maintained.—Filial duties.—The laws of matrilineal filiation.—The two principles of social continuity reflected in the two units: family and real kindred, and in the two principles of village life: male and female.—The sub-clan as a unit in law, economics and rank.—The chief of Omarakana and his military rival of Kabwaku.—Chiefs and headmen.—Several sub-clans side by side in one village.—The gradations of rank: minor chiefs, sub-chiefs, notables and the rank and file.

Sec. 10. THE MOVEMENT OF WEALTH IN THE TROBRIANDS AND THE RÔLE OF AGRICULTURE THEREIN (*Pp.* 40–48). The chief orders a new basket.—Payments by 'solicitary gift', 'maintenance gift' and 'clinching gift'.—How this scheme of barter works.—The ceremonial exchange of fish for food.—Fishing for mortuary distributions.—Partners in exchange.—Ordinary barter.—Other transactions: an enterprise inaugurated by gifts of food.—Direct barter of manufactured articles.—Absence of a common measure of value.—Preponderance of payments in food.—The value of trade tobacco and its limitations.—Other European goods.—Financing a specialist.—Measures of equivalence.—No native money or currency.—The gardener's point of view.—Both industrialist and agriculturalist mainly aiming at the acquisition of food.—The economic position of the headman or chief.—His dues from palms and pigs.—How the chief uses his wealth.—The constructive side of economic give and take: industrial and political organisation.—Food in religious ceremonial.—The all-pervading influence of magic.

Part Two

Gardens and their Magic on a Coral Atoll

CHAPTER I

GENERAL ACCOUNT OF GARDENING

Gardening and Fishing, the two main food-providing activities.—Their mutual dependence (*P.* 52).

Sec. 1. THE SEASONAL RHYTHM OF GARDENING (*Pp.* 52–55). Agricultural activities as rhythm and measure of seasonal sequence.—Other pursuits subordinated to gardening.—Accessory character of the lunar calendar.—Correspondences in time-reckoning.

Sec. 2. THE SEVERAL ASPECTS OF TROBRIAND AGRICULTURE (*Pp.* 55–57). Magic, law and social structure in gardening.—Agriculture, the core of tribal economics and the foundation of social order, notably of chieftainship.—Aesthetics of gardening.

Sec. 3. A WALK THROUGH THE GARDENS (*Pp.* 57–60). The character of the soil and the landscape.—The area under cultivation.—Varieties of gardens.—The standard plots.—The magical corner.—The garden as the setting of economic and sociable life.—Output in gardening.—The chief's rôle in agriculture.

Sec. 4. THE PRACTICAL TASKS OF THE 'GOOD GARDENER' (*TOKWAYBAGULA*) (*Pp.* 61–62). The four main stages of gardening.—Types of labour and its character.—Technique and equipment.

Sec. 5. THE MAGIC OF GARDEN (*Pp.* 62–64). Magic, public and official, closely interwoven with work.—No confusion between magic and work.—Outline of gardening cycle showing inter-relation and distinctness of magic and work.—The rôle of standard plots in gardening.

Sec. 6. THE GARDEN WIZARD (*Pp.* 64–68). The hereditary nature, mythological foundations, and spiritual filiations of the magician's office.—Combined gifts to the ancestral spirits and to the magician.—The acolytes of the magical service.—The garden wizard's taboos.—The magician as leader, expert and supervisor of work.—The organising function of magic.

Sec. 7. THE GLORY OF THE GARDENS AND ITS MYTHOLOGICAL BACKGROUND (*Pp.* 68–75). Pride and fame of Kiriwina.—The legends of Tudava and Gere'u.—Text of one version of the Tudava myth.—Correlation of legend with conditions of fertility and garden craft.—Other versions of the Tudava story.—Tales about garden origins from the surrounding islands.—The supremacy of Kiriwina, the agricultural production of other districts and the corresponding local mythology.

Sec. 8. THE POWER OF MAGIC AND THE EFFICIENCY OF WORK (*Pp.* 75–80). Two sources of fertility: natural and magical.—Native knowledge of soil, crops, technical processes; statements and terminologies.—The distinction between magic and work in the mind of the native.—Private magic.—Black magic, wielded by individuals or, officially, by the chief.—The types and conditions of labour in agriculture.—The division of functions within the family.

Sec. 9. THE PLACE OF GARDENING IN TRIBAL ECONOMY AND PUBLIC LIFE (*Pp.* 80–83). The beauty of the gardens.—Delight in perfect work and finish.—Economic function of the diverse crops.—Accumulated food in connexion with rank, prestige, and social organisation.—The dangers of wealth and success in gardening.—What the gardens mean to the Trobriander.

CHAPTER II

THE GARDENS OF OMARAKANA: EARLY WORK AND INAUGURATIVE MAGIC

Omarakana, the capital of the most fertile district.—Its hereditary rulers (*P.* 84).

Sec. 1. SOME PERSONALITIES OF THE GARDENING CIRCLE (*Pp.* 84–86). To'uluwa, the chief, and Bagido'u, the magician.—Bagido'u's helpers.—The "Good Gardeners".—Bagido'u's share in this book.

Sec. 2. THE STANDARD OF TROBRIAND AGRICULTURE (*Pp.* 86–87). Omarakana, the paragon to the other communities.—The significance of detail for the natives and the ethnologist.—The beginning of the agricultural year.—The main and early gardens.

Sec. 3. *KAYAKU*—THE CHIEF AND MAGICIAN IN COUNCIL (*Pp.* 87–93). The garden magician's summons to the garden council.—

Proceedings at the garden council.—The counting and apportionment of the plots.—Apportionment of standard plots.—The number of plots cultivated by a gardener.—The dowry plots (*urigubu*) and the personal plots (*gubakayeki*).—The enumeration of plots following the native system of reference.—Garden council in a southern village.—The cutting of the boundary belt and the opening up of the plots.—Quarrels over land.

Sec. 4. THE GRAND INAUGURAL RITE: THE STRIKING OF THE SOIL (*Pp.* 93–102). This rite representative of garden magic in general.—The garden magician's preliminary address.—The two expeditions for fish and magical herbs.—Mixing of the magical ingredients. —The feast of fish.—Preparing magical axes.—Oblation to spirits.—Charming the axes.—The social setting of the magical performance. —Imprisoning the magical virtue.—Ceremonial procession to gardens.—Proceedings on the standard plots: the spell of the bad wood; the spells of fertility; the magical striking of the soil.—Extension of the magic to all remaining plots.

Sec. 5. THE WORK OF CUTTING THE SCRUB (*Pp.* 102–105). Communal cutting of the standard plots.—Toils and amenities of the work.—Quarrelling about gardens.—Keeping the pace in successive garden activities.

Sec. 6. DIGRESSION ON MAGICAL INGREDIENTS (*Pp.* 105–106).

Sec. 7. THE MAGICIAN'S TABOOS (*Pp.* 106–107).—Fasting.—Forbidden plants and animals.—Taboos on new crops.

Sec. 8. VARIATIONS IN THE INAUGURAL RITE (*Pp.* 107–109). The simple rite.—The more complicated form.

CHAPTER III

THE GARDENS OF OMARAKANA: PREPARING THE SOIL AND PLANTING THE SEED

The purpose and significance of burning the gardens (*P.* 110).

Sec. 1. THE SECOND GRAND INAUGURAL ACT: THE RITUAL BURNING OF THE GARDENS (*Pp.* 110–115). Overlapping of magic and work at this stage.—Act of the wholesale burning described. —Behaviour of the performers and the villagers.—A magistrate makes a mistake.—Preparation and proceedings of the burning in detail.—The rite of the "dog's excrement".—The double rite of the fourth day; planting of the yam; planting of the second taro; "the spirits' hut".

Sec. 2. DIGRESSION ON NATIVE IDEAS ABOUT THIS MAGIC AND A THEORETICAL INTERPRETATION (*Pp.* 115–120). The sympathetic significance of the torches.—Why these are charmed at harvest.—Imprisonment of magical virtue.—Magical vilification.—The home of bush-pigs.—Black magic to attract them.—Footnote on sexual taboos.—Unexplainable details.

Sec. 3. THE WORK OF FINAL CLEARING: *KOUMWALA* (*Pp.* 120–123). The need of final clearing.—Pedantry and aesthetics.—The artificial chessboard made of sticks.—Its practical and psychological uses.—Preliminary planting.—Fencing.

Sec. 4. THE CORNER-STONES OF THE MAGICAL WALL (*Pp.* 123–132). The shape and character of the corner prism, *kamkokola*.—Doubts as to its functional significance.—The main planting of taytu a distinct activity.—Proceedings in preparation for the ceremony.—Collecting poles during tabooed periods.—Ceremonial erection of the *kamkokola*.—The "magical wall".—Its aesthetic value to the natives.—Second rite of the *kamkokola*.—The magical herbs.—The function of this rite.

Sec. 5. THE PLANTING OF TAYTU (*Pp.* 132–136). Simplicity and brevity of the central act of Trobriand agriculture.—Implements.—The technique of tilling with the digging-stick.—Native skill.—Communal labour.—Planting cries.—Obscene allusions.

CHAPTER IV

THE GARDENS OF OMARAKANA: THE MAGIC OF GROWTH

The new cycle in work and magic (*P.* 137).

Sec. 1. THE TURNING-POINT OF GARDEN WORK (*Pp.* 137–139). Henceforth taytu the centre of interest.—Training the vines round poles.—Protection of the gardens from pests.—Thinning out the tubers.—Weeding, women's work.—Harvesting and its successive stages.—The magic of growth.—The remaining inaugurative acts to weeding, thinning and harvest.

Sec. 2. THE MAGIC OF GROWTH—THE STIMULATION OF SHOOTS AND LEAVES (*Pp.* 139–149). The process of growth as seen through the natives' eyes.—The forces of nature and of magic.—The awakening of the sprout.—Preparation of the small supports.—Making the sprout emerge.—Women at weeding.—Its inaugural rite.—Magical induction of leaves.—Magical production of luxuriant foliage.—Magic for the formation of a canopy of leaves.

Sec. 3. THE MAGIC OF GROWTH—THE STIMULATION OF ROOTS AND TUBERS (*Pp.* 149–152). Reference to the Chart of Time-reckoning.—Breaking forth of tubers in magic.—The rite of "fetching back" the taytu.—"Anchoring" of the roots.—Thinning out the new tubers and the inaugural ceremony of this.—The eating of the new tubers.

Sec. 4. PRIVATE GARDEN MAGIC (*Pp.* 152–157). Private magic, individually owned, inconspicuous.—Private magic of seed-yams; of boundary poles; of digging-stick.—Faint influence of fairy-tales on growth of crops.

Sec. 5. COMMUNAL LABOUR (*Pp.* 157–158). Mutual dependence of gardeners.—Communal labour to help the chief.—Other forms of communal labour.—Co-operation between villages.

CHAPTER V

HARVEST

Harvest, the most important, the most joyous of gardening events (*Pp.* 159–160).

Sec. 1. FAMINE AND PLENTY (*Pp.* 160–164). Contrast of hunger and plenty.—Hunger as result of consecutive drought.—The tradition of past famines.—Endocannibalism from living memory.—The effects on chiefs and commoners of bad harvests.

Sec. 2. THE PRELIMINARY GARNERING (*Pp.* 165–167). The several harvests.—The magic of the pearl shell inaugurative of the early harvest.—Some subsequent rites.—Public display of food offering to spirits.—Lifting the magician's taboo.—The technique of harvesting.

Sec. 3. THE RITUAL OF REAPING THE MAIN CROPS (*Pp.* 167–171). The ceremonies inaugurative of the main harvest.—Remarks on the function of garden magic.—The rite and spell of *okwala*.—The principal harvesting ceremony.—Torches for next year's burning ceremony medicated.—Another lifting of the magician's taboo.

Sec. 4. WORK AND PLEASURE AT HARVEST (*Pp.* 171–176). Technique of harvesting and cleaning the taytu tubers.—The *kalimomyo*, garden arbour, centre of visits and social intercourse.—Classification of taytu in heaps.—Envy and dangers of sorcery behind the mask of friendly admiration.—A commoner's humility in fear of a chief.

Sec. 5. THE BRINGING IN OF THE CROPS (*Pp.* 176–181). Festivity of carrying in the crops.—Payments to the carriers.—The social setting.—The counting of yams by basketsful.—Arrival of carriers at the village.—Display of crops on the central place.—Gleaning.—The harvest of sweet potatoes.—A fallow garden.

Sec. 6. *BURITILA'ULO*—THE COMPETITIVE CONTEST IN HARVESTED WEALTH (*Pp.* 181–187). Undercurrent of jealousy and animosity leading to contests of wealth.—The initial quarrel between two villages.—The rules of the contest.—A native account of an actual case.—Mustering the crops and building the receptacles.—Presentation of the food.—The return gift.

CHAPTER VI

THE CUSTOMARY LAW OF HARVEST GIFTS

The Sociological and Economic framework behind the gay and picturesque activities of harvest (*P.* 188).

Sec. 1. DUTIES OF FILLING THE STOREHOUSE (*Pp.* 188–196). The main principle of the harvest transaction stated.—Complications: (*a*) ratio of brothers to sisters; (*b*) redistribution as *kovisi* and *taytupeta*; (*c*) "untying of the yam-house"; (*d*) gift—to husband or wife? (*e*) rôle of sons; (*f*) terminological confusion: *urigubu* as marriage contribution, as chief's tribute, as duties to sister's husband.—Tributes to the paramount chief and to chiefs of lesser rank.—Further complications: (*g*) the filling of the storehouse by kinsmen (and not relatives-in-law); (*h*) the filling of the storehouse, only a part of *urigubu* duties; (*i*) the return gifts, *youlo* and *vewoulo*; (*j*) own taytu (*taytumwala*).—Distinction between the dowry and personal share in plots and harvest heaps.—Direct and comprehensive statement of the rules of the harvest transaction.—Digression on terminology.—Diagram of harvest transactions.

Sec. 2. HUNGER, LOVE AND VANITY AS DRIVING FORCES IN THE TROBRIAND HARVEST GIFT (*Pp.* 196–210). Apparent absurdity and unfairness of the Trobriand harvest arrangement to the stupid European.—The puzzle of motive.—Need of sociological digression in order to understand the driving forces of the *urigubu*.—Economic reciprocity as basis of Trobriand marriage.—The matrilineal principle in Trobriand kinship.—The native theory of procreation.—The two groupings corresponding to "family" in the Trobriands represented by the two males: husband and wife's brother.—The necessity of a husband to a Trobriand woman: as guardian of her procreative life; as legitimiser of the offspring.—Fatherhood as derived from husband to wife relationship.—Duties towards the household incumbent on the wife's brother or nearest matrilineal kinsman.—The paternal sentiment: its strength and foundation; the customs and institutions arising out of it.—The avuncular sentiment.—*Urigubu* gift as the result of the double male guardianship in the Trobriand family.—*Urigubu* as the endowment of the family by its real head given to its resident guardian.—Personal motives for the *urigubu* gift: sentiments of lineage, vanity and attachment.—Why the *urigubu* cannot be a commercial transaction.—Why *urigubu* contributes to the stability of marriage.—How *urigubu* contributes to political organisation.

Sec. 3. THE THEORY AND PRACTICE OF THE HARVEST GIFT (*Pp.* 210–217). Filling a storehouse in Yalumugwa.—*Urigubu* in smaller villages.—The big competitive *urigubu* in Omarakana.—Its origin in a strange custom named *kiliketi*.—A native text about *kiliketi* and *kayasa*.—Account of the ceremonial bringing in of crops at Omarakana, 1918.—A quarrel during the *kayasa*.—Entertainment of carrying parties and refreshments on the return journey.—Building of a new *bwayma*.

CHAPTER VII

THE WORK AND MAGIC OF PROSPERITY

Taytu as the foundation of wealth.—Storehouse as centre of interest and activities.—The type of storehouse correlated with the class of crops (*Pp.* 218–219).

Sec. 1. THE MAGICAL CONSECRATION OF THE STORE-HOUSE (*Pp.* 219–222). The magic of prosperity performed by the garden magician.—Function and meaning of this magic.—The first act.—Symbolism of spell and rite.

Sec. 2. THE FILLING OF THE *BWAYMA* (*Pp.* 222–223). Work and technique of filling the storehouse.—The sociabilities of this performance.—Distribution of *kovisi* and *taytupeta* gifts.

Sec. 3. THE SECOND ACT OF *VILAMALIA* MAGIC (*Pp.* 223–225). The rite and spell of "piercing the village".—Description of an actual performance.—Absence of ceremonialism in Trobriand magic.

Sec. 4. THE OBJECT AND FUNCTION OF THE *VILAMALIA* (*Pp.* 226–228). Discrepancy between the character of spell and rite and the native comments on these.—Analysis of the two magical acts, showing that they are directed at storehouse and crops.—Native comments defining this magic as directed against human appetite.—Native ignorance of the nutritive rôle of food.—Virtue in abstention from food.

Sec. 5. THE FUNCTION OF THE STOREHOUSE (*Pp.* 228–232). Contrast between the show storehouses and small domestic ones.—Show storehouses a privilege of rank.—Differences of storing.—The various types of crops.—Various uses of crops and their removal from the store-house.—Incidental uses of the storehouse for sociabilities.—Protection of stored taytu from smells of cooking.—Ownership of storehouses.

Sec. 6. THE MAGIC OF HEALTH, WEALTH AND PROS-PERITY IN OBURAKU (*Pp.* 233–239). Differences in prosperity magic between a northern and a southern village.—Spell and rite of the shell trumpet.—Drowning of the shell trumpet.—Substances collected for the second ceremony and their sympathetic character.—The spell and the rite.—Comparative analysis of the two magical systems recorded.—*Vilamalia* also used in times of hunger, sickness, disaster.—The forces of nature and the figments of magic.

CHAPTER VIII

STRUCTURE AND CONSTRUCTION OF THE "BWAYMA"

Relevance of technological studies for sociological purposes.—Gaps in the material and unsatisfactory treatment in present work (*Pp*. 240–241).

Sec. 1. THE FORM OF THE *BWAYMA* AS CONDITIONED BY ITS FUNCTION (*Pp*. 242–243). Structural features imposed by cultural conditions.—Distinctions of rank in storehouses.—Storehouses better made than dwellings.—Instruction for following the diagrams.

Sec. 2. THE PREPARATION OF MATERIALS (*Pp*. 243–245). The implements used.—The preparatory work.—The boards, beams, arches and gables made ready.—The carving and bending of front-boards.

Sec. 3. THE CONSTRUCTION OF THE STOREHOUSE (*Pp*. 245–248). Preparing the beams for the log-cabin.—The work and payment of specialists.—The three main structural elements: foundations, log-cabin and roof.—No magic connected with building or repairing.—Laying of the foundation.

Sec. 4. THE LOG CABIN (*Pp*. 248–252). The manner of building.—Parallel between storehouse and canoe suggested by the terminology.—The use of scaffolding.—The framing logs.—Floor of magical importance but badly constructed.—The interior compartments and their sociological significance.

Sec. 5. THE ROOF (*Pp*. 252–255). The rods, poles and boards, and their putting together.—The three ways of placing the roof in position.—Thatching the storehouse.—Flooring the projecting gable ends.—*Bwayma* terminology in the magic of prosperity.

Sec 6. THE STRUCTURE OF THE SMALLER *BWAYMA* (*Pp*. 255–257). Difference between show *bwayma* and domestic store.—The seven types of small store.—The best *bwayma* used only for harbouring and displaying of crops.—Accessibility of inferior *bwayma*.

Sec. 7. SUMMARY OF THE STRUCTURAL, SOCIOLOGICAL AND ECONOMIC CHARACTERISTICS OF THE *BWAYMA*: LINGUISTIC TERMINOLOGY (*Pp*. 257–259). *Bwayma* as receptacle for food and medium for display.—The correlation between structure, economics, aesthetics and the social position of owner in the show *bwayma* and the inferior ones respectively.—Inconsistencies of terminology.

Sec. 8. NOTE ON PROPORTIONS (*Pp*. 259–260).

Sec. 9. TECHNICAL TERMINOLOGY OF THE *BWAYMA* (*Pp*. 260–262).

CHAPTER IX

A COMPARATIVE GLANCE AT TROBRIAND GARDENING

The system of Omarakana as representative of Trobriand gardening (*P.* 273).

Sec. 1. THE ESSENTIAL UNITY OF TROBRIAND GARDENING (*Pp.* 273–278). Example from early notes on gardening in the western district.—Critical analysis of these notes.—The system of garden magic particular to every village.—General character of garden magic and the main ceremonies similar throughout.—Each system distinguished by its own spells and magical substances, minor rites and taboos.—Some comparative remarks on the various systems.

Sec. 2. A PUBLIC CEREMONY IN THE GARDENS OF KUROKAYWA (*Pp.* 278–289). A ceremony in the sacred grove of Ovavavile and its spell.—The *kayaku.*—First inaugural ceremony and its spells.—The ritual burning and the spells.—The cleaning of the gardens.—The extensive ceremonial of the *kamkokola* rite.—Description of the earlier proceedings.—The magician's harangue; preparation of the larger poles; erection of the *kamkokola*; magical consecration of the *kamkokola*; uses of the ceremonial stone axe; the spell; private magic witnessed.—Description of the main ceremony: Preparation for the ceremonial distribution; men's and women's work; the gathering in the garden; the distribution; the spirits' share; the men's share; chanting over the magical herbs; the magician's helpmates; placing the magical herbs on the *kamkokola.*

CHAPTER X

THE CULTIVATION OF TARO, PALMS AND BANANAS

Sec. 1. THE GARDEN SYSTEMS OF THE SOUTHERN DISTRICT (*Pp.* 290–295). Barren stretches on the southern coast.—Fishing communities.—Intermarriage and reciprocal gifts of vegetable food and valuables.—Greater importance of taro in the south.—Similarity of gardening and garden magic throughout the Trobriands.—Minor differences between the systems of the north and south.—A garden rite in Oburaku.—The four main rites of Sinaketa.—Harvest customs of Vakuta.

Sec. 2. *TAPOPU*—TARO GARDENS (*Pp.* 295–300). Period of growth shorter for taro than yams; taro less dependent on seasons.—The concatenation of the yearly cycle.—No special taro magic in Omarakana.—Private magic.—No standard plots.—One plot shared by several men.—Special magic at the burning.—Driving the bush-pigs away.—The second burning and the *kamkokola*.—Planting the taro tuber.—Weeding the plots.—Native scarecrows.—Growth magic.

Sec. 3. THE MAGIC OF PALMS (*Pp.* 300–310). Coconut and betel-nut, their importance and their magic.—How trees are owned.—Uses of the fruit.—Method of cultivation: fencing in the young tree.—The cultural rôle of betel-nut.—The *kaytubutabu* ceremony to secure plentiful nuts for a distribution.—Trimming the palms and preparing the pole and collecting herbs for the rite.—The bands tied round the pole.—The *kaytubutabu* spell.—Planting the pole and placing the herbs.—Taboo on nuts, noises and fires, not to shock the nuts.—The 'breaking of the sprout'.—The 'breaking of the tree-tops'.—'Bringing down the first nuts.'—The function of this magic.—New decrees undermining old custom.

Sec. 4. FRUITS OF THE WOOD AND OF THE WILD (*Pp.* 310–312). Ownership in fruit-trees.—Wild fruits, nuts and leaves.—Methods of collecting and seasons of ripening.—Fruits of the bush and swamp.—Industrial plants.

Sec. 5. MINOR TYPES OF CULTIVATION AND SECONDARY CROPS (*Pp.* 312–316). Bananas and their magic.—Mango and bread-fruit.—European fruits recently introduced.—Holes in the coral-ridge.—Gardens on the sea-board.

CHAPTER XI

THE METHOD OF FIELD-WORK AND THE INVISIBLE FACTS OF NATIVE LAW AND ECONOMICS

Rôle of Observation and Construction in Field-work.—The invisible facts of cultural reality.—The institution of land tenure as an invisible fact; practical value of its study (*Pp.* 317–318).

Sec. 1. A PRELIMINARY DEFINITION OF LAND TENURE (*Pp.* 318–320). Direct questions unsatisfactory.—Mistaken approach by preconceived ideas.—Land tenure as the relation of man to soil in the widest sense.—How man uses his soil and builds his institutions round it.

Sec. 2. AN ANTHROPOLOGICAL EXPERIMENT IN DETECTION (*Pp.* 320–324). A challenge to the reader.—All the relevant data concerning land tenure as contained in the preceding chapters.—The chaos of unorganised facts.—An autobiography of mistakes illustrated from writings on the Mailu.—How the clan should be studied.

Sec. 3. AN ODYSSEY OF BLUNDERS IN FIELD-WORK (*Pp.* 324–330). Questions in pidgin.—Conflicting answers and conclusions.—The title of *toli*.—Next step: approach in native still unsatisfactory.—The approach through "objective documentation".—Result of all these approaches: the list of disconnected claims to land: (1) chief, (2) headman, (3) magician, (4) head of sub-clan, (5) minor sub-clan, (6) village community, (7) individual member of community, (8) actual gardener, (9) gardener's sister.—The incompleteness of the list.—The disconnected claims: harmonious or conflicting?

Sec. 4. FUNCTIONAL ANALYSIS OF THE TITLES TO LAND (*Pp.* 331–340). Value of the land to the producer.—Importance of claim 8.—The permanent subdivisions of the soil and its economic exploitation.—Relation of the various claims in the process of production.—Value of the land to the consumer.—Distinction between 'producer' and 'consumer' in relation to land tenure justified: the arbour and the yam-house.—The emergence from this analysis of four fundamental doctrines underlying claims to land: A. First Emergence, B. Law of Marriage, C. Magical Leadership, D. Rank.—These doctrines not codified by the natives themselves, but based on native theories.—Claims and practices based on these doctrines.—The several aspects of land tenure: sociological, legal, economic and mythological.—A synoptic summary of the argument.

CHAPTER XII

LAND TENURE

The four doctrines and their setting (*P.* 341).

Sec. 1. THE DOCTRINE OF FIRST EMERGENCE AS THE MAIN CHARTER OF LAND TENURE (*Pp.* 341–351). The right of citizenship based on First Emergence and matrilineal descent.—Concrete data concerning spots of emergence.—Emergence of the four clans in animal form.—Assortment of lands around each hole of emergence.—Fundamental validity of claims derived from Emergence.—Sub-clan as the sociological result of Doctrine A.—The prepayment for the immediate exercise of hereditary rights.—Seniority of lineages.—Villages and village clusters with several sub-clans.—Supremacy of one sub-clan, its leader acting as village headman.—Titles and duties of such a headman; the native view on this subject.—Members of a sub-clan related to their land until death and beyond.—The resultant unity of the sub-clan.—Magic transmitted within the sub-clan.—Exceptions to this rule.—Review of Doctrine A and its complications: citizenship and the mother earth.—The non-co-operating, non-cohabiting sub-clan established by this doctrine.

Sec. 2. THE LAW OF MARRIAGE IN ITS TWO-FOLD EFFECT (*Pp.* 352–358). Marriage: patrilocal, exogamous and matrilineal.—The rôle of *urigubu* (marriage gift) in marriage.—The patri-potestal co-operative family unit.—*Urigubu* and the woman's economic position.—The woman's claims to land implied in *urigubu*.—The family as a productive unit.—Division of labour within it.—The family and its rights to land.—The garden team consisting of a number of families.—The leading sub-clan in village gardening.—Several garden teams in one compound village.—Organisation of gardening under the leadership of the magician. —The garden team as a changing unit.—Importance of the doctrine of magical leadership in co-operative gardening (Doctrine C).

Sec. 3. RANK AS A PRINCIPLE OF TERRITORIAL OCCUPATION (*Pp.* 358–369). Doctrine A matrilineal.—Doctrine B patri-potestal.—Father and uncle: their behaviour and motives.—A dynamic adjustment between these two conflicting forces.—The claims of rank.—High rank father allowed to keep his sons by him.—A chief's sons and the difficulties of favouritism.—Prominent 'aliens'.—Rank as supporting the matrilineal principle: sons of a high rank mother.—A noble lineage in a commoner village.—How titles and dignities are taken over by the new immigrants culminating in the acquisition of garden magic.—The spread of the Tabalu.—Mobility and ubiquity of high rank sub-clans.—A short history of migrations.—Omarakana: the Tabalu stronghold.—Surrendering the rights of first emergence.—Dormant claims of the autochthonous sub-clan.—Alien garden magic used in Omarakana.—Privileges and offices gradually acquired.—Contribution of rank to land tenure.

Sec. 4. LAND TENURE AT WORK (*Pp.* 369–375). Apparent digressions into sociology justified.—Technicalities in the use of land: titles to ownership reviewed; the constitution of the gardening team.—The garden council.—Apportionment of plots.—The economic and legal character of leasing plots.—The leasing of a whole field.—The legal force of public apportionment at the garden council.—Legal foundations of the various claims in land tenure restated.—The safeguarding of land and produce against theft: protective magic.—The shame of stealing.—Rare encroachments on land.

Sec. 5. SUMMARY AND THEORETICAL REFLECTIONS ON LAND TENURE (*Pp.* 376–381). Land tenure: the relation of human beings to their soil.—Comprehensive view of land tenure.—The historical process at work.—An objective picture of land tenure.—Matrilineal and patri-potestal principles in ownership.—The study of legal and mythological claims.—The fallacy of communism *versus* individualism.—The human being and the machine.

Part Three

Documents and Appendices

Doc. I. *DODIGE BWAYMA* IN YALUMUGWA (*Pp.* 385–391). Plan of central place in Yalumugwa.—Clans, citizenship, rank.—Rules of ownership of yam-houses.—The headman.—List of yam-houses.—Analysis of gifts.—Table of recipients and donors.—(Pedigrees 1, 2, and 3).

Doc. II. COMPUTATION OF THE HARVEST GIFT PRESENTED AFTER THE *KAYASA* IN OMARAKANA IN 1918 (*Pp.* 392–405). Plan showing contributions to the *kayasa*, 1918.—Comparison of past and present contributions to the paramount chief.—Categories of the harvest gift.—Contributions to the main *bwayma*.—Number of chief's wives.—Sociological analysis of gifts.—*Urigubu* as the prototype of tribute to chief.—Table 1: *Urigubu* given to other residents in Omarakana.—Table 2: To'uluwa's harvest gift.

Doc. III. THE DECAY OF THE CHIEF'S HARVEST GIFT (*Pp.* 406–412). Initial mistakes in field-work.—Misconception of "tribute".—Early analysis of who fills the chief's *bwayma*.—Correction of analysis.—Legal distortion: *urigubu* a marriage gift, not tribute.—Chief's wives as each representing a tributary village community.—The list computed in 1916 compared with Document II.—Only part of chief's *urigubu* stored in his own yam-house.—Chief's sons' contribution.—Plots allotted to chief's wives.—Add sons' *urigubu* to chief's annual income.

Doc. IV. A FEW REPRESENTATIVE EXAMPLES OF AVERAGE *URIGUBU* (*Pp.* 413–418). (Pedigree 1).—Tokulubakiki's lineage, his number of plots, the distribution of his crops.—Who fills his *bwayma*.—Gifts to his mother's clanswoman, and others.—Errors in early notes.—Karisibeba, his lineage.—His number of plots.—His indirect contribution to *kayasa*.—How his *bwayma* is filled.—(Pedigree 2).—Bukubeku: cessation of *urigubu* on death of wife.—Motives for remarriage.—Namwana Guya'u.—His lineage; his number of plots; who fills his *bwayma*.—Tovakakita, his lineage; his modesty as a commoner concerning his *urigubu*.—What *urigubu* he receives and gives.—Rough estimate of number of baskets per heap.

Doc. V. LIST OF SYSTEMS OF GARDEN MAGIC (*P.* 419).

Doc. VI. GARDEN MAGIC OF VAKUTA (*Pp.* 420–426). Character of information received in Vakuta.—List of fields.—The garden council.—Oblation to spirits.—Some magical substances.—The inaugural rite.—Cutting the scrub.—The ceremonial burning and its rites.—Taboos on work.—The clearing of the ground.—The erection of the *kamkokola*.—Charming over seed taytu.—Growth magic more complex than in

Kiriwina.—Details thereof.—The harvest and its magic.—Limitations of Vakuta information.—Comparative table of garden magic in Omarakana and Vakuta.

Doc. VII. GARDEN MAGIC IN TEYAVA (*Pp.* 427–429). The magician's address and the distribution of plots at the garden council.—The inaugural ceremony.—*Kamkokola* rite.—Weeding rite.—The harvest and its ceremonies.

Doc. VIII. THE GARDEN LANDS OF OMARAKANA (*Pp.* 430–434). (Map).—Omarakana, the political capital of the Trobriands.—The assortment of lands: the grove; the water-hole; the roads.—Principles of land tenure recapitulated: the two autochthonous sub-clans; the in-married Tabalu.—The paramount chief; his personal dwelling; his court and visitors.—The three parts of the village and their respective inhabitants.—High rank and lower rank villages.—Rights to wood, wild fruit, roads, water-holes, uncut jungle.—Holes in the *rayboag*.—The garden land subdivided into fields.—Plots as units of measurement in land tenure.—Survey of fields.—Limitation of data.—Synoptic table of fields and plots.

App. I. ANALYSIS AND CHART SHOWINC THE ORGANISING FUNCTION OF TROBRIAND GARDEN MAGIC (*Pp.* 435–451). (Chart of Magic and Work).—Importance of magic.

A. Some General Characteristics of Trobriand Magic :—Magic in gardens and fishing.—Private magic in hunting.—Limitations of building and manufacturing magic.—The magic of canoes.—The magic of health and success.—Sorcery.—Native views about origin and nature of magic.—Spell and rite.—Magic: an attribute of man.—Performance of spell as monopoly of magician.—How magic is transmitted.—Status and taboos of magician.—Magic performed by headman and chief.

B. Garden Magic as a System of Rites running parallel to an Organised Pursuit :—Systematic and independent magic.—Association between technical pursuit and its magical correlate.—How magic supplements work.—Explanation of chart: specimen readings.—Practical and theoretical import of chart.

App. II. CONFESSIONS OF IGNORANCE AND FAILURE (*Pp.* 452–482).

Sec. 1. "*Nothing to Say*":—Possible reasons for lacunae.—Insufficient statement of gaps in former books.—Remedy here adopted.

Sec. 2. *Method of Collecting Information:*—Length of residence.—Interval for digesting material.—Linguistic equipment.—Advantage of living in a native village.—First impressions.—Three aspects of field-work.—Separation of these in plan of book.—The three stages in the study of native life.—Gradual deepening of knowledge as documented in field notes.—Magic and work.—"Survey work."—Discovering the meaning of *urigubu*.

Sec. 3. *Gaps and Side-steps:*—Mistaken over-emphasis on the inaugurative function of ritual.—Resulting oversight of growth magic.—Neglect of taro gardening.—Deficiency of quantitative data.—Botanical ignorance.—Difficulties in time sequence.—Technological ignorance.—Photography: inadequate and haphazard.—Overlooking the everyday in language and photograph.

Sec. 4. *Some Detailed Statements about Errors of Omission and Commission:*—

Chapter I: 1. Chart of Time-reckoning; 2. *Kaymata* and *Kaymugwa*; 3. Sociology of the *Leywota*; 4. The Ever-nascent Myth; 4*a*. Chief Working with his Wives.—Chapter II: 5. Black Magic; 6. *Kayaku*; 7. Number of Plots Cultivated; 8. Visits of the Spirits.—Chapter III: 9. Rest Days; 10. Black Magic of Bush-pigs; 11. Taro Magic in Taytu Gardening; 12. The Function of Small Squares; 13. Crops on the *Kaymata* and *Kaymugwa* respectively; 14. The *Kamkokola* Rite as Magic of Planting; 15. Construction of Fence and *Kamkokola*.—Chapter IV: 16. Observations on the Growth of Taytu; 17. Growth of Taytu as due to Magic; 18. Disproportion between Magical and Technological Information; 19. Private Garden Magic.—Chapter V: 20. Stories about Famine; 21. The Sociology of the First Fruit Display; 22. The Symbolism of *Tum*; 23. Information about *Buritila'ulo*.—Chapter VI: 24. Taytu for *Urigubu* and for own Consumption; 25. Filling a Man's *Bwayma* after his Wife's Death; 26. No Arbours on *Gubakayeki*; 27. The Chief's *Urigubu* and Tribute; 28. The one-sidedness of Documentary Evidence; 29. Estimate of Sociological Information in this Chapter.—Chapter VII: 30. *Malia-Mana?* 31. *Binabina* Stones; 32. Theory and Practice of the *Vilamalia* Magic; 33. Filling of Small *Bwayma*; 34. The Consumption of Crops.—Chapter VIII: 35. The Study of the Uses of *Bwayma*.—Chapter IX: 36. Methodological Problems; 37. The Value of Sampling.—Chapter X: 38. The Inclusion of Gaps; 39. Proportion of Taro to Taytu in the South; 40. Lack of Time and Opportunity; 41. The Neglect of Taro Gardens; 42. Hearsay Evidence; 43. Decay of Custom under European Influence; 44. Scattered Data on Minor Subjects.—Chapter XII: 45. Assessment of this Chapter; 46. Holes of Emergence; 47. Maps of Village Ground; 48. Transactions at *Kayaku*; 49. Stealing of Produce.

INDEX, *page* 483.

PART I

INTRODUCTION: TRIBAL ECONOMICS AND SOCIAL ORGANISATION OF THE TROBRIANDERS

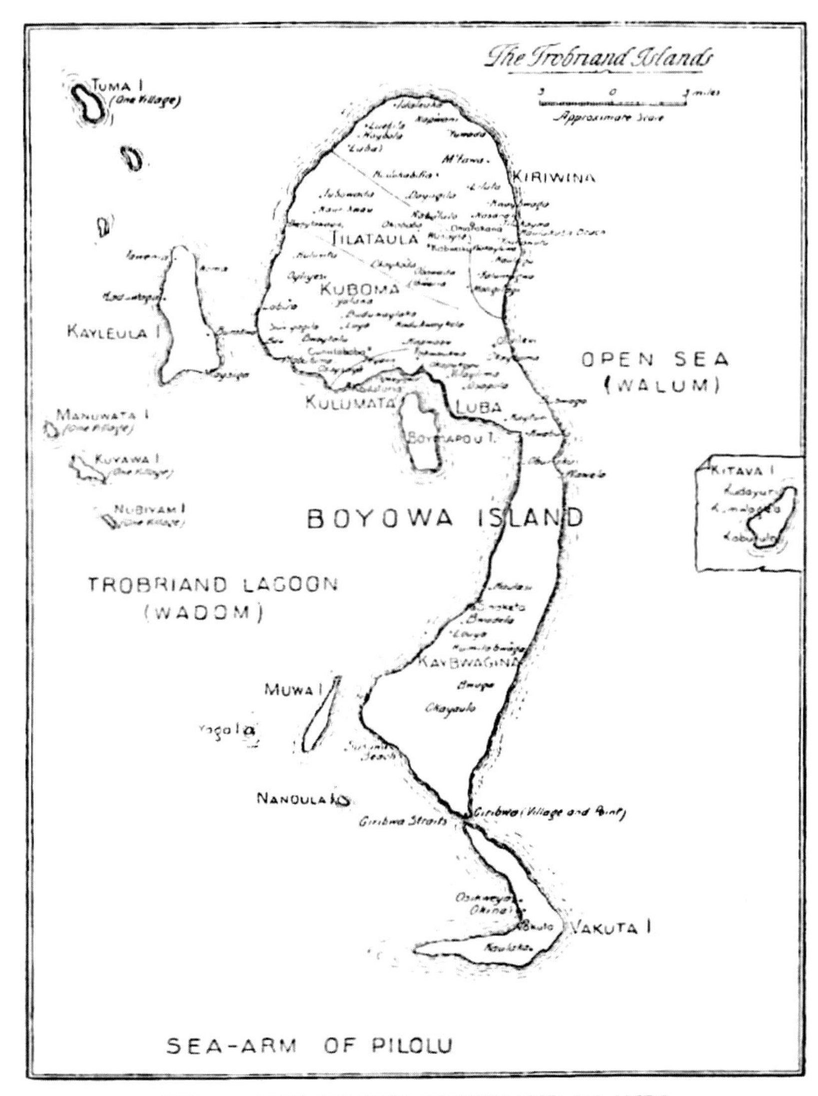

FIG. 1. MAP OF THE TROBRIAND ISLANDS.

Reproduced from the author's "The Sexual Life of Savages" by kind permission of the publishers Messrs. Routledge & Sons, Ltd.

TRIBAL ECONOMICS AND SOCIAL ORGANISATION OF THE TROBRIANDERS

1. THE SETTING AND SCENERY OF NEW GUINEA GARDENS

THIS will largely be a study of human effort on tropical soil, of man's struggles to draw his sustenance from the earth in one part of the exotic world, in the Trobriand Islands off the east end of New Guinea. Nothing is perhaps more impressive to an ethnographer on his first pilgrimage to the field than the overwhelming force of vegetable life and the apparent futility of man's efforts to control it. This contrast is brought home with great force when, on your first voyage along the south coast of New Guinea or through the archipelagoes in the east, you are surveying almost at a glance the character of this enormous expanse of tropical country. Chains of hills succeeding each other; deep transversal valleys which often afford a glimpse right into the heart of the country; the foreground, at times rising in an almost vertical wall of vegetation or again sloping down and extending into alluvial flats—all these reveal the strength of tropical jungle, the tenacity of lalang steppe, the impressive solidity of the undergrowth and tangled creepers. But to perceive man or even traces of him and his works you have to be a trained ethnographer. To the experienced eye the blot of withered vegetation on the waves of living green is a little village, with huts built of sere wicker work, thatched with bronzed palm-leaves, palisaded with dried timber. Here and there on the slope of a hill a geometrical patch, brown at harvest-time with the foliage of ripe vines or earlier in the year covered with the lighter green of sprouting crops, is a village plantation. If you are lucky you might even pass at night a constellation of smouldering fires where the bush has been cleared and the trees and brush are being burnt down. But the more you concentrate your attention on such almost imperceptible symptoms and strain your imagination to interpret them, the more you realise what little imprint man has made as yet on this soil, how easily his efforts are obliterated, how everything which he has made his own comes directly as a gift of spontaneous growth. Nature here seems not yet to have been subdued by man and fashioned to serve his purposes. Man, on the contrary, is but a part of her scheme sheltering precariously under what the jungle has yielded, clad in dried leaves, subsisting on that which, year after year, he wrests from the virgin forest and which, after a few years, returns to it again.

If you were to settle in one of the hamlets and follow the work and interests of the natives, the perspective would change considerably. You would find everywhere that agriculture is a businesslike procedure, that it is not merely a highly skilled and technical enterprise but also an important ceremonial of the tribe; that the whole territory is well marked, legally defined and more or less appropriated to individuals or groups.

If you were to sail further, surveying the various cultures and exploring the different parts of the country and islands, you would sooner or later come upon the flat coral archipelago of the Trobriands, which lies some hundred and twenty miles directly north of the easternmost tip of New Guinea. There you would recognise at once that you were in a region where the relations of man to nature are entirely different. You would realise at first sight that the soil is valued more highly, that it is mapped out very definitely and used more effectively than anywhere in the land of mountainous forest, sago swamp or lalang steppe. Even during a casual visit to the Trobriand Islands, the ethnographer would be struck by the density of the population, by the extent of the gardens, by their variety and thoroughness of cultivation. He would also find that relatively little of that territory is left to nature and its spontaneous growth. In the villages again it would be easy to see that more than half of the buildings are storehouses, and that produce is accumulated, stored and handled in a manner which makes it evident that man here in no way leads a hand-to-mouth existence, but that he depends on what he has achieved and has made into a solid foundation of wealth.

In the following chapters we shall have to repair to the yam gardens of the Trobrianders and to their taro and banana plantations. We shall take part in their work and follow their harvest joys and amusements. We shall scour the coconut groves and enter the magician's house to watch him at his spells and ritual. In all this we shall have to follow two lines of approach: on the one hand we must state with as much precision as possible the principles of social organisation, the rules of tribal law and custom; the leading ideas, magical, technological and scientific, of the natives. On the other hand we shall try to remain in touch with a living people, to keep before our eyes a clear picture of the setting and scenery. In order to achieve this it will be necessary, before we plunge into our special subject, to give a general introduction to the Trobrianders, to their land, to their sea and their lagoon.

Some of you may be already acquainted with the natives of our archipelago.[1] You may have followed me on the pilgrimage which I had to make several times from one of the white settlements along the south coast, and through the archipelagoes of the east end of New Guinea which are inhabited by the "Southern Massim" —a term coined by Dr. Haddon and describing a Papuo-Melanesian culture of which you will find a comprehensive picture in the third part of Seligman's *Melanesians of British New Guinea*.

I shall not retrace this pilgrimage in detail here. The scenery of the south coast and the east end; the charming scattered settlements of the natives; some of the customs of these cannibals, head-hunters and bloodthirsty warriors, I have described in my *Argonauts* (Ch. I). There also I have given an outline of the culture of some of the immediate neighbours of the Trobrianders—those who live on the foreshore and lofty inaccessible slopes of the d'Entrecasteaux group, and on the scattered rocks of the Amphletts. I have also dwelled on the contrast between the two types of scenery and the two types of culture.

"Leaving the brown rocks and the dark jungle of the Amphletts we sail North into an entirely different world of flat coral islands; into an ethnographic district, which stands out by ever so many peculiar manners and customs from the rest of Papuo-Melanesia. So far we have sailed over intensely blue, clear seas, where in shallow places the coral bottom, with its variety of colour and form, with its wonderful plant and fish life, is a fascinating spectacle in itself—a sea framed in all the splendours of tropical jungle, of volcanic and mountainous scenery, with lively watercourses and falls, with steamy clouds trailing in the high valleys. From all this we take a final farewell as we sail North. The outlines of the Amphletts soon fade away in tropical haze, till only Koyatabu's slender pyramid, lifted over them, remains on the horizon, the graceful form which follows us even as far as the Lagoon of Kiriwina.

"We now enter an opaque, greenish sea, whose monotony is broken only by a few sandbanks, some bare and awash, others with a few pandanus trees squatting on their air roots, high in the

[1] Notably those who have read Professor Seligman's *Melanesians of British New Guinea* (1910), or the three introductory chapters to my *Argonauts of the Western Pacific* (1922). Their domesticity, their family life and their love-affairs have been described in my *Sexual Life of Savages of North-Western Melanesia*, while their childhood is discussed in my small volume *Sex and Repression*. For the contiguous area of New Guinea we have first and foremost the excellent book of Doctor Fortune, *The Sorcerers of Dobu*, which should be consulted by everyone who is interested in the region; also the book on *Rossell Island* by W. A. Armstrong, the accounts of Mr. F. E. Williams (*Orokaiva Society* and *Orokaiva Magic*) and D. Jenness and A. Ballantyne, *The Northern D'Entrecasteaux*.

land. To these banks, the Amphlett natives come and there they spend weeks on end, fishing for turtle and dugong. Here is also laid the scene of several of the mythical incidents of primeval Kula (the inter-tribal trade to which a great deal of time, effort and ambition is devoted). Further ahead through the misty spray the line of horizon thickens here and there, as if faint pencil marks had been drawn upon it. These become more substantial, one of them lengthens and broadens, the others spring into the distinct shapes of small islands, and we find ourselves in the big Lagoon of the Trobriands, with Boyowa, the largest island, on our right, and with many others, inhabited and uninhabited, to the North and North-West.

"As we sail in the Lagoon, following the intricate passages between the shallows, and as we approach the main island, the thick, tangled matting of the low jungle breaks here and there over a beach, and we can see into a palm grove, like an interior, supported by pillars. This indicates the site of a village. We step ashore on to the sea front, as a rule covered with mud and refuse, with canoes drawn up high and dry, and passing through the grove, we enter the village itself."[1]

2. THE HABITAT AND PURSUITS OF THE TROBRIANDERS

From now on we shall dwell among the Trobrianders. This archipelago lies, as we know already, due north of East Cape. Every anthropologist also knows that the inhabitants are Melanesians. They show greater affinities, however, with some of the more distant oceanic populations in physique, culture and institutions than with their near Papuan neighbours on the mainland of New Guinea. They have a developed chieftainship, they are expert sailors and traders, their decorative art forms the glory of many an ethnographic museum.

The Trobriand archipelago which faces you on the map (Fig. 1) is a coral atoll, or more correctly a part of the Lusançay atoll. The group which concerns us consists of one big island, two of fair size—Vakuta and Kayleula—and a number of smaller ones surrounding a basin or lagoon. This latter is very shallow, parts of it are not navigable even to the native canoes, but it is crossed with deeper channels. It is open to all winds, giving no shelter whatever from the north-westerly monsoon or the strong southerly winds, and affording only a little protection near the shores of the main island from the south-easterly trade wind. To the east, at a distance

[1] *Argonauts of the Western Pacific*, pp. 49-51.

of about one hundred miles, lies the second large centre of the Northern Massim culture—Woodlark Island. The Northern Massim—the name is again Dr. Haddon's—are the second branch of the Eastern Papuo-Melanesians. Between Woodlark Island and the Trobriands there is a bridge of five small islands—Kitava, Iwa, Gawa, Kwaywata and Digumenu—also inhabited by people of the same culture. We shall meet them again in the mythology of gardening.

In our detailed descriptions we shall, however, dwell almost exclusively on the main island of the Trobriands, with only brief references to the contiguous areas. On this large island, called by the natives Boyowa or, after its principal province, Kiriwina, we shall find several types of scenery, soil and agriculture. The northern part, a wide circular expanse of land, harbours most of the fertile soil. Only the narrow coral ridge which runs along its northern and eastern border remains almost completely outside cultivation and is covered with patches of primeval jungle. But this never attains to full tropical luxuriance, and some economically important plants, such as the sago palm, the lawyer cane, and the bamboo, do not grow there and have to be imported as raw material from abroad. Some portions of the land in the interior are also useless because they are too swampy; while, in the west, large stretches on the coast are covered with mangrove, which grows on a brackish swamp awash at high tide. In the southern part of the island the dead coral crops up, especially at the extreme end, leaving large tracks of country uncultivable and uninhabited. The brackish swamps of the southern portion extend deeper inland and the villages are placed either on the lagoon, where fishing makes their existence possible, or on one or two fertile spots inland.

3. FIRST IMPRESSIONS OF TROBRIAND GARDENS

Since our subject here, the systems of native gardening, forms only part of the economic life of the tribe, albeit its principal part, we shall have to consider the exploitation of natural resources as a whole. The brief outline of tribal economics here given, forms an indispensable background to the detailed descriptions of agriculture.[1]

The description of the territory just given, taken in connexion

[1] I should like to state that this, as with the brief account already published in the *Economic Journal*, 1921, on "The primitive Economics of the Trobriand Islanders", is but a preliminary sketch of the subject. I am now working on a full account of Trobriand fishing, hunting, industries and inland trade. Their overseas expeditions, ceremonial exchanges and inter-tribal trade have already been described in *Argonauts of the Western Pacific*.

with the map of the Archipelago (which shows incidentally that for a South Sea tribe the Trobriands have a very dense population), the realisation that these natives have a high level of cultural ability, of political and economic organisation—enables us roughly to assess the type of their production and industrial development. The fertile humus covering the wide expanse of dead coral lends itself obviously to an intensive cultivation of useful plants, i.e., since we are in the South Seas, of yam, taro, sweet potato, banana and coconut. The open lagoon, teeming with submarine life, would naturally invite an enterprising and intelligent population to develop effective fishing. The industrious and compact settlements would lead us to anticipate excellence in arts and crafts. Differences in habitat and opportunity might well be expected to produce special centres of industry and systems of internal trade. Again, the absence of certain indispensable raw materials—stone (dead coral is useless for any industrial purpose), clay, rattan, bamboo, sago—would suggest an extensive trade with the outside world. The absence of primeval jungle indicates that hunting cannot be of any importance and the search for wild produce can play only a subsidiary part.

This rough estimate is indeed correct almost in every essential. The Trobriander is above all a cultivator, not only by opportunity and need, but also by passion and his traditional system of values. As I have already said elsewhere: "Half of the native's working life is spent in the garden and around it centres perhaps more than half of his interests and ambitions. In gardening the natives produce much more than they actually require, and in any average year they harvest perhaps twice as much as they can eat. Nowadays this surplus is exported by Europeans to feed plantation hands in other parts of New Guinea; in olden days it was simply allowed to rot. Again, they produce this surplus in a manner which entails much more work than is strictly necessary for obtaining the crops. Much time and labour is given up to aesthetic purposes, to making the gardens tidy, clean, cleared of all débris; to building fine, solid fences; to providing specially strong and big yam-poles. All these things are to some extent required for the growth of the plant, but there can be no doubt that the natives push their conscientiousness far beyond the limits of the purely necessary. The non-utilitarian element in their garden work is still more clearly perceptible in the various tasks which they carry out entirely for the sake of ornamentation, in connexion with magical ceremonies and in obedience to tribal usage."[1]

Fishing comes next in importance. In some villages situated on

[1] *Argonauts*, pp. 58, 59.

PLATE 2

THE CHIEF'S POLYGAMOUS FAMILY

Seated on the platform is the Chief To'uluwa; on his right, standing, his son Gilayviyaka. At the chief's feet can be seen some of his wives: in the middle Isupwana, on her right the chief's son Dipapa, and next to him Bokuyoba, then Bomawise, and to the right against the sapling, Ilaka'isi (Part I, Sec. 3)

PLATE 4

FISHING CANOE ON THE LAGOON

All large-scale fishing is done in most southern lagoon villages with nets. The large triangular nets here seen are held out (see next plate) close to the water. The fish are driven into a seine net, and in attempting to jump over this fall into the triangular net (Part I, Sec. 5: cf. also Sec. 4)

PLATE 3

SEA-GOING CANOE UNDER SAIL

These large canoes are mainly employed on overseas expeditions to the eastern archipelagos from Kiriwina, and south to the Amphletts, Dobu, and D'Entrecasteaux Islands from Sinaketa and Vakuta. For details of the overseas expeditions, compare *Argonauts of the Western Pacific* (Part I, Sec. 4)

the lagoon it is the main source of sustenance and claims about half of their time and labour. But while fishing is prominent in some districts, agriculture is paramount in all. Were fishing made impossible to the Trobrianders by a natural or cultural calamity, the population as a whole would find enough sustenance from agriculture. But when the gardens fail in times of drought, famine inevitably sets in. Hunting is hardly an economic pursuit. From time to time you see a native walking out of the village spear in hand, and he tells you that perhaps he will be able to kill a small wallaby or a bush-pig. Bird-snaring has a little more importance. But every time I saw the natives eating a wild-fowl I found that it had been shot by some white trader and passed on from a distant village. Collecting of food from the bush in times of drought, the catching of crabs and molluscs in mangrove swamp and lagoon, are much more substantial contributions to the tribal larder. Transport and trade are well developed. The inland barter of fish and vegetable food is an institution which controls a great deal of their public life.

Thus, in brief, we find that the environmentalist's predictions are substantially correct. But there are a great many things referring to work and its organisation, to the production and distribution of wealth and to its consumption, which cannot be inferred from oecological indications. The environmentalist will foresee nothing of the great importance of magic and of political power in the organisation of gardening. In the distribution of produce he cannot anticipate the extremely complex way in which kinship and relationship by marriage impose obligations and place the Trobriand household economically on a two-fold foundation (cf. Chs. V and VI). Nor could he guess the intricate manner in which mother-right combined with patrilocal marriage complicates the system. The contrivances and customs which allow these natives to accumulate large quantities of food, and the legal system which concentrates wealth in the hands of a few leaders who can then organise enterprises on a tribal scale, have to be observed and stated from experience.

Let us then survey the various food-producing activities, the arts and crafts and trades one after the other. We will begin with the gardens.

On my arrival I certainly was both impressed and fascinated by the life in the gardens, by its bucolic beauty and richness, as well as overwhelmed by the complexity of agricultural events. I arrived in the Trobriands early in June, 1915, and after a few days on the coast settled in Omarakana, the chief's residence and the principal village of the archipelago. The harvest in most of the surrounding villages was in full swing; in the capital it had just started or was about to start. There is no time at which the Trobriand gardens show to better advantage or the natives' interest in produce displays itself with greater intensity; no season at which so many threads in the fabric of gardening are interwoven (Chs. I, II, V and VI). All the village is in the garden and in a way the garden comes to the village. Thus at one moment no one is to be seen among the deserted houses, but old men and women at work and small children at play. Then one party after another rush in with the crops and fill the whole settlement with yams, baskets, chatter and frolic, and with the importance of gardening (cf. Pls. 60–64). It was at that time that I received the first inkling that the Trobriander is above all a gardener, who digs with pleasure and collects with pride, to whom accumulated food gives the sense of safety and pleasure in achievement, to whom the rich foliage of yam-vines or taro leaves is a direct expression of beauty. In this, as in many other matters, the Trobriander would agree with Stendhal's definition of beauty as the promise of bliss, rather than with Kant's emasculated statement about disinterested contemplation as the essence of aesthetic enjoyment. To the Trobriander all that is lovely to the eye and to the heart, or—as he would put it more correctly—to the stomach, which to him is the seat of the emotions as well as of understanding, lies in things which promise to him safety, prosperity, abundance and sensual pleasure.

Walking across the country at that season, you would see some gardens in all the glory of their green foliage, just turning to gold (cf. Pls. 21 and 31). These would be some of the principal plantations of yams, which have matured later than the majority. Then again you would see some of the next season's gardens being started (cf. Pl. 20), and from time to time you would pass through a flat stretch of broad green leaves—the taro gardens (cf. Pl. 112). During my first rough inspection of the gardens round Omarakana I was astonished by the bewildering variety of garden scenery, garden work and garden significance. In one place a harvest was going on, men and women cutting vines, digging up roots, cleaning them and stacking them in heaps; in some of the taro plantations women were weeding; men were clearing away the low bush with axes in parts of the garden, in others they were laying out the ground in small squares like a chessboard, the purpose of which at first eluded my most assiduous inquiries in Pidgin (cf. Pls. 26 and 38).

It was at that time also that, on one of my first days, I was an unwitting witness of the grand garden council which will engage our attention several times during the following narrative (Chs. II, XI and XII). I saw the chief, his heir and nephew Bagido'u, who was also the magician, together with all the notables in assembly, discussing matters which my interpreter failed to translate. Soon after that the chief himself took me for a morning's walk and work in the gardens. I was impressed that he, like the humblest of his subjects, worked day after day on his own land, wielding a powerful digging-stick, for he was among the tallest and strongest of the Trobrianders. Like everyone else he would plant the taytu, tuber after tuber, seeking for each an appropriate piece of soil. At harvest he would work with the same minuteness and precision, breaking up the soil, taking out the taytu with his own hands and cleaning it, just as carefully, lovingly and patiently as anyone else (Ch. V, Sec. 4). Usually he was accompanied by one or other of his wives; the strong, healthy Isupwana (cf. Pl. 81), the beautiful young Ilaka'isi (cf. Pl. 82), or the first wife of his own personal wedding, Kadamwasila, or the oldest of his wives, whom he had inherited from his elder brother, Bokuyoba (cf. Pl. 2). It was to To'uluwa and his wives that I was indebted for the first few lessons in the technology of gardening. It is useful in field-work to show some practical interest in a pursuit and to put up some show of manual competence, to offset that theoretical and yet personal curiosity which is so apt to offend native susceptibilities.

At that season the various types of labour characteristic of garden work could also be observed side by side. In the taking out of the tubers, as well as in the cleaning and preparing of the new gardens, usually the whole family, husband, wife, children and dependents, work together (cf. Pls. 24 and 36). Even small children are often given a toy digging-stick and a miniature axe with which to play at garden work, and they begin to garden seriously at a surprisingly early age (Ch. I, Sec. 3; and Part IV, Div. 5).

At planting and the cutting of the scrub communal labour preponderates. Living in the village among the natives it is impossible not to notice it. Summoned by the chief and treated by him in advance to some food, betel-nut and tobacco, the assembled men go out in semi-festive attire, with frolic, customary cries and jokes, and work for half a day or so. In smaller villages the men work communally for each other, cutting or planting one or two plots in turn. At times a man will work alone on his own ground, or a woman will continue weeding her own plot, or more correctly that of her husband.

Hardly less evident would be another important side of Trobriand gardening—magic. It certainly is not esoteric. It is a public duty of the garden magician, its existence and even its details are known by everyone, and since magic is regarded as both inalienable and normal, there is no secrecy about it. With all that, I can imagine that an ordinary visitor or trader or missionary might live for a long time among the natives before he discovered its existence. For it is extraordinarily simple, devoid in most cases of any ceremonial, and in the carrying out of a magical act the performer looks very much like an ordinary native bent on practical work (Chs. III, IV, VII, IX and X). The rites are simple and direct, and only the fact that some of the spells are chanted aloud in the field or in the village would lead you to inquire what the man was really after. Mostly the magical act is carried out by the magician alone, or with those who help him or who have some right or interest in the magic (Ch. I, Sec. 6; cf. also Pls. 23, 29, 33-35, 52 and 100). There are only one or two really ostentatious and public ceremonies of gardening (Ch. IX, Sec. 2; cf. also Pls. 103-108).

Early in my experience I already recognised that tribal life revolved round the gardens. In the distribution of tasks and interests, through the cycle of seasons, agriculture always takes precedence. The districts rich in produce are on the whole politically dominant as well as economically the most wealthy. In each village the garden magician—which office is held either by the headman or his heir or a near relative—is either the most or the next most influential person. Garden produce is the foundation of wealth throughout the area. It is distributed according to a complex system of marriage endowment and political tribute, and in this form it is about the most significant fact in Trobriand sociology. The few villages where fishing plays a considerable part still depend economically on their agricultural neighbours. This last point brings us to the subject of economic provinces or districts.

4. ECONOMIC DISTRICTS: THE TILLERS AND CRAFTSMEN

We know already that in the Trobriands, besides a rich development of trade and industry, two principal food-providing activities are practised—agriculture and fishing. But they are not practised universally throughout the archipelago, for there is a specialisation corresponding roughly to two districts. A glance at the map (Fig. 1) will show that the broad expanse in the north is thickly populated with villages scattered all over the circular area, whereas in the south there is a collection of villages near the western coast which run on in a continuous line to the lagoon settlements of the north. These two constellations of villages, the evenly distributed batch in the north and the semicircular belt of lagoon settlements, correspond to the occupational difference between agriculture and fishing. But besides these two main distinctions, further differences can be found between every one of the several districts—differences which are partly political, partly sociological and partly, what is of special interest to us here, economic.

Thus in the north we have three central provinces; Kiriwina, to the north-east, Tilataula in the middle, and Kuboma to the south-west. The first two depend almost exclusively on agriculture. Kiriwina is the politically dominant, socially most exalted and economically perhaps the richest province, and the paramount chief of the whole area has his residence in Omarakana. The villages at the northern end of the island count as part of Kiriwina; and among these are Laba'i and Kaybola, the only two fishing settlements in this district. They specialise in two types of fishing only, shark and mullet, which though strictly seasonal have yet some economic importance. When a large shark is caught off Kaybola—for this is the place where shark fishing is known in magic and practice—the whole district will have its fill of this pungent fish. Again, when rich shoals of mullet appear at the full moon off Laba'i, and are caught in large quantities with air-nets (Pl. 5), tribute will be sent to the paramount chief and to lesser chiefs, and fish will be plentiful all over Kiriwina. There is a special magic connected with this fishing, chartered by mythological tradition and localised ritual, and carried out by the respective headman of each community, with taboos and ceremonial. The villages are bound to give tribute of their catch and in turn are presented with counter-gifts from the recipient communities.

I am unable to provide numerical data of any value as to the relative importance of fishing *versus* agriculture in such communities as Laba'i or Kaybola. I think it is desirable to substantiate certain statements by quantitative indices. But in most cases it is not possible to do this in the anthropological field, and faked or pretended numerical data are worse than useless. Shark was caught and distributed several times during my stay, so that I was able to taste of this delicacy in Omarakana, but there is no doubt that this occupation is to a large extent in abeyance, so that present-day observations are no indication of what the trade was when it flourished. If you ask the magician how many sharks were caught yearly in olden days, he would shut his eyes and clap his fists together time after time, meaning scores and scores—two or four hundred perhaps. With another man the number will fall to twenty or may rise to eight hundred. The natives aim at giving an impression rather than stating a fact. Numbers and the amount of food produced or fish caught are a matter of local glory. Ask someone in the interior, who has no personal interest in the quantity of shark landed in Kaybola, and he will tell you that probably six to twelve were caught in the year—no doubt an understatement. The amount of mullet caught off Laba'i is even more variable. I was told that at times more fish came from there to Omarakana than was produced by exchange with the coastal villages. At other times hardly any shoals would turn up throughout the whole year.

In Kiriwina most of the villages have a "sea-front" on the eastern shore, where a large canoe for overseas trips, and several small canoes for fishing or coasting, are beached. In these villages a man would go out with his fish-hook or a group of people with a seine and make some haul on the reef. This was an amateur pursuit which enabled the villagers to obtain a little fish now and then during the calms in autumn and spring, more for pleasure of the sport and delicacy of the relish than for business. When I say amateur pursuit, I mean that there was no official magic, no season for fishing, no communal organised expeditions, no obligation to outside communities or to their own people.

The next province, Tilataula, did no fishing whatever. They would be described by a native expression meaning "real land-lubbers" (see Part V, Div. V, § 13). If Kiriwina, as the brilliant, exalted, aristocratic province, might be called the Athens of the Trobriands, Tilataula, strong in military arts, hard working and sober, could be called the Sparta of the island. They themselves are proud of their agriculture and of their frequent victories over their more aristocratic but less militaristic neighbours. The chief of Kabwaku, the capital of Tilataula, used to wage war occasionally against the Paramount Chief, to whom in one way he was subject, but in another a rival and a dangerous antagonist. Economically these natives concentrate on gardening; they have no canoes either for fishing or for overseas expeditions, and they are not skilled in any art except one, the polishing of stone.

This art is practised by specialists in several centres, notably Obowada, Kaurikwa'u and Okobobo. A large coral boulder of a special grain, cut out from the live reef, is planted before the hut of the craftsman who also procures some very fine sand from a special beach. The art requires some skill and patience, and the traditionally handed on determination to remain for some weeks on end bent over the polishing apparatus. These craftsmen worked mainly for chiefs of rank—the Tabalu of Omarakana, the Toliwaga of Kabwaku, the sub-chiefs of Liluta, Yalumugwa or Kwaybwaga.

While working they are maintained by their employer with periodic gifts of food.

But it is only when we move further west to the district of Kuboma that we find really developed industries on that stonier soil which produces distinctly less brilliant gardens than its eastern neighbours. The most fertile grounds in Kuboma belong to the village of Gumilababa—the residence of the divisional chief. He also is a Tabalu. For—to anticipate a piece of sociological information—the Trobrianders are divided into four main totemic clans: the Malasi, Lukuba, Lukwasisiga and Lukulabuta. Each of these clans again is divided into sub-clans and the sub-clan is by far the more important social unit (cf. Ch. XII, Secs. 1 and 3). Each sub-clan has what might be called an index of rank: and within this hierarchy the sub-clan of Tabalu is universally acknowledged as the highest. Although there is no precise equivalence in the manner of expression, the language of the Trobrianders conveys both the idea that a man is a Tabalu or a Mwauri or a Tudava, even as we speak of a Campbell, a Cameron or a MacDonald; and also that there is *the* Tabalu, that is, the reigning chief of Omarakana; *the* Toliwaga, the head of the province of Tilataula; *the* Mwauri, who resides in Liluta, and so on. Besides the main Tabalu, however, there are also subsidiary lines, for this most aristocratic sub-clan is scattered over several capitals. (See Ch. XII, Sec. 3.)

Returning to Kuboma, however, this district interests us here really because of its industrial character. We might feel tempted to speak of the inhabitants of Kuboma as the industrial caste of the Trobriands; for neither in ancient Greece nor even on the Mediterranean can we find any exact parallel. They are not like Phoenicians or Jews, primarily traders, but rather industrialists and craftsmen; and, as in any strict caste system, their high manual ability does not give them rank but rather places them among the despised. This refers especially to the most admirable of all Trobriand craftsmen, the inhabitants of Bwoytalu. This village, which shares with its neighbours of Ba'u the reputation for the highest efficiency in sorcery,[1] can certainly show the best results in carving; it is traditionally cultivated there and both for perfection and quantity of output is unparalleled in the region. From time immemorial its people have been the woodworkers and carvers of eastern New Guinea. And they still turn out wooden platters, hunting- and fishing-spears, staffs, polishing-boards, combs, wooden hammers and bailers in large quantities, and with a degree of geometrical and artistic perfection which any visitor to an ethno-

[1] That is, the art of injury or killing by magic.

graphic museum will appreciate. They also excel in plaited fibre work and in certain forms of basketry. During the wet season, when some other communities are busy preparing overseas expeditions, or engaging in festivities and ceremonial distributions, or (generations ago) indulging in war, the men of Bwoytalu will day after day sit on one of their large covered platforms (cf. Pl. 98) rounding, bending, carving and polishing their masterpieces in wood. It is a wholesale manufacture for trade and export. There is no magic whatever connected with their work, but from childhood skill is drilled into every individual, the knowledge of material, ambition and a sense of value. No other community can or tries to compete with them.

The other villages, Yalaka, Buduwaylaka and Kudukwaykela specialise in the production of quick lime for betel chewing. The last named village used also to produce the burnt-in designs on decorated lime pots, which can still be admired in an ethnographic museum and form undoubtedly one of the high-water marks of South Sea art. Unfortunately this industry is now dead. Plain lime pots, gaudily and as a rule vulgarly bedecked with cheap European trade beads, have completely superseded the beautiful native product. The inhabitants of Luya are the main producers of the finely plaited basket work made of lalang grass, chiefly used for the three-tiered basket, the widower's cap and small handbags. These are traded even now all over the archipelago, indeed over the whole Kula district. Some of the villages, notably Ba'u, Bwoytalu and Wabutuma, also practise fishing and specialise in catching, by means of a multi-pronged spear, that despised fish, the stingaree.

5. ECONOMIC DISTRICTS: THE FISHERMEN

Moving in our general economic survey, we come to the lagoon district of the north—Kulumata. In the large compound village of Kavataria we find again a Tabalu in residence, and in two other neighbouring settlements chiefs of the same rank have also become naturalised (Ch. XII, Sec. 3). But this district is not one political unit under the sway of one headman as is the case with the three preceding ones.

The natives are fishermen who treat their calling as a serious and important pursuit. Since in this they are closely akin to some of the southern villages, let us cast our eye on the map again and consider the other fishing districts. There we find Luba, the complex of villages situated on what might be called the waist of the main island. Here the new capital Olivilevi, founded a few generations

PLATE 5

CATCHING FISH IN THE AIR

This picture, taken in the village of Laba'i, shows the manner in which the triangular nets are used. Compare also previous plate (Part I, Sec. 5)

PLATE 6

SCENE IN A LAGOON VILLAGE

Village street in Kavataria, a shore village on the lagoon. The seine nets used for fish driving (cf. also Plates IV and V) are seen drying on the left. Men and women are busy at their daily tasks (Part I, Sec. 6)

PLATE 7

THE BEACH OF TEYAVA

"At times the whole community (of a fishing village) is seen setting forth or returning from a fishing expedition", carrying their punting poles and perhaps also agricultural produce from their neighbouring plantations (Part I, Sec. 6)

ago as an offshoot of Omarakana, and its neighbour Okayboma are mainly agricultural. The other villages, however, from Okopukopu down to Oburaku, depended chiefly on their fishing. Further south still and separated by a somewhat prolonged stretch of unoccupied land, we find the large village of Sinaketa surrounded by a few smaller settlements, and south of these, situated towards the eastern shore, three villages. These latter are mainly agricultural. Sinaketa, on the other hand, is an important fishing centre as well as the seat of at least one dominant industry—the production of red shell-disks used as ornaments and tokens of value (cf. *Argonauts of the Western Pacific*, pp. 371 to 374, and the Plates L, LI and LII).

Now concentrating our attention on the fishing villages we find that each of the coastal settlements has a type of fishing of its own. As we know, Bwoytalu and its neighbours, though they net fish, collect molluscs and catch crab, are primarily interested in the spearing of stingaree. Kavataria, the large settlement in the centre of Kulumata, has an importance specially due to the presence in their portion of the lagoon of a number of coral outcrops with cavities and shelters which afford the best opportunity for catching fish by means of a poisonous root (see Part V, Div. II, § 4). One of the sub-clans resident in Kavataria has a family tradition explaining this. Their ancestors emerged from underground in one of the small islands lying between the Trobriands and Woodlark to the east; but later on they migrated to Kavataria and brought with them the coral outcrop, together with the skill in fishing and the magic by which the fish were attracted to the outcrop—a magic which has been long since lost. Needless to say, an adept at historical interpretation of myths and legends would interpret this story—and even be correct in doing so—as meaning that this type of fishing was diffused into the Trobriands from the eastern archipelagoes. The coral patches are now owned individually, at times leased, and they are worked often and worked hard. The great importance of this fishing is that it is possible to make a catch in weather and under conditions in which no other type of fishing is practicable. It is remarkable that no magic whatever is now practised in connexion with this industry. By the ease of their work and their relative independence of weather, the natives of Kavataria are monopolists, in that they can provide fish when no one else can supply it. In the exchange of fish for vegetables, which plays an important rôle in Trobriand economic life, they exact about double the usual price from the other party, but in return offer punctual and reliable delivery.

The communities further east, Teyava and Osaysuya, Tukwa'ukwa and Oyweyowa, also fish, but are inferior in their effectiveness not only to Kavataria, the premier fishing centre, but also to Oburaku and Okopukopu. The last named places have no coral outcrops in their portion of the lagoon and all large-scale fishing they do by nets and beating (cf. Pls. 4 and 5). For this they need a calm day, favourable movements of the shoals and, of course, an organised communal enterprise. When successful their yield is large and they give a better measure in exchange; but their partners may very often have to wait for a long time and even in the case of a successful expedition the yield may be fitful. Moving further south we come to the only village situated on the eastern shore—Wawela. The inhabitants do some odd fishing on calm days, but their speciality is knowledge of native astronomy and meteorology, or more correctly, of native time-reckoning (cf. Fig. 3 and Ch. X). Economically they depend on their gardens and to a considerable extent also on the rich coconut plantations on their beach.

The large settlement of Sinaketa, comprising some seven or eight component villages, is important in that from here, and from the large village of Vakuta on the adjacent island to the south, some of the main sailing expeditions are made to the Amphletts and to Dobu, where the ceremonial Kula exchange takes place and also some straightforward inter-tribal trade. In olden days, and to a certain extent even now, the natives of Sinaketa and Vakuta used also, on their expeditions, to fish for the spondylus shell out of which the shell-disks were manufactured (cf. *Argonauts*, Ch. XV, Secs. 2 and 3.) This industry, as we shall see presently, has in its economic character a great deal in common with the production of large polished blades which we have observed already in Tilataula. There is a third industry of the same type and that is the making of arm-shells (cf. *op. cit.* Ch. XXI, Sec. 4). This was above all the speciality of the one district not yet mentioned—the small island of Kayleula to the west of Kuboma. To this district also belong the smaller one-village islands of Manuwata, Kuyawa and Nubiyam. But Kavataria, the village of the coral outcrops, runs a near second in the production of arm-shells. The two centres, Kavataria and the villages of Kayleula, used also to practise a special offshoot of the Kula ring, in which the exchange of useful industrial articles as well as foodstuffs played a greater part than it does in the main circuit of the Kula. (*Op. cit.*, Ch. XXI, Sec. 3.)

One more subject must be briefly discussed here although it stands a little outside the proper Trobriand economics. The small pearl-shell, called by the natives *lapi*, has been fished and collected from time immemorial, as providing the natives with the most important edible mollusc. When in opening the shell the natives would find a large, beautifully rounded off pearl, they would throw it to the children to play with. Under European influence a new industry has blossomed. By the wise legislation of the Papuan government, European traders are allowed only to purchase pearls from the natives and must not carry out or organise any diving on their own account. To five communities—Kavataria, Teyava, Tukwa'ukwa, Oburaku and Sinaketa—this has, for the last quarter of a century or so, proved what to the natives seems an incredibly large source of income. In many ways it has produced a revolution in native economics. This provides the anthropologist with some interesting sidelights on native habits and ideas; and first on their "conservatism", their strict adherence to tradition and usage.

In spite of the fact that pearl diving opened up a prospect of untold wealth and upset the whole balance of power, it is only those communities where the *lapi* shell was fished of old which continue the industry. Neither the expert and skilled fishermen of Vakuta and Kayleula, nor the crafty and intelligent stingaree fishers of Bwoytalu, nor, least of all, any of the land-lubbers, take to pearling. Technically they could do it perfectly well, as the pursuit is simple. As regards organised effort and their ability to appreciate wealth and trade, they are all on the same level. They abstain partly because the communities in traditional possession of the industry would object and would have a moral—in native opinion also a legal—right to prevent poaching; but more because it is felt that it would be unjust as well as unseemly to encroach upon established rights.

Secondly, pearling gives the anthropologist an insight into the difficulty of creating a demand. The only foreign article which exercises any purchasing power on the natives is tobacco. And even this has its limits; for a native will not value ten cases of trade tobacco as ten times one. For really good pearls the trader has to give native objects of wealth in exchange—arm-shells, large ceremonial blades, and ornaments made of spondylus shell-disks. Again attempts to send certain South Sea material abroad and there to produce imitation valuables has completely failed. An enterprising firm of stone-cutters (I cannot say whether it was in England, Holland or Germany) made an attempt some thirty or forty years ago to produce large stone blades of European schist or slate and to flood various districts of the South Seas. These articles were discarded by the natives as dirt. My friend M. Brudo had one or two pieces of the original stone from Woodlark Island polished in Paris.

It was not accepted by the natives either. So nowadays each trader keeps a retinue of native workers who polish large axe-blades, rub spondylus shell into the shape of small disks, occasionally break up and clean an arm-shell—so that for savage ornaments civilised "valuables" may be exchanged. The Trobriander indeed shows and expresses as readily his contempt for the European's childish acquisitiveness in pearls as a duchess or Parisian cocotte would show for a necklace made of red shell-disks.

The third and perhaps most important sidelight thrown by modern pearling on Trobriand habits and ideas is that the greatest bribery and economic lures, the personal pressure of the white trader and competitive keenness on wealth, cannot make the native give up his own pursuits for those foisted upon him. When the gardens are in full swing "the god-damn niggers won't swim even if you stuff them with *kaloma* and tobacco", as it was put to me by one of my trader friends. When a fishing expedition has to be carried out under tribal contract with an agricultural community "nothing will make the bloody cows do an honest piece of work on the *lapi*." I have calculated roughly that, measuring in standard yam baskets or in their present-day equivalent, sticks of tobacco, the average fisherman can earn about ten to twenty times as much in an ordinary day's pearling as he does by successful fishing. But he will disregard this; he will disregard the gambler's excitement, his day-dreams and ambitions, and go out in order to provide two or three strings of fish for the taro and yams which he has received. Obedience to tradition and the sense of tribal honour make him invariably put his gardens first, his fishing for exchange second, and pearling last of all.

This survey of the various economic districts was indispensable, though it may have been somewhat tedious. It has provided the reader with a full though piecemeal description of the economic data necessary as a background to our special subject. In fact, barring details of technique, descriptions of magical ritual and of the customary pomp and routine of work, we have given here a complete summary of Island economics.

6. WHAT INDUSTRIAL SPECIALISATION LOOKS LIKE IN MELANESIA

Our survey has shown us that agriculture is not merely most important integrally, that is, for the tribe as a whole, but that everywhere it is the main food-producing activity in that everywhere it takes precedence over all other work. We have also seen that fishing, which is unknown in about half the villages, takes the second place even in those centres where it is most intensively practised. As regards industries it is important to make it quite clear that they do not play a part comparable in any way to that of agriculture or even fishing. When we speak about division of labour by districts, we do not mean anything similar to specialisation in a modern industrial community. All villages exercise every art and craft, except such as have only a local utility. Inland villages do not build canoes or make nets.

Apart from that there is no article which cannot be produced not only in wood-carving Bwoytalu but also in the agricultural capital, Omarakana; which cannot be made in a small insignificant hamlet such as Giribwa in the extreme south or Moligilagi in the east, as well as be imported there from the centre of basketry, Luya, or the centre of the lime-pot industries, Kudukwaykela. If you were to penetrate into a Trobriand village on a rainy, sultry day during the agriculturally slack season, or on any day after sunset, you would find every second male in the community busy: carving a lime spatula, or polishing a nose-stick; tinkering at an axe or adze, reshafting or reshaping its handle; making a water-bottle out of a coconut, or finishing off a lime pot. Women again, unless engaged in cooking or domestic work, would be trimming their fibre skirts, sewing mats, blanching pandanus leaves or plaiting baskets. All these crafts are practised and the more specialised articles, such as wooden dishes, clubs, spears, fine baskets, are produced and reproduced everywhere.

Specialisation in this sense obtains in the following classes of article: wooden platters, combs, trimming- and pounding-boards, lime gourds and quick lime, finely plaited basketry, lime pots with burnt-in designs, shell-disks, arm-shells, large polished stone blades, though slightly inferior specimens can be and are manufactured occasionally everywhere. In some cases, notably in the manufacture of shell-disks, arm-shells and polished stone, the industry depends on the supply of raw material. In the case of shells this has to be fished out in the few places where it is found, and is accessible, therefore, only to the three sailing and fishing communities. Again, the stone imported from Woodlark Island is naturally held up in some of the eastern communities and worked there. One or two other villages have some local advantage in obtaining material, e.g. in the swamps of northern Kuboma grows the fine grass necessary for the baskets made in Luya, while the mangrove bogs round Ba'u and Bwoytalu supply especially suitable wood for carving. On the whole, however, it is tradition, communal pride and the organisation of industry which determines the intensive work in industrial centres.

The concrete appearance of such "industrial communities" at times fulfils the ethnographer's expectations and at times is surprisingly, even disappointingly, unimpressive. When instead of going to an average Trobriand village you enter Bwoytalu especially but also Yalaka or Luya, particularly during the time of monsoon rains or inter-seasonal calms, you would find in the central place groups of men sitting on covered platforms, working, reciting fairy tales, competing with each other in a friendly way, exhibiting the finished article and admiring it. Bwoytalu, in fact, gives a picture of a Papuo-Melanesian industrial community in its most developed form. In Sinaketa and Vakuta also you would, at certain times, find clear signs of shell-disk manufacture. In these villages, if you chanced to come in a year when many ornaments had been produced, you might witness the interesting organised competitive exhibition of these, set in a festive framework (cf. *Sexual Life of Savages*, p. 34.) Obviously in the fishing villages nets are often drying (cf. Pl. 6), canoes are being prepared, and at times the whole community is seen setting forth or returning from a fishing expedition (cf. Pl. 7) or can be watched from the beach pursuing their industry (cf. Pl. 5).

I have given this brief summary of the data scattered through my description of districts partly to place these facts of primitive economics in their correct perspective, since in using such words as "agricultural", "fisheries", "industries", "specialisation" it is easy to introduce spurious implications; and partly because even an ethnographer, unless he has met this particular problem before, is liable from his reading and from his museum impressions to approach a centre of South Sea industries such as the Trobriands with somewhat exaggerated expectations. That I know from my own experience in field-work. When you have been contemplating articles from the Trobriands all over the museums of Europe, Australia, the United States—the civilised world, in short; when in travelling through Melanesia and New Guinea you pick them up as prized pieces of perfection hundreds of miles from their place of origin, your curiosity is worked up. You unavoidably anticipate some technical plant, some regular system of production with markets, agencies and channels of distribution.

Your first surprise comes when you find that the production of, say, the lime pots or three-tiered baskets is practically restricted to one or two villages. When you enter the mere cluster of ramshackle huts called Yalaka or Kudukwaykela, and instead of your guild of artists or artisans headed by a South Seas captain of industry you find that half a dozen men, youths and boys carry on the tradition, a shock of surprise is inevitable. The wonderful pottery of the Amphletts—a manifold widely distributed monument to the industry and skill of the Papuo-Melanesians—rests, as I found, on the shoulders of seven old women who then had just three or four young apprentices. A severe epidemic of influenza might destroy the whole traditional line. This seems actually to have happened in the lime-pot industry through the untimely death of three or four men in Kudukwaykela, a generation ago. It is this, of course, combined with the regression in birth-rate, a change of taste and the consequent loss of interest, which has made the boys slack in learning crafts. In olden days the output of even a small settlement could have been, must have been, indeed, enormous. Even if we estimate the time for the production of a masterpiece of a lime gourd or wooden dish at about a week and assume that the men worked for less than a third of their time, some fifteen specimens must have been produced by a craftsman each year. Ten or twenty craftsmen thus could produce anything from 150 to 300 objects. The production of course was limited in the case of some articles by a scarcity of raw material. The volcanic tuff for the stone blades had to be imported from Woodlark Island, gourds had to be grown every year, suitable wood for carving had to be obtained by felling large trees. In the Amphletts clay was obtainable in one spot only and that at a considerable distance from the villages. When I was there, in two villages there was no clay at all and in the third there was very little.

Such handicaps and limitations of time show in what sense we can speak about "industrial activities" in the Trobriands. If we compare these with the breadth of opportunity, the intensive and surplus production which obtains in agriculture, it will help us to see gardening in its correct perspective.

7. THE VILLAGE AND WHAT HAPPENS IN IT

Since we are about, however, to supply the concrete setting for our picture of native economics, let us stop here for a moment and examine more closely the local setting of all this activity. We have been listing various villages but we have as yet no clear idea of their lay-out or of the details of their structure.

The visitor has to get ashore in one of the coastal villages—Sinaketa or Vakuta, if he intends starting from the south, or in Kavataria or Tukwa'ukwa if he begins with the north. These villages—an attractive palm grove with here and there a house showing among the trees and a few canoes drawn up on the beach (cf. Pl. 7), or being paddled into the offing (cf. Pl. 8), do not gain by near inspection. The beach itself is very often littered with rubbish, shells, fish-bones and, nowadays, tins and cotton rags. European influence also shows itself in pieces of corrugated iron disfiguring the thatch, in oil cans gradually replacing the graceful coconut water-bottles. The coastal settlements also appear at first sight chaotic, with huts irregularly placed and storehouses scattered among the palms without much system. But after a close inspection of one of them, or preferably of several, we find that in reality each coastal settlement consists of a cluster of smaller villages. We find also that every small village has definite constituent parts: a central public place surrounded with one or two rings composed of stores and dwellings. Where these are kept separate, the inner ring is formed by the storehouses and the dwellings line the outer side of a circular street (See Fig. 2).

Walking inland we find that the villages vary in size, in structure and in the degree to which the buildings are finished and decorated. If we enter one of the district capitals, say Gumilababa or better still Omarakana, we see a wide central place of a good shape, and large, well-built, and at times decorated yam-houses. But we shall also pass through small inland villages with some twenty huts and storehouses surrounding a diminutive central place; and compound settlements where several small hamlets are clustered in the same large grove. Whatever the village it is marked from a distance by a large clump of tall trees; at times this looks like a piece of untouched jungle, at times the coconut palm predominates. Such clumps of trees are sometimes found to be groves without any habitation—either sites of now deserted villages or places where the jungle is left standing because of a mythological association or some special magical taboo. Every village, whether compound or simple, whether large or small, is a unit of settlement; that is, it is surrounded by its own territory, agriculturally exploited; it has its own water-hole, fruit-trees and palm groves, and in most cases its own access to the seashore. In the chapters which follow we shall become better acquainted with the principle on which villages are divided into component hamlets or sections (Ch. XII, Sec. 3), the principles of land-tenure (Ch. XII), and the structural elements of a village, especially the way in which the yam-houses are distributed and built (Chs. VII and VIII).

The central place in the village is the seat of most public life. Here also many of the garden events which we shall recount in the

PLATE 8

FISHING CANOES IN THE CREEK OF TUKWA'UKWA

Young men and boys from a village often set out on a short pleasure trip to some neighbouring village, or out on the lagoon. The photograph shows the sea-front of Tukwa'ukwa, the village grove of palms passing into the mangroves growing on the neighbouring swamp (Part I, Sec. 7)

PLATE 9

MEN PRACTISING A DANCE

During the dancing period, which takes place in the moon of Milamala, men have a full-dress rehearsal each afternoon, and dance ceremonially in the evening, late into the night. Drummers and singers stand in the middle. Men at times wearing women's petticoats, as in this photograph, at times carrying decorated shields, move rhythmically in a circle. The main art of dancing consists in the rhythm of movement, above all in the complicated shaking of the dancing shield or pandanus streamer (Part I, Sec. 7)

PLATE 10

VILLAGE STREET DURING THE SEASON OF MILAMALA

During these festivities, held for the Reception of Departed Spirits, women can be seen preparing taro and yam dumplings which will presently be cooked in boiling coconut oil in the large pot to the left. To the left is the row of houses, to the right the inner ring of yam-houses (Part I, Sec. 7)

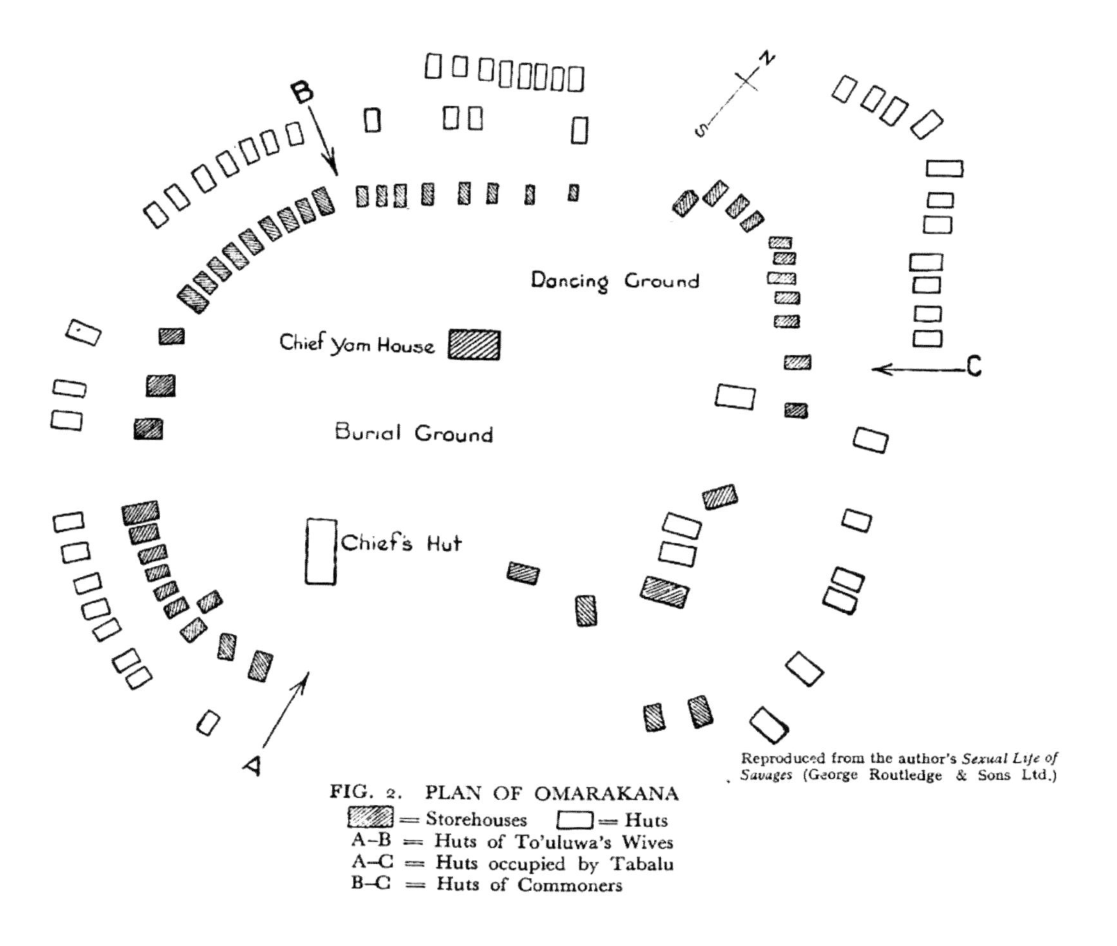

FIG. 2. PLAN OF OMARAKANA

▨ = Storehouses ▭ = Huts
A–B = Huts of To'uluwa's Wives
A–C = Huts occupied by Tabalu
B–C = Huts of Commoners

Reproduced from the author's *Sexual Life of Savages* (George Routledge & Sons Ltd.)

following chapters will take place. We shall assemble there for the annual garden council (Chs. II, Sec. 3, and XII, Sec. 4). At harvest we shall witness the display of new crops and later on we shall follow the ceremonial erection of yam heaps and the filling of the stores (Chs. V and VII). On this place an especially interesting form of magic—the magic of plenty—will be performed (Ch. VII). There also we shall see men forgathering for communal work, or to start on their expeditions to bring fish for one of the great inaugural ceremonies.

It is astonishing how many of the various events of public life which always happen in the central place refer to gardens, or at least involve the use of garden produce. Glancing at the Chart of Time-reckoning (Fig. 3) we see that the half to the right side of Column 5 is taken up with garden activities. For even the ceremonial (Col. 10) is in the first place directly dependent on harvest. The first three moons and the last three are noted in Columns 1 and 10 as occupied by social, sexual and ceremonial life. If we project these entries into their actual setting in the village, they will always be framed by decorations of food: large heaps of yams on the central place, prismatic receptacles filled with produce, festoons made of coconuts and bananas, taro and yam tubers (cf. Pls. 1, 11 and 16). The spirits of the deceased are greeted and gladdened at the feast of *Milamala* by the sight of accumulated raw vegetables and fruit (cf. Pl. 16). The memory of those dead during the last year or two is celebrated with ceremonial distributions of uncooked food. In fact, no sooner have the harvest displays of yams been stacked away in the full yam-houses than these are tapped and some of the food moved again.

If during the last year no one has died in the village—that is no one of importance or rank—a dancing period is opened by a large distribution of food, followed by cooking and eating (cf. Pl. 9). From then on day after day the men will be dancing the whole afternoon and evening, the women, in new and especially gaudy petticoats, will be cooking an unusual amount of food; platters of cooked food will be exchanged from one house to another and constant gifts of uncooked food will pass from hand to hand (cf. Pl. 10).

The dancing period, like most festivities or enterprises in the Trobriands, is punctuated by communal distributions and feasts. There is usually a specially large one at the beginning or in the middle, and several at the end. The headman of the community—whether he be also a district chief or only the local *primus inter pares* does not matter—acts throughout as master of ceremonies. He supplies the food. The members of the community by receiving it are put under an obligation to co-operate in whatever ceremonial activity is in progress: to participate in the dancing or the display of valuables, in competitive games or public singing.

If a member of the community has died during the previous year, there would be no dancing. Probably one or two ceremonial distributions of food would take place instead. Such distributions are among the most usual and characteristic ceremonial acts in the Trobriands (cf. Pls. I and II).

They attain their most elaborate and quantitatively biggest form at mortuary commemorative feasts. Roughly speaking, the principles, sociological, economic and dogmatic, of such ceremonies are as follows. The eldest male of the deceased's kindred acts as master of the ceremony. From his kinsmen and from his wife's relatives and, if he is a notable, from those who normally give him tribute, he receives on such an occasion considerable gifts of food and to this he adds to the extent perhaps of half the contents of his storehouse. On lesser occasions all this food is then placed on the ground in circular heaps (cf. Pl. II); on greater ones into prismatic receptacles (cf. Pl. I). Each heap is apportioned to a person who stands in a definite social relationship to the deceased. As a rule most food is given to the deceased's children and to their kindred, for these are the people who have performed the bulk of the mortuary duties. The distribution is made by a ceremonial calling out of the name of each person as his share is allotted.

After this the gifts, which—let it be noted—consist in uncooked tubers and fruit, are put into baskets by each recipient, taken away to his own village, which is usually at a distance, and there unloaded in the house. With a few insignificant exceptions it is not put away in the storehouses, but distributed by the recipient among his kindred and relatives-in-law on both sides. Every community round the centre of distribution will be thus provided with an abundance of food which must be consumed in especially lavish meals. Plenty in the sense of glut will obtain in the district in which the mortuary distribution has been held.

I have expatiated on these facts because it is important to note that "feasts" or "festivities" in this part of Melanesia consist not of joint eating on a large scale, but of big agglomerations of uncooked food which are redistributed to individuals. At the same time such a redistribution does not merely mean that food changes hands, often several times, within the space of two or three days—it means also that in the whole district more food is prepared and consumed on a lavish scale. As a rule such a distribution will be accompanied by the killing of a pig or two, often as many as a dozen. Or else fish will be provided from one of the coastal villages (cf. Pl. 12). This again means that more tubers are consumed for, as the natives put it, "eaten with a relish" (i.e. with fat, flesh or fish) "the staple food slides down more easily".

When a communal cooking of taro or yam is arranged, the food is not consumed communally; a group of men squat round a pot and each dips his shell spoon into it (*Sexual Life of Savages*, Pl. 86, p. 372). The festive character of the meal does not, therefore, lie in a communal sharing of each other's company in the act of eating —the tendency is to eat in small groups, usually within the family —but much more in the common enjoyment of seeing large quantities of food accumulated and displayed. The substance of the feast consists in giving and receiving and handing on, and in satisfying the obligations of kinship by contributing first of all to the pooling and then to the redistribution of the produce.

To return to the sociological scheme of mortuary distributions, this is so complex and involved that I was only able with the greatest effort and by the collection of a number of concrete instances to establish the fundamental principles of the transaction. The master of ceremonies must be assisted by all those who have kinship obligations to him, but who have not carried out the mortuary duties towards the deceased. On the other hand, he has to give to all those who have in some way or other been discharging duties during the dead man's illness and after his death. Very often it happens that a man will be seen on one day contributing to the distribution because he is a relative-in-law of the master of ceremonies, and after the distribution will be seen returning home with practically the same amount of food, because he was related to the deceased as a classificatory son.

The contemplation of agglomerated food, its handling and its movements constitute, as we have said, the substance of the collective ceremony; but the meal itself, eaten in small groups and families, or it may even be alone, forms also an essential part of this though it is never a collective act. Full satisfaction of appetite, glut and surfeit, are to the native bound up emotionally with the vision of accumulated food. They are the substance of *malia*—prosperity, fullness, satiety—a concept which we shall constantly meet in the following chapters. The idea of "eating till one vomits", of surfeit, of eating with the addition of meat and other relishes, is something which fits naturally into the scheme of the "moons of plenty" which are the months of harvest. This motive we shall find expressed in Magical Formulae, in customs and traditional institutions, as well as actually on the lips of natives (cf. Ch. VI, Sec. 3, Part V, Div. V, § 3, and Part VII, M.F. 16 and 25).

Glancing back at our Chart (Fig. 3,) on the left we find in the subsequent moons, 3, 4, and 5, such entries as preparations in sea-trading communities for overseas expeditions (Col. 3), for shark fishing and fishing in the open sea (Col. 2). If we watched such events as they actually happen in and around the village and on the seashore we would find that once more food is part of the ceremonial setting of canoe-making and that there are important sociological and economic implications in the presents given. Considerable gifts in food would be given by the master and owner of the craft to the specialists in canoe-making. The man who orders large axe-blades, arm-shells or shell-disks in preparation for the Kula, also has to maintain the craftsmen (see above, Sec. 4). The bulk of the activities at this season is not exclusively agricultural, or even directly connected with gardens. It would be a one-sided picture, drawn in a false perspective, that led us to imagine that gardening is the only thing in the Trobriands. But exchange of agricultural produce, gifts in raw or cooked food, constitute the economic side of most of the preparations for overseas trade. In the same way, if we studied the shark and mullet fishing on the north-eastern coast, we should find that while substantial gifts are made to the inland villages whenever there is a good haul, these gifts are sooner or later repaid in garden produce.

During the latter months of the monsoon season and during the calms, there would be less public display of yam food. For by that time the stores are getting low; interest in the gardens is manifested in watching the new crops and trying to estimate the chances of a "plenty" or "scarcity" harvest. Soon some of the early crops grow ready for harvesting. The preliminary ceremonies take place and the displays of the first fruits (Ch. V, Secs. 2 and 3). And this starts the harvest which again fills the whole village with agricultural produce. Then we should see new stores and new dwellings being constructed, and gifts and counter-gifts would be exchanged: basketsful of yams, bundles of taro and bunches of banana or sugar-cane.

Taking our stand on the central place of one of the inland villages we have watched the round of the seasons. Public life consists during the early time of the year in various festivities, some of which, such as dancing, games and singing competitions, are merely framed with distributions of food. Others, such as the mortuary ceremonies, consist very largely in the handling of produce. Throughout these activities the people are busy in the evening repairing or making things or at some craft, and in the industrial centres the men work day after day on their speciality. But it can be said that throughout the year in one way or another garden produce is their mainstay, and the public display of crops forms part of all ceremonial. Small wonder, since agriculture is the dominant food-producing activity and the principal personal interest of every native.

8. WHAT HAPPENS IN THE HOUSE AND TO THE HOUSEHOLD

So far we have lingered on the central place and watched the events of public life. Walking from the inner to the outer ring we would have to pass between two storehouses. Facing one of these on the other side of the street we would see its owner's dwelling, where he lives with his wife and children, the rule being in the Trobriands: one living house—one family. Only seldom do remoter relatives stay with them. Usually even an old couple have a hut to themselves; a widow or widower lives alone, while bachelors share a small communal house (cf. *Sexual Life of Savages*, Ch. III, Sec. 4). The house, the store, the intervening part of the street, the clear space behind where the refuse heap is to be found, all these are the domestic territory of the family. Behind the house the wife cleans and prepares the food for cooking, or works on some of her purely feminine activities, such as skirt-making, mat-sewing, or mending some domestic implement. There also small children often play. The cooking is done either in the house on the hearth stones or, when an earth oven has to be used, behind the house, or at times in front of it. The front is the social centre of family life. There in the evening they sit together, eat, and chat; there they are visited by their neighbours.

Entering the house we would find a stuffy but not offensive interior, scantily furnished and illuminated only by such light as creeps through the small door. The house is no more than an arched thatched roof placed directly on the ground (Pl. 94). Against the walls are bunks, usually two along the back and one along the side (cf. *Sexual Life of Savages*, Pl. 8). By spreading a few pandanus mats, these are made ready for the night. On one of those low bunks we shall watch the magician charming over his herbs and substances. (Ch. II, Sec. 4). Towards the free wall you invariably find the fireplace, consisting of three hearth stones—to the Trobriander in a way as to the Westerner the symbolic centre of domesticity. Here cooking is done on wet days; and on one of the stones, the garden magician will place the offering to the ancestral spirits at one of the most important ritual acts of his system. Higher up above the bunks there run one or two shelves on one of which our magician keeps the magical torches between harvest and the burning of the gardens. On the shelves you would find as a rule the cooking-pots in which food is prepared, the wooden platters from which it is eaten, and a small provision of tubers, the main supply of which, as we shall see, is kept in the storehouse. Only when tubers are received at a ceremonial distribution will they be placed for immediate consumption within the house. Perishable vegetables, such as taro, bananas, fruit, are also kept in the house.

Following the round of daily life we would find that at the busy seasons for agriculture the family set out early for the gardens. There is no cooking of food in the morning. Some cold yams or taro are left over from the previous evening's cooking, and a meal of these is described by a special word meaning "food left over for cold eating". The main meal—and there is only one in the day—is prepared at sunset after the people have returned from their daily work, and consists of small yams, taytu, or the larger variety, *kuvi*; or else of some taro, occasionally seasoned with fish or wild fowl or pork. These may be roasted on embers or boiled in a small pot or, more rarely, baked in the earth oven. During the day, whether the people be out in the gardens or return for the hot hours to the village, no cooked meal is served. Fruit, such as mangoes, breadfruit or bananas, is eaten, or a green coconut broken, its milk drunk and its flesh eaten. Or when there is a good supply of fresh fruit, some of it may occasionally be roasted.

The large clay pots are used only for communal cooking (cf. Pl. 10); that is, when cooked, not raw, food is distributed. This is done in payment for communal work and sometimes it may figure as part of the gift at the other distributions already mentioned.

The natives distinguish between first "staple food",[1] a term which includes all agricultural produce, but refers principally to the small yams (see Part V, Div. II, § 8); secondly 'dainty' or 'lighter food', that is wild fruit, sugar-cane and breadfruit; and thirdly 'relish', that is all forms of protein food: pork, fowl, fish, edible grubs, and shell-fish. Though the staple food is the foundation of their prosperity, and when accumulated symbolises for them plenty, wealth

[1] The expression "staple food" has a native equivalent *kaulo*. In order not to confuse the reader, I have eliminated native words almost completely in this introduction. Their existence, however, is marked by placing the English equivalent in inverted commas. For the analysis of "equivalents" see Part IV, Div. II.

and a succession of joyful tribal events, the natives show a certain lukewarmness about eating it alone. We shall enter into this more fully when we touch upon their attitude towards eating and its physiological importance to the organism (Ch. VII, Sec. 4). They only dimly realise that food possesses nutritive value. They know that absence of 'staple food' means famine which they profoundly dread. But the main importance of eating is that it is a lively pleasure and this is enhanced and prolonged if some relish is added to it.

Another interesting rule of conduct towards food is founded on what might be described as the antagonism between cooking and accumulated food. Cooking, as we have seen, is always done in or around the dwelling. But in villages with an inner ring of storehouses and where these are constructed with open log-work, there is a definite taboo on cooking within this ring. Thus the chief's personal residence which always stands within the inner ring, or a bachelors' house which is also often placed there, must not be used for cooking. There is a special native word for 'dwellings where cooking is allowed'. In those villages where stores and dwellings stand side by side, the stores are invariably covered. The only open stores standing near the cooking-house are the toy stores of children (Ch. VIII, Sec. 6).

We have seen that agriculture is of prime importance in village life and that it enters, either directly or indirectly, into almost every activity. But really to understand its position we must firstly gain a clearer idea of the structure of Trobriand society; and secondly of the economic principles underlying the constant give and take, the movements of wealth, the produce and other objects changing hands. Whether it be a mortuary distribution, a tribute to the chief, an offering at harvest, the exchange of fish for vegetables, the production of objects of value—everywhere we find in the Trobriands the interesting phenomenon that ceremonial and industrial activity is accompanied by the exchange of presents and payments. A closer inspection will reveal some significant principles of this tribal exchange. We shall see that what appears at first sight a free, spontaneous gift is usually woven into a network of reciprocal obligations in which services and contributions are mutually exchanged. Independently of this we shall also find, what concerns us here more particularly, that in most, practically in all Trobriand exchanges—whether they be highly ceremonial and apparently disinterested or as nearly commercial as they ever are—agricultural produce plays not only a conspicuous but a unique part, not matched by any other article of native wealth.

PLATE 11

A SMALL MORTUARY DISTRIBUTION OF TUBERS

The yams, which have been brought into the village by kinsmen and friends of the master of ceremonies, are displayed on the central place in circular heaps. "Each heap is apportioned to a person who stands in a definite social relationship to the deceased" (Part I, Sec. 7)

PLATE 12

SMOKING OF FISH FOR A FESTIVE EATING

Fish caught in large quantities will be smoked and then distributed to be eaten with yams. The quantity here photographed represents one man's share. Fish thus smoked will last for a day or two before it goes bad (Part I, Sec. 7)

PLATE 13

SPECIALIST AT WORK

A master carver decorating a prow-board, such as is seen on the canoe in Plate III (Part I, Sec. 10)

9. THE CONSTITUTION OF TROBRIAND SOCIETY

We must then first introduce some order into the sociological implications which have been running through the foregoing descriptive sketches. We must get a clearer idea of what is meant by "chiefs", "notables", "matriarchy" or "matriliny", "patrilocal marriage", "clans", "sub-clans", "the tribe", "village community", "household" and "family". I have not given any of these words or expressions any special or private meaning. *Notes and Queries in Anthropology* or an ordinary dictionary of the English language will give the plain common-sense meaning of each word and it is in such sense that I have used them. But, on the one hand, in every culture the "family" or the "clan" have a slightly different specific shape and possess certain characteristics which are their own. On the other hand, some of the words, notably perhaps such words as "clan", "matriarchy", "family", are often used in current anthropological literature with a great many hypothetical implications which might therefore be read into them by the specialist or the amateur.

But first let me indicate one or two salient characteristics of Trobriand society which would strike forcibly even a casual visitor and which have their deeper foundation in the social organisation of these natives. The social principles which would thus impress everyone because they come freely to the surface and which the expert sociologist would list as main attributes of Trobriand society, are rank, matriarchy, or more correctly matriliny, and totemism, or more correctly division into clans with animal associations. Put briefly, rank in the Trobriands consists in personal prestige and titles. A man of rank has to be addressed as chief; he is styled either *guya'u* (full chief) or *gumguya'u* (sub-chief) though the latter title would never be used as a term of address. Rank also entitles people to the wearing of certain ornaments, minutely specified and divided into at least four or five grades. Thus the Tabalu have the highest range of insignia, such sub-chiefs as Kwoynama, Burayama and Mwauri would be entitled to wear a slightly lower range of shell ornaments and decorations, the Toliwaga of Kabwaku and their peers would have still a different insignia, slightly less aristocratic. Other sub-chiefs would have the right to certain distinctive ornaments, the commoners would have to be very soberly adorned, and the pariahs of Bwoytalu would have, especially in the presence of people of higher rank, to go without any shell ornaments at all. The most poignant means of differentiation, however, refers to taboos. This is both a burden and a mark of distinction. Roughly speaking, the higher the rank, the more stringent the taboo. The disgust assumed by people of rank with regard to the 'pariahs' of Bwoytalu and Ba'u, culminates in the revulsion which the clean-eating members of a high sub-clan feel towards people who partake of such abominations as the bush-pig, the stingaree, and certain other kinds of fish eaten in Bwoytalu, but forbidden in Kiriwina. Custom here, however, is as usual very complicated and extremely contradictory. Thus the chiefs of Sinaketa eat bush-pig and stingaree and are accepted as almost equals by the chiefs of Omarakana, while the same habit in a member of the Bwoytalu sub-clan makes him profoundly disgusting. This is not the place, however, to enter into more detail concerning the psychology of this complex attitude, which any European, however, ought to be able to understand from his own cultural setting.

Rank also entails a definite ceremonial, the main principle of which is that elevation must be commensurate to rank. The chief's head must not be overtopped by anybody. When commoners are moving about, he must be seated on a high platform; if he stands, they must bend. When commoners have to pass a chief who happens to be seated, the chief has to rise—an uncomfortable prerogative strictly adhered to in the Trobriands. If, for any reason, he remains seated or squatting, commoners have to creep on their bellies. I have myself witnessed all the people present in the village of Bwoytalu dropping from various elevations to the ground, as if mown down by a hurricane, at the sound of the long drawn *O guya'u!* announcing the arrival of one of my Omarakana friends who came to visit me in that village of low rank. In Omarakana also I have frequently seen people of low rank approaching the chief in a crouching position, dragging their hunting-spears behind them as was due and proper.

I shall not dwell on certain further inconsistencies in Trobriand ideas about rank, about the origins of clans and of sub-clans. I have described elsewhere the myth which tells how the totemic ancestors of the four main clans, Malasi, Lukuba, Lukwasisiga, and Lukulabuta came out of the hole of Obukula, near Laba'i; how the dog, the Lukuba, who originally had the highest rank, lost it in favour of the pig, the Malasi. In this myth the general precedence in rank of the four main clans has its charter. The Malasi coming first, the Lukuba second, the Lukwasisiga third and the Lukulabuta last.[1]

The high position of women is equally patent. They take part in a great many public ceremonies. A woman of rank has to a large extent the same privileges as a man of rank, and will be treated with

[1] Cf. Ch. II of *Myth in Primitive Psychology* for further documentation.

a similar show of deference by men who are commoners. The freedom, influence and independence of women is clearly manifested in their surface behaviour. The sociologist will readily get on the track of the matrilineal constitution of their society.

The visitor to the district, especially if he knew a little of the language, or else were present at some public ceremony such, for instance, as the mortuary distribution, would also soon find that society is divided into a number of groups. If you speak the language you will soon find that when two strangers meet, one of the first questions is as to which of the four large divisions of the tribe the other belongs. Even the European is invariably questioned on this point and finds it useful to adopt himself into one of the divisions. For to the native every human being must be either a Malasi or a Lukuba, a Lukwasisiga or a Lukulabuta. A great many of the animals and plants are also associated with such a clan. But the division is not at all exhaustive. That is, the majority of birds and insects, fish and plants, are not subdivided.

To put more precision into these surface observations, we would have to examine not only what is queer and sensational, such as the physical elevation of chiefs, the existence of subdivisions with animal names and associations, and the counting of descent, inheritance and succession in the mother line, but also some of the fundamental forms of human grouping which are less obtrusive, because they are not specifically Trobriand or Melanesian or savage, but which are nevertheless very important. That the household plays a large part in tribal life, and that the household consists of the family, that is of the husband, wife, and their children, is such a fundamental fact of Trobriand sociology. The fact again that all the inhabitants of a local community, that is of a village, do a great many things together on their village soil, jointly cultivate it within the same enclosure, carry out their distributions, enterprises and festivities in common—this is another of the "obvious" facts which, like many obvious things, escapes attention because it does not impress us at once as something queer and exceptional.

It will be best, perhaps, to give a brief outline of all the types of social grouping found among our natives—an outline which will at the same time indicate the sense in which various terms are used in the present book, and give two or three of the outstanding principles of social organisation, of the legal system and of customary usage.

Every individual is born first of all as a member of a household. The child is suckled and looked after by the mother, who, during the first month or so, is isolated, being assisted in her early maternity cares by her mother and sisters. Later, father and mother share in most of the nursery duties. Child-birth, especially of the first child, often takes place in the house of the mother's parents, that is the infant's maternal grandparents. Then mother and child move back to the husband's hut which, as often as not, is situated in another community. Marriage in the Trobriands is patrilocal, that is, the bride moves into her husband's village where the new couple usually build a new house for themselves. There they carry on a joint economy—husband, wife, and later on the children, eating from a common larder, having their food cooked for themselves, and for themselves only, by the mother, eating from the same pot or platter. They all sleep together while the children are small. At puberty the boys move into a bachelors' hut. The family also do a great deal of work together, especially in the gardens. The early education is given to the children by the parents, the mother instructing the girls, the father training the boys. The unity of the family group is not destroyed when the sons leave the parental house for the bachelors' hut, and later on when they leave the father's community and join that of the mother's brother (cf. Ch. VI, Sec. 2). The sons, as long as they remain in the same village, always eat in the parental house; the daughters remain at home till marriage. The sons have to provide for the parental house and later on they have to contribute to their sisters' households when these get married. As long as the father lives, both sons and daughters have varied personal duties towards him in illness or economic distress, when he needs support in a quarrel or when he is in danger from personal enemies. If he dies, his children, more particularly his sons, will have to carry out the complicated and unpleasing mortuary duties. The family thus is an important unit and resembles very much the ordinary patriarchal family, especially on the surface.

If we were to study, however, the methods of tracing descent and the systems of mutual duties, obligations and services, we would find that matters are more complicated than they appear. Filiation —if I may use a word covering all traditional and genealogical continuities, that is, covering descent, inheritance, and succession —is matrilineal. The children are regarded as of the same body and blood as the mother. They inherit her totemic identity; that is, they belong to her sub-clan and clan. Her daughters succeed her in any positions and privileges which she may possess in virtue of her rank or of her relation to the headman of a community. Her sons succeed to her brother. The inheritance of material goods and privileges is also in the mother line. Connected with this we have the curious fact that a man at maturity has to change residence from his father's village to that of his mother's brother.

We have then obviously two principles of social continuity: matriliny, which is the dominant legal rule, makes position, possessions and social identity pass from a group of brother and sister to the children of the latter. The patriarchal constitution of the household and the very strong attachment between the father and his children produce a tendency for many things to be handed over by a father to his children. The anthropologist who is sufficiently cautious not to read into his facts any evolutionary theories or historical reconstructions will not readily jump to the conclusion that here we have a "system of matriarchy in decay, with patriarchal institutions budding". He will simply observe that two independent principles of filiation can exist side by side, and that they produce a somewhat complicated picture, the study of the working of which is our first duty. In this book we shall have occasion to watch one aspect of this two-fold system of filiation at work, when we describe and analyse the harvest gifts, which are among the most important economic and legal institutions of the Trobrianders (Ch. VI, Secs. 1 and 2).

When we speak about the family in the Trobriands, we have to remember, however, that instead of one simple unit we have two groupings: the household with the father at the head, which is the dominant social unit of daily life and which is an important subdivision in the local community, i.e. the village; and the "real kindred", that is the group which consists of the mother and her children, and her brother who is the legal head of that group (Ch. VI, Sec. 2).

If we walk through any local settlement in the Trobriands, we find that a village consists of a simple arithmetic sum and geometrical uxtaposition of a number of households. This again is the obvious or surface phenomenon; and here the male dominance and what might be almost described as the patriarchal principle prevails. For at the head of each village community stands the oldest male of the oldest lineage in the sub-clan of highest rank. But here also working behind the scenes is the matrilineal law, introducing another important group which is neither the village community nor the totemic clan. We know already that descent is matrilineal; that is, every person traces his kindred, from mother to mother, to one common ancestress. The final ancestress is a person who was not born of another woman but who came on earth from underground (Ch. XII, Sec. 1). All the descendants from such a common ancestress form what the natives call 'kindred', but for which I shall use the term 'sub-clan'. Each of such sub-clans, of which there may be perhaps thirty or forty or fifty—I was not able to make an exhaustive list of them—belongs to one of the four dominant clans. On the surface, in conversation, even in folklore, the clan is the more apparent unit. In reality, in law and in economics, in actual behaviour and in sociological implications, the sub-clan is by far more important. As we shall see, the sub-clan is a very powerful unit in the mythological foundations of land-tenure (Ch. XII, Secs. 1 and 3), in the constitution of the village (Ch. VI, Secs. 1 and 2, and Ch. XII) and in magic (Ch. II, Sec. 1, Ch. IX, Sec. 1, and Ch. XII, Sec. 3). The sub-clan also is the unit to which rank is attached. I have listed some ten or twelve sub-clans belonging to the Malasi clan. One of them, the one who rules in Omarakana and other capitals (cf. above, Sec. 4), is universally acknowledged as being the group of highest rank. But to the same clan belong the lowest of the low, the inhabitants of Bwoytalu and Ba'u. This clan therefore is compounded of the highest and the most despised sub-clans.

And this brings us to the problem of chieftainship. In this book, chieftainship, rank, headmanship, play so conspicuous a part that a clear idea must be given of the meaning of these words. And in this connexion it is extremely important to note that in matters of apportionment, of rank, of political constitution, of privileges and the relative position of the various leaders in the various districts, tradition and history have introduced an accumulation of concrete detail in prerogative, hereditary right, claims of rank, and reciprocal dependencies. A brief outline of chieftainship must therefore be rather rough and ready and cannot do justice to the complexity of the subject. Take for instance the unquestionably paramount chief of Omarakana in his relation to his military rival, the chief of the province of Tilataula, resident at Kabwaku. The paramount chief has the higher rank. He is the wielder of the rain and sunshine magic; that is, master of tribal fertility. His personal prestige, the aura of power and dignity around him, are incomparably higher and more august. At the same time, he could be beaten and driven out of his home on occasions by his rival. This rival again has the right to certain ornaments which the chief would not, but also could not, use. This rival also would not when meeting his paramount chief bend very low. A full description of their relationship would be further complicated by the fact that the military rival would give an entirely different account from his point of view of the relationship than would the chief, and that the opinion of third parties would differ from one district to another. Again, the position of the other chiefs of highest rank is complicated: they belong to the same sub-clan as the paramount chief—a fact which no one disputes—

and yet their power and prestige are so much less that it is beyond comparison. Here also we would find a queer paradox; certain food taboos and abstentions are to the paramount chief of Kiriwina and his kindred the very quintessence of their rank. Yet the same abstentions are observed by his low-rank subjects in the capital but completely disregarded by his higher rank cousins on the lagoon coast (see above, Sec. 7).

Bearing, therefore, in mind that, as in every human order, and especially in all primitive institutions built up on a small scale, every single rule has its many exceptions which very often contradict one another and seem to overrule the rule, it is yet possible to lay down the following essentials: chieftainship is a combination of two institutions; rank and headmanship of a village community. Each village has its headman. He is the eldest of the males in the senior lineage of the dominant sub-clan. This sounds a little complicated, but as will be seen (Ch. XII, Sec. 3) every village consists of one or more sub-clans. Usually rank decides which of these will be dominant, and the head of it is the leader of the village community. He will interest us more particularly in the following chapters, because it is either he or his deputy who carry out the official garden magic on the joint garden enclosure, and on behalf of all the members of the gardening team (cf. Ch. I, Sec. 6). This leader, when his village is a capital, that is, a village of high rank, is also the chief of the district (cf. above, Sec. 4) and in the case of one district at least—that of Kiriwina—this man is also the paramount chief in the whole area. His writ does not extend beyond his province. On the other hand some of his economic prerogatives extend well beyond it. The recognition of his supremacy is universal and his fame runs through many archipelagoes where his language is no more spoken nor the customs and ways of his people understood. Besides this paramount chief there is the powerful headman of Kabwaku, capital of Tilataula, and the headmen of the other districts enumerated above. When I speak, therefore, about *the* chief, or *a* chief, this does not refer only to the first and foremost among them. When I speak about a minor chief or sub-chief, I usually mean either a head of a less important district or else a headman of rank whose influence does not extend beyond his own village community. By the term notables, which I am using often, I mean the kinsmen of a chief or, in a community of low rank, the old men, especially those who either practise an important form of magic, or have acquired an outstanding position by virtue of personal skill in some industry or the knowledge of some pursuit, such as sailing, fishing, or gardening. In a village of high rank the sons of a chief, who of course under matriliny are not his kinsmen, the heads of minor sub-clans and the brothers of the chief's wives, who might be invited to reside there, belong also to the group of notables. The rank and file consists, in an aristocratic village, of the members of the lesser sub-clan, of people who reside there because they render specific services to the chief and of those who, because of the smaller importance of their mothers in the chief's polygamous household, are pushed into the background. In villages of low rank, the distinction between notable and ordinary villager depends, as we have said, largely on age and personal capacity.

I have spoken here mostly about chieftainship. The character of the Trobriand population as a whole, what might be called their tribal constitution, is known as much as is necessary for the present book from our survey of districts and from our analysis of the position of the chief. Our description of the village and our analysis of headmanship has given a preliminary grip on the sociology of the village community, and this will become more plastic and precise as we watch the members of the village at their garden work and discuss the constitution of the village in relation to land-tenure. Clans and sub-clans are now sufficiently known for the purpose of the present volume (but cf. Ch. XII, Sec. 3).

10. THE MOVEMENT OF WEALTH IN THE TROBRIANDS, AND THE RÔLE OF AGRICULTURE THEREIN

We are fully aware already that a constant flow of wealth, whether this be agricultural produce, fish, objects of daily use or "valuables", is characteristic of Trobriand economics, of public and ceremonial life. The variety of types of exchange and the complexity of the transactions is so great and bewildering that it is not easy to present the facts briefly and yet without mutilating their reality and omitting some of their essential features.

Let us approach the matter with one or two concrete examples. Soon after I arrived in Omarakana in 1914, the paramount chief, To'uluwa, became aware of the need for a new three-tiered basket. Such baskets are properly produced only in the village of Luya, as we know. Their use to a man of high rank is double: on the one hand their three-fold construction symbolises wealth and plenty. But furthermore the three low cylinders can be placed one into the other, so that only the contents of the topmost are visible. And this is very useful, as the chief can put away into the lower compartments his stores in tobacco or betel-nut. If these were exposed to the public gaze, he would, on the principle of *noblesse oblige,*

A CEREMONIAL FOOD OFFERING IN THE EXCHANGE OF TUBERS FOR FISH

Canoes from an inland village have arrived on the beach of Oburaku—one of the main fishing centres—with a cargo of tubers. This is stacked into square receptacles and carried into the village where they are ceremonially offered by each man to his partner. The picture gives a characteristic view of the palm grove on the beach and the main swamp stretching further south (Part I, Sec. 10)

PLATE 15

ARRIVAL OF A FISHING FLEET

"The haul of fish is taken directly from the canoes, where they have already been made into strings." This picture presents, incidentally, also a scene of direct barter. "Whenever a rich haul in fish is anticipated . . . people from inland may bring yams, taro, or bananas, and exchange them directly for fish" (Part I, Sec. 10)

PLATE 16

THE DISPLAY OF AGRICULTURAL PRODUCE DURING THE MILAMALA

"In the festivities which follow harvest, when the spirits of the dead return to the village to be present at the dancing and feasting, they enjoy the display of food" (Part I, Sec. 10)

have to distribute them among the surrounding people. The necessity for a three-tiered basket, therefore, might have been functionally connected with my arrival in the village, because I used to supply the chief with about half a stick of tobacco a day, and usually with a small bunch of betel-nut.

In order to get this basket, the chief sent one of his daughters to Luya; she went there accompanied by her friends and carrying a basket of yams. After having received it, a specialist from Luya started on the three-tiered basket. Two or three weeks later he brought the unfinished object to show it to the chief and ask him whether he was satisfied with what was being done. He received then another basketful of taytu, and a few rings of tortoiseshell. After about a month's interval from the initial gift, he brought the finished article and received a third basketful of yams.

This was a typical example of purchase of an article by a 'solicitary gift', a 'maintenance gift' and a 'clinching gift'. The article, it will be noted, was made to order. This type of exchange, combined with an order to produce and with occasional intermediary gifts, is characteristic of a great deal of Trobriand barter. We know already that this is the way in which valuables are produced (see above, Sec. 4). This is also the manner in which a specialist would be invited to construct a canoe or to carve a decorative board.

One of the main characteristics of this type of transaction is that agricultural produce always figures prominently in the soliciting gifts and in the maintenance gifts. The production of a valuable—for instance, the carving of a prow-board (cf. Pl. 13) or of a gable-board for a storehouse or dwelling; of an ebony spatula or walking-stick—would entail the feeding of the worker by the chief or notable for whom he was producing the object.

This type of barter, therefore, is closely connected with industrial production. It consists, in fact, in giving initiative to production; in making it economically possible through the maintenance of a specialist or specialists; in reducing the need of one section to work on their gardens and allowing them to follow their craft while others produce food for them. I hardly need to repeat here that this only happens on a small scale. I have said already that everybody is a gardener in the Trobriands, and I have precisely stated in what sense we can speak about specialisation. At the same time, it is important also to make the point clear that all industries are fed by agriculture and that this feeding happens in a very direct manner.

The scheme of barter by solicitary gift and an ensuing exchange of services or goods runs, as some of my readers may know from other works on the Trobriands, right through several of their economic systems. Thus all the ceremonial gifts linked up in the vast institution of the Kula are based on the exchange of an 'opening gift' and 'return gift' with an additional play of 'solicitary gifts' of which the natives distinguish at least four classes (cf. *Argonauts of the Western Pacific*, Ch. XIV). Then come intermediary gifts and finally a 'clinching gift'. In the exchange of fish for food there are several ways or techniques. One of them, the most ceremonial and honourable, consists in an 'opening gift' in which the agriculturalists usually, I think invariably, take the initiative. Food will be brought into the coastal village in elegantly decorated receptacles (cf. Pl. 14). Such an opening gift of food puts the fishermen under an obligation which cannot be waived nor subordinated to other interests (see above, Secs. 4 and 5) to return the gift by its equivalent in fish. The fishermen have to wait for a spell of favourable weather and, having given notice to the inland villages that an expedition will take place, they go out to make the haul. The inland community then arrive on the beach about the time when the fishermen are expected back. Sometimes they have to wait for a couple of hours. The haul of fish is taken directly from the canoes where they have been already made into strings (cf. Pl. 15). Each man takes his strings and runs as fast as he can back to his village community.

As mentioned already, this exchange of vegetable food for fish plays a conspicuous part in tribal life. On certain occasions, as, for instance, at the first inaugural rite of garden magic (Ch. II, Sec. 4), fish must be procured by each inland community. On such occasions the reliable fishing of Kavataria is usually resorted to. Again, when a rich harvest has been garnered in Kiriwina or Tilataula, and either big mortuary distributions are being prepared or a ceremonial season of dancing or games arranged for, an inland community will send a substantial present of vegetables to the coastal village. In payment for this, the fishermen will be bound to go out and supply the return gift. Since festivities are arranged in preference at full moon, the fishermen very often have to do their best in order to procure at least some fish on a certain date. If they cannot do it, the inland distribution may have to take place without fish, which will then be procured for an intermediary ceremony or a final feast.

Such ceremonial exchange of fish for vegetables is based on a permanent system of partnership, whole communities being linked up with each other, and in each community every man having his individual partner. There is a rough equivalence between the measure of food, which is the bell-shaped standard basketful, and the measure of fish, which is a string of fish some two to three kilograms in weight. Besides the ceremonial exchange, there are other forms of barter. When a rich haul in fish is anticipated, or even after it has been reported, people from inland may bring yams, taro or bananas, and exchange them directly for fish. At times, the fishermen take the initiative and ask for vegetable produce. This is especially the case when they need some taro or some yams for seed. But in such cases they will never bring their fish to the inland villages. The produce will be brought straight down to the canoes and exchanged then and there for fish. In the transfer and handling of goods, agricultural produce either takes the initiative or else is directly traded on the spot. On the whole, this system of barter is based on the obvious craving of the inland villagers for albuminous food and of the corresponding need on the coast for more agricultural crops than they can produce.

Another type of transaction, less commercial perhaps than the previous ones, conforms also to the pattern of an initial gift in crops followed by counter-services. Whenever an enterprise is started, the organiser—chief, headman or notable—will arrange for an abundant supply of food which will be distributed to the participants in the enterprise and then eaten. We discussed this in connexion with the manners and social groupings at meals. Such anticipatory payment in food takes place on every occasion of communal labour, not only in the gardens but also in the piecing together of a canoe, in the building of a storehouse or the thatching of a hut (Ch. VIII). Large supplies of food also have to be provided by the master of ceremonies at a 'competitive enterprise', at the beginning of the dancing season, or other festive periods. When in olden days a chief of a district summoned his "allies and supporters" to help him in a war, his also would be the duty of looking after the commissariat.

Besides the circumstantial exchange by gift and counter-gift we have already met one act in which objects are exchanged immediately—in the non-ceremonial barter of fish for food. This occurs also in the case of some of the manufactured articles. When a great number of wooden dishes or combs or lime pots or baskets and plaited armlets have been produced in one of the industrial centres the villagers will take their wares and visit places rich in agricultural produce, or fish, or possessing some of the articles imported from overseas. They show their goods and ask for what they want. They will directly mention the quantity and then bargaining will inevitably take place till both sides agree on the fair amount. On other occasions some of the seafaring people, starting on an overseas expedition, may go to an industrial community in order to acquire some goods for trade overseas. Here also "trade with bargaining" would take place. There is a native word for "trade with bargaining" and the forms of exchange here described have a much more definitely commercial character. At the same time, it must be remembered that in this "trade with bargaining" the range of possible equivalents is very limited, and that the equivalence is fixed fairly narrowly by tradition. Thus a small wooden dish from Bwoytalu would be, in nine cases out of ten, exchanged for a basketful of yams. In the tenth case the wood-worker might wish to get some turtleshell ear-rings in a coastal village, or in olden days a stone blade on the eastern coast, or he might be prepared to give two wooden dishes for a small cooking-pot, usually obtainable in Kavataria or on the island of Kayleula. But the most striking feature of all trade in the Trobriands is that *de facto* in most exchanges garden produce figures as one of the elements. When the industrialists turn east to Kiriwina or Tilataula, or go south to the non-industrial parts of Kuboma or Luba, their main interest is in food. A basketful of yams is always useful and the quality of the food which can be obtained in Kiriwina is the best. And in food, especially that used for ceremonial purposes and gifts, size and quality count for very much.

Again, when the members of an overseas trading community, let us say, Sinaketa or Vakuta, prepare for their expeditions, and visit, as we have seen, the industrial centres, they may take, as well as their garden produce, coconut and betel-nut, certain articles from overseas which they have kept for this purpose from a previous expedition. But food always preponderates. For while food is always welcome and always fetches its price, the exchange value of any other article depends very much on whether it is wanted by one or other of the potential purchasers at that moment. There is an interesting counterpart to this in modern conditions, when among all the European importations there is one and one only which can be used as an almost universal medium of exchange. This is, as we know already, trade tobacco. I pointed out above (Sec. 5) that even this has its quantitative limits; that is, the purchasing power of this narcotic will not increase in direct ratio to its quantity. When it is a matter of exercising a powerful economic incentive, only native objects of value can be used. But in matters of small exchange, that is in everything which an ethnographer might need from a native, either as specimens or services, tobacco has become now universal currency. The white traders, who as we know are mainly after pearls, find it necessary to keep in their stores an assortment of other goods: steel blades for axe or adze, knives, belts, mirrors, calico, hurricane-lamps, kerosene, rice and sugar. When a native needs one of these articles, he wants that and nothing else. The trader who lacks it is at a disadvantage. But if the trader has to take the initiative—unless it is the matter of a very large pearl—he can best obtain the native's commercial response with tobacco.

In this tobacco has completely paralleled, though not replaced, the one and only article which, under the old conditions, approached currency—a basketful of taytu, the staple variety of yam. Yams measured by basketsful constitute always the initial or solicitary gift. They figure as maintenance gift in what might be called feeding or financing the work of a specialist. Large quantities of yams are indispensable for the starting and carrying on of any tribal enterprise. In ordinary barter, a basketful of yams is the only commercial unit which functions as a measure of value. A basket of yams has another equivalent in a bundle of taro. But there are two differences between taro and yams. Firstly taro ripens in small quantities throughout the seasons, so that there is always a steady supply but never any great abundance; and secondly it does not keep. Thus, though a bunch of taro is technically equivalent to a basket of yams, it never attains anything like the importance of the latter. But in neither case would it be correct to speak about food being currency, still less money. As a rule, in every type of article exchanged, there is a limited range of counter-articles, and there is a specific measure of equivalence established. Sometimes this can be reduced to and expressed in basketsful of yams. At times, however, it would be impossible to do so. Thus while a small wooden dish is worth a basketful of yams in Kiriwina, and eight coconuts in some of the southern villages, this does not establish the equivalence of eight coconuts to one basketful of yams. There are communities which would exchange one basketful of yams for four coconuts and others where sixteen coconuts would be readily offered. Again, in some of the more traditionally fixed forms of exchange we might find that a large valuable object, such as a ceremonial axe-blade, a string of shell-disks, or a pair of large arm-shells, has cost a man some one hundred basketsful of yams; but if you wanted to purchase a valuable with tobacco you would have to pay perhaps a thousand or two thousand sticks, that is about ten or twenty times its price in yams. In fact it would be very doubtful whether you could make this exchange at all, except on the very rare occasion when the person in possession of the valuable needs a large quantity of tobacco for a distribution. The explanation of this lies in the fact that the first transaction, that of valuable for yams, conforms to the traditional obligatory type of exchange. In the selling of a valuable for tobacco, the native's greed or his need of smoking supplies for a large ceremonial distribution would supply the incentive. In other words, there is no regular market, hence no prices, hence no established mechanism of exchange, hence no room for currency—still less for money.

In brief, it would be as incorrect to speak about currency or money or a medium of exchange in the Trobriands as it would be to apply to their economics the concept of capital and interest, or imagine that they have specialisation of industry and labour comparable to ours, or to imply that their trade is based on a system of markets where haggling determines the price. There is nothing of the sort in the Trobriands, and the paramount part played in their exchanges by agricultural produce has been clearly defined.

We have looked at matters from the point of view of the industrialist who offers his wares to members of neighbouring communities. The average man in an agricultural district—and in a way he is a "standard Trobriander"—produces enough vegetable food for his household; though actually he only keeps half of it, giving the other half to his sister's household and receiving a corresponding amount from his wife's brother (Ch. VI). Therefore of garden produce he has as much as he needs. The only things which he must obtain by exchange are some of the raw materials not to be found in his district, protein food in the form of fish, which is a very welcome addition to his overwhelmingly starchy diet; and finally articles of some special kind or excellence. Now for every one of these things he is able to pay in food. Occasionally, when trading with his western neighbours, he may be asked for a finished or unfinished piece of stone (this refers to olden days); or when trading eastwards, he may offer a cooking-pot, some rattan or feathers, or some other importation from the d'Entrecasteaux Archipelago. We see therefore that both the industrialist and the agriculturalist are mostly driven to exchange manufactures for food, and only very seldom and in so far as they act as middlemen are other objects introduced into the exchange.

There is one more actor in this game whom we have to consider. How does the movement of wealth and the forms of exchange look from the point of view of a chief or of an important headman? As we shall see, the chief through his plurality of wives is able to accumulate a considerable proportion of agricultural produce (Ch. VI, Secs. 1 and 2, and Doc. III). In addition to this, he also receives tribute from his subjects. He has also a number of important economic monopolies. Thus in Kiriwina, the paramount chief is the titular owner of all the coconut and betel-nut palms in the district, which means in practice that he is given annually a small proportion of the nuts from each palm—enough to make his quota mount to the large figure of several thousand nuts per annum. He also is the only person who is allowed to keep pigs within the whole district of Kiriwina. This means again that every pig in the district is 'apportioned' to an individual who on the killing has to give part of it to the chief. In reality pigs used to be killed almost exclusively in the chief's village, some part of them left there, and only what remained taken to the home of the actual keeper. A chief of minor rank, an important headman, enjoys similar privileges within his own village.

But the real importance of this lies in the fact that the chief on the one hand has the power to accumulate agricultural produce and to control the live-stock and the palms of the district, while on the other hand, he is the one who has both the right and duty to use this accumulated wealth effectively. At the chief's bidding, objects of wealth are produced, canoes constructed, large storehouses and dwellings built. Again he is the organiser of big enterprises. Finally the wealth which he accumulates, or such wealth as is produced by means of this accumulated food, he can use to organise war, to pay a sorcerer for killing a man by witchcraft, or even to pay a man for spearing an offender. The actual way in which a chief in the Trobriands exercises his power is largely economic. For every service received he has to pay, and the wherewithal for these payments he obtains through the duty of most of his subjects to produce for him. Yet in the long run, all the wealth accumulated by him flows back to his subjects. This pooling and reapportionment, however, is not a mere idle play of changing hands. In the course of it some wealth becomes transformed into more permanent objects, and again a great many events and institutions in tribal life are organised by this process of concentration and redistribution. It is this process which allows of such industrial specialisation as exists. It is this process also which makes wealth an instrument of political organisation.

It is not possible further to elaborate here the concepts underlying Trobriand economics, or enter more fully into the relations between agriculture and industry, economics and the ceremonial life of the tribe, social organisation and the various forms of give and take; nor yet to follow the astonishing complexities of such an institution as the mortuary distribution of food. Enough has been said to provide the reader with all that he needs to know about Trobriand organisation in general and also to supply the necessary background for the chapters which follow. We have seen that all ceremonial life is framed in displays and repartitions of food and that gardening ramifies into other aspects of culture. Take religion for instance. In the festivities which follow harvest, when the spirits of the dead return to the village to be present at the dancing and feasting, to enjoy the display of food and valuables, and to partake of the cooked dishes of food which are exposed for them (Ch. I, Sec. 6 and Ch. IX, Sec. 2)—in this native All Souls' Day, or as the Trobriander would put it, All Souls' Moon, agriculture plays an essential part (cf. Pl. XVI).[1] Again in mortuary distributions, agricultural produce, moral duties, and pious remembrance of the departed one, are all inextricably bound up. The spirits of the dead come again into touch with the gardens through garden magic, though here the belief in their "real" presence is not quite so clear (cf. Part VII, M.F. 1, D.).

The all-pervading influence of magic is nowhere better shown perhaps than in agriculture. But I shall not dwell further on it here, because it is sufficiently prominent in the chapters which follow, and also I have dealt with it specifically in another place.[2] In the fact that it is connected with all other aspects of Trobriand culture, gardening does not differ from any other important activity or institution. But in that it gives a tone and initiative in many ways to tribal life, that its produce is the foundation of native wealth and the root of political power and of law and order, that in all the exchanges it plays a dominant part, and provides a leverage, agriculture occupies quite a special position in Trobriand life. We shall now proceed to its description and analysis.

[1] Compare also my article on the "Spirits of the Dead" in *J. R. A. I.*, 1916.

[2] Cf. *Argonauts of the Western Pacific*, Chapters 17 and 18, and for a theoretical analysis of the subject "Science, Religion and Magic", in *Science, Religion and Reality*, edited by J. Needham, Sheldon Press, 1925. In the present book besides recurrent references I deal with magic specifically in Chapter I, Sections 5 and 6, in Appendix I and in Part VI.

PART II

GARDENS AND THEIR MAGIC ON A CORAL ATOLL

FIG. 3. CHART

1	2	3	4	5	
OTHER ACTIVITIES	FISHING	TRADE and KULA	WIND SEASONS	MOONS	
				European Months	Nativ Moon
Social and Sexual Life (cf. also Column 10)			South-East Trade Wind	August	1 MILAMAL
				September	2 YAKOS
	Pearling—an occupation associated with lagoon fishing; Shark Fishing; Fishing in the Open Sea; Lagoon Fishing (Mullet)	Preparations in Oversea trading Communities	Calms	October	3 YAVATAKU
Period of Monsoon and Garden Pause: In non-KULA Communities, Industrial Activity: Carving in BWOYTALU, Basket work in LUYA and YALAKA, Producing of Nets.				November	4 TOLIYAVA
			North-West Monsoon	December	5 YAVATA
		Sailings from the Trobriands to the East and South		January	6 GELIVILA
				February	7 BULUMADU
					8 KULUWOT
	Fishing in the Open Sea	Sailings to the Trobriands	Calms	March	9 UTOKAKAS
				April	10 ILAYBISILA
Social and Sexual Life		Preparations for Canoe Building begin	South-East Trade Wind	May	11 YAKOKI
				June	12 KALUWALA
				July	13 KULUWASAS

NOTE:—Alternative Names for certain Months:—GAYGILA = TOLIYAVA

TIME-RECKONING

6	7	8	9	10
AGRICULTURAL SEASONS	KAYMATA (Main Gardens)	KAYMUGWA (Early Gardens)	TAPOPU (Taro Gardens)	CEREMONIAL
GEGUDA — MALIA	Pause in Gardening; GABU			MILAMALA in KIRIWINA — Dancing; KAYASA; Ceremonial Distribution
	KOUMWALA and Early Planting	SOPU		MILAMALA in VAKUTA
	KAMKOKOLA and SOPU			
MOLU	SOPU and Making Fences	ISUNAPULO	Planting (Dry Soil) — Second Cycle	
	Growth Magic (1); KAVATAM and Weeding			
	ISUNAPULO		Second Cycle	
MATUWO	Growth Magic (2)			
	BASI	HARVEST	ISUNAPULO	
MALIA	OKWALA and TUM; Harvest begins		Planting (on DUMYA) — First cycle	KAYTUBUTABU Season
	KAYAKU; Harvest continues	KAYAKU; TAKAYWA	First cycle	Dress in Preparation for MILAMALA
	Filling the BWAYMA; VILAMALIA	GABU		MILAMALA in KITAVA
GEGUDA	YOWOTA; TAKAYWA	KOUMWALA; Early Planting: KUVI, Taro, etc.	ISUNAPULO	MILAMALA in SINAKETA, LUBA and Western District

KATUBUGIBOGI = YAVATAM. OBWATAYOUYO = YAKOKI.

CHAPTER I

GENERAL ACCOUNT OF GARDENING

Gardening and fishing then—briefly to sum up what has just been said—are the two principal sources of livelihood to the Trobriander. Hunting provides him with hardly any sustenance. His domestic animals, pig and fowls, afford a pleasant seasoning to his food on festive occasions, and the fruits collected from the bush help him out in times of scarcity. But neither collecting nor fishing nor domestic animals are sufficient when gardens fail. A drought or a destructive blight on the crops inevitably mean hunger (*molu*) for the whole tribe; and this, the most dreaded of calamities, though it happens but rarely, is remembered for centuries. A year of good harvest on the other hand means prosperity (*malia*), that is satisfaction, festivities and, incidentally, village brawls and fights; in short, all that makes life worth living. We know also that fishing and gardening are closely interrelated and that agriculture forms the backbone of tribal economics.

1. THE SEASONAL RHYTHM OF GARDENING

The cultivation of gardens gives the full rhythm and measure of the seasonal sequence in the year. The chart of native time-reckoning (Fig. 3) brings out clearly the dependence of the gardening cycle on the seasons, and of the various phases of tribal life on gardening.[1] To the natives the annual revolution of the seasons is defined by the cycle of agricultural activities. Even the name for 'year' is taytu, a small species of yam, which is the staple crop of the district. And this etymology is not far-fetched or antiquated to the native, to whom the past year is literally "the time of the past taytu" or shortly, "the past taytu". Immersed as he is in his garden work, he reckons time in terms of crops: the past crop, two crops back, three crops ahead, and so on.

Again the year is subdivided for him into the season when the gardens are unripe (*geguda*) and into that when they begin to mature (*matuwo*. Chart of Time-reckoning, Col. 6). Within both subdivisions there are periods of intensive and exacting labour and times when work can be slackened; but, during the time that gardening is really important and the whole community are busy at

[1] Cf. Note 1 in App. II, Sec. 4, for methodological comments concerning the construction of the chart, and certain inadequacies of it.

it, nothing else is allowed to interfere with this work. (See Cols. 1–3, which show only such tribal activities, during moons 2, 3 and 4, as can be fitted in with agriculture.) If the crops are not ready, the overseas expeditions, *kula* (Col. 3) which have such a fascination for the Trobriander, are postponed, until all that can be done by man is finished. Sex interest, dancing, festivities are made subordinate to agriculture; even warfare (as far as I was able to ascertain, for it has been completely suppressed) was not allowed to break out during the three moons (2, 3 and 4, Cols. 1 and 10) of intensive work after the gardens had been burnt and the clearing, planting, fencing and erection of supports were in progress. The festive, ceremonial and love-making season falls between harvest and this period of work, while fighting and sailing take place during the slack time of gardening.

The correlation of the season of rejoicing and dancing, or else of mortuary festivities and distributions (cf. Part I, Sec. 7) with its phase of garden work is shown on our chart. It is not very easy, however, to express it clearly there because, in the first place, the central moon of ceremonial life, *Milamala*, varies according to four districts (see below). In the second place, when the festive season is extended by agreement, the gardens are started later. Thus, though there is no confusion in the seasonal correlation, the exact adjustment is fluid. Roughly speaking, the longest time between harvest and the new gardens is about four moons, while it may, on occasion, extend only over two.

Gardening seasons thus constitute the real measure of time. The native who wishes to define a period or to place an event will always co-ordinate it with the most important, the most rigidly maintained, and the most characteristic index of that period; that is, with the concurrent gardening activity. He will say: this happened *o takaywa*, during the cutting of the scrub; *wa gabu*, in the burning, the period when the cut and dried scrub is being burnt; *wa sopu*, in planting time; *o pwakova*, during weeding; *wa basi*, during the removal of the surplus tubers; *o kopo'i*, during the preliminary taking out of taro and yams; *o tayoyuwa*, during the harvest proper.

Gardening activities are correlated with the sequence of moons, for which the natives also have names. Thus the moon of *Milamala* or festive period, in which the spirits of the departed visit their native villages, usually coincides with the pause between the cutting and burning. This correlation is shown in Column 5 of our chart. Here we find the thirteen names for native moons which correspond, not rigidly, but with a high degree of approximation, to our counting of months within the year. Since the prevailing winds (Col. 4) fit precisely into our calendar, the correspondence of Columns 4 and 5 is close. Column 6 shows the native distinction between the seasons of plenty and hunger (*malia* and *molu*), indicating the supply of crops, and the seasons of ripe and unripe (*matuwo* and *geguda*) gardens which marks the development of the crops. Garden activities (Cols. 7, 8 and 9) are through this correlated to moons and seasons, though this correlation may, as we know, vary within a moon or even more.

Let us follow this in more detail. Glancing at our Chart of Time-reckoning, we can see from Column 4 that about the month of September or October the dry season of the trade winds comes to an end. During the calms which follow and later during the monsoon, there is a considerable rainfall. Now the gardens must be cut, burnt and cleared some time before the wet season really sets in, nor must the sprouting and growth of the plants fall too late in the rainy season. On our chart, the typical or ideal sequence of activities in relation to moons and seasons is depicted. Thus the *Milamala* moon is made to coincide with the pause in work on the main gardens (*kaymata*); the planting begin with the early rains of the monsoon and the harvest towards the end of the second calms which precede the onset of the trade winds; while the garden council (*kayaku*) takes place towards the end of the native year. If, as occasionally happens, the natives decide to extend the dancing period for a moon after that of *Milamala*, their inaugural activities would be postponed. In such case, however, they would usually work communally and compress the early stages into a shorter period, so that the planting is not substantially delayed.

Another complication which is not shown on the chart is the difference in moon-reckoning in four distinct districts. In Kitava the visits of the spirits are received in June; in the southern part of the main island, from Okopukopu to Olivilevi southwards, as well as in the western districts, that is Kuboma and Kulumata, in July; Kiriwina receives them in August, and Vakuta in September or even October. In the last-mentioned district, the moon of *Milamala* coincides with the appearance of the palolo worm on the fringing reef, which is also called by the natives *Milamala*. I think that the gardening activities are synchronised throughout the district, in spite of the differences in moon-reckoning. In terms of our chart we can say that the correlation between seasons, European months and economic and agricultural activities remains fixed, and that it is only Column 5 that moves, according to the naming convention of the district, while the entries in Column 10 indicate the differences of the festive period.

But it is the gardening activities that really matter and that determine the sequence of time—the moons have a subordinate importance. When precision in dating is required, the natives have to refer coming events to such and such a moon, and such and such a day within a moon, and for this purpose their lunar calendar is necessary. But even then they usually have first to place the moon within the gardening activities to which it belongs and then only use the moon names as a means of more detailed and precise definition.[1]

The chronological sequence of the years is also defined by gardening activities. The natives have a name for every *kwabila* (field, or division of garden land) and since the gardens are made successively on such fields they are able to associate a past event with the name of the two or three fields which were put under cultivation in that year. For, as we shall see in the chapter on land-tenure, every field has its proper name. Thus when they are asked about a past date, they will enumerate the names of the sites on which gardens have been made in each of the preceding years, till they arrive at the right combination of names, and thus they are able to count the years for several decades back.

2. THE SEVERAL ASPECTS OF TROBRIAND AGRICULTURE

In the Introduction I have shown in a concrete manner how agriculture permeates tribal life. Now let us see how other aspects of Trobriand culture enter into the system of gardening. Any observer who has lived, worked, and conversed with the natives, would be impressed by the sheer bulk, complexity, and abundant detail of their gardening occupations, and the number of extraneous and supererogatory activities which cluster about these.

Gardening is associated with an extremely complicated and important body of magic, which, in turn, has its mythology, traditional charters and privileges. Magic appears side by side with work, not accidentally or sporadically as occasion arises or as whim dictates, but as an essential part of the whole scheme and in a way which does not permit any honest observer to dismiss it as a mere excrescence (cf. App. I).

Agriculture also has its legal aspect. When we come to the dis-

[1] I have dealt in another place with the lunar and seasonal calendar (*Journal of the Royal Anthropological Institute*, Vol. LVII, 1927). I have there brought out the subordinate part played by the native moons in their calendar reckoning and the paramount importance of seasonal gardening activities. I also tried to account for the fact that the moon names and distinctions are more clearly and definitely laid down in the case of the first eight or nine moons—that is, during the time that gardening is actually in progress (cf., however, Note 1 in Sec. 4 of App. II).

tribution of plots for cultivation, we shall see that a complex system of privileges and claims and duties is involved, accompanied by semi-ceremonial transactions which the natives by no means treat as trivial or irrelevant.

Again the sociology of garden-making is intricate. The parts played by the chief or headman of the community, by the official garden magician, by the owners of the soil, by those who lease garden plots, by those who benefit from the harvest, dovetail and intertwine into a complex economic and social network which constitutes the land-tenure of these natives (cf. Chs. VI, XI, and XII). Another important aspect of culture into which gardening enters directly is social organisation, notably the kinship system and political power. The natives usually harvest a large surplus over and above what is necessary to nourish them; and this surplus figures in tribute and the marriage gift. When we study the distribution of yams at harvest (Chs. V to VIII) we shall see that the best produce is always given by the gardener to his sister and her husband, and, owing to the system of polygamous marriage which is definitely the privilege of rank and chieftainship, a large proportion of such matrimonial gifts finds its way to the storehouses of various chiefs and notables. The whole institution of chieftainship is founded on the large tribute in staple crops which the chief receives from the maternal kinsmen of his wives (cf. Part I, Sec. 10 and Ch. VI, Sec. 2). The enormous quantity of yams thus placed at his disposal, he in turn has to distribute, partly in the financing of feasts and tribal enterprises and partly in maintaining a number of industrial workers who produce objects of permanent wealth for him (Part I, Secs. 4, 6 and 10). Gardening, and effective gardening at that, with a large surplus produce, lies at the root of all tribal authority as well as of the kinship system and communal organisation of the Islanders.

Finally, among other apparently extrinsic elements, we find a surprising care for the aesthetics of gardening. The gardens of the community are not merely a means to food; they are a source of pride and the main object of collective ambition. Care is lavished upon effects of beauty, pleasing to the eye and the heart of the Trobriander, upon the finish of the work, the perfection of various contrivances and the show of food. We shall also see how other incentives besides mere greed and anxiety are brought into action by joint family farmings and competitive display. A further complexity is added to Trobriand gardening by the diversity of crops, the various kinds of gardens, and the differentiation of plots according to their magical, aesthetic and practical function.

The theoretical synthesis of all these elements—the meaning and

PLATE 17

THE CULMINATING ACHIEVEMENT OF GARDENING

"At two stages of harvesting the natives stack the neatly cleaned yams into conical heaps; cover each heap with an arbour, and allow it to remain a few days, or even weeks, to be admired by villagers and visitors" (Ch. I, Sec. 9; cf. also Chs. V, Sec. 4, and VI, Sec. 3)

PLATE 18

ON THE ROAD CROSSING THE JUNGLE

". . . between two green walls of low, dense jungle of recent growth . . ." (Ch. I, Sec. 3; cf. also Ch. V, Sec. 5)

PLATE 19

ON THE ROAD THROUGH A HARVESTED GARDEN

". . . men carry in oblong valise-shaped baskets, or else they shoulder the very big, long yams" (Ch. I, Sec. 6; cf. also Ch. V, Sec. 6)

function of magic, the part played by elegance and aesthetic finish, the relations between the privileges of kinship and the influence of myth—are subjects which we shall be discussing in the following chapters. At present, so as not to lose our way through the detailed accounts which follow, we must lay down a few general principles.

3. A WALK THROUGH THE GARDENS

We have gathered something of the work and care lavished upon gardens through our somewhat desultory visit to them in the Introduction. The landscape of the Trobriands is not at first sight beautiful. We are on a flat, even, coral foundation, covered for the most part with fertile black soil, interspersed with patches of swampy ground, and of drier, stonier soil. Round the northern and eastern shores of the main island and of Vakuta there runs a low, irregular, coral ridge, named by the natives *rayboag*, which is covered with primeval forest. The remainder is almost entirely under intermittent cultivation, so that the bush, cleared away every few years, cannot grow to any height. When you walk across the country, therefore, you either move between two green walls of low, dense jungle of recent growth, or you pass through gardens. Glancing at Plates 18 and 19 we see parties of men and women carrying tubers—a sight typical of the district at harvest-time, when food is constantly being transported from gardens to villages, or, again, when there is a competitive food display between two villages (Ch. V, Secs. 5 and 6). Women carry in bell-shaped baskets on their heads; men in oblong, valise-shaped baskets or, in the case of the very big long yams, they shoulder their burden. On Plate 18 we see a party on a road through the low jungle, and on Plate 19 another passing through a harvested garden.

The gardens are certainly the more attractive part of the landscape. We pass over completely cleared ground which leaves an open view into the distance, where the horizon is broken by an occasional clump of trees marking the site of a *boma* (sacred grove) or one of the numerous villages; or else our eye travels towards the jungle on the coral ridge or sweeps across the green lagoon between the islands. The gardens shown on Plate 20, which was taken towards the end of the clearing and after most of the preliminary crops had been planted, illustrates such a view. These crops can be seen already growing, the tall tufts of sugar-cane, the young heart-shaped leaves of the taro, and here and there an early yam vine of the large variety (*kuvi*) climbing round the stems left standing after the cutting and burning (Ch. II, Sec. 5, Ch. III, Sec. 1). In the foreground we see the poles already laid which divide the gardens like a chessboard into squares. In the background the scrub, which in this case was almost at the foot of the coral ridge, rises behind the fence. A group of men are seen at work.

Or again we traverse a yam garden in full development, reminiscent somewhat of a Kentish hop-field and unquestionably more attractive. The exuberant vines climb round tall stout poles, their full shady garlands of foliage rising like fountains of green, or spilling downwards; producing the effect of abundance and darkness so often referred to in native spells (Pl. 21). Even the gardens already harvested, in which here and there a banana tree is left growing and the old crop of sweet potatoes still continues, have their charm, the charm of an old untidy orchard. In the marshy districts we might pass by a taro garden with its array of scarecrows and wind rattles, a new stout fence encircling the low flat surface of broad green leaves. We would meet a somewhat different picture in the south, where patches of fertile soil are scarce and small gardens are often wedged in between jungle, mangrove swamp, and stony coral outcrops. Plate 22 shows such a taro garden. The new plants are seen growing among large heaps of stones; a small *kamkokola* stands near the well-built fence. On the other side of the fence is the site of an old garden. Walking along the coral ridge we would, from time to time, come across a more or less deep hole in the dead coral filled with black humus, and planted with the large variety of yam which grows specially well in this soil, the vine trained round one or two supports and spreading over the rim.

On a rough computation I estimated that about one-fifth or perhaps one-quarter of the total area is under tillage at any one time. The cultivation of this area is very varied because, in the first place, the natives have two entirely different types of garden—the *tapopu*, exclusively planted with taro, and the gardens in which the yam predominates; and in the second place, these latter are of two kinds, the earlier, *kaymugwa*, and the main gardens, *kaymata*. The *kaymugwa* are made on a smaller scale and planted with very much more mixed crops than the *kaymata*, which are almost all taytu. With the various stages of decaying, flourishing and harvested crops, the occasional plantations of banana and sugar-cane, and the cultivated holes in the coral ridge, the complexity of gardening and its claim on human attention and labour become patent.[1]

A closer inspection of the gardens reveals other interesting details. For instance, some plots are much more carefully worked than

[1] On the relation between the early and the main gardens, see Note 2 in Sec. 4 of App. II.

others. These are usually the most advanced, and the surrounding fence, the vine supports and certain large magical structures called *kamkokola*, display a better finish and are of bigger proportions. Usually these plots will be the first we meet on entering the gardens from the village. They have a special name, *leywota*, and we shall designate them throughout as "standard plots". They are generally cultivated by some important persons and play a leading part in magic and in gardening.[1] They are in a way representative plots as the work on them has to be done with full aesthetic finish and the maximum of perfection; on them no magical rite may be omitted, and some ceremonies are performed on them only, though meant indirectly to benefit the rest of the gardens. These plots are the pride of the community and the focus of all magical activities.

Thus even an occasional visitor would find the Trobriand gardens not only attractive but intriguing in their detail. The ethnographer finds them, even during his preliminary explorations, full of interest and significance. The obviously non-utilitarian, geometrical erections at the corners of each plot, the *kamkokola*, promise well for what might be called the magical or esoteric dimension in gardening. A careful inspection of the corners on which the biggest *kamkokola* stand, the "magical corners" as we shall call them, would reveal further details: a small construction of sticks, *si bwala baloma*, "the house of spirits" as it is called by the natives; a miniature *kamkokola* made of slender sticks; a group of special plants leaning against the *kamkokola;* some herbs inserted into it; and again, a strand of tough grass wound round the pole. Sooner or later the ethnographer discovers that these are traces of magical activities, and indeed, in any of his walks he might come across the magician leaning over a *kamkokola* and reciting his spells. On Plate 23 which was taken during an actual performance, we see a clump of plants in the magical corner. The large *kamkokola*, as well as the miniature replica, can also be seen, together with a row of large poles, ready to be put up as yam supports (cf. Ch. III, Sec. 4). Starting from such visible signs, he is led gradually to discover the world of mythology and magic, the ideas of value and the sentiments of a sociological nature which surround gardening.

When he walks with the natives through a well-tilled fully developed garden in a year of plenty, he realises that to the Trobriander the whole garden oozes prosperity (*malia*). When he watches the natives at their communal work during the preliminary clearing or the planting of the seeds, when he accompanies a family at some other stage of gardening and spends the whole day with

[1] See Note 3 in Sec. 4 of App. II.

them in their open-air work, he comes to understand how much of sociable life centres round the gardens and gardening. Spice is given to routine by competitive efforts—at times on a tribal scale, at times in a much smaller way as between families or individuals.

To possess a good and showy garden is not only a matter of pride, it is also a privilege. Only chiefs, or those who make their gardens for a chief, are allowed to have absolutely first-class gardens. For men of lesser rank to be too successful would entail serious consequences to themselves.

A man who had no gardens would be an outcast, whereas a man who for one reason or another is no good at gardening is an object for contempt. Everybody has to make gardens, and the more garden plots a man is capable of tilling, the greater is his renown. The average number that a strong, grown-up married man can manage with the help of his wife is three to six. A boy or youth would make one or two; exceptionally strong men eight to ten. But we shall return to the question of work and its division between the men and the women; and to the question of land tenure and the right of each man to cultivate as many plots as he needs (cf. Sec. 8 of this chapter; and Chs. XI and XII).

The age at which boys begin to make their own gardens is exceedingly early. A small boy in Omarakana, by name Bwoysabwoyse, honoured me with his friendship and often visited me, his favourite place to dispose his person being a five-pound tin of biscuits, whence he watched the proceedings in the tent. Even on such an unmonumental basis he looked diminutive, and he could hardly have been more than six years old. When walking through the gardens once, I was told that we were crossing the plot of Bwoysabwoyse. I looked upon this simply as a joke, and it was only after I had received various corroborating statements and had myself seen him and other small boys at garden work, that I was convinced that such tiny children actually did make their own gardens. The heavier labour is, of course, done for them by their elders, but they have to work seriously for many hours at cleaning, planting and weeding, and it is by no means a mild amusement to them, but rather a stern duty and a matter for keen ambition.

In gardening, then, we have a big department of tribal life. It has its spiritual depth in magic and in the mystical powers displayed solemnly and publicly by the hereditary officiating magician of the community. His office again is backed by a mythology closely connected with native ideas of the original association between man and the soil from which his ancestors have sprung (cf. Ch. XII, especially Sec. 1).

4. THE PRACTICAL TASKS OF THE GOOD GARDENER, *TOKWAYBAGULA*

Let us now pass from our territorial survey and follow the seasonal round of garden work. This falls into four main divisions. First comes the preparing of the soil by cutting down the scrub and burning it after it has dried.[1] The second stage consists in clearing the soil, planting, erecting the yam supports, and making the fence.[2] The third stage has for the most part to be left to nature; the seeds sprout, the vines climb upwards round the supports, the taro plants develop their big leaves and their roots; while human intervention is confined to weeding, which is done by women, and a preliminary pruning or thinning out of the tubers and training of vines by the men.[3] Meanwhile the magician is at work, casting spells favourable to growth. Finally, after the crops have matured, we come to the last stage, the harvest.[4] Apart from the magic of growth just mentioned, each new type of work is inaugurated by a magical rite, and these form a series which correspond to the sequence of practical activities.

Garden work is never done in heavy rain or in windy and what to the natives would be cold weather. During the intolerably hot hours of the day, at the season of calms, the gardeners usually return home or rest in the shade. Whether for communal or individual or family work, the farmers generally go early to the gardens, return between ten and eleven to the village, and then start out again, perhaps after a light meal and a siesta, to work from about three or four o'clock till nightfall. Since some of the gardens directly adjoin the village and the most distant are not more than half an hour's walk away, there is no difficulty in interrupting and resuming work at the convenience of the moment.

The technical efficiency of the work is great. This is the more

[1] Looking at the Chart of Time-reckoning (Fig. 3) we see in column 7 (main gardens) that the first stage falls in the thirteenth moon, and in column 8 (early gardens) in the eleventh. Taro gardens are more complicated because the cycles are shorter (cf. Ch. X, Sec. 2) and we have two periods of plenty and preparation, falling about the third and fourth moon, and about the eighth and ninth.

[2] Column 7 of the chart (main gardens) shows that this stage occupies moons 2, 3, and 4; column 8 (early gardens) moons 13, 1 and 2; column 9 (taro gardens) moons 4 and 9.

[3] Column 7 (main gardens) moons 5 to 8; column 8 (early gardens), moons 3 to 6 or 7. This stage is not noted in column 9 (taro gardens), but would fall in moons 6 and 7, and 11 and 12.

[4] Column 7 (main gardens), moons 10 to 12; column 8 (early gardens), moons 8–10; column 9 (taro gardens), moons 8 and 13. Taro is also harvested on the early gardens in the fourth moon, and on the main gardens in the sixth moon.

remarkable because the outfit of the Trobriand farmer is of the most rudimentary nature. It consists of a digging-stick (*dayma*), an axe (*kema*), an adze (*ligogu*) and, last but not least, of the human hand, which in many of their activities serves as an implement and often comes into actual contact with the soil. The digging-stick is used for turning up the soil at planting and thinning, at harvest and weeding. Axe and adze play an important part in the cutting of the scrub, the thinning out of the tubers and at harvesting. Skill with the hand is important during clearing, planting, weeding, thinning and at harvesting. These then are the tasks and the tools of a "good gardener" (*tokwaybagula*)—one of the proudest titles which a Trobriander can enjoy.

But besides hard work, and a technical skill based on a sound knowledge of the soil and its properties, of the weather and its vicissitudes, of the nature of crops and the need of intelligent adaptation to the soil, another element enters into Trobriand gardening which, to the natives, is as essential to success as husbandry. This is magic.

5. THE MAGIC OF THE GARDEN

It may be said that among the forces and beliefs which bear upon and regulate gardening, magic is the most important, apart, of course, from the practical work.

Garden magic (*megwa towosi* or simply *towosi*) is in the Trobriands a public and official service. It is performed by the garden magician, also called *towosi*, for the benefit of the community. Everybody has to take part in some of the ceremonial and have the rest performed on his account. Everybody also has to contribute to certain payments for magic. The magic being done for each village community as a whole, every village and at times every subdivision of a village has its own *towosi* (garden magician) and its own system of *towosi* magic, and this is perhaps the main expression of village unity.

Magic and practical work are, in native ideas, inseparable from each other, though they are not confused. Garden magic and garden work run in one intertwined series of consecutive effort, form one continuous story, and must be the subject-matter of one narrative.

To the natives, magic is as indispensable to the success of gardens as competent and effective husbandry. It is essential to the fertility of the soil: "The garden magician utters magic by mouth; the magical virtue enters the soil" (Text 36, Part V, Div. VII, § 2). Magic is to them an almost natural element in the growth of the gardens. I have often been asked: "What is the magic which is done in your country over your gardens—is it like ours or is it different?"

They did not seem at all to approve of our ways as I described them, saying that we either do not perform any magic at all, or else let our "misinaris" do the magic wholesale in the *bwala tapwaroro*—the house of the divine service. They doubted whether our yams could "sprout" properly, "rise up in foliage" and "swell". In the course of one such conversation, held in Omarakana with Kayla'i and Gatoyawa, I jotted down the following pointed comment on our method (Text 81, Part V, Div XI, § 9): "The missionaries state: 'We make divine service and because of this the gardens grow.' This is a lie." It should be noted that the native word for 'lie' covers anything from a purely accidental mistake, a *bona fide* flight of imagination not pretending to be anything else, to the most blatant lie. The natives do not accuse the missionaries of deception, but rather of a certain feeble-mindedness or, as Professor Lévy-Bruhl would put it, of a prelogical mentality when it comes to gardening magic.

I am afraid that converted natives who act as missionary teachers have *towosi* magic surreptitiously chanted over their gardens. And white traders married to native women have, under the pressure of public opinion and of the wife's influence, to engage the help of the local *towosi* to chant over their gardens; so monstrous did it appear to everybody that a cultivated patch of soil should go without the benefit of magic.

The round of gardening opens with a conference, summoned by the chief and held in front of the magician's house, to decide where the gardens are going to be made, who will cultivate such and such a plot, and when the work will be started.[1] Directly in connexion with this, the magician prepares for the first big ceremony, which is to inaugurate the whole gardening sequence, while the villagers procure a quantity of special food, usually fish, to be offered as a ceremonial payment to the magician. A small portion of this gift is exposed in the evening to the ancestral spirits, sacrifically and with an invocation;[2] the bulk is eaten by the magician and his kinsmen. Then he utters a lengthy spell over certain leaves which will be used on the morrow. Next morning the magician and the men of the village go to the gardens and the inaugural ceremony takes place. The *towosi* strikes the ground and rubs it with the charmed leaves—acts which symbolise in speech and sentiment the garden magic as a whole. This rite officially opens the season's gardening as well as its first stage: the cutting of the scrub. Thereafter each stage of practical work is ushered in by the appropriate ceremony. After the cut scrub is sufficiently dried, he imposes a taboo on garden work,

[1] Cf. App. 1, Comparative Table of Magic and Work, for the correlation between practical work and magic.

[2] Cf. Note 8 in App. II, Sec. 4.

ritually burns the refuse, and introduces the planting of certain minor crops by a series of ceremonies extending over a few days. Later on, a sequence of rites inaugurate successively the main planting of yams, the erection of vine supports, weeding, preliminary thinning out, and finally of harvesting. At the same time in a parallel sequence of rites and spells, the garden magician assists the growth of the crops. He helps the plants to sprout, to burst into leaf, to climb; he makes their roots bud, develop and swell; and he produces the rich garlands of exuberant foliage which intertwine among the vine supports.

Each rite is first performed on one of the standard magical plots, the *leywota*. This is important from the practical point of view, because the men who cultivate these plots are bound to keep time with the rhythm of magical ritual and not lag behind. At the same time they must also be worked with special care. They are scrupulously cleared and cleaned, perfect seed tubers are selected, and since they are always made on good soil, they represent not only a very high standard of garden work but also of gardening success. Thus, in punctuality, quality and finish of work, and in perfection of results, these plots set a definite pattern to all the others, and this excellence is mainly attributed to the influence of magic.

6. THE GARDEN WIZARD

The *towosi* or garden magician is an hereditary official of every village community. As a matter of fact, the position of *towosi* coincides with that of the Chief or the head-man, if not in identity of person, at least in the principle of lineage. In native mythology and legal theory, it is always the head of the kinship group owning a village who is the garden magician. This man, however, frequently delegates his duties to his younger brother, his matrilineal nephew, or his son. Such handing over of the office of garden magician was especially frequent in the lineage of the paramount chiefs of Omarakana, on whom the duties of charming the gardens weighed too heavily.

The mythological system of the Trobrianders establishes a very close connexion between the soil and human beings. The origins of humanity are in the soil; the first ancestors of each local group or sub-clan—for these two are identical—are always said to have emerged from a certain spot, carrying their garden magic with them (cf. Ch. XII, Sec. 1). It is the spot from which they emerged which is usually, though not always, the sub-clan's soil, the territory to which it has an hereditary right.[1] This hereditary ownership of the

[1] At times a sub-clan obtains rights of ownership in a district to which they have migrated. Cf. Ch. XII, Sec. 3.

PLATE 20

VIEW OVER A NEW GARDEN

". . . completely cleared ground . . . the horizon broken by an occasional clump of trees marking the site of a sacred grove . . . or the jungle on the coral ridge" (Ch. I, Sec. 3; cf. also Ch. III, Sec. 3)

PLATE 21

A YAM GARDEN IN FULL DEVELOPMENT

"The exuberant vines climb round tall stout poles, their full shady garlands of foliage rising like fountains of green" (Ch. I, Sec. 3; cf. also Ch. IV, Sec. 2)

PLATE 22

A TARO GARDEN

"In the south small gardens are often wedged between jungle and stony coral outcrops" (Ch. I, Sec. 3; cf. also Ch. X, Sec. 2)

soil—mythological, legal, moral and economic—is vested in the head-man; and it is in virtue of these combined claims that he exercises the function of garden magician. "I strike the ground," as I was told by Bagido'u, the proudest garden magician of the island, "because I am the owner of the soil." The first person meant, "I, as the representative of my sub-clan and my lineage."

We shall see in our study of magical texts (Part VII) that the traditional filiation[1] of garden magic is kept alive by every officiating magician. In some of the spells he has to repeat the whole series of the names of those who have wielded the magic before him. At one or two stages of his magic, he offers a ceremonial oblation, consisting of a minute portion of cooked food taken from the substantial present he has received, to the spirits of his predecessor. Such presents from the community are the expression of their gratitude and their submission to him rather than a commercial gift. They are the recognition of his services, and in this spirit they are offered to him and to his forerunners. This ritual offering of food, which is an integral part of the magical proceedings, is called *ula'ula*.

The members of the community, however, usually offer the magician other presents as well. At the beginning of the gardening cycle he is usually given small gifts of food, such as coconuts or bananas; or else he may accept a bunch of betel-nut or such objects of daily use as baskets, axes, mats, spears or cooking-pots. This type of gift, called *sousula*, is meant to repay him for the hardships undergone in the exercise of his calling. As one of my *towosi* friends explained to me, putting it in the concrete form characteristic of native utterance: "When I go about making magic in the gardens, and I hurt my foot, I exclaim: '*Wi! Iwoye kaygegu; gala sene si sousula.*' " "Oh! (the object) has hit my foot; not very much their *sousula* payment (i.e. they don't give me enough to repay me for all my hardships)."

Again from time to time the magician receives a present of valuables called *sibugibogi*: a large ceremonial axe-blade, belts or ornaments of shell-disks or a pair of arm-shells. This gift is usually offered after a bad season to propitiate him, or else at an especially good harvest to express gratitude.

In the carrying out of his duties, the magician is usually helped by some younger men: his younger brothers and his sisters' sons are his natural successors, whom he will have in due course to instruct in magic, teaching them the spells, telling them the substances to be used, advising them how to carry out the ritual and what per-

[1] For the meaning of the term 'filiation' and the account of the principles of inheritance, see Part I, Sec. 9.

sonal observances they have to keep. Of this instruction, the most difficult is the learning of the formulae. Even this, however, does not require much special training, for garden magic is a public ceremonial, the spells are heard often by everybody, while the ritual is well known and anyone is able to tell you exactly what observances the magician has to keep. Those who have to inherit garden magic and practise it, and are therefore more interested in it, will be acquainted with every detail early in their life. They are the magician's natural help-mates and acolytes. Whenever the ceremony is cumbersome, they take part in it; or they repeat on other garden plots the rite which the chief magician performs on the standard plots. And they assist him often in the collecting of ingredients or preparing of magical mixtures and structures.

Besides these, he has non-official helpers among the younger people and children, who carry some of his paraphernalia, assist him in putting up certain magical signs and do other such minor services.

I have just mentioned the magician's taboos. These consist almost exclusively in the abstention from certain foods. In no circumstances may he touch the meat of certain animals and fish, or eat certain vegetables. Generally these are sympathetically connected with the substances which he uses in his ritual or with the aims of his magic. The magician is also not allowed to partake of the new crops until after the performance of a special ceremony, which consists as a rule in an offering to the ancestral spirits. A third type of abstention is the fast which he has to keep on the days on which he performs any ceremony (cf. Ch. II, Sec. 4).

From all this it can be seen that a garden magician's office in the Trobriands is no sinecure. Not only does he have to carry out a series of inaugural rites, following closely the practical work of the gardens, not only does he stimulate the growth of the plants in his spells of encouragement; but he also has to observe a system of by no means easy abstentions and fasts, and last, but not least, to carry out a considerable amount of practical work and control.

The garden magician is regarded by the community as the garden expert. He, together perhaps with his elder kinsman, the chief, decides what fields are to be cultivated in a given year. Later on, at each stage, he has to find out how the work in the gardens stands; how the crops are sprouting, budding, ripening, and then he has to give the initiative to the next stage. He must watch the weather and the state of the cut scrub before the burning. He has to see whether the gardens are sufficiently advanced before he performs the planting magic, and so at every stage. And when he finds that people are lagging behind, or that some of them, by neglecting a communal duty, such as the fencing of the garden plots, are endangering the interests of the whole community, it is his function to upbraid the culprits and induce them to mend their ways and to work energetically.

Time after time, as I sat in my tent reading or looking over my notes, or talking to some of my native friends, I would hear the voice of Bagido'u of Omarakana or Navavile of Oburaku or Motago'i of Sinaketa rising from somewhere in front of his house. In a public harangue, he would accuse such and such a one of not having completed his share of the fence, thus leaving a wide gap in the common enclosure through which the bush-pigs or wallabies could enter; and now that the seeds were in the garden and beginning to sprout, the wild animals would soon be attracted and might do a great deal of damage. Or again he would announce that the cut scrub was practically dry and that the burning would be inaugurated in three or four days. Or again he would impose one of the public taboos on work, saying that as in a few days the large *kamkokola* would be erected, everybody must stop all other work, and bring in the long stout poles necessary for the magical structure and for the final yam supports.

Thus the *towosi* exercises not merely an indirect influence on garden work, by giving the initiative and inaugurating the successive stages, by imposing taboos, and by setting the pace, but he also directly supervises a number of activities. In order to do this he has constantly to visit the gardens, survey the work, discover shortcomings, and last but not least, note any special excellencies. For public praise from the *towosi* is a highly appreciated reward and a great stimulus to the perfect gardener, the *tokwaybagula*.

The natives are deeply convinced that through his magic the *towosi* controls the forces of fertility, and in virtue of this they are prepared to admit that he should also control the work of man. And let us remember, his magical power, his expert knowledge and his traditional filiation to his magical ancestors are reinforced by the fact that he is the head-man, or, in a community of rank, a chief of high lineage, or a nephew or younger brother of such. When the office is in the hands of the chief's son, he again only holds it as the delegate of the rightful head of the community (cf. Ch. XII, Secs. 2 and 3). Furthermore, the acts of magic are an organising influence in communal life: firstly because they punctuate the progress of activities at regular intervals and impose a series of taboo days or rest periods; and secondly because each rite must be fully performed on the standard plots, and these plots must be perfectly prepared for it, whereby a model is established for the whole village (cf. App. I). Magic therefore is not merely a mental force, making for a more highly organised attitude of mind in each individual, it is also a social force, closely connected with the economic organisation of garden work. Yet magic and technical activities are very sharply distinguished by the natives in theory and in practice—but to this point we shall have to return presently.

7. THE GLORY OF THE GARDENS AND ITS MYTHOLOGICAL BACKGROUND

To each village community this magic, as already mentioned, is a very precious possession and a symbol of its social integrity as well as of its standing in the tribal hierarchy. A village of high rank always leads in gardening. This is natural because, as we shall see (Ch. XII, Sec. 3), the most noble sub-clans have settled in the most fertile territory—that of Kiriwina. The renown of this district, especially for its gardening excellence, extends over the whole area of the Northern Massim. The first evening on which I arrived at the village of Dikoyas in Woodlark Island, some eighty miles from the paramount chief's capital, I was told, in glowing language, about the wealth of Kiriwina, and that the culture hero, Tudava, who gave gardens and taught gardening to all men in and around Woodlark Island, had come out of the soil in Kiriwina; also that Kiriwina was the first island to exist, and the first where gardens had been made. Thus, more than a year before I came to Kiriwina by a very roundabout way of several thousand miles, I was made to feel that agriculture originated in the Trobriands; that gardening, the knowledge of how to do things, the knowledge of proper conduct, of totemic origins and the totemic identity of man, had spread from Kiriwina, eastwards and southwards; that Kiriwina still remained the most fertile, wealthy, and aristocratic place in the world.

The legend of Tudava is known all over the district of the Northern Massim. Everywhere in Woodlark Island, in the d'Entrecasteaux Archipelago, in the Amphletts, in the Marshall Bennett group, stories are told of how the culture hero was a native of Kiriwina, and was the first to institute gardening and garden magic. In one version of the legend, also obtained in Woodlark Island, I was told that Tudava was the first man to come out of the ground in Kiriwina. After him the other men came out. As each man emerged, Tudava gave him his totem. When he first came out, there was no land except Kiriwina. He threw a large stone into the sea and there arose the island of Kitava. Then he went there and threw other stones, and the islands of Iwa, Kwayawata, and Digumenu came into being. Then he made the district of Madawa (part of Woodlark Island), then the rest of Woodlark Island, and Suloga where the big stone quarries are. Then he came to Nada or Nadili (Laughlan Islands) where some people had been before him, for these islands already existed.[1] The other islands which were made by Tudava were all peopled by men from Kiriwina, who came over in canoes. When he went to the Laughlans, the people wanted to kill him, so he went away.

In another version I was told that there were two men at that time, one of them was Tudava and the other was Gere'u. Gere'u came out before Tudava from the ground in Kiriwina. He had a sister called Marita. They both came over to Woodlark Island. Tudava came to the district of Wanuma on the northern side, Gere'u came to the Kropan side (southern side). Tudava then came across to see what Gere'u was doing. Gere'u had a big garden, and Tudava asked him: "Who makes this garden?" "I make it myself," answers Gere'u. "You make such a garden as big as three or four men could make?" Gere'u says: "Yes, I make it alone." "How do you make your garden?" asks Tudava. "I cut one small tree," Gere'u tells him "and all the saplings fall down. I cut a large tree and all the big trees fall down. I make a small fire and everything burns. I break one stick and plenty of sticks come. I start to make the fence and the fence makes itself. I plant one taro and many others grow. I plant one taytu and plenty of taytu grows. I plant one yam, and lots of yams grow."

Tudava answers: "Oh, this is no good. This quick work is no good. Supposing we make too much food, man will not work. It is better that a strong man should make a big garden and a weak man a small garden. A weak man will go and fish and exchange fish for taytu."

Then Tudava told Gere'u how to make gardens and gave him magic—the magic which Tudava had was not so strong as the magic of Gere'u, but the magic of Gere'u is lost and the magic which people now know is the magic of Tudava, and this is the reason why people cannot make such big gardens now.

Gere'u had also a cooking-stove made of stones underground. Tudava looked into it and found that Gere'u baked snakes and iguanas and opossum and bush-pigs and rats and fish. Tudava told him: "The rat is no good, throw it away. The snake is no good, throw it away. The iguana is no good, throw it away. You keep pig and opossum and fish." And that is the reason why people now only eat fish, opossum and pig, and do not eat rats and snakes and iguanas.[2]

[1] Such contradictions as the one contained in the two sentences "there was no land" and "these islands already existed" occur in all myth, including our own, that of Christianity.

[2] I am including this incident in order to give as complete a version of the myth as I received, though it has no direct bearing on gardening. Incidentally it suggests that Gere'u represents a more ancient cultural order, and that Tudava was a reformer.

Gere'u followed his advice and afterwards went to Misima and left plenty of betel-nut there. He then went to Du'a'u, where his canoe broke on the reef. There Gere'u and Marita were drowned and transformed into stones, but the big yams came over to Du'a'u, while the taytu drifted over to Kiriwina. This is why in the north of Fergusson Island there are very good big yams, and in Kiriwina there is plenty of taytu.

Thus we have two cultural heroes, one of them representing an old system of gardening which in the myth is endowed with an even stronger magic; while the other, Tudava, remains the lawgiver whose word has established the order which at present obtains.

I was also told in Woodlark Island that garden magic largely consists of spells given by Tudava, and that his name is always mentioned in magic. The first garden magician received the spells from Tudava himself and the formulae are still handed down in the mother-line. We shall find the name of Tudava and the name of Gere'u's sister, Marita, in one or two of the spells of Kiriwinian garden magic.

Another version of the legend I picked up in Vakuta, the southern island of the Trobriand Archipelago. There I was told that Tudava was the first to show men how to make gardens properly, that he was the first to make garden magic, to institute garden taboos, and tell people what to plant and what not to plant. After having finished his work in Kiriwina, he sailed east and came first to Kitava, the nearest island, plainly visible from the eastern shore of Kiriwina. Let me give in free translation the full text of the myth as I obtained it from Mbwasisi, the garden magician of Vakuta (cf. Text 96, Part V, Div. XII, § 40)[1]:—

"(i) In Kitava, Tudava endowed the place well; he went to the village, he planted big yams, taytu, taro, arum. The countryside was made bright. (ii) He made Kitava indeed a very good (agricultural) country. Not in the slightest bitter (the crops), yams planted round the village very big, also in the bush and on the coral ridge all very good. The very home of the large yams. (iii) He went round the village from one end to the other. He stopped as he had already finished the island of Kitava. Already the countryside of Kitava was made bright and the task was over.

(iv) (He said) 'I shall sail, I shall go to Iwa.' There he anchored and went ashore; he planted yams in the village, he planted taytu.

[1] Comparing this free rendering with the interlineal translation and commentary (cf. loc. cit., §§ 40, 41) the reader will find that the words and phrases enclosed in brackets are implied by the context though not explicitly stated in the text.

Then he went ashore so that he might go to the bush and plant there. (v) They said: 'Tudava, thy canoe has drifted away. Get to thy canoe and pull it ashore.' (vi) And (comments the narrator) look, he did not plant in the bush. There it is bitter. Only in the village (did he plant and there it is sweet). (vii) He approached (his canoe). 'Oh no, I shall sail. It is enough that I have already planted in the village.'

(viii) He sailed to Digumenu. The people of Kwaywata came, the masters of the island. They drove him off (shouting): 'This is our island. Do not you settle (here), O Tudava.' (ix) They drove him off. He told them: 'In this place Digumenu I thought I should plant yams, I should plant taytu, I should plant bananas. (But) because you have driven me away already, I shall give you only coconuts, and I shall sail to Kwaywata.'

(x) He sailed, he went to Kwaywata. He made fast the canoe and anchored it; he went ashore and planted taro, taytu and yams in the village. But when they attacked him, he moved over to the island of Gawa.

(xi) He went ashore at Gawa, he planted yams in the village, he planted bananas, arum, taytu; he planted taro. (xii) Afterwards (he tried) perhaps to go to the bush, so that he might plant, so that he might make good all the countryside. They attacked him.

(xiii) He got into the canoe, he sailed and came to Bovagise. There at Bovagise he planted taro, he planted sago and taytu. He got to like the village of Bovagise, he gave fish, and then he sailed to Wamwara.

(xiv) He went (there), he remained, he spoke: 'Men of Wamwara, I shall plant yams, taytu and taro; I shall plant all the countryside till there is nothing left. I shall get all your lands in order.' (xv) They consented, and he planted their countryside, and when it was over, he got into his canoe and sailed and went to Nadili.

(xvi) He anchored, he anchored indeed, (but) they rose up and attacked him. (He fled and) he broke off one end (of the island); it remained one (separate) island (made by) the bottom of the canoe of Tudava. (xvii) He went to the open sea, running away. They went round so that they might kill him. He broke off one piece of land (from) another, (making) a sea passage in between. (xviii) He broke off three places (islands): one by name of Obulaku, one Bugwalamwa, one Budayuma. (xix) He spoke: 'You are very bad, people of Nadili. I should have put order into your lands so that they might be good. I should have given you yams, taytu, bananas, taro. But since you have wronged me, I shall give you coconuts.'

(xx) He sailed, he went away to a foreign land beyond Nadili.

He went, he settled there. While he was there in the land beyond Nadili, a man was fishing with a large shark-hook for shark. (xxi) The shark which he had caught swam and went towards the strange land and there went ashore. Tudava spoke: 'Who art thou?'

(xxii) 'Oh, I, I belong to Nadili. I fish for shark with shark-hook. It pulls me. I come to thy village.' They sat down together. (xxiii) Tudava spoke: 'You are my friend. Let us two garden together.' They gardened. (xxiv) After the moon was over, he (the man of Nadili) spoke: 'I shall go to my village.' Said Tudava: 'Let us lash thy canoe so that thou might fill it with thy staple food.' (xxv) He spoke (the man of Nadili): 'Well, I shall go to-morrow.' Tudava filled his canoe and said: 'It is done. Come here, I shall magic thy wild ginger.' (xxvi) He (Tudava) charmed the wild ginger. He wrapped it up, making two bundles. He (the man of Nadili) said: 'Thou remain and I shall paddle away.' (xxvii) Tudava stood up and said: 'While you paddle on the sea, chew one bundle of wild ginger, ritually bespit thy village, so as to make it clear (on the horizon); then turn round, chew the second bundle, bespit my place that it might disappear, so that while I remain here, no one should see me.' (xxviii) He (the man of Nadili) paddled off; he ritually bespat Tudava's island, it disappeared. He ritually bespat Nadili, and it appeared clearly. (xxix) He paddled, he was already on the shallow water (near his village). His canoe capsized because of the breakers; its contents spilled; all the food went into the sea and his boars' tusk pendant sank. It was the boar's tusk pendant of Tudava, who gave it to the man of Nadili as his valuable. We see it flashing through the sea. (xxxi) All his personal effects fell down into the sea. The man went to the village and remained there."

The really significant part of this myth ends with Tudava's exploits in the island of Nadili, as it is called by the Trobrianders, or Nada—to use its own local name—the Laughlan Island of our map.

We have in this myth a justification of why, as is really the case, the best large yams (*kuvi*) of the whole district grow in Kitava, while Iwa has good crops round the village, but the rest of the island is stony and cannot be used for gardens. Here, as can be seen, it is due to the accident of Tudava's canoe drifting away. The island of Digumenu, which is but a sandbank, can bear only coconuts and a few aromatic herbs such as *kwebila* and *sulumwoya*. These are not mentioned in the myth, but were specified to me in a commentary. The hostility of the people of Kwaywata who are the owners of the sandbank, and garden it from their own island, is responsible for the poverty of the soil, both on the sandbank and on the island of Kwaywata. The same explanation accounts for the agricultural poverty of the island of Gawa, or Woodlark Island (Muruwa or Muyuwa), where he was received well in one district, but driven off in another; while in the Laughlan Islands, a low atoll with sandy soil, only coconuts grow, because of the hostile behaviour of the inhabitants.

The last incident, the friendly intercourse of the culture hero with one fisherman from the Laughlans, leads us up to what might be named the token of truth: the visible mark left by past events, at the bottom of the sea. The petrified boar's tusk can be seen through the water on a reef on the other side of the Laughlans. To a certain extent it also explains why the people of Nada have no garden crops. For although Tudava gave his friend all the agricultural produce, this was lost again when the canoe foundered. The very end strikes us as singularly pointless, as often happens in native myths. The penultimate verse was spoken in a different tone, and given obviously as a narrator's commentary on the events. Another legend which I obtained from a few natives of the d'Entrecasteaux Archipelago and which is obviously a variant of the story which I recorded in Woodlark Island about Gere'u and his sister Marita, tells how a mythical hero and his sister sailed from Misima to Du'a'u on Normanby Island. All the betel-nut they left behind, and this is the reason why Misima is so rich in this coveted stimulant. They only took away taytu (small yams) and *kuvi* (large yams). In leaving the canoe the sister dropped her petticoat, and the brother had to turn away not to see her nakedness, so that he did not notice an approaching squall. This capsized the canoe and upset all the large yams into the water; but the taytu remained in the canoe, which was charmed by the man and drifted north to the Trobriands. This is the reason why *kuvi* (large yams) grow well on Normanby Island and taytu is plentiful in the Trobriands.

Another legend concerning a different overseas expedition of canoes laden full with crops explains why certain parts of Fergusson Island, notably the northern shore round the slopes of Koyatabu, are so fertile. The poor districts of the region are Dobu, which can never boast of a great surplus, the Amphletts, the Lusançay Islands at the north-west rim of the Northern Massim, and the Laughlan Islands or Nadili. All these places had in one way or another incurred the displeasure of the original mythical dispensers of plenty, and therefore they have now to rely on others, or else to eat coconut, fish and wild fruit.

The mythological hero, Tudava, is well known in the whole district; and in Kiriwina itself, of course, his story is told—or rather a great many stories, for there is a whole cycle of Tudava myths, legends and even fairy-tales. But remarkably enough that part of the legend which expresses so tellingly the gardening supremacy of Kiriwina over the outlying districts is not known in Kiriwina itself. The data which I have given here came, as I have mentioned, from Woodlark Island and from the extreme southern island of Vakuta. I was also told about the cultural rôle of Tudava, as the first maker of gardens and of garden magic, by natives coming from the outlying islands of Kayleula, Simsim and Kitava. I have heard it recounted again in the Amphletts by the natives there. Travelling though the d'Entrecasteaux group I found that the natives of Du'a'u, Dobu and Goodenough Island knew it well. But the natives who are glorified by it do not seem even to have heard of it!

The reason for this apparent anomaly is, I think, that the Trobrianders, notably those of Kiriwina and Tilataula, take their supremacy in agriculture for granted. They have good reason for doing so, and no one ever challenges their claim. They have tangible evidence of their competence and wealth. It is obvious to everyone, whether he be a native of Kiriwina or a visitor from some neighbouring village, a *kula* partner from Kitava or Woodlark Island, the Amphletts or Dobu, that the storehouses of Kiriwina are unparalleled, their gardens are the biggest and best in the district, and the total amount of yams harvested there yearly the greatest. In short the Kiriwinian does not need to tell a story about his past wealth; he can point to the present with pride and assurance. Still less does he need to justify his poverty as do some natives; he does not suffer from it.

With all this it would be incorrect to assume that there is no mythological foundation to gardening in Kiriwina. In the first place they have a brief mythology which merely asserts that the local or naturalised sub-clan whose ancestors emerged from a given territory controls the magic of fertility of that territory. Such brief mythological affirmations define traditionally and legally safeguard the claims of the sub-clan to their lands.

In the second place there exists an ever growing and constantly renewed tradition of good gardening. This consists of the stories told about the achievements of one village or another; of the accounts of specially glorious harvests (Ch. V, and Doc. II); of the results of institutionalised food competitions (*buritila'ulo*, cf. Ch. V, Sec. 6), in which one village is "beaten" by another. But all this is more a glorification of the present than a reference to miracles in the past. To a certain extent the historical records of occasional famines, offset by accounts of plenty, function in the same way (Ch. V, Sec. 1).[1]

[1] See Note 4 in Sec. 4 of App. II.

To return to the wider cycle of myths, the gist of the stories about the origins of gardening is an explanation of the excellence of some places and the poverty of others. They contain a legendary charter of gardening in general and of the differences in local fertility and custom. In most of them we have a moral reason for Tudava's preference for certain islands: where he was well received their gardens blossomed, yams swelled inside the fertile ground, taro flourished, and garlands of taytu shaded the soil with foliage, and so they continue till the present day. Where the inhabitants threatened and drove him off—a compound idea expressed by the native word *bokavili*—there the soil remained barren or swampy, good only for coconut palms and wild fruits of the bush, the two commodities which will grow anywhere and everywhere on these islands. The main island of the Trobriands, notably the district of Kiriwina where gardens originate and whence Tudava went off on his wanderings, has always remained the supreme centre of gardening.

8. THE POWER OF MAGIC AND THE EFFICIENCY OF WORK

In order to appreciate this mythological cycle of ideas, we must keep in mind the relation of magic to practical work as this is conceived by the natives. The short myths of first emergence have in the Trobriands a very close connexion with magic, since this latter has always been brought by the ancestors from underground (Ch. XII, Sec. 1). The gift of fertility bestowed by the mythical founders and wielders of magic on the richest districts of that region is without exception conceived in a two-fold manner, magical and natural. The natives realise that on sandy, brackish and stony soil neither yams nor taro, and still less taytu, could ever grow. If you ask whether one could start any plantation or garden on the precipitous slopes of the Amphletts, on the barren sands of the Laughlans or on the windswept, brine-drenched fragments of the Lusançay atoll, they will answer, no, and explain why most plants cannot thrive there in perfectly reasonable, almost scientific language.

At the same time they attribute the supreme fertility of some districts, the prosperity which dwells there permanently and the beautiful expanse of successful gardens to the superiority of one magical system over another. Thus from native commentaries on the above myths, it becomes clear that the culture hero is, on the one hand, always supposed to bestow fertile soil, sound seedlings, instruction in gardening skill, the knowledge of how to handle crops and protect them from blights and other dangers; and, on the other hand, it is understood that he brings with him a powerful system of magic. The two ways, the way of magic and the way of garden work—*megwa la keda, bagula la keda*—are inseparable. They are never confused, nor is one of them ever allowed to supersede the other. The natives will never try to clean the soil by magic, to erect a fence or yam support by a rite. They know quite well that they have to do it by hand and in the sweat of their brow. They also know that no work can be skimped without danger to the crops, nor do they ever assume that by an overdose of magic you can make good any deficiencies in work.

Moreover, they are able to express this knowledge clearly and to formulate it in a number of principles and causal relations. They have a sound knowledge of the soil and of the crops; in fact they distinguish between six or seven types of soil and know well which variety of crop is best adapted to swampy, heavy soil, to black humus, and to the light and stony ground of the dry regions. Thus, although there is no word corresponding to 'waste land', the natives know and can explain that no garden is possible on coral stone;[1] nor on the soft mud of swamp or marsh; nor in the black humus drenched with salt water on which mangroves grow; nor on sand. They will also state that *pwaypwaya*, the real soil or earth, is to be found only where the bush (*odila*) grows, and that after having cut the trees and shrubs you can plant crops. The soil suitable for cultivation is classified into black, heavy, fertile humus, the best soil for taytu, *kuvi* and taro alike; light red soil, on the whole inferior to humus, and not suitable for taro, but giving good results in a wet year with yam crops; soil found in and near the *rayboag*, similar to the previous, but heavier and less dry, specially good for the large variety of yams; swampy soil, which can be used for taro gardens and, in very dry years, even for yams; poor stony soil which is quite unsuitable for taro, but can be used for hardy varieties of yam. The natives know also that the black humus which fills the holes in the *rayboag* (coral ridge), is good for the large yams, *kuvi*.

As to the varieties of yam, taro and taytu, they have literally hundreds of names for each of them.[2] Some names distinguish what might be properly regarded as real botanical varieties, others decribe characteristics of size, shape, perfection and so on. The natives will explain intelligently why it is necessary to have the soil well cleared and weeded. They have got a clear thory as to why taytu must be planted rather deeply and covered with a hillock. They have clever devices for keeping off pigs, and whenever their fence is temporarily damaged they will put sharp pointed stakes on the other side of

[1] For the native terminology which documents these distinctions, see Part V, Div. I, §§ 11 and 12.
[2] Cf. Part V, Div. III, §§ 9–22.

it, so that the pigs are caught in jumping over the fences. They have several types of scarecrow and bird rattle. All these practical devices they handle rationally and according to sound empirical rules.

Nor does this distinction between work and magic remain implicit and unexpressed. I was always able to ask whether it was the way of magic or of gardening, and received unambiguous answers very early in my work. I was told, for instance, that the large, bulky structures, the *kamkokola,* were matters of magic: "*Megwa wala; gala tuwayle si koni wa bagula*—only magic; no other task (is incidental) to them in the garden." I was told also that the spirit houses (*si bwala baloma*), the miniature fences, the tufts of grass bound round the *kavatam* (yam support) and certain horizontal sticks placed on the *kamkokola* were all purely magical. It was, on the other hand, explained to me that ashes fertilise the ground; that deep planting is advisable in dry seasons; that stones must be removed from the soil; that weeds choke the crop, and so on. To the natives, therefore, the aims of magic are different from the aims of work. They know quite well what effects can be produced by careful tilling of the soil and these effects they try to produce by competent and industrious labour. They equally know that certain evils, such as pests, blights, bush-pigs, drought or rain, cannot be overcome by human work however hard and consistent. They see also that, at times and in a mysterious way, gardens thrive in spite of all anticipations to the contrary, or else that, in a fairly good season favoured by good work, the gardens do not give the results they should. Any unaccountable good luck over and above what is due the natives attribute to magic; exactly as they attribute unexpected and undeserved bad luck to black magic or to some deficiency in the carrying out of their own magic.

Briefly, magic, performed officially by the garden magician under ceremonial conditions, by means of rite and spell and with the observance of taboos, forms a special department. Practical husbandry, on the other hand, carried out by each one with the aid of his hands and common sense, and based on the recognition of the causal relation between effort and achievement, constitutes another department. Magic is based on myth, practical work on empirical theory. The former aims at forestalling unaccountable mishaps and procuring undeserved good luck, the latter supplies what human effort is known naturally to bring about. The first one is a sociological prerogative of the leader, the *towosi*; the second is the economic duty of every member of the community.

I have so far spoken exclusively of public magic and, when making clear the distinction between magic and work, I referred only to the rites and spells of the public garden magician, the *towosi*. It will be well to state here at once that private garden magic exists in the Trobriands. I obtained one or two formulae which will be adduced in one of the following chapters, together with a fairly complete account of its rather meagre ritual and limited influence. Private magic is performed chiefly over seed yams at the time of planting, and over the digging-stick and axe at the time of the thinning of the tubers. It consists of simple spells directly chanted over the object to be charmed. It never integrates with the work in the gardens as public garden magic does.

Another form of garden magic which must be mentioned here is the evil and malicious magic, *bulubwalata*, which is supposed to be carried out by neighbours in order to injure the garden. Unfortunately I only got on the track of this towards the end of my stay in the Trobriands, and, as in most cases of black magic, it would have required a long time and a great deal of patience to make sure whether it even exists. It would have taken even longer to obtain details of it, and I was not successful in either task. My impression is that *bulubwalata* is a mere myth as far as the gardens are concerned; that people are suspected of doing it but that they never live up to the suspicion; that unskilful and unsuccessful gardeners would impute it to neighbours of their own village or some more distant place, and that these would return the suspicion and the ill-will (cf., however, Ch. III, Sec. 2).[1]

The true black magic of the gardens is the magic of rain and drought, the magic which regulates the conditions of fertility. This is not a surreptitious form of sorcery, however: it remains in the hands of the chief; he wields it openly and officially as an expression of his anger and as a means of collective punishment and enforcement of his will. The wielding of rain and drought magic is, as a matter of fact, one of the most dreaded and coveted privileges of the paramount chief of Omarakana (cf. Ch.V, Sec. 1).

Turning now to the practical side: garden work is done in the Trobriands by everybody, man and woman, chief and commoner, chief's principal wife, chief's own sister, as well as the humblest spinster (cf. Part I, Sec. 3). Nor is the garden magician excluded from work unless, as was the case with Bagido'u, the garden magician of Omarakana, he is ill. He will work on quite as many garden plots as everyone else, he will carry out all the activities with the same vigour and he will not be able to shuffle any of his burdens on to another man. The garden magician and the chief were indeed

[1] Also Note 5, App. II, Sec. 4.

always expected to be specially effective gardeners. The lesser chiefs of Liluta, Kwaybwaga and Mtawa were always assumed to be good gardeners who would vie with one another and with their commoner subjects. There is no specialisation in garden work; nor is the specialist exempt from it. Even the magician is not excused by his activities on behalf of the whole community from doing his own share in the work.

The most important distinction is that between a man's and woman's part in gardening. A woman never gardens in her own right. She is never styled "owner of a garden" or "owner of a plot". She never works independently but must always have a male for whom and with whom she works the soil, and this refers also to women of the highest rank whose husbands are necessarily of a lower rank than themselves. An unmarried girl will simply assist her mother. On marriage she works on her husband's garden land. The family, that is husband, wife and their children, is the smallest co-operative unit in Trobriand gardening. Between them they cultivate the several plots allocated to them at the garden council (cf. Ch. II, Sec. 3). On Plate 24 we see such a family group in a garden, completely cleared and ready for planting the preliminary crops. In this joint work there is a regular division of labour. The man cuts the scrub; man and woman clear the ground and prepare it for planting; the man does the planting; the woman weeds. The man has to train the vines and to thin out the roots; while finally harvesting is done by men and women together. The renown of good gardening, the praise and other emoluments of ambition go to the man and not to the woman. She only shares in them vicariously. It is rather the quantity of produce which a woman receives for her household from her brother that redounds to her credit.

Certain complications arise, however, from the fact that work is done differently by an unmarried man or woman; that the sharing of garden work between a chief and one of his several wives cannot be done on the same principle as between a monogamous couple. When in olden days the paramount chief had some fifty wives and plots to correspond, he could only do roughly one-fiftieth of the male's share for each. It will be best to take one type of status after another and see how the rights to a garden, the work in the garden, and co-operation appear:—

(1) A bachelor or widower has to do all the work for himself, including weeding which, however, might be done by some related woman out of kindness. (2) An ordinary man with one wife and perhaps with children will do only the male's work, assisted by his sons, as long as these are small, while his wife and daughters carry

out the female part. The grown-up sons may take up their own plots in the father's community if they continue to reside there for some time, or else they make their own gardens in their maternal community after they have moved there. (3) The head-man or notable with two or more wives will usually have as many times the normal portion of plots as he has wives. On this greater number of plots he is not able to do all the work himself. On the other hand, through his rank he can command and pay for communal work at cutting and planting. Such a man again would usually retain his sons for a longer time in the village, and these also would assist him, each working on his mother's portion. (4) The paramount chief with a large compound establishment of some forty to sixty wives (at the time of my stay in the Trobriands they were reduced to fourteen) would cultivate some two or three plots for each wife. Obviously his own share in the work would amount to very little. As far as I can judge, he would work with one or two, or at the most three wives. On Plate 25 we see the males of the gardening team at work; in the foreground are the slanting poles of a *kamkokola*. The rest of the work would be done partly by the sons of each wife, partly by communal labour. At cutting and at the planting of the main crops, communal labour would invariably be used (cf. Ch. IV, Sec. 5).

9. THE PLACE OF GARDENING IN TRIBAL ECONOMY AND PUBLIC LIFE

We have seen in the Introduction what a fundamental part agriculture plays in Trobriand economics: we can now gauge how the magical superstructure, the intermixture of legal principles and of aesthetic ideas, of work and knowledge supply the institution of agriculture with many other facets besides the economic one.

The gardens are, in a way, a work of art. Exactly as a native will take an artist's delight in constructing a canoe or a house, perfect in shape, decoration and finish, and the whole community will glory in such an achievement, exactly thus will he go about the laying out and developing of his garden. He and his kinsmen and his fellow-villagers as well, will be proud of the splendid results of his labours.

A considerable amount of energy is spent on purely aesthetic effects, to make the garden look clean, showy and dainty (cf. Part I, Sec. 3). The ground before planting is cleared of stones, sticks and débris, with a meticulousness far beyond what would be strictly necessary on purely technical grounds. The cleared soil is divided into neat rectangles about 4 to 10 metres long, and 2 to 5 metres broad by means of sticks laid on the ground (see Pl. 26). These

PLATE 23

THE MAGICAL CORNER

"The magician leaning over the *kamkokola* just before reciting his spells. . . ." (Ch. I, Sec. 3; cf. also Ch. III, Sec. 4)

PLATE 24

A FAMILY GROUP IN THE GARDEN

"The family . . . is the smallest co-operative unit in Trobriand gardening. . . ." (Ch. I, Sec. 9; cf. also Ch. XII, Sec. 2)

PLATE 25

THE MEN OF THE GARDENING TEAM PLANTING TAYTU

"At cutting and planting communal labour would invariably be used. . . ." The slanting poles of a part of a *kamkokola* are seen in the foreground (Ch. I, Sec. 8)

rectangles have little practical purpose, but much value is attached to the proportions and quality of the sticks which mark their boundaries. There are the purely magical constructions already mentioned, the *kamkokola*, and much effort is expended to make them look imposing. Pride is taken in selecting strong, stout and straight poles as supports for the yam vine. During all the successive stages of the work, visits are exchanged and mutual admiration and appreciation of the aesthetic qualities of the gardens are a constant feature of village life.

A considerable amount of pleasure in well-accomplished work and the social pressure embodied in the imperative: "It is the right, honourable and enviable thing to have fine-looking gardens and rich crops"—these are the psychological elements which we shall find expressed in many features of gardening, harvesting and of the general economic conditions.

A fuller insight into the way in which the proceeds of their gardening are utilised will show why the natives devote so much work, attention and aesthetic care to their gardens. Only after we have seen in detail how the crops are taken out of the soil and stored; how they are displayed several times in the process—cleaned, counted and adorned; how they are cajoled by special magic to remain stored and not to stimulate the appetite of greedy human beings; how they are redistributed, renamed and classified by sociological categories, only then shall we be able to appreciate the value of the crops for the Trobriand farmer (cf. Chs. V–VII).

Here certain distinctions between the economic functions of crops may be pointed out. Taro, which as we shall see figures almost more conspicuously in magic than does taytu or *kuvi*, becomes far less important in the handling and in tribal economics. The explanation is very simple. Taro has to be eaten as soon as it is dug up, because it is perishable (cf. Part I, Sec. 10). *Kuvi*, the large, bulky, unwieldy yams, can be stored for a long time, but they never keep quite as well as the small yam, taytu, and they are far less palatable, and therefore less appreciated by the natives as staple food. On the other hand, their size and their fantastic shape make them a valued object for decoration and display (cf. Pls. 63, 65 and 74). Taytu, the staple food, is to the natives *kaulo*, vegetable food *par excellence*, and it comes into prominence at harvest and after. This is the sheet-anchor of prosperity, the symbol of plenty, *malia*, and the main source of native wealth.

The presence of food means to the natives the absence of fear; security and confidence in the future. But it means more than that. It means the possibility of dancing and feasting, of leisure for carving and canoe-building, the opportunity for pleasant overseas expeditions, for visiting and social intercourse on a large scale. And to those natives who are pre-eminent as gardeners, chiefs or magicians, it also means *butura* (renown) and the satisfaction of vanity.

The native attitude towards food will become clear in our descriptions of harvesting, which to the Trobriander is a quite definitely festive occasion and one of pleasant though hard work. In the manner in which the food is handled and displayed, in the custom of admiring and counting the yams and the yam heaps, in the filling of the painted yam-houses, in the magic of *vilamalia*, the ritual of prosperity—in all this we shall constantly be faced with the emotional appeal of food, or rather of accumulated food. Thus at two stages of harvesting the natives instead of continuing directly with the business in hand, interrupt this in order to stack the neatly cleaned yams into fine conical heaps, cover each heap with an arbour, and let it remain for a few days or even weeks to be admired by villagers and visitors. Such a heap (cf. Pl. 17) is a token of the gardener's achievement, of the wealth of the community, and of the fertility of the soil.

Accumulated food is to them a good thing—its absence is not only something to be dreaded, but something to be ashamed of. There is no greater insult than to tell a man that he has no food, *gala kam*, "no food thine"; or that he is hungry, *kam molu*, "thy hunger".[1] (cf. also Ch. VI, Sec. 3). No one would ever ask for food, or eat in a strange place, or accept food unless in obedience to traditional usage. The giving of food is an act of superiority; and generosity is the highest of privileges, as well as an appreciated virtue.

Food is displayed on all occasions—at death and at dancing, at marriage and at mourning feasts. It is presented to the ancestral spirits at their annual return, and it is given to them as a ceremonial tribute (cf. Part I, Secs. 7 and 10).

Rank controls, as already indicated, the accumulation of food. Only men of rank can have large, decorated storehouses with wide open interstices. Commoners must store their yams in covered *bwayma* (storehouses: cf. Ch. VIII).

The intricacies of the sociology of food distribution; the obligations associated with kinship and relationship-in-law and tributes due to the chief, already touched upon (Part I, Sec. 9), will be further discussed in Chapter VI (Secs. 1 and 2) and in Chapter XII (especially Secs. 1, 2 and 3). In our account of harvesting, we shall see how great is the amount of labour involved in handling and

[1] *Kam* in the two expressions are accidental homophones.

transporting the produce in accordance with the complex native system of apportionments or distributions of the harvested crops.

Food, as we know, is used in the production of other utilities and its accumulation in the hands of the chiefs makes possible certain enterprises on a tribal scale (Part I, Secs. 4, 6 and 10). In connexion with this it is important to remember that the glory of gardening, the renown which attaches to a *tokwaybagula* (efficient husbandman) is always subordinate to the rules which make accumulation of food a privilege of rank. You may earn the reputation of a good gardener, but you must devote your energies to contributing to the yam-houses of the chiefs or head-men. No commoner must become too rich or work for anyone else but those really in power. Ill-health or even death by sorcery rather than renown would then reward his labours.

The crops harvested each year, especially the crops of small yams (taytu) are, then, the economic foundation of public and private life, of most institutions and pursuits in the Trobriands (cf. Part I, Sec. 10). With this in mind we shall be able to understand the mixture of aesthetic pleasure and passionate emotion, of hard work and delightful activities, of magical mysticism and really strenuous labour, which we shall encounter in the following chapters. We shall understand the minute care with which the Trobriander tills the earth, erects the strong vine supports, builds his magical structures, divides his garden plots into a chess-board with long slender poles. We shall be able to appreciate the sympathetic influence which these works of aesthetic supererogation exercise on the growth of the gardens—an influence kept still more alive by the constant performance of magic.

In pursuit of his duties to the chief, his relatives-in-law and kinsmen, and anxious also to garner his own share, the average farmer works with a will and with pleasure. If he is really efficient he will earn what he needs for his household, supply everyone with the share due to them, and gain a legitimate fame for himself. But he can never indulge in the ambition of surpassing his rivals and of producing personal wealth—he cannot do that with impunity.

Over and above individual effort, there is the power of *towosi* magic, the ordinary garden magic of the community; the influence of *vilamalia*, the magic of prosperity, and last but not least, the supreme magical power of the paramount chief, *tourikuna*. This magic finally decides on excessive sunshine, which means drought, or the adequate rainfall, which brings fertility to the islands.

CHAPTER II

THE GARDENS OF OMARAKANA: EARLY WORK AND INAUGURATIVE MAGIC

In the heart of Kiriwina, the richest and most fertile district of the Trobriands, lies Omarakana, the capital of the Tabalu, the highest and most powerful sub-clan in the whole nation of the Northern Massim (cf. Part I, Sec. 3). We are already aware that Kiriwina is surrounded with a mythological halo of superiority in gardens and garden magic; that it is the most aristocratic, the most wealthy and hence the most powerful province, and that its wealth and power are in the hands of the paramount chief, *the* Tabalu of Omarakana.

The glorious pedigree of this sub-clan leads them back to the original Malasi, the brother and sister or, as we ought to say in a matrilineal community, sister and brother, who emerged when mankind was first coming out of the soil. They originated on the north-west shore near the village of Laba'i, which is the mythological centre of the whole district.

According to tradition, they moved to Omarakana, and before the splendour of their rank and the power of their office, the original children of the soil had to give way (cf. Ch. XII, Secs. 1 and 3).

1. SOME PERSONALITIES OF THE GARDENING CIRCLE

The Tabalu have come, therefore, to be regarded as the owners of the soil, and as such they have the right and the duty of acting as garden magicians. The system of *towosi* garden magic named *Kaylu'ebila*, which is publicly and officially performed over the gardens of Omarakana, is in the hands of the supreme chief. At the time when I visited the district (1915–18), To'uluwa, the last chief ever allowed to wield the full powers of chieftainship, was still alive. He was a conspicuous presence in the village, sometimes squatting on the ground in front of his hut or storehouse, sometimes perched high on his *kubudoga* (raised platform), as in Plate 27, so as to allow his subjects to move about freely (cf. Part I, Sec. 9). He was a shrewd, well-balanced man, but his pride had been broken by the European invasion, and he had retired from most of his offices. He also had a bad memory and was not well versed in magical lore. The garden magic he had therefore delegated to his matrilineal nephew and direct successor, Bagido'u, the heir-apparent to the chieftainship.

Like all dignities and positions, the office of garden magician, which, let us remember, is inherent in the chieftainship or headmanship of a community, is hereditary in the female line, that is, from elder to younger brother and from maternal uncle to nephew (cf. Part I, Sec. 9). Sometimes, however, and not rarely at that, a chief would delegate the office to his son, especially if the latter belonged to a sub-clan of higher rank. The Kwoynama, a sub-clan of the Lukwasisiga, whose village is Osapola, is the most suitable for supplying husbands and wives to the Tabalu. Sons from such a marriage would usually have a special place in the capital and would often carry out among other important offices that of the garden magician. Thus Bagido'u's predecessor was his own father, Yowana, who was the son of Purayasi, a Tabalu chief, and of Vise'u, a woman of the Kwoynama sub-clan, and therefore belonging to the latter sub-clan. He in turn married Kadubulami, a Tabalu woman, so that his sons became heirs to the chieftainship. He had taught the magic of gardens to Bagido'u, who with the consent of his maternal uncle was carrying it out in Omarakana.

Bagido'u was a man of outstanding ability and intelligence (cf. Pl. 28). He, like most of the aristocrats of the Trobriands, had a great deal of personal dignity, extremely good and retiring manners; could be won by personal considerations rather than payment in tobacco; and once he took an interest, even in a white man, became something of a friend. As an informant, he was very easy and effective owing to his personal misfortune: he was, at the time when I met him, obviously in a fairly advanced stage of consumption, had to remain a great deal in or around his house, and was very willing to talk. Since he was a repository of native tradition, intelligent, gifted, and with an excellent memory, he was one of the main sources of my information on magic. Moreover, in spite of his illness, he still kept the whole conduct of gardening and garden magic in his own hands and personally performed most of the main ceremonies (Pl. 29).

Besides these two principal actors in all garden ritual, we shall become acquainted with the main helpers of Bagido'u: his two younger brothers, Towese'i and Mitakata, both of whom had already learned most of the formulae and were acting as Bagido'u's acolytes; and Yobukwa'u of the Kwoynama clan, a son of the paramount chief, who was a personal friend of the garden magician and had also learned from him the spells and the details of the ritual.

Apart from these outstanding personalities, the community consisted of the classes typical of every village of rank (cf. Part I

Sec. 9); that is to say: firstly of citizens of rank, the matrilineal relatives of the chieftain; secondly of those sub-clans who belonged to the dispossessed original owners—the sub-clan Burayama (Lukwasisiga clan) of high rank, who now rule a part of Kwaybwaga, and the sub-clan Kaluva'u, commoners of the Malasi clan; and finally of the *vilomugwa*, that is, men of low rank who live in the village not by right of citizenship but as vassals or servants to the chief. Three or four of these latter were distinguished by their efficiency in gardening; they were styled *tokwaybagula*, perfect gardeners.

It is clear that the garden magic and garden work of Omarakana should be the principal object of our study, and thanks to the good offices of Bagido'u, this is possible. For he not only allowed me to be present at every rite in the garden, but usually advised me some days beforehand, explained the rationale of most of his arrangements, invited me to his own house while he was reciting spells in the solemnity of actual performance, dictated them to me with unusual patience and capability, and helped me to translate them —by no means an easy task for him or me. Thus we can follow the whole of each ceremony in detail, hear all the spells and benefit by the comments of one of the best minds in the Trobriands.

2. THE STANDARD OF TROBRIAND AGRICULTURE

The Omarakana garden magic and work, the most elaborate and perhaps the most famous of all systems, is a paragon and a pattern to the other communities. Therefore the following account will serve as an illustration of gardening in general. The various systems differ but little from village to village, and such differences as there are we can assess in the comparative survey in Chapters IX and X, and Documents V to VII. The details into which we shall presently have to enter are not given merely from pedantry or love of accumulated fact. It is only when we follow the natives in their work and consider from their point of view everything which they are doing —that is, follow as minutely as they carry out themselves all that is essential to them—it is only then that we can really integrate native behaviour into native significance, and appreciate the values which surround the gardens for them.

The annual feast of *Milamala* is the Trobrianders' new year. This feast is definitely bound up with their gardening work, and takes place in the month of *Milamala* which in Kiriwina roughly corresponds to August (Chart of Time-reckoning, Fig. 3). It is preceded by the harvest as well as by the inauguration of the new gardens, and is succeeded by the burning, clearing and planting of the *kaymata*. The early or preliminary gardens, the *kaymugwa*, may be started, and usually are started, about one or two months before the main gardens. When I was in Omarakana in 1915, the scrub had already been cut on the site of the early gardens in June, while the main harvest was still in progress.

The earlier gardens are made within a separate enclosure, and, as a rule, on fields adjoining the settlement. In fact they often nestle within a bay of the village grove, as can be seen on Plate 30 where a few palms show that the plot has encroached on such a grove. Each man cultivates only one or two plots in the earlier gardens as against three to six in the later, the main gardens (*kaymata*). The two kinds of garden comprise practically the same crops, though the main gardens are more exclusively devoted to taytu while in the earlier there is a greater variety. They differ principally as to the season in which they are made. The magic performed over the earlier ones is identical with that of the main gardens, but the full set of ceremonies is performed on the main gardens only. Roughly speaking, most of what you will read in this and the following chapters (II to VII inclusive) refers both to the *kaymugwa* and the *kaymata*, the earlier and the main gardens. It would be confusing as well as tedious to duplicate the account and interrupt the narrative by constant jumps from one type of garden to the other, with a reservation here and a parallel there.[1]

3. *KAYAKU*—THE CHIEF AND MAGICIAN IN COUNCIL

Some time in the moon of *Yakoki* Bagido'u waits for a day or two of fine weather, and, good practical meteorologist that he is, fixes a day with To'uluwa, the chief, for the ceremonial deliberations. One fine evening he gets up, and standing by the fireplace in front of his hut, he addresses the village, announcing in a short harangue that the *kayaku* (garden council) will take place to-morrow.

"Good! To-morrow To'uluwa will hold his *kayaku*. It is time to begin our gardens. Last year's crops were good, we filled our *bwayma* (yam-houses), we ate our *kaulo* (vegetable food), we made big *sagali* (ceremonial distributions); we gave plenty of food to all the villages. This year we shall make even better gardens; the *bwayma* (yam-houses) must be filled. You old men will all come to-morrow. We shall sit on the *baku* (central place), we shall talk; we shall find which *kwabila* (land division) we shall garden; we shall

[1] A glance at the Chart of Time-reckoning will show the seasonal differences in the gardens.

count all the *baleko* (plots), who makes one *baleko*, who another. Good, come to-morrow to my house, to-morrow we shall chew betel-nut, we shall drink green coconuts, we shall suck sugar-cane and we shall talk. We shall count our gardens."[1]

Kayaku[2] is the name for any social gathering, conversation or council, in which the natives sit down together, discuss business or just talk socially. But the *kayaku*, the one by which a season is defined when a native refers to the period *o kayaku* ("at the time of the council sitting"), is the village council at which gardening business is discussed before the new gardens are started. It takes place in moon 11 (see the Chart of Time-reckoning), and is invariably held in front of the house of the *towosi* (garden magician). In Omarakana the men would forgather first of all in front of the chief's large personal dwelling (*lisiga*). Then, together with the chief, they would walk the few steps to Bagido'u's house, and there, squatting on the ground in a semicircle, they would wait for the chief and the garden magician to open the proceedings. Dagiribu'a, Bagido'u's wife, would hand round green coconuts, at times even some cooked yams on platters, while bunches of betel-nut would be distributed by the chief. Then To'uluwa would speak, telling them in the usual circumstantial manner that they had gathered there to settle all about the gardens.

"Last year we made gardens on Ibutaku and Lomilawayla. Two years ago we used Duguvayusi, our biggest *kwabila* (land division). We have not made gardens on Sakapu and Obwabi'u for a long time, but these are a long way off. Shall we strike Tubuloma, Kavakaylige, and Ovabodu this year?"[3]

To this Bagido'u, who had settled the matter beforehand with the chief, and for whose sake the very near lands had been selected as he was ill that year, would answer:

[1] This and the next speech quoted are not free translations of recorded native texts. During the early stages of my field-work I was not able to note down speeches rapidly in the vernacular. The quoted speech was, as a matter of fact, given to me in pidgin English by my earliest informant Gumigawaya. As always, I noted "untranslatable" (cf. Part IV, Div. 2) native terms in the vernacular, as they are given here. Throughout the following chapters all texts which are free translations of a recorded native original are numbered and referred to Part V, where an interlineal translation and commentary will be found.

[2] The linguistic analysis of this term and of associated words, and a few texts concerning *kayaku* noted in the vernacular, will be found in Part V (Div. V, §§ 17–24).

[3] For the proper names of fields, consult Doc. VIII. The legal character of the *kayaku* proceedings will be better understood by consulting the data given on land-tenure in Chs. XI and XII, and also Doc. VIII.

PLATE 26

AESTHETICS IN GARDENING

"The cleared soil is divided into neat rectangles by means of sticks laid on the ground" (Ch. I, Sec. 9; cf. also Ch. III, Sec. 3)

PLATE 28

BAGIDO'U IN FRONT OF HIS STOREHOUSE

". . . A man of outstanding ability and intelligence. . . ." (Ch. II, Sec. 1)

PLATE 27

THE PARAMOUNT CHIEF SUPERVISING HIS HARVEST TRIBUTE

"He was a conspicuous presence in the village, sometimes squatting on the ground . . . sometimes perched high on his raised platform (as here seen), so as to allow his subjects to move freely about" (Ch. II, Sec. 1; cf. also Part I, Sec. 9)

"Yes, O Chief, I want to strike Tubuloma, Kavakaylige and Ovabodu. The *odila* (bush) has grown well there, it is a wet year, and it will be good for these lands, which are all *galaluwa* (black, heavy, but dry soil) and *butuma* (light, red soil). Let us make our gardens there."

DIAGRAM OF A FIELD (*KWABILA*)

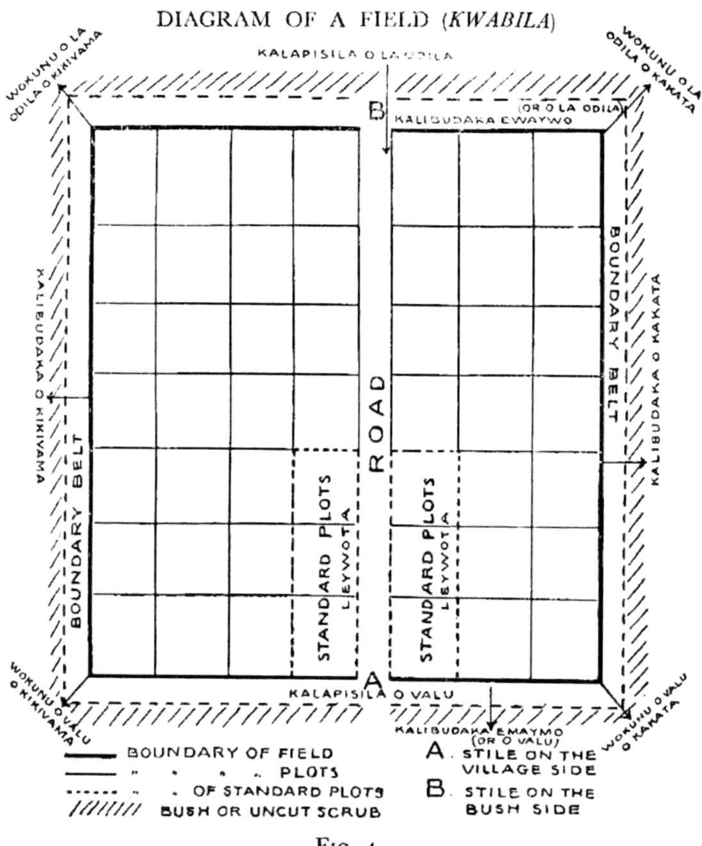

FIG. 4

Such speeches, telling to all what everyone knows and to which everyone has to consent, are a feature of Trobriand public life.[1]

After the selection of the site has been made, there comes what is called the counting of the garden plots. In studying land-tenure (Chs. XI and XII) we shall see that each field or garden portion (*kwabila*) is subdivided into smaller plots, *baleko*. These are in a way

[1] Cf. also the account of *kayaku* in Vakuta and Teyava, Docs. VI and VII—especially the latter. Cf. also Note 6 in App. II, Sec. 4.

individually owned, and as each field has its name, so have also most plots, though not all of them. The counting, or rather enumeration (*kalawa*), of the *baleko* (plots) is conducted by the chief or the garden magician. It can be done even if the names of the plots are not remembered. The typical arrangement of a garden can be seen from the preceding diagram (Fig. 4).

It must, of course, be kept in mind that, like all diagrams, this one represents a schematic arrangement of a more complex reality. As will be seen from the Map of Omarakana Garden Lands (Fig. 13), which reproduces both form and dimension with a high degree of approximation, many of the large fields are clear rectangles (e.g. fields 8–12) while others, such as the large field 1, and the smaller fields 2, 3, 4 and 5 are partly rounded, partly trapezoid. Without exception, however, the main points of orientation—the two stiles, the four corners, and the intermediate sides—are found (see also below in this section).

The Map also shows that certain large fields such as Duguvayusi, Opikwakula and Waribu are crossed by a road. Other adjoining fields again, such as Kavakaylige and Ovaboda, or Ibutaku and Sakapu, lie symmetrically on either side of the road, their boundaries usually meeting across the road, so that when fields 9 and 8, or 9 and 10, or 3 and 2, or 3 and 4, are gardened in the same year, the road naturally runs through them. Where the road strikes the garden, a stile will be made in the fence. This place is called *kalapisila o valu* (cf. Fig. 4). It is at this point that the four or six standard plots, the *leywota*, are always placed. The chief or garden magician would ask who is going to cut the main standard plot on the right hand. "I", answers the man who has been given the honour of gardening this choice spot.[1] The chief would then allot ceremonially the remaining plots belonging to the *leywota* type; each of these plots has its own name.[2]

After each standard plot has been allotted, usually to a man of rank or to a renowned gardener, the chief proceeds to distribute in the same manner the remaining plots.

As we have already said (Ch. I, Sec. 3) each man makes several plots, the number depending on his status, on the obligations he has to fulfil, and on his strength and industry. The average number cultivated is three to six. A very efficient gardener might cultivate up to ten plots, even if such a number was not absolutely necessary. Thus Mitayuwo of Omarakana normally made eight *baleko*. He

[1] Cf. also Note 3 in App. II, Sec. 4.

[2] These names and their linguistic analysis are given in Part V, Div. VI, § 29.

was a particularly strong and intelligent man, capable of working for long hours without being tired (I used him often as a carrier) and capable also of long stretches of strenuous mental attention, as I was able to judge by using him as my informant. A man called Kalumwaywo was another *tokwaybagula* and vied with Mitayuwo in efficiency; and Kawatalu, whom I never met, was said to be making twelve *baleko* each year with the help of his wife; but he was quoted as the greatest gardener (cf. also Note 7 in App. II, Sec. 4). But strict necessity demands that a man should make one and preferably two plots for the support of his own household, and at least one to satisfy his *urigubu* obligations, that is, the duty of filling his sister's husband's yam-house. Plots cultivated for own consumption are called *gubakayeki*, and those for sister's husband *urigubu*. A man would usually have all his *baleko* together.

The chief or garden magician, following some topographical scheme, enumerates one plot after another. Many plots have their names; the unnamed plots would be defined by their position in the garden. Each side of the garden can be described by reference to the village and the bush respectively, and to the road passing through it, as well as by its place to the right or left. Therefore a very clear system of co-ordination is provided by following the garden side (*kalibudaka*) to the corner (*wokunu*), then proceeding again along the side (*kalibudaka*) to the next corner, and so on. They would speak about the "hither garden side", *kalibudaka emaymo*, that is the garden side which you strike coming from the village; or simply, "garden side of the village", *kalibudaka o valu*. Then there would come the "corner of the village side to the right", *wokunu o valu o kakata*; then the right garden side, *kalibudaka o kakata*. Then comes the "thither corner or angle", *wokunu ewaywo*, also called "the bush angle on the right", *wokunu o la odila o kakata*. The thither garden side, *kalibudaka ewaywo*, otherwise called, *kalibudaka o la odila*, "bush garden side", leads to the "bush stile", *kalapisila o la odila*; then the corner in the bush on the left, *wokunu o la odila o kikivama*; the left garden side, *kalibudaka o kikivama*; and lastly the "corner of the village side to the left", *wokunu o valu o kikivama* (cf. Fig. 4).

Thus the garden site is clearly subdivided to the natives, who can define the position of most plots by reference to the sides, corners, stiles and central road. They know their fields, moreover, very well, gardening each of them every few years, and "the counting in the village" done at the *kayaku*, allows them usually to allot every plot to its prospective cultivator without uncertainties or error.[1]

[1] Cf. also Ch. XI (especially Sec. 4) and Ch. XII which will make these details clearer on the legal and economic side.

This system of orientation would not, however, be applicable to very irregularly shaped fields. Although in Omarakana, as we know, there are no such fields, villages in the south, especially those near coral ridge or mangrove swamp, with fewer and smaller patches of fertile land, have a good many. For these fields the individually named plots and topographical singularities would be taken as points of orientation. A concrete example, recorded during a *kayaku* at Sinaketa at which I was present, may be interesting (Text 28 *c*, Part V, Div. V, § 21). The lead was taken by To'udawada, chief and magician of the component hamlets.

(i) *Chief:* "Who will make the plot at the stile?"
His son: "I."
Chief: "And the next one?"
A Commoner: "I."
After three or four plots had been enumerated *seriatim*, the chief continued:
(ii) *Chief:* "Good. Let us proceed to the mangrove swamp. Who follows?"
Certain Commoners: "I . . ." (claiming successively the plots adjacent to the swamp).
(iii) *Chief:* "Already we have arrived at the garden corner. Who will make his plot round the garden corner?"
A Commoner: "I."
(iv) *Chief:* "Who will take over the plots at the garden side?"
Certain Commoners: "I . . ."
(v) *Chief:* And the corner which returns from Bwadela?"
A Commoner: "I."
(vi) *Chief:* "Who will garden the plot named Ogayasu?"
A Commoner: "I."

The last question was repeated with a number of plots which had individual names: Ogayasu, "where the *gayasu* plant grows"; Okaybu'a, "where the coral outcrop stands"; Wabusa, "where the *busa* tree grows"; Omwaydogu, "where the mangrove tree stands".

Afterwards I walked over the field. It was narrow and irregular in shape, but had one or two well-marked corners, one of which was at its extreme end on the road to the neighbouring village of Bwadela.[1]

But it still remains to prepare the gardens for the magical ceremony which is soon to follow, and to ascertain on the spot whether the allotment has been done correctly and to the satisfaction of everybody. For this purpose the men go on the same or the next day

[1] For other *kayaku* texts, see Doc. VII.

to the chosen fields, and cut a narrow belt all round the prospective garden, making thus a clear boundary space.

After the boundary belt has been cut round the new garden, in-roads are made into the encircled site in such a manner that free access may be had to every plot. This marks the final recognition by each man of the *baleko* which he is going to cultivate, and is what the natives call: *takalawa baleko o buyagu, takalawa mokita*, "we count the plots on the garden site, we count them truly". Thus this process of penetration, *sunini*, has a two-fold purpose; it seals the arrangements made during the gardening council and it makes each plot accessible so that magic can be performed on it.

As we shall see in our analysis of land-tenure, there is not much scope for quarrelling over land, since there is plenty of room for everyone to cultivate as much as he likes or can. But there are preferences; and quarrels over gardens are by no means infrequent, traces of which are to be found in Trobriand folk-lore (see below, Sec. 5). Such quarrels never occur during the garden council in the village in the presence of the chief and the magician. In fact they seldom occur at the stage now described, but only at the *takaywa*, the cutting of the scrub. Then they sometimes do quarrel about who is going to cut this or that plot.

4. THE GRAND INAUGURAL RITE: THE STRIKING OF THE SOIL

We now pass from these legal and ceremonial preliminaries to the magical rite, *yowota*, which inaugurates the gardening cycle. If so far To'uluwa, the chief, has been more prominent than anyone else, he now makes way for Bagido'u, his matrilineal nephew, to whom he has delegated the duties of garden magician. Henceforth Bagido'u will address the village; and it will be his part to fix the dates, and perform most of the ritual acts.

The first ceremony comprises an offering of food to the ancestral spirits, several magical spells, and also a very complicated rite in the gardens. It inaugurates not merely the first attack on the gardens the cutting of the scrub; but also the season's work as a whole, by the sacrifice to the spirits, the striking of the ground, and the rubbing of the soil. These last two rites, especially the striking of the ground with the *kaylepa*, the magical wand, are the representative acts of garden magic as a whole; in common speech they define the functions of garden magician. Thus instead of saying, "So-and-so is our *towosi*, our garden magician", "So-and-so charms our gardens"—they will usually say "he strikes our soil", *iwoye da pwaypwaya*, or "he strikes our garden", *iwoye da buyagu*.

Shortly after the *kayaku* (garden council) and a few days before the magical ceremonies begin, Bagido'u, at the hour of the evening meal, soon after sunset, addresses his fellow-villagers:

"Soon we shall strike our soil, old men. To-morrow you shall go to Kavataria (or some other coastal village). You will arrange for the fish. The day after to-morrow our partners will go out fishing and you will bring the ceremonial offering to our spirits. I shall go and collect herbs. The day after that, we shall strike the soil. We shall begin our gardens so they may grow high and swell underground."[1]

The harangue must be pronounced at least two or three days before the actual ceremony, for an arrangement with the coastal villages has to be made for the *wasi*, exchange of vegetable food for fish (cf. Part I, Sec. 10).

On the morning of the day before the ceremony, the village is early astir. Two parties are setting off for a day's expedition. Most of the young men of the village prepare their baskets, fill them with yams, tuck aromatic herbs into their armlets, anoint their bodies and put on some paint. Presently they set out for the coastal village with which they have previously arranged the date for the exchange. At the same time the coastal villagers have set sail early in search of fish.

Bagido'u, meanwhile, and perhaps his two brothers with one or two of their friends, are starting in the opposite direction. They are going to the eastern shore and to the adjoining coral ridge, the *rayboag*, to collect the magical herbs and other ingredients of the complicated mixture which will be prepared for the next day. They also have to take baskets, for among the thirteen elements required are large bunches of leaves, clumps of soil taken from the nest of the bush-hen, chunks of hornets' nest, and bits scraped off squat coral boulders. The *towosi* and his friends are usually first back in the village. There they sit down in front of his house and, without much ceremony, proceed to the preparation of the mixture. I have been present several times at these proceedings, in Bagido'u's house or in front of it, and have even lent a hand myself. Two mats are spread: the raw materials are deposited on one and shifted to the other as they are made ready for use. All the leaves have to be torn up into small bits; the earth, the pounded chalk and the broken up hornets' nest are mixed separately and strewn over the chopped leaves, which then look like a lightly peppered salad before it is dressed with oil and vinegar.

The magician is usually just about to finish his work when, from

[1] Cf. Note on p. 88.

the western outskirts of the village, a shrill *tilaykiki*, an intermittent yell, is heard, and panting, screaming, racing one another, the men with the offering rush in, and throw down the strings of fish at the magician's feet, with the words, *kam ula'ula da towosi*, "thy sacrificial oblation, O garden magician". Usually they add some such words as, "make our gardens good", or "offer it to the spirits —may they bring prosperity to our village".

Soon after the women gather round the house of the magician, who forthwith distributes the fish so that the whole village can have a festive supper. The men are resting after a very tiring day, for they have had to run on the way back, and usually most of them have their legs bled by means of the miniature bow and arrow arrangement called *gipita*. Speed is an essential element in the ritual, and it is also necessary so that the fish shall not arrive in a too decayed state. In spite of this, the village reeks with putrefying fish on such ceremonial evenings, so that a white man finds it difficult to remain there. But the ancestral spirits are quite fond of the smell, as are their living descendants.

After the magician has had his share of fish and the men have rested and eaten their meal, the first magical act of the whole system takes place. All the men bring their axes to the magician, who lays them on a mat spread upon the bunk along the side opposite to the fireplace (cf. Part I, Sec. 8). To each axe a piece of dried banana leaf, about six inches by four, has been tied. The main part of the leaf lies flat against the blade while the other part is left free. After the spell has been uttered the free flap will be folded over; but first some of the magical mixture must be inserted, between the leaf and the cutting edge of the axe, and mixture and blade be left open, so that the voice of the *towosi* and the magical virtue which it carries can penetrate into the blade and the herbs. After the axes have been placed on the mat, the magician takes some cooked fish and puts it down on one of the three hearthstones in his house with the following words:—

FORMULA 1

"*Here, this is our oblation, O old men, our ancestral spirits! I am laying it down for you, behold!*

"Here, this is our joint oblation, O Yowana, my father, behold!

"To-morrow, we shall enter our gardens, take heed! O Vikita, O

---

[1] The word "myth" is here used in the sense of the native word *liboguo*, that is, "sacred tradition", "charter of magical ritual and social order". Compare also my *Myth in Primitive Psychology*. For the difficulties which had to be overcome in

Iyavata, fountain-head of our myth[1] and magic, banish the pests, the insects and grubs.

"I shall open for you, O pests, the sea-passage of Kaulokoki.

"Your sea-channel Kiya'u! Drown, begone!"

This act as well as the exhortation recited during it is described as "the shredding of the spirits oblation". The word "to shred" refers here to the tearing off of small bits of fish and placing them on the hearthstone. The words of this incantation are spoken in a slow, solemn, persuasive voice. It is not regarded as a *yopa* (spell), and it is not uttered in the usual singsong of the magical formulae. The natives conceive of it as being spoken directly to the ancestral spirits by the *towosi*, as from one man to another.[1]

The blessing of the spirits being thus invoked, and their presence in a vague, mystical way established, the magician now proceeds to the charming of the axes. He places a second mat over the axes, only leaving a sufficient aperture for the breath carrying his words to penetrate between the two mats, and then he utters in the clear, melodious singsong characteristic of magical incantations, what is perhaps the most important formula in the whole system of Omarakana garden magic, the *vatuvi* spell:

FORMULA 2

I. "Show the way, show the way,
Show the way, show the way,
Show the way groundwards, into the deep ground,
Show the way, show the way,
Show the way, show the way,
Show the way firmly, show the way to the firm moorings.

"O grandfathers of the name of Polu, O grandfathers of the name of Koleko, . . . Takikila, . . . Mulabwoyta, . . . Kwayudila, . . . Katupwala, . . . Bugwabwaga, . . . Purayasi, . . . Numakala; and thou, new spirit, my grandfather Mwakenuwa, and thou my father Yowana.

"The belly of my garden leavens,
The belly of my garden rises,
The belly of my garden reclines,

---

translating these spells and a justification of the way in which I have done this, see Part VI. For fuller commentaries on the spells—their sociological and dogmatic setting, manner of recitation and so forth—and for their linguistic analysis, see the commentaries which follow each native text in Part VII. The formulae are there given in the same order in which they occur here, and with the same numbers.

[1] Cf. also Note 8 to App. II, Sec. 4.

PLATE 29

BAGIDO'U AT THE FIRST BURNING

"He personally performed most of the main ceremonial" (Ch. II, Sec. 1)

PLATE 30

SITE OF AN EARLY GARDEN

"They are as a rule made on fields adjoining the settlement" (Ch. II, Sec. 2. Note the stems left standing and the boundary poles; Ch. III, Sec. 3)

PLATE 31

THE CENTRE OF CEREMONIAL ACTIVITIES

"The point where a stile will lead over the fence into the garden—a point which will be the centre of most magical activities throughout the whole cycle" (Ch. II, Sec. 4)

PLATE 32

AN EXUBERANT PATCH OF VEGETATION

"A number of creepers, of which the leaves are torn up and mixed, are used (in garden magic) because their foliage closely resembles that of the taytu, and grows luxuriantly and to an enormous height" (Ch. II, Sec. 6)

The belly of my garden grows to the size of a bush-hen's nest,
The belly of my garden grows like an ant-hill;
The belly of my garden rises and is bowed down,
The belly of my garden rises like the iron-wood palm,
The belly of my garden lies down,
The belly of my garden swells,
The belly of my garden swells as with a child.
I sweep away.

II. "I sweep, I sweep, I sweep away. The grubs I sweep, I sweep away; the blight I sweep, I sweep away; insects I sweep, I sweep away; the beetle with the sharp tooth, I sweep, I sweep away; the beetle that bores, I sweep, I sweep away; the beetle that destroys the taro underground, I sweep, I sweep away; the marking blight, I sweep, I sweep away; the white blight on taro leaves, I sweep, I sweep away; the blight that shines, I sweep, I sweep away.

"I blow, I blow, I blow away. The grubs I blow, I blow away; the blight I blow, I blow away; insects I blow, I blow away; the beetle with the sharp tooth, I blow, I blow away; the beetle that bores, I blow, I blow away; the beetle that destroys the taro underground, I blow, I blow away; the white blight on taro leaves, I blow, I blow away; the marking blight, I blow, I blow away; the blight that shines, I blow, I blow away.

"I drive thee, I drive thee off, begone! The grubs I drive, I drive off, begone! The blight I drive, I drive off, begone! Insects I drive, I drive off, begone! The beetle with the sharp tooth, I drive, I drive off, begone! The beetle that bores I drive, I drive off, begone! The beetle that destroys the taro underground, I drive, I drive off, begone! The marking blight on taro leaves, I drive, I drive off, begone! The white blight, I drive, I drive off, begone! The blight that shines, I drive, I drive off, begone!

"I send thee, I send thee off, begone! The grubs I send, I send off, begone! The blight I send, I send off, begone! Insects I send, I send off, begone! The beetle with the sharp tooth, I send, I send off, begone! The beetle that bores I send, I send off, begone! The beetle that destroys the taro underground, I send, I send off, begone! The white blight on taro leaves, I send, I send off, begone! The marking blight, I send, I send off, begone! The blight that shines, I send, I send off, begone!

"I chase, I chase thee away, begone! The grubs I chase, I chase away, begone! The blight I chase, I chase away, begone! Insects, I chase, I chase away, begone! The beetle with the sharp tooth, I chase, I chase away, begone! The beetle that bores, I chase, I chase away, begone! The beetle that destroys the taro underground, I chase, I chase away, begone! The white blight on taro leaves, I chase, I chase away, begone! The marking blight, I chase, I chase away, begone! The blight that shines, I chase, I chase away, begone!

III. "I split for thee thy sea-passage of Kadilaboma, O garden blight.

"Laba'i is thy village, Ituloma is thy coral boulder. Sail on a *de'u* leaf, that is thy boat. Paddle with a coconut leaf rib.

"I shall stow you away, begone! Bubble away, begone! Disappear like a whirlwind, begone! Be lost, begone! Thou art my sister, keep off me! Be ashamed of me, get off me! Begone, slink away! Slink away bending.

"Show the way, show the way.
Show the way groundwards, into the deep ground.
Show the way, show the way.
Show the way firmly, show the way to the firm moorings.

"O grandfathers of the name of Polu, O grandfathers of the name of Koleko, . . . Takikila, . . . Mulabwoyta, . . . Kwayudila, . . . Katupwala, . . . Bugwabaga, . . . Purayasi, . . . Numakala, and thou, new spirit, my grandfather Mwakenuwa, and thou my father Yowana.

"The belly of my garden leavens,
The belly of my garden rises,
The belly of my garden reclines,
The belly of my garden grows to the size of a bush-hen's nest,
The belly of my garden grows like an ant-hill,
The belly of my garden rises and is bowed down,
The belly of my garden rises like the iron-wood palm,
The belly of my garden lies down,
The belly of my garden swells,
The belly of my garden swells as with a child.
I sweep, I sweep, I sweep away!"

Thus in the hearing of all the ancestral spirits, the magician recites the long sacred spell over the axes. I have quoted it here in full, but without reproducing all the repetitions; for, especially in the middle part, the magician will repeat his assertions and exorcisms over and over again, "I blow off, I drive off, I sweep away"—adding one pest, one blight, not always in the same order but somewhat at haphazard. The more important the occasion, the longer the repetitions. The spell here quoted will be recited as we shall see several times during the course of garden magic. When I listened to it in Bagido'u's hut, the proceedings took a good three-quarters of an hour, from the first exhortation to the spirits to the tying up of the magical virtue round the axe-blades, which ends the ritual.

There is no special taboo to be observed during this ritual. When I asked to be present my request was granted without further ado. The usual rule of good manners forbidding anyone to enter a house without very good reason, especially the house of a chief or garden magician, results in the spell being heard by a very small audience; but the rest of the villagers sitting outside over their festive meal of fish and taro pudding realise that it is being recited. As a matter of fact, since this spell is uttered in the usual loud singsong of magic, the immediate neighbours and all those who pass near the hut hear it.[1]

The charm ended, the magician takes off the top mat and immediately proceeds to fasten the open flap of the leaf round each axe-blade. By this means the magical virtue becomes imprisoned round the blade. To keep it there even more safely, the top mat is carefully replaced so that it enfolds all the axes, which remain on a top bunk or shelf till the next morning.

Early next morning all the men of the village again assemble in front of the magician's hut, who, unable to sleep for the excitement of the occasion and the preparations before him, has himself risen very early. He has not yet touched food, however, for on this day he must observe fast until all the magical proceedings are over. Each man receives his axe and hangs it over his shoulder in the usual Trobriand fashion. Then the whole company, headed by the chief and the magician, march in Indian file slowly and with due solemnity to the gardens. They are in what might be called semi-festive attire: an hibiscus flower or two in the hair, pandanus petals or green aromatic herbs in their armlets, black paint over one half of the face, or a few lines drawn with the crimson-red of crushed betel-nut mixed with lime. They have all washed at one of the water-holes and are anointed with coconut cream. They follow the road until they come to the point where a stile will lead over the fence into the garden—a point which will be the centre of most magical activities throughout the cycle (see Pl. 31). So far, of course, there is no stile and no garden, and only the boundary belt cuts the road at this place. The magician now comes forward holding his *kaylepa*, which in Omarakana is a short stick without any special ornamentation, but sacred since it is the hereditary wand of office of the garden magician. Bagido'u takes this stick in his left hand and his axe in the right and steps into the near corner of the

[1] See note, p. 95. The sociological, ritual and dogmatic context of this important spell are fully treated in Part VII (M.F. 2, A, B and D). Als oits structure and mode of recitation (C and E) which are typical of the fully developed Trobriand magical formula.

principal plot of the gardens, the one on the right nearest the village, and adjoining the village stile (*kalapisila o valu*—see Fig. 5). With a vigorous stroke he cuts down a small sapling; then lifts it in

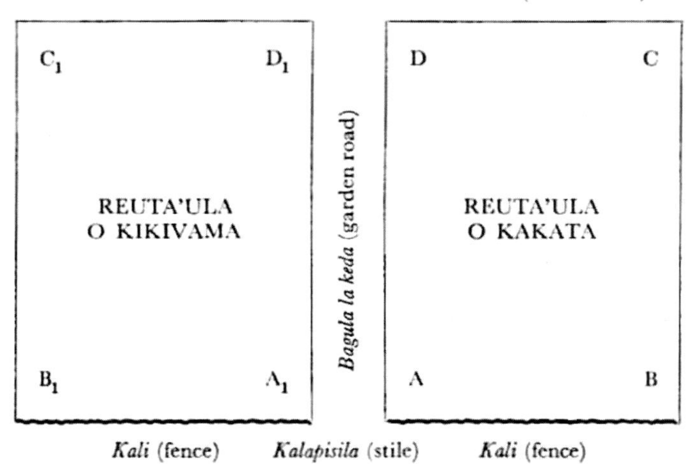

All the corners are called *nunula*.

A A_1 = *mukuvalu*—the "village corners". A is the main magical corner of the whole garden.

B, B_1 / C, C_1 / D, D_1 = *nukulaodila*—the "bush corners".

Fig. 5

his right hand and recites over it and, as it were, into it, the following spell called *kaygaga*:—

Formula 3

I. "This is our bad wood, O ancestral spirits!

"O bush-pig, who fightest, O bush-pig from the great stone in the *rayboag*, O bush-pig of the garden stakes, O bush-pig drawn by evil smells, O bush-pig of the narrow face, O bush-pig of the ugly countenance, O fierce bush-pig.

"Thy sail, O bush-pig, is in thy ear, thy steering-oar is in thy tail.

II. "I kick thee from behind, I despatch thee. Go away. Go to Ulawola. Return whence you have come. It burns your eyes, it turns your stomach."

The "bad wood", the sapling over which the charm is spoken, is then thrown across the boundary belt into the *yosewo*, the uncut jungle. It stands for all the evil influences, but above all for the bush-pig. It conveys a magical message to them, as I was told (cf. Ch. III, Sec. 2).

Then Bagido'u cuts a sapling standing next to the first one, inserts this *kayowota*, as it is called, into the ground, and squats or rather sits down by it. This attitude has a ritual and mythological meaning. According to tradition, the Kaylu'ebila garden magic, the magic used in Omarakana, was originated by two women, Iyavata and Vikita (cf. above, M.F. 1). It is in their honour that Bagido'u now rests on the ground female fashion. For no man ever sits with his buttocks on the ground, but has to squat without touching the soil.

The magician, swaying his whole body to right and left, and the stick with it, recites another short spell, the spell of *kayowota*:—

FORMULA 4

"I cut my garden; with my charmed axe, I make the belly of my garden blossom. My garden rises, it stands up on that side; it rises, it stands up on this side."

Without rising, the *towosi* tears up a handful of weeds and, while he rubs the ground with a swaying motion, he recites the *yowota* spell:—

FORMULA 5

"Who sits down and blesses on all sides within the tabooed grove of Yema?

"It is I myself, Yayabwa, with Gagabwa—we sit down and bless on all sides, we anoint it with coconut cream, we make the taytu vine grow up quick and straight in the tabooed grove of Yema.

"The belly of my garden lifts,
The belly of my garden rises,
The belly of my garden reclines,
The belly of my garden grows to the size of a bush-hen's nest,
The belly of my garden grows like an ant-hill,
The belly of my garden rises and is bowed down,
The belly of my garden rises like the iron-wood palm,
The belly of my garden lies down,
The belly of my garden swells,
The belly of my garden swells as with a child."

This rite, which is described as *talala*, "we make to flower", is meant to impart fertility to the soil and make the garden flourish with exuberant crops. As Bagido'u put it to me (Text 41, Part V, Div. VIII, §9): "After I rub the soil ritually, it becomes soft and fertile like the *dumya*." As the *kaygaga* spell was negative, so this is the positive spell which will promote the growth of the gardens throughout the whole cycle. More specifically the planting of the *yowota* sapling has the effect of making the crops sprout and grow, and the rubbing of the ground makes it soft and good.

The ceremony is not quite at an end yet; Bagido'u rises to his feet and, with his magical wand, strikes the soil, saying the *kaylepa* spell:—

FORMULA 6

> "I strike thee, O soil, open thou up and let the crops through the ground. Shake, O soil, swell out, O soil, swell out as with a child, O soil."

The men have been watching the ceremony in silence, standing in a close group on the small clearing which is always made at the *kalapisila*, on the road and on the boundary belt. After the magician has completed his share of the performance, they cease to be passive and, uttering the long-drawn scream, *tilaykiki*, each runs to his own garden plot, grasping his medicated axe in his right hand. There the ceremony is repeated in miniature. The magician does not go with them, but his acolytes, his younger brothers and helpers follow the men. Each man cuts two sticks on his *baleko*. One he has to throw out of his *baleko* into the uncut bush, or, if his *baleko* does not adjoin this, carry it to the boundary belt and throw it across. The other, the *kayowota*, he plants in the ground. Afterwards he has to cut the little piece of scrub on that part of the *baleko* where later the *kamkokola* will be erected. Then one of the acolytes recites one formula only on each plot, the formula of *yowota*, rubbing the soil with weeds pulled out of the ground.

This compound ceremony inaugurates the cutting of the scrub, to which the men now proceed energetically; the work is done in part communally and in part individually, each man on his own plot (cf. Ch. IV, Sec. 5).

5. THE WORK OF CUTTING THE SCRUB

The *leywota* are first attacked. Whenever these, as often happens, are worked by the chief, or directly for him, communal labour is summoned. The work to be done consists in cutting down the small trees, uprooting bigger weeds, and trimming such of the saplings and larger trees as can be used later on for vine supports. A number of good and straight stems are put aside for future use as boundary poles (*tula*) and fence stakes (see Ch. III, Sec. 3). The scrub, however, even though it be only three or four years old, is rank and, before the Europeans' advent when the natives had to use stone implements, the cutting of it must have required a considerable amount of skill. Communal labour is therefore highly appreciated, and emulation as well as other customary amenities serve to take away from the tedium of the work. Thus the men engaged on it assemble first in the village, where they are given some slight refreshment—green coconuts, betel-nuts, baked yams—some of which they eat on the spot and some in the gardens (cf. Ch. IV, Sec. 5). They start work simultaneously, standing in a row and advancing into the scrub as they cut it. From time to time they utter the *ta'ukuwakula* scream, and then fall to again with renewed energy.

When it is felt that work is becoming slack, or when the heat is too great, or when the bush has been denser and more difficult to deal with than usual, one of the men, generally a notable, will call out *uri yakala towamata*—"my treat to-day will be the giving of a taro cake". Or of course it may be some other food, at times even a pig. The work would then be immediately stopped, and the men would answer in chorus, "*O yakala, O yakala, O yakala, O, O, O* . . ." and all would go back to the village and eat the taro cake or the coconuts or squat round chewing betel-nut and talking.

Thus is communal work made amusing on a chief's plot. On ordinary plots every man, accompanied perhaps by his son or his nephew, cuts down his own scrub; and here severe quarrels may break out if two people want to cut the same plot, or if they cannot agree on the boundary line, though this is usually pretty clear. Or again, a man may regret having chosen an inferior plot at the *kayaku*. He will then pretend that his choice was different and start cutting the plot he now covets.

Whatever the cause, quarrelling frequently occurs. Even during my stay in the Trobriands I was told that there were long *yakala*, native litigations, arising out of disputes at cutting. Nowadays such cases are not infrequently brought before the white Resident Magistrate. I understand that Dr. Bellamy, who was for a long time A.R.M. in the Trobriands, but unfortunately left the district soon after my arrival, having enlisted for the Great War, invariably refused to give judgment on cases of this sort. His successor, who had no experience in administrative work, adopted a method natural to the European but fatal in a matrilineal community. He would enquire whose father had cut the disputed plot in olden days, a question which, under maternal descent, was beside the point and usually admitted of no answer, since the fathers of both litigants probably belonged to other communities (cf. Ch. XII).

There are several references in native folk-lore to quarrelling about gardens. One cat's-cradle or string figure shows two men quarrelling while cutting, and in one or two fairy-tales the same motif occurs.

As a commentary on the myth of Kudayuri, I was told that the quarrel which ended in the murder of the elder brother, Mokatuboda, arose over the cutting of a plot. As it was put by my informant (Text 97, Part V, Div. XII, § 42): "(i) People came and said to Mokatuboda: 'Your younger brother is cutting a garden plot.' (ii) 'Indeed. I shall go and shall see.' He went. (iii) 'What art thou doing?' His younger brother answered: 'That is all right; keep off; let me cut my own garden plot.' (iv) 'No, indeed, thou art a youngster; I am your senior and this is my garden plot.' (v) They quarrelled: already the younger brother is roused to anger; he hits his brother and kills him."[1]

Really good gardeners never lag behind in work. A few days after the inaugural ceremony their plots are cut, and the scrub is lying in the sun to dry in readiness for burning. Some are more dilatory; but it is essential that a little of the scrub should be cut on every plot, so that all can participate in the ceremonial burning. When the majority of plots are cut there is invariably a long pause in garden work, as the scrub must be fairly dry before it burns well, and during this respite the festivities connected with the previous harvest, the *milamala*, begin.

This is a suitable place to make a few remarks about the co-ordination of work which obtains throughout the gardening cycle. After each inaugural ceremony, the practical work is started immediately on the standard plots and also on those of really good gardeners. The standard plots must be completely finished as regards their respective stages—that is, they have to be completely cut, cleared, planted with the preliminary crops and so on, before the next rite inaugurates the following stage. As to the other plots, on every one of them some part of the work corresponding to the inaugural ceremony must be done, but such work need not be finished. Thus at times the magician is performing some of the later rites on a garden in which one or two of the plots (*baleko*) are not yet completely cut. Speaking about the lazy men who lag behind, one of my informants

[1] Cf. for Kudayuri myth, *Argonauts of the Western Pacific*, pp. 311–316. In the narrative I was told that the brother killed Mokatuboda as he sat in the village, but on asking the real reason why he killed him, I received the comment here given. In another locality I was told that he was really killed because he made rain magic, so that a small cloud formed just over his garden and gave it rain and fertility, while the surrounding plots, including those of his younger brother and maternal nephews, remained parched.

told me (Text 28a, Part V, Div. V, § 18): "They would not catch the time-reckoning! This man here [referring to a good gardener whose plot was up to time] he would be well ahead, he would have planted his plot, he would be abreast of the reckoning, his crops would be already sprouting." The phrase "to catch a time-reckoning" corresponds to the concept of being well up to time.

6. DIGRESSION ON MAGICAL INGREDIENTS

Let us pause with the gardeners and make a digression on certain points connected with garden work, which have been left unexplained, so as not to interfere with the sequence of events.

First of all, we must enumerate the several ingredients of the magical mixture. It is a tedious business, which brings out a prosaic, pseudo-scientific side of magic, but it throws an important light on the structure of magical belief.

The explanations here given of the why and wherefore of the substances are a reproduction of what the garden magician, his acolytes and one or two other natives told me, for neither their use nor their significance is secret. The spells are not secret either; they are chanted in a sufficiently loud voice in the house and certainly quite loudly in the garden. They have been heard by everyone, and most natives know them by heart. But no one save the garden magician or his accredited acolytes would dare to utter them. I could not get any unauthorised native even to check them with me, or to help me in the translation.

In enumerating his materials, Bagido'u would always first mention that very ordinary substance *yoyu*, coconut leaves. I obtained the list from him several times, because it is never quite certain whether the memory of a native will not fail, but each time, in 1915 as well as in 1918, the coconut leaves came first. They are of the dark green colour which the taytu leaves should have if they are to be strong and healthy.

The second ingredient, the leaf of the areca nut, is used for exactly the same reason. The next three substances which I find in conspicuous places on my list are all used because of their size and shape. *Ge'u* is the native name for the enormous mounds scraped together by the bush-hen for brooding purposes. A chunk of caked earth is taken from such a mound, carried to the village in a *vataga*, an oblong basket, and then crumbled up in the fingers; it is used so that the taytu may grow and swell up, like one of these mounds. For the same reason some chalk is scraped off with a mussel-shell (*kaybomatu*) from the large coral boulders or outcrops of coral found in the *rayboag*. The name for these, *kaybu'a*, refers to their massive and spherical shape, which is the proper shape for the maturing taytu; they also, as the natives put it, "grow deep into the ground as good taytu should do". *Kabwabu* are the large round nests which hornets make in the ground; the taytu should be as bulging and large as one of these nests.

A number of creepers, of which the leaves are torn up and mixed, are used because their foliage closely resembles that of the taytu, and grows luxuriantly and to an enormous height (Pl. 32). Taytu with a rich foliage will also have good tubers. I only have the native names for these creepers and have not been able to identify them botanically. They are: *youla'ula* (a creeper with white flowers), *yokwa'oma*, *ipikwanada* and *yokunukwanada*. These last two are probably varieties of wild yam.

Ubwara is a small plant growing in the bush, with long tubers which are white and beautiful to look at, as the natives put it. Its leaves are mixed with the rest in the hope that the taytu in the garden will also produce beautiful white tubers. The white petals of an especially fragrant species of pandanus, *kaybwibwi*, are used partly because of its scent—"the taytu should have a pleasant smell like that of *kaybwibwi* after we take it out of the ground"—partly because the bunches of taytu tubers should be as thick and large and long as the aerial roots of the pandanus. *Kubila*, a plant with lovely scented flowers, is also used to infect the taytu with its pleasant smell. *Sasoka*, a tree with very big, round and bulky fruit, and the leaves of the *wakaya*, which is the largest variety of banana grown in the Trobriands and has a massive trunk swelling out near the ground, are both incorporated to influence the size of the taytu. The leaves of the *wakaya* are not mixed with the rest, but are wrapped round the blade of the axes, so as to enclose the mixture.

The meticulous care with which all these substances are brought and prepared; the clear and well-established doctrine which justifies their use, and the rational interest shown in their application, are very characteristic of Trobriand magic.

7. THE MAGICIAN'S TABOOS

The same type of doctrine underlies also the taboos which Bagido'u has to keep. It has been already mentioned that on the day of a ceremony he must fast completely till the ceremony is over, after which he can eat his fill. He has also permanently to abstain from certain foods; some of which are associated with the substances used in his magic. Thus Bagido'u may never touch the flesh of the ordinary bush-hen (*kwaroto*) or its eggs. Neither may he eat *mulubida*, the smaller species of bush-hen. He must not eat the *wakaya* bananas, nor the tubers of the *ubwara*. He is not forbidden, however, either coconut or betel-nut. If he broke any of the food taboos associated with his magic, the taytu would not grow properly. His magic would become "blunt".

Sina, a bird with black plumage, must not be eaten or else the rain would be frightened. The magic would not produce wet weather, and this, though not its direct end, inevitably accompanies every important ceremony as part of its *kariyala* (magical portent). The cuttle fish, *kwita*, if eaten would have the same undesirable effect, highly pernicious to the gardens. This fish, which squirts out its black fluid into the sea, is mystically associated with rain-clouds. As Bagido'u explained to me, the reason (*u'ula*) for this taboo is that the animals are black. Should he partake of their flesh, the rain-clouds (*bwabwa'u*, lit. the black things) would not follow the magic, there would be no *kariyala* and the gardens would die.

A number of other fish are not allowed to the magician, some of them because they are of dark or black colour, some of them because they live in the coral outcrops of the reef. Why these latter are tabooed I was not able to ascertain. The following are the native names of the forbidden fish:—*yabwa'u, milabwaka, mamila, sekela, siga'u, mawa, bayba'i, madolu, lum'gwa.*

As mentioned already, Bagido'u must not eat the new taytu taken out of the soil during the thinning out, the *basi*. These are named *bwanawa*, as distinguished from the *taytuva'u* taken out at harvest proper. To this taboo we shall return again in discussing the harvest.

8. VARIATIONS IN THE INAUGURAL RITE

One more technical detail must be mentioned. In the description of the inaugural ceremony (see above, Sec. 4), I gave an outline of the simplified rite such as I witnessed in Omarakana in 1918. But a more complex form is usual in other villages, though Bagido'u, who was at that time a very sick man, invariably performed the simplified *yowota*. The simplified rite is called *burakema*, the more complex and circumstantial one, *bulukaylepa*.

This latter lasts four days and here is a brief outline. On the day after the offering to the ancestral spirits, the garden magician goes alone into the gardens, and rubs each plot with a handful of weeds, uttering the *yowota* spell. This ceremony is called *vapopula digadaga*. Next day the *towosi*, perhaps accompanied by one or two acolytes, goes to the principal *leywota*, and with an ordinary axe, unmedicated, cuts down some trees at the magical corner (*o nunula*). On the third day, the same is done by the acolytes on the plots of the other *leywota*, while the *towosi* collects the magical mixture and brings it to the village. In the evening the axes are medicated as described above. Then on the fourth day all the men of the village go with the magician and the chief to the gardens. There the ceremony is performed very much as described above: the *kaygaga*, the bad stick, is cut and thrown away; the *kayowota* is cut, inserted into the ground, swayed and charmed. The only difference is that the rubbing of the soil, which has been done already on the first day, is not repeated. But the ground is hit with the *kaylepa* and the ordinary spell is uttered. The magician then goes round the gardens, each *baleko* is hit and over each the spell is spoken.

I failed to ascertain exactly on which occasions the simpler rite, and on which the more complex one is performed. As far as I can judge, the simpler rite is used when the magician does not feel very well, or when there is no special anxiety about the gardens; that is, after there has been a run of successful gardening seasons.

In the following text, which I obtained rather late in my field-work, there is contained the distinction between the two forms of magic: *burakema* and *kaylepa*. I obtained it in Obweria from Modulabu, the garden magician, in connexion with a garden council and the opening rites which I had witnessed there (Text 28d, Part V, Div. V, § 23). "(i) We forgather in council—she (the magician's wife) would bake the yams and bring them, so that we might eat them while sitting in council. (ii) We enumerate (the garden plots following the) fence; we make the rounds till we make the two ends meet, we adjourn. (iii) We reopen the council (next day): we count (the plots right) inside the garden until we have finished. (iv) We present the oblation (to the spirits)—(and then) the garden magician will perform the rite called 'to cause the breaking forth of the bracken'. (v) Whether he might perform the *burakema* rite—which is also called 'to make flower the gardens'; whether he might walk round with the magician's wand—both rites are *yowota* (inaugural garden magic). (vi) It depends on the garden magician's inclination, whether he desires to make the *burakema* rite or whether he desires to strike (the garden site)."

This corresponds to what we find in Document VI on Vakuta, as a very clearly marked distinction in the proceedings. At the same time, it is probable that the division into two sittings was just an accidental matter. The proceedings on the first day extending too long, the *kayaku* was broken up.

In comparing verses (iv) and (v) we see that the partial rite *vapopula digadaga* may be carried out on the same day on which the *ula'ula* (oblation to spirits) is given. In (v) and (vi) we find the statement that it depends entirely upon the magician's personal inclination whether he will perform one or the other of the two rites.

With this we have finished our digressions on the technicalities incidental to the beginnings of Omarakana garden magic, and we can return now to a consecutive account of the further stages.

CHAPTER III

THE GARDENS OF OMARAKANA: PREPARING THE SOIL AND PLANTING THE SEED

WALKING through the *buyagu*, the garden site, a week or ten days after the first inaugural ceremony, we see most of the plots covered with cut weeds, branches and saplings, and only here and there a solid wall of tangled greenery, where a plot has been left untouched. There is that part of the moon of *Milamala*, when the spirits are being received in the village, feasted, and then expelled, for the refuse to dry. Soon after the *yoba*, the driving away of the spirits, in the second half of the moon of *Milamala* or early in the moon of *Yakosi*, the gardens are ready for burning. The burning of the gardens (*gabu*) has a double function. In the first place, the mass of faded and dried greenery is best cleared by burning; and in the second place, the ashes form an excellent manure when washed into the soil by the rain. As Motago'i, one of my best informants, put it (Text 29 Part V, Div. VI, § 14): "If we were not to burn the leaves on the soil, it would become sterile and not fertile." I did not often come across this realisation of the fertilising value of ashes, but, besides Motago'i, one or two others, including Bagido'u, were ready to admit that ashes were favourable to the growth of plants.

The burning, however, does not clear and clean the fields sufficiently for the planting. Some of the weeds do not burn easily, and some will sprout immediately after being burned, even though only a small shoot has been spared by the flames. Thus the burning is immediately followed by the *koumwala*, or cleaning up of the rubbish, and usually directly after the *koumwala* comes the planting of several of the crops: taro, sugar-cane, peas, pumpkins, large yams, and bananas. The planting of the staple crop, the small yam (taytu), occupies a somewhat special position.

1. THE SECOND GRAND INAUGURAL ACT: THE RITUAL BURNING OF THE GARDENS

This complex set of gardening activities, to which we shall return presently, is inaugurated by a four-fold magical ceremony. At this point the practical and magical activities are perhaps more intimately associated and intertwined with one another than at any other stage.

The wholesale burning is itself a magical act performed by the magician and his helpers on the first day of the ceremonial series.

On the second day, the magician ritually starts the practically more effective second burning of the small weeds and accumulated refuse. On the third and fourth days he performs ceremonies which consist of a magical planting of taro and yams respectively. Thus each of the four inaugural rites of magic is correlated with one or the other aspect of the complex activities on which the gardeners now embark.

At this time of the year fine weather may generally be expected to last; for we are in the middle of the dry season of the south-east trade winds. Bagido'u, with his fine meteorological appreciation, decides on a date that will allow him to expect a run of a few fine days, a desirable though not indispensable condition. Then, one evening, he will address the village:

"To-morrow I put the *kaboma* (garden taboo) on our fields. You remain in the village; I with my younger brothers will go and make the wholesale burning (*vakavayla'u*). You remain in the village. Do not do any garden work. Just remain and rest.[1] I shall go, I shall burn the garden."[2]

Next morning Bagido'u will produce some magical torches, *kaykapola*, from one of the shelves which run under the thatch of his roof. Enfolded in a mat, these have been lying there since the last harvest a few months before. At that time he prepared them for the present occasion, charming them over with a spell which we know already—the long spell beginning with the words: *Vatuvi, vatuvi*—"Show the way, show the way." He takes the mat with the torches inside it and, escorted by some of his acolytes, repairs to the garden.

When I accompanied Bagido'u on the *vakavayla'u* ceremony of September 11th, 1915, the chief himself went with us, rather in attendance on me than on the garden, I suspect. The chief's eldest wife, Bokuyoba, came also, and about half a dozen men, among them Towese'i and Mitakata, the two younger blood-brothers of Bagido'u. Elsewhere, at one or two other ceremonies of this kind, I saw the garden magician in company with two or three men who were helping him. At my first *vakavayla'u*, the chief, To'uluwa, himself took one torch, his eldest wife another, Molubabeba, the chief's maternal first cousin, another, Bagido'u yet another and, standing to the windward of the garden, they applied the torches simultaneously to the cut scrub (Pl. 34. See also Pl. 29). It was dry and blazed away easily. After the fire was under way, Towese'i and Mitakata took a torch each and went round the boundary belt kindling the cut scrub at intervals, at least once on each garden plot. In Omarakana, in spite of all the greatness of its magic and dignity of the ritual, an ordinary match-box was produced in the

[1] See also Note 9 in App. II, Sec. 4.

[2] Cf. footnote, p. 188.

gardens, from which the torches were lit. In some other villages, however, the fire has to be either made by friction in the garden, or carried from the village by means of fire-sticks lit at the magician's hearth.

On our return to the village, we observe that a somewhat festive though by no means highly ceremonial or religious atmosphere prevails; the majority of the people have remained there, for the most part just sitting about and chatting. There is no taboo on attendance at such ceremonies—even women may be present, as we have seen, and may handle the magical torches—but naturally only those specially authorised to go would be there.

But though this ceremony has few trappings, it is regarded as very sacred and important, indeed indispensable. On one occasion—it was I think some ten years before my arrival in the islands—one of the resident magistrates, not aware of the necessity for magical burning, was tempted, in passing a cut and dried garden, to put a match to it. It was in a specially dry year, he told me, and the bush had been very tall and rank, so that the idea of a gigantic bonfire attracted him. He lit the garden, and soon the night was ablaze with high-towering flames. In a few minutes he was surrounded by a crowd of natives, half despairing, half angry, who then explained to him for the first time their magical theory of gardening, besought him never to do anything of this sort again, and prophesied that the gardens would suffer. And indeed this was a year of drought—hence the beauty of the bonfire—so that the prophecy did come true, and the white man for once contributed by his act to the strengthening of native belief. For although this occurred years before I came to the Trobriands and in an entirely different district, the natives of Omarakana would quote this case to me as proof of the truth of their doctrine and of the danger of European interference with their custom.

Back in his house, Bagido'u sits down with his assistants and eats his first meal for that day. Soon after, he proceeds to the preparation of the next act in the sequence: the second burning, called *gibuviyaka*, literally "the big burning", though in fact it is a burning carried out in detail. He goes out in search of two kinds of leaves: to the seashore for acacia leaves (*vayoulo*), and afterwards to the swampy marches, to the *dumya*, for lalang grass (*gipware'i*). Some time before, a few budding coconut sprouts have been cut and have by now dried, thus yielding the easily inflammable material of the torches.

Next morning the magician places the acacia leaves, the lalang grass and the torches beside each other on one mat, covering this with another. A piece of dried banana leaf is tied round the end of

PLATE 33

VIEW OF A GARDEN AFTER THE RITUAL BURNING

This picture gives a good view of the garden littered with sticks, with the small trees left for yam supports, after cutting and burning; and of the surrounding jungle with here and there a big tree (in the centre of background). Some of the plots are still uncut (Ch. III, Sec. 1)

PLATE 34

THE BURNING OF THE GARDEN

". . . standing to the windward of the garden they (the magician's helpers, on this photograph Bokuyoba and Molubabeba) applied the torches to the cut scrub" (Ch. III, Sec. 1). Note the rolled up grass petticoat worn by Bokuyoba as a protection against the sun. Her head was shaved for mourning

PLATE 35

THE SECOND BURNING

"Mitakata performed the ceremony on the other *baleko*" (Ch. III, Sec. 1)

each torch by means of some lalang grass, leaving a flap open which is not folded over and tied down until after the spell has been spoken. The procedure is similar to that used with the axes (cf. Ch. II, Sec. 4). The function of the leaf is to catch and imprison the magical virtue of the spell. The spell recited over the torches and acacia leaves is again the *vatuvi* (M.F. 2 of our series).

Immediately after this spell has been recited, the magician and his acolytes set out for the gardens to carry out the actual rite of *gibuviyaka*. On the occasion on which I saw and photographed the rite, Bagido'u personally performed it only on the *leywota*. His younger brother Mitakata, accompanied by Molubabeba (a Tabalu), Gumguya'u and Ka'ututa'u, ritually carried out the two-fold ceremony on all the other plots.

They carried the acacia leaves, which had been tied with lalang grass into as many bundles as there were plots and wrapped up in a large folded mat. They went first to the *leywota*. There the assistants quickly performed a sort of preliminary *koumwala*, gathering sticks, weeds and branches and piling these in front of Bagido'u. Bagido'u took a bundle of acacia leaves out of the mat, and put it into the heap. After the *leywota*, Mitakata performed the ceremony on the other *baleko* (Pl. 35; cf. also Pl. 33). The heaps are called *lumlum*, a word to my knowledge used only with reference to these magical piles.[1]

After his return Bagido'u cannot yet sit down to his morning meal. He is once more called by his duties to the garden. Accompanied again by one or two helpers, who this time carry a mat full of medicated torches, he goes back to the *leywota*. With no preliminary spell, he lights the torch by means of a match and sets fire to the *lumlum*, and again the ceremony is performed on the other plots by Mitakata. Returning to the village in the afternoon, he can break his fast; this ends the second day's ceremonies, which are to be continued on the morrow.

The rite of the third day, called by the strange name *pelaka'ukwa* (lit. "dog's excrement"), consists in the planting of a taro. Bagido'u goes to the *leywota*, accompanied by helpers who carry large bunches of taro tops; that is, of taro leaves attached to the top of the root from which the edible bulbous part has been cut off. It is by putting such a taro top into the soil that this vegetable is planted (cf. Ch. X, Sec. 2). He takes the biggest and most promising taro top and, holding it in his right hand, puts his face close to the cut surface and speaks into it the following charm:—

[1] Compare, for the light thrown by linguistic analysis on this word, Part V, Div. VII, § 26.

FORMULA 7

"This is our dog dung, O ancestral spirits!

"O bush-pig, who fightest, O bush-pig from the great stone in the *rayboag*, O bush-pig of the garden stakes, O bush-pig, drawn by evil smells, O bush-pig of the narrow face, O bush-pig of the ugly countenance, O fierce bush-pig.

"Thy sail, O bush-pig, is in thy ear, thy steering-oar is in thy tail.

"I kick thee from behind, I despatch thee. Go away. Go to Ulawola. Return whence you have come. It burns your eyes, it turns your stomach."

In comparing this with the spell of the bad stick (M.F. 3) we see that they are identical, save for the first line. Also, by the magical tradition of Omarakana, this spell is spoken in a soft persuasive voice into the cut surface of the taro, and not chanted.

This ritual taro is planted in the corner of the *leywota* which adjoins the "stile of the village side"—the corner which will gradually become the main "magical corner" of the gardens (Pls. 23 and 31).

The magician's younger brothers now proceed to all the other garden plots, and repeat the ceremony.

On the fourth day a double rite is performed exactly on the spot where the first taro has been planted on the previous day. In the first of these rites, which is called *kalimamata*, a seed tuber of a special variety of yam, called *kwanada*, is planted. Bagido'u again squats down at the magical corner. He takes the *kwanada* tuber in his right hand, holds it close to his mouth and speaks the following spell into it:—

FORMULA 8

"Blaze up, flare up towards the village!
Spread fast, move fast towards the bush!"

This short spell is spoken, not in the soft persuasive voice, but loudly as if giving a peremptory order or perhaps inciting to some decisive action. The spell is obviously an exhortation to the vegetables to grow and spread on both sides of the garden.

Immediately after this, the garden magician takes a taro top from one of his assistants and chants it over, this time in the ordinary singsong of magical formulae:—

FORMULA 9

"Swell there, O taro, swell there, O taro. Swell here, O taro, swell here, O taro, O stout taro. It comes on quickly, the immovable taro."

The words of the spell explain the intention of the rite, which is to make the taro grow in size and in quality.

After the spell, Bagido'u performs this rite: he collects a few small twigs, some twenty to forty inches long, and constructs or rather sketches out a miniature hut: he plants four verticals, with forks at the top into which he then places four horizontals, and these he covers with a little sketchy roof by laying a number of horizontal sticks across the top rectangle. This miniature hut he surrounds with a miniature fence, made by planting a few forked sticks and placing horizontals on them.

The whole construction is called *si bwala baloma*, the spirits' house. Bagido'u could not explain to me the meaning of it. He was quite clear, however, that the spirits had really nothing to do with it, for, as he told me, the *milamala* festival is over and the spirits have returned to the spirit-land, Tuma.[1] But the little erection is supposed in some way to make the future fence strong and effective. The whole ceremony of planting the second taro and constructing the spirits' house is called *bisikola*.

These four rites, that is to say, the rite of the second, that of the third, and the two rites of the fourth day, should be performed on every garden plot. When I watched them in Omarakana, the master magician, Bagido'u, was seriously ill and every effort tired him, so that he only did the rites on one or two plots. I think actually the *gibuviyaka* was done on all the six *leywota*, but the planting rites only on one plot. The remaining plots were dealt with by his younger brothers. In other villages, however, the garden magician would personally officiate on every *baleko* and the whole thing would not take long, because the spells are very short and the rites simple.

After the fourth day the taboo on work is lifted, and each man with his family goes to his own plot and carries on with the cleaning (*koumwala*) and the planting (*sopu*).

2. DIGRESSION ON NATIVE IDEAS ABOUT THIS MAGIC AND A THEORETICAL INTERPRETATION

But here we must pause and, even as an ethnographer, after he has watched a ceremony and taken rapid notes, is forced to call the magician and his acolytes, with perhaps two or three of the most intelligent informants as well, to his tent to go over the things he has seen and heard and obtain native comments, so we also will sit down in front of Bagido'u's hut and listen to what he has to say on certain details of his magic.

[1] Cf. Note 8 in App. II, Sec. 4.

He would confirm without any hesitation the general conclusions which we might have drawn from observing the rites and translating the spells, viz. that the compound ceremony of *gabu* is the inaugural magic of burning, cleaning and planting. He would tell us that before the magician sets fire to the cut gardens, no one would dare to burn them; that the three planting rites are necessary to start the planting of the minor crops, though the planting of the taytu proper cannot begin in Omarakana before the next big ritual act, the *kamkokola*.

Bagido'u could also give us more detailed information on the magic of the torches. He would send a small boy to the top of a coconut palm to fetch a young sprouting coconut leaf, the *kaykapola*, which dries and burns extremely well and makes an excellent torch. Now, as we remember, the torches for the first wholesale burning are prepared not immediately before their use, but at the previous harvest. Bagido'u on that occasion also uses young coconut leaves and lalang grass. The coconut leaf, he would tell us, is chosen because the taytu leaf should have the dark deep green colour and shiny surface of the coconut palm; the lalang grass because it has sharp points and good taytu has sharp spines on the tubers. The sharper these are, the better the tuber.

When we enquire from Bagido'u why the charming of the *vakavayla'u* torches for the first burning has to be performed at harvest, three or four moons before they are used, he has no authoritative answer but gives rein to speculation: "At harvest there is plenty of taytu. I perform my *vilamalia* magic, the taytu is strong and good, I charm over the *kaykapola*. The next garden crops will be strong and plentiful."[1] The explanation conveys a feeling that at harvest-time prosperity is in the village and that it is good to carry over the past year's prosperity to next year. But this explanation contains a good deal of the ethnographer's inference.

One thing, however, is perfectly clear, viz. that the magic power contained primarily in the spell must be conveyed right into the substances to be affected, and that if it is not going to be used immediately, it has to be imprisoned there by tying up and covering with mats. In every rite so far described we find that the magician's mouth is directly turned on the object he charms, and the spell is spoken at close range; if possible some receptacle—such as the cavity formed by one mat over another, and the open flaps of a leaf—is constructed which conducts the magician's breath and prevents it from being dispersed (cf. also Ch. VII, Sec. 6).

Two spells (M.F. 3 and M.F 7) were identical save for the first

[1] Cf. footnote, p. 88.

lines, and these had a definite point in common; both refer to objects which were, in the very act of ritual treatment, verbally disparaged—I mean the spells of the "bad stick" and the "dog's excrement". Whether this magical form of vilification is meant to protect the gardens and the taro by giving them a bad name, I cannot say on native authority, but it seems very likely that some such traditional attitude is embedded in the spells.[1]

This explanation of the two pejorative acts is confirmed by a parallel from another part of the island which I insert here where it best fits. The rite of *pelaka'ukwa* is performed throughout the whole district of the Trobriands. In the southern half of the island it is the main magical rite of the taro gardens and it is directly meant to protect these gardens from bush-pigs. It is very interesting in the present context that in Sinaketa, the principal village of the south where I obtained the account of this magic, the *pelaka'ukwa* is performed not over a taro top but over a stone. This stone is charmed, apparently, with a formula in which bush-pigs are mentioned and exorcised, and then it is thrown over the fence into the *yosewo* (uncut bush). So that here a stone is labelled as "dog's excrement", and thrown to the bush-pigs to disgust them with the garden. I obtained, moreover, the following statement from Motago'i, my best Sinaketan informant (Text 78, in Part V, Div. XI, § 2).

(i) "When they see the stone (they) go away to a place called Tepila, the home of the pigs. (ii) As the magicians charm over the garden-site (with this spell), they order the bush-pigs away, and these go to Tepila. (iii) One home of the bush-pigs is Tepila, the other Lukubwaku, a place on this side of Giribwa (the straits which separate the main island of the Trobriands from Vakuta, the second largest island, and the village lying on the straits). (iv) When the magicians drive away these pigs, they go to Lukubwaku; (v) instead, there would come the pig of the fence-stake, that pig who has four ears and two tails. They order them away, all the other pigs go to Lukubwaku. This on the other hand (the pig of the fence-stake) would keep watch over the fence because it is the pig of the stake. (vii) The *towosi* know this magic because it (the pig of the stake) is their pig. (viii) They (the pigs) really live in their village, as is known in old men's memory from long, long, long ago."[2]

[1] This suggestion was made to me by my friend, Dr. Obrebski, who from his extensive knowledge of Slavonic magical charms, tells me that such expressions are characteristic of our own, Polish and Slavonic, magic and have there the function attributed in the text to the Trobriand spell. Cf. also Frazer's *Golden Bough*.

[2] The linguistic commentary on this text will be found to contain additional information of ethnographic interest.

It is unfortunate that I obtained this statement too late to be able to follow it up.[1] It was just a few weeks before I left, and I was overwhelmed by other subjects of enquiry arising out of actual happenings round me, so that I could not investigate the belief of the bush-pig of the fence-stake in the northern part of the island. But I do not hesitate to give this as a general belief in the southern part, because a statement of Motago'i's was usually worth those of twenty other informants. The interesting points in it are the *bulukwa gado'i*—"the pig of the fence-stake" with four ears and two tails—and the two homes of bush-pigs. That somewhere in the densest and least accessible part of the *rayboag* there is a home to which the bush-pigs can be driven back by magic and whence they issue if they are invoked by sorcery, is a widespread belief. Tepila must be somewhere in the north because this name is known through the district. Lukubwaku is, according to Motago'i, an actual spot through which he himself and many other people have passed, though when human beings go there bush-pigs are never to be seen. It lies apparently in the long stretch of *rayboag* at the south-west end, where the island extends club-like and is completely covered with uncultivable, inhospitable rock, seldom ever visited by human beings. The most important black magic is apparently the one which attracts the bush-pigs. And here again I received from Motago'i, on the same occasion, the following statement (Text 79, Part V, Div. XI, § 4):—

(i) "People might quarrel: 'Later on I am going to bewitch your garden.' (ii) He (the man who threatens) would then charm over a stone, and throw it across the fence. Pigs would come from Lukubwaku in great numbers. They would eat the staple food till it was finished. That would be the result of the bewitching of the garden. (iii) If no one bewitches a garden, the bush-pigs would not come."

Returning to the first part of the statement which I obtained about the black magic of gardens: the rite here described, characteristically enough, consists also of the throwing of a stone; only, as I was told in commentaries, the stone is not this time thrown from within out of the garden, but from outside into the garden. Motago'i did not know the formula and had not even a vague idea about its contents. The fact that if you want to drive away pigs you throw a stone from the garden into the bush, while if you want to attract them you throw the stone from the bush into the garden, would make us surmise that the evil spell might be something like the reverse of the positive magic, and that the stone may symbolise magically a specially attractive taro.

[1] See Note 10 in App. II, Sec. 4.

The belief that only sorcery lets loose the pests and plagues which torment man is very deeply rooted. It has got a significant parallel in the belief that sorcery and sorcery alone is the ultimate cause of all which threatens human health and welfare and produces the accidents to human life. Neither crocodiles nor a falling tree nor death by drowning ever come of themselves, they are always induced by black magic.

Since we are here discussing various beliefs, "superstitions" and ideas connected with the safety and cleanliness of gardens, two important taboos must be mentioned. The first is that which prohibits sexual intercourse in or close to the gardens. If Text 79, quoted above, is compared with its original in Part V (Div. XI, § 5), it will be seen that it continues as follows: "People would sometimes enquire: 'Why do pigs come to this garden day after day?' And the answer would be: 'They fornicate near the garden.' " As has been stated elsewhere (cf. *Sexual Life of Savages*, pp. 231, 383, 415), gardens and garden work must be kept clear of love-making. The act of sex must never be committed within the enclosure. The usual native phrase for having intercourse in or near the gardens is *isikayse tokeda*, "they sit down upon the belt of bush adjoining the garden", though the more specific phrase "sit down on the boundary sticks" is used in the following native comment on this taboo (Text 34, Part V, Div. VI, § 47): "It is taboo to sit down on the boundary sticks. If a man does this, he will get elephantiasis (of the testicles). Women likewise would get elephantiasis (of the labia)." A special taboo forbids men to approach women while they are engaged on weeding and clearing.

The second taboo concerns human excrement; the gardens must be kept clear of all such matters. There must be no defecation or micturition in them or the soil would be polluted and the crops blighted. There is, also, a very strong repugnance against defecation near human habitations.[1]

Let us now return to the magical system of Omarakana. The reason why so much magic is devoted to taro and so little to taytu, which is the principal crop; the meaning of certain verses in the magic; the reason why one spell is sung and another spoken; these and many other questions must have arisen in the mind of the reader, as they did in mine when I witnessed the ceremonial in Kiriwina. But no answer is forthcoming save the eternal: *tokunabogwo ayguri*—"it has been ordained from of old", "it is the tradition, it is the usage,

[1] Cf. Part V, Div. VI, §§ 15 and 29; and *Sexual Life of Savages*, pp. 383 and 415.

and we do not understand it", "we have no *u'ula* (reason, lit. foundation) to give for it".[1]

3. THE WORK OF FINAL CLEANING, *KOUMWALA*

After this digression, we must take up again the thread of our narrative. The compound ceremony of burning (*gabu*) has inaugurated a new stage of work which the natives call by the generic term *koumwala*, "the clearing of the ground from rubbish". But this stage also comprises an additional burning of the scorched and partly carbonised plants, a preparatory planting and, to a certain extent, fencing. This latter has to be immediately taken in hand, because as soon as some crops are in the soil they have to be protected from the bush-pigs and wallabies.

A glance at Plate 43, taken before clearing, if compared with Plates 26 and 38 on which we see the ground finally cleared, shows that after the two magical burnings the ground is still littered with leaves, twigs and sticks. All this must be cleared away, while the roots, the most dangerous part of the old scrub, have to be dug or torn out of the ground. Apart from the danger of the old scrub sprouting again, the *koumwala* (clearing) is necessary because it would be difficult to find the really suitable patches of fertile soil under the litter with which it is still covered.

The natives as a matter of fact carry the clearing of the ground to an almost pedantic perfection, especially on the standard plots, the *leywota*. There is no doubt that aesthetic motives play an important part in this work, as a glance at the neatly uncovered, well swept and garnished and ornamentally subdivided plots on Plates 26, 36 and 38 will show. The digging-stick and axe are used occasionally for digging and cutting out roots, but the human hand is the principal implement at this stage of the work. This is illustrated by Plate 36, where a whole family are seen engaged in picking up débris from their garden plot.

The twigs, leaves, roots and sticks are all collected in heaps and fired. Often on calm days the gardeners will accumulate the refuse till towards sunset, when they set fire to all the heaps at the same time. The stuff, usually still green and somewhat damp, smoulders long into the night. Walking through the gardens at such a time one receives the impression of a gigantic deserted camp with its diminutive fires scattered in regular constellations over a vast area.

Each year during the *koumwala*, the fields are cleared of the small stones which always stray on to the fertile soil. These are put either

[1] See also Note 11 in App. II, Sec. 4.

PLATE 36

A FAMILY GROUP CLEARING THEIR GARDEN

"The human hand is the principal implement at this stage . . ." (Ch. III, Sec. 3)

PLATE 37

A STONY GARDEN IN THE SOUTH

"The stone heaps rise to large conical mounds, so close to each other that the tillable soil runs only in valleys" (Ch. III, Sec. 3)

PLATE 38

A PLOT SUBDIVIDED INTO SQUARES

The squares are here "specially small, as they were made on a very fertile *kaymugwa* plot near the village" (Ch. III, Sec. 3) on the longitudinal heap which forms the boundary between two *baleko* (garden plots), and is called *kakulumwala* (cf. Pl. 25 in the foreground, and Pls. 40 and 44 from left bottom to right top).[1] In districts where the ground is very stony, large heaps called *tuwaga* have to be made all over the field. In Kiriwina such heaps are almost non-existent (Pls. 20, 23 and 26) or small and scattered far apart (Pls. 33 and 43). But in the western district, in Kuboma, Kulumata and in the south, the *tuwaga* heaps rise to large conical mounds which stand so close to each other that the fertile, tillable soil runs only in valleys between each stony hillock (see Pl. 37).

During the clearing, a special arrangement is made characteristic of Trobriand gardens. The garden plot is subdivided into small squares by means of even, straight sticks simply laid on the ground. The size of these squares can be gauged from Plates 26, 36 and 38. On Plates 26 and 38 they are specially small, as they were made on a very fertile *kaymugwa* plot near the village. As will be remembered, during the cutting of the scrub all good and straight stems were put aside so that they would escape being burned at the *gabu* (magical burning). These sticks are now used to make the *tula*, which name is given to the boundaries of the small squares. The square itself is called *gubwatala*. When there are not enough sticks remaining from the cutting, new ones are of course provided from the bush. As we shall see presently, the network of *tula* bounding the small squares is connected with the system of high poles erected at the corners and on the boundaries of each garden plot.

Like everything else, the boundary poles (*tula*) and squares (*gubwatala*) are done with extra care and neatness on the standard plots. The best straight sticks are selected, the squares carefully laid out and made specially small and neat. A well cleaned *baleko*, showing the dark soil meticulously free from stones and rubbish and with the *tula* elegantly laid out, is to the native a very pleasant sight; the owner enjoys showing it and takes due pride in the results of his labour.

There is no doubt that the *tula* add to the elegance of a garden plot. At first sight, however, it is difficult to see how they influence the economic or technical side of gardening, but there is no doubt that they fulfil indirectly such a function. In the first place, when a man works a plot in the early gardens he often has to subdivide it for different uses. He may have to keep part of the crops for himself, give some to his sister, and some to his mother. For this purpose the *gubwatala* are convenient. During my first visit to Omarakana, the

[1] A corner in which two garden boundaries meet can be clearly seen on Plate 48.

paramount chief, To'uluwa, had several *baleko* (garden plots) under cultivation, but not as many as he had wives. In order to share the produce among them equitably, he allotted to each of them the same number of *gubwatala*.

The divisions play an important part in planting, by enabling the gardener to calculate his distances, and measure his time more easily. As we watch him at work, a gardener will begin by distributing several handfuls of seed yams on each square. "How could we do the planting properly", a native will say, "if there were no *tula*?" "The work in the gardens where there are no *tula* is bad, in gardens which are divided into *gubwatala* work is good."[1] Or again (Text 33, Part V, Div. VI, § 45): "We lay down the boundary sticks so that the garden work might go quickly. We plant one garden square till it is finished. We change to another square—we plant, we plant, and already it is finished." Later on at weeding, which the women often do communally, the division again exercises its psychological influence on economic efficiency. Each woman will take a pride in having her square cleaned better than that of her neighbour. "When the *tula* are there, the work goes quickly, it is pleasant to do it; when no *tula* are there, or when they lie crooked, work goes slowly, we do not want to work."

Thus the subdivision of the plots into the small squares enables the native to portion out his task. It allows him to measure and systematically divide his work into what he has done and what he has still to do. He has not got a big indiscriminate stretch of labour before him but a series of appointed tasks of which he can see the end at the beginning.[2]

As soon as the *tula* are ready, some preliminary planting is done. In each square (*gubwatala*) one or two banana seedlings are planted, a couple of taro tops, two or three sticks of sugar-cane, and a few tubers of the larger yam, *kuvi*. An assortment of such young plants can be seen on Plates 25 and 38. Here the division into squares obviously helps in ordering the work and securing the right distribution of crops. In the later gardens, the *kaymata*, such mixed crops are, as we know, of minor importance, but in the early gardens (*kaymugwa*) they take up a considerable proportion of the ground.[3]

All this work is done by the owner himself, helped by his wife and children, or, if he is not married, by his mother or other female relatives (see Pls. 24, 36 and 37). Communal work is never used in connexion with the *koumwala*. A chief may request a few of his

[1] Cf. footnote, p. 88. [2] Cf. also Note 12 in App. II, Sec. 4.
[3] Cf. also Note 13 in App. II, Sec. 4.

kinsmen and dependants to lay the boundary poles and do some of the preliminary planting on his plots. The men seen on Plates 25 and 38 are an example. But this is not, sociologically speaking, communal work even of the type called *kabutu*, which will be described with other forms of communal labour in Chapter IV (Sec. 5). It is rather a case of a minor *corvée* incidental to the relation between subjects and chief.

Normally a whole family goes into the garden and camps there for the day. A fire is lit, mats spread on the soil, the small children play round the camp, while bigger ones aid in the work. The clearing of refuse from the ground is more especially women's work, whereas the laying out of the *tula* and the preliminary planting is done by men; but there is no strict division of labour according to sex.

As soon as the crops are in the ground, the fence (*kali*) has to be erected. The sticks (*gado'i*) left over from cutting or brought from the bush are planted vertically in two parallel rows at a distance of some five to seven centimetres from each other. The space between them is filled out with horizontal sticks. The verticals are lashed together with bast about the middle of their height and then more horizontals are laid, and the verticals are lashed again at the top. The average fence is about 1 to 1·20 metres high. Wherever a path strikes the gardens, a stile (*pisila* or *kalapisila kali*) is made, as can be seen on Plates 31 and 39. The structure of a fence can also be seen on Plate 22, where the stile appears on the left. Everybody contributes his share to the common fence, by making the portion which runs along the outer boundary of his own *baleko* (garden plot); for one fence encloses the joint gardens, protecting all the crops from bush-pig and wallaby. Such a self-contained garden unit is diagrammatically represented on Fig. 4, and a general impression of it can be obtained from Plate 20 where, in the background, a fence is seen dividing it from the bush.

4. THE CORNER-STONES OF THE MAGICAL WALL

After the gardens are cleaned, covered with a network of squares and partially planted, there comes another rite. It is one of the three or four principal ceremonies of the garden: it has its period of taboo; it consists of a series of rites lasting for two to three days; it entails a great deal of manual work on the part of the gardeners and even of the magician himself; and it changes considerably the face of the garden. The *kamkokola*, the structure erected on each of the four corners of every garden plot, has a special aesthetic value to the natives. Its sight rejoices the heart of the Kiriwinian even as it would that of a modern Cubist. A glance at Plates 23, 40 and 41 will supply the reader with the essentials of this structure. Later on we shall see that these corner structures are connected with the system of cross poles, *tula*, which divides each plot into small squares, and with another type of geometrical erection which borders the plot. The chief *kamkokola* (magical prism) of each garden plot stands at what we have already got to know as the "magical corner" of the garden; that is to say, the corner in which at previous ceremonies the magician has planted a taro and, later on, a *kwanada* yam; where he has built the little house of the spirits; where, in short, he has performed a series of rites connected with the big magical burning, and where he will have to return, once more at least, during the ceremonies of harvesting.

The ceremony which we are about to describe is perhaps less definitely inaugural than the two preceding magical acts. The erection of the magical prism is undoubtedly connected with the putting up of the stout yam supports, the *kavatam*, round which the vines of the taytu will presently climb. It is also connected, though in a less definite manner, with the planting of the taytu. The planting of subsidiary crops has, as we know, already been done in each plot; but on account of local variations in different villages, and the uneven progress of gardening, I could never ascertain by personal observation whether it is allowed to plant any of the staple crop, the taytu, before the *kamkokola* (magical prism) ceremony is performed.[1] In Omarakana I was repeatedly assured by Bagido'u that the real *sopu* or *sopu malaga*, that is the planting of the taytu or main crop, cannot begin until each of the four corners of every garden plot, or at least of the standard plots, the *leywota*, is garnished with the magical prism. But in other villages I have myself seen taytu being planted before the *kamkokola* ceremony.

But two things stand out clearly. In the first place the *kamkokola* ceremony is connected with the planting of the taytu, it is a planting rite—whether invariably inaugural, as it probably is in the Omarakana system, or not, I do not know. If not, it is so to speak a consecration of planting already in course of completion. My native informants were emphatic that this ceremony was connected with planting, for, where it is not performed, they said, planting would not be successful. Its spells, the ritual and the structure itself, however, associate the *kamkokola* much more directly with the yam poles.

In the second place the planting of taytu, though it is called by the same term, *sopu*, as any other planting, is, to the natives, an activity

[1] See also Note 14 in App. II, Sec. 4.

entirely apart. The word *sopu*, when it is used to define a stage in gardening, never refers to the first preliminary planting but to the main planting only. This main planting, whether it be done immediately before or only after the *kamkokola*, entails a different manner of working: the preliminary planting is never done except by individual labour (or chief's *corvée*), each family working its own plot; whereas for the planting of taytu, communal labour is usual. During such communal planting of taytu and also, though in a lesser degree, when individual labour is used, certain customs are observed—such as traditional yells and songs, forbidden during any other garden work; competitive bets; and one or two minor taboos.

After the standard plots and a number of others have been properly cleared, Bagido'u decides that the time has come for the next magical step. One evening he will thus adress his fellow-villagers:—

"The *koumwala* (clearing) is now finished. The fences are made. There are one or two places where the bush-pig can still jump in. Setukwa ought to make his fence higher, and Gumigawaya must still finish his piece. We do not want the bush-pig or the wallaby to eat our sugar-cane or prune our taro tops. The time has now come for the planting of taytu. We must make *sopu* (planting). Our taytu will grow high, it will climb up; we want the large supports, the *kavatam*. You, my friends, must bring the *lapu* (stems) to make *kavatam*, to make *kamkokola*. To-morrow I shall put the *kayluvalova* (taboo-stick) into the ground, I shall put a stick into each *baleko* (plot). Your gardens will be tabooed (*kaboma*). You must not do any work. You men, you all go to the *rayboag*, to the *momola* (seashore), and bring long straight sticks, so that our gardens look beautiful, and our taytu grow tall and strong."[1]

Such would be the harangue in which the garden magician of Omarakana, upbraiding some men for their laziness and remissness, indicating the stage at which the gardens were at present, exhorting to good work, would announce his intention to begin the new series of rites, the *kamkokola* rites.

Next day Bagido'u starts on his first round. In each *baleko* he inserts a slender stick not much higher than a metre and a half. This is the *kayluvalova*, the stick which indicates that henceforth a taboo is placed on the gardens and that all energies must be devoted to one task and one task only, the bringing of the stout poles, the *lapu*. There is no spell said at the inserting of the *kayluvalova*, no rite performed, no magical virtue embedded in the object; it is simply a sign that the gardens are under a taboo.

[1] Cf. footnote, p. 88.

The *kayluvalova* is inserted in every *baleko*. In the standard or show plots, the *leywota*, it is put up with special care and usually two slanting sticks are placed against it, making a miniature *kamkokola*, as can be well seen on Plates 23 and 41. The natives call it sometimes *si kamkokola baloma*, "the *kamkokola* of the spirits". On other plots a smaller stick would be put without the slanting cross sticks. On the day on which I accompanied Bagido'u during his preliminary walk, I observed that no garden plot was omitted. We first went to the *leywota*, all finished, cleared and planted, the clean red-brown soil showing through the green leaves of taro, sugar-cane and banana which had already taken root and were growing. The fresh greenery and the brown soil looked well against the glistening fence which, with its new wood, seemed as though it were made of golden bars. Then we went to the other plots, some already cleared and subdivided, others half through with the *koumwala*, others again full of uncut scrub, *kapopu*, as the natives call it. But even in these last, right at the apex of the magical corner where at least a little of the scrub had been cut and all previous ceremonies had been performed, Bagido'u would insert a slender stick.[1]

After his return to the village, he would usually announce briefly that all the plots had been tabooed and that now was the time to begin collecting the *lapu*. Let us follow the men who are streaming out of the village in every direction, generally in small groups of three, four and five. The lazier, weaker, or less important ones go just round the corner to the nearest patch of uncut bush and make the best selection they can from the stouter saplings, which they cut and bring to the village. But most men, above all the owners of the *leywota* and those who pride themselves on their gardens, the real *tokwaybagula*, have a much more strenuous task to accomplish. Going across the plain, they climb the coral ridge and scour the jungle for suitable trees. It takes some time to find these, as I know from personal experience. The *rayboag* is covered with high jungle, with big banyan trees, ironwood, wild mango, Malay apple and pandanus, often entangled and encircled by creepers; but straight, light and easily cut trees are not very plentiful.

After such a tree has been found, it has to be cut, trimmed, and carried to the village. The average *lapu* is about three to five metres, but I have seen them as long as six or even a little more. The carrying back is the most difficult task. To see the men panting under the weight of an exaggeratedly large *lapu*, rest, and then plod on again—at times obliged to hide it away in the bush in order to get back to a village for a meal or for the night—makes one realise the great

[1] See also Note 15 in App. II, Sec. 4.

value which these natives attach to the aesthetic element, to magic, and to competitive renown. All this lasts some three to five days. When a sufficient number of *lapu* (straight poles) have been brought to garnish the gardens with magical prisms (*kamkokola*), yam poles (*kavatam*) and lateral triangles (*karivisi*), the ceremony proper begins.

One evening Bagido'u harangues the villagers again. He tells them that enough poles have been brought; he moralises a little, reminding them that good and strong *lapu* make a good garden, and he announces that on the morrow the magic will begin and that the taboo on all work still has to be observed. The ceremonies take in Omarakana, as in most communities, exactly three days.

The first of them is devoted to preparations and preliminary magic. What happens in fact is almost a replica of the magic that precedes the very first garden rite. Again Bagido'u goes in the morning to collect the thirteen magical ingredients, which are pounded and torn into the standard mixture of Omarakana garden magic. A little of the mixture is wrapped round the cutting edge of the blades of two or three axes in the manner described in Chapter II (Sec. 4). These, together with the remainder of the mixture, are enclosed between two mats and charmed by the *towosi* in his hut with the same formula used at the first inaugural ceremony (*vatuvi* spell, M.F. 2).

On the next day the ceremonial erection of the magical prisms, the *kamkokola*, takes place. This act is not as a rule accompanied by any great festivity and a few people only take part in it; indeed, but for one or two details, it would not be easy for a casual observer to see that anything more than a purely technical activity is being carried out.

I have seen *kamkokola* erected in Omarakana and in other villages, and the proceedings are almost identical. When I first saw it done in Omarakana, Bagido'u, the chief magician, was not able to come, being feeble and tired as usual, and reserving his strength for the next day. So his brothers, Towese'i and Mitakata, together with three other men, performed the rite. In following the technical proceedings it is well to keep in mind the magical setting, the structural aspect, and what might be called the sociology of this act.

The putting up of the heavy and cumbrous structure and all the work associated with it has to be done by the chief magician himself or by those directly delegated by him, in this case his brothers and acolytes, Towese'i and Mitakata. These two had to dig a deep hole, about 75 cm. to one metre in depth; they used a digging-stick, scooping out the soil with the hand. Then the vertical *lapu*—the pole which changes its name into *kamkokola* as soon as it is erected—was rubbed with the magical mixture charmed over on the previous evening. No spell was uttered during the rubbing. Some of the mixture was then put into the hole, the pole was planted by being lifted and thumped into place, and finally the soil was packed round it and stamped down. Then two more *lapu*, as a rule lighter though not shorter, were placed at right angles to each other, one end resting against a fork of the vertical *lapu* and the other resting on the ground, in the characteristic position shown on Plate 40 and others. From whatever side you look at the *kamkokola*, you always see one or two triangles; the whole outline of the structure is that of a prism overtopped by the end of the main pole. The word *kamkokola*, which in its wider sense embraces the whole structure, is more particularly the name for the vertical pole, while the slanting sticks are called *kaybaba*.

This was the whole ceremony, but they had to go to every *baleko* and on each erect one *kamkokola* in the magical corner. So it can be seen that the magician's office is not a sinecure even as regards purely manual labour. The other three prisms on each *baleko* are built by the owner. His also is the task of constructing the *karivisi*, or decorative triangles which run along the side of the *baleko*.

The *karivisi* also consists of a vertical pole with one, two or three slanting sticks, *kaybaba*, leaning against it (cf. Pls. 42 and 103–108). This structure does not play any part in magic, but fulfils an aesthetic function and acts as a yam support. On Plate 42 it can be seen that the *kaybaba* of a *karivisi* joins a *tula* at the point where it touches the ground. The *kaybaba* of a *kamkokola* often also joins one of the boundary *tula* of the *baleko*. Such a *tula* is in turn connected with the *kaybaba* of a neighbouring *karivisi*, and so on. The *tula* which cut across the *baleko* terminate each in one of the *kaybaba* which run up on to a *karivisi*. Thus the whole system of *tula*, *karivisi* and *kamkokola* is interconnected; they divide the *baleko* into smaller portions and surround it with a complete magical structure. Every one of the uprights of this system serves as a support to a vine; the upright of the *kamkokola* itself supporting the tuber which was ceremonially planted during the *bisikola* ceremony.

Thus, at one stroke, the garden is provided with a third, the vertical dimension. What a few days previously gave an impression of a flat bare piece of ground, is now incased in a glittering framework of strength and loftiness—for almost invariably the *lapu* are stripped of their bark. The aesthetic appeal of this to the natives is shown in the zeal of their work, their pride in achievement and in the comments of the *towosi*. I felt, as though they themselves were feeling it, that they had buttressed the garden about with a magical wall of which the *kamkokola* were the corner-stones. My phraseology and even my

PLATE 40

KAMKOKOLA

Tokulubakiki making an anticipatory demonstration of how the magical rite will be performed in the neighbouring village (Ch. III, Sec. 3; cf. also Ch. IX, Sec. 2)

PLATE 39

STILE IN A GARDEN FENCE

On this photograph the structure of the fence, as well as of the stile, its height and its relation to path and garden can be seen (Ch. III, Sec. 3)

PLATE 41

BIG AND MINIATURE KAMKOKOLA

The structures here seen are on a *kaymata* plot, a village grove in the background. At the foot of the big structure is a miniature "spirits" *kamkokola* (Ch. III, Sec. 4)

PLATE 42

KARIVISI

This structure "consists in a vertical pole, with one, two, or three slanting sticks leaning against it" (Ch. III, Sec. 4)

"empathy" may be purely subjective, but the native attitude to this aspect of gardening is, none the less, an interesting compound of the practical and the mystical. They know empirically that tall strong verticals are of value; but the fact that the higher and more luxuriant the taytu vine, the richer its development underground, engenders in them an aesthetic love of height and strength for their own sake, which is expressed in their careful choice of poles. Finally, I have no doubt that this aesthetic appreciation merges into a mystical feeling that the height and strength of their vertical system, and above all of the *kamkokola*, has a stimulating effect on the young plants.

Let us return from this rather theoretical digression to the final magical act of the *kamkokola*, which takes place upon the last, that is, the third day of the sequence. On this day Bagido'u, ill though he felt, went himself into the garden, accompanied by his acolytes, Towese'i and Mitakata, and a few other men. They carried the charmed axes for him and several baskets full of magical herbs which they had collected earlier in the morning. This time the mixture consisted of the leaves of two plants: *yayu* (casuarina), used because of its rich foliage and deep green colour, which it should impart to the taytu; and *youlumwala*, a bush plant with large tubers, used that the taytu may produce edible roots of a similar size.

Arrived at the *leywota*, Bagido'u has to squat down close to the *kamkokola*. Laying a bunch of each kind of leaves on the ground, he charms them over. He faces the *kamkokola*, his back to the outer bush, and in a loud voice, allowing his breath to go freely over the garden plot, he utters a magical spell. The charm which he had said two days previously in his house, he had condensed on to the medicated substances and imprisoned by means of two mats. But this time his voice and the magical virtue therein contained is made to flow over the fields and penetrate the soil. This is characteristic of all the spells chanted in the field, such as those occurring in the *gabu* ritual which we have met already, and those we shall find in the magic of growth which follows. This is the spell which Bagido'u chants over the "leaves of covering", as the mixture of *yayu* and *youlumwala* is called:—

FORMULA 10

I. "Anchoring, anchoring of my garden,
Taking deep root, taking deep root in my garden,
Anchoring in the name of Tudava,
Taking deep root in the name of Malita,
Tudava will climb up, he will seat himself on the high platform.
What shall I strike?
I shall strike the firmly moored bottom of my taytu.
It shall be anchored.

II. "It shall be anchored, it shall be anchored!
My soil is anchored,
My *kamkokola*, my magical prism, shall be anchored.
My *kavatam*, my strong yam pole, shall be anchored,
My *kaysalu*, my branching pole, shall be anchored,
My *kamtuya*, my stem saved from the cutting, shall be anchored,
My *kaybudi*, my training-stick that leans against the great yam pole, shall be anchored,
My *kaynutatala*, my uncharmed *kamkokola*, shall be anchored,
My *tula*, my partition-stick, shall be anchored,
My *yeye'i*, my small slender support, shall be anchored,
My *tukulumwala*, my boundary line, shall be anchored,
My *karivisi*, my boundary triangle, shall be anchored,
My *tamkwaluma*, my light yam pole, shall be anchored,
My *kayluvalova*, my tabooing-stick, shall be anchored,
My *kayvaliluwa*, my great yam pole, shall be anchored.

III. "It is anchored, my garden is anchored,
Like an immovable stone is my garden.
Like the bed-rock is my garden.
Like a deep-rooted stone is my garden.
My garden is anchored, it is anchored for good and all.
Tudududu . . . . . .
The magical portent of my garden rumbles over the north east."

The magical portent (*kariyala*), mentioned at the end of this spell, refers to the natural event or convulsion which, in native belief, is a by-product of magic. Usually it is lightning and thunder, sometimes a violent wind or a slight earthquake. Generally speaking, each type of magic has the same *kariyala* for each of its rites. Thus thunder and lightning accompany all acts of garden magic. Considering that this magic begins towards the change of the seasons from trade-wind to monsoon, and that most of its ceremonies take place during the early part of the monsoon season when there is thunder and lightning every day, the miracle is not hard to account for. The natives explain any occasional absence of the magical portent by saying that the magic was not properly performed on that occasion. Once or twice I was told that the thunder and lightning were there only very distant, or again that they occurred in the middle of the night when everybody was asleep.

After having uttered this spell over the "covering leaves" (*kavapatu*), Bagido'u inserts them into the ground, right against the pole of the *kamkokola*. He then rises, takes the axe, which on a previous evening had been charmed over with the *vatuvi* spell, from his shoulder and strikes the pole with the words:—

FORMULA 11

"Our *kamkokola* ours, the garden magicians'. It is strong and hard. Our *kaybaba* ours, the chiefs'. . . ."

This rite must be performed on every plot, but only on the four to six standard plots, the *leywota*, need the magician do it in person. This ends the ceremony of the *kamkokola*, one of the most important magical acts of gardening, and certainly the most mysterious and ill-defined. As I have mentioned previously, the magical function of this rite seems to be most clearly inaugurative to the erection of the vine supports. The *kamkokola* itself is such a support, it is connected with the system of *tula* and of *karivisi* which latter again serve as poles for the taytu to wind round. The *kamkokola* is in fact the first big vertical pole introduced into the garden, though some verticals are supplied by the remnants of the cut jungle. It is also a finishing off, so to speak, of the surface work in the garden; it is the last decorative touch given to the magical corner, and it artistically completes the *tula* system (cf. Pls. 105 and 108).

On the other hand, it is only after the *kamkokola* that communal planting is done, and, as I have mentioned already, in some villages at least taytu may not be planted before the *kamkokola*, or so I was told by competent informants. Also I find the following statement quoted in my notes:—

"The big *kamkokola* will make the taytu grow high and be strong. Without *kamkokola* we cannot plant taytu."[1]

Again I was told that when the *kamkokola* is bad, the whole garden will be bad. Several informants assured me that the *kamkokola* frighten the bush-pig and that they make the fence strong. "When no *kamkokola* are put up in a garden, the bush-pigs come, they destroy all crops."[1]

Immediately after the *kamkokola* ceremony there is a small rite called *vakalova*. I think it is performed a day or two after the striking of the *kamkokola*, but I am not certain since I never witnessed it. Some *kotila*, a plant growing in the *rayboag*, mixed with earth from the bush-hen's nest, *ge'u*, is made into a small heap in the magical corner and burned there. The mixture has been charmed with the following spell:—

FORMULA 12

"O *nabugwa* taytu,
O *nakoya* taytu,
O *teyo'u* plant,
Boil in the belly of my garden,
Go on boiling in the corner of my garden."

[1] Cf. footnote, p. 88.

This rite is made specifically so as to prevent *leria* (the pests) from coming into the gardens. It is an act of exorcism and fumigation.

Perhaps it would be best to consider the *kamkokola* ceremony as marking the turning-point in the making of the gardens, the point where most of the preparatory activities are finished. The fences are completed, the network of sticks drawn on the ground, the preliminary planting ended; the real planting has been inaugurated and will be rapidly finished, and with this the growth of the gardens can begin. The *kamkokola* might thus be regarded as the ceremony completing all the preparatory activities and ushering in the real growth and development of the gardens.

5. THE PLANTING OF TAYTU

The whole activity which the natives call *sopu* and which for each individual plant lasts but a few minutes, comprises the whole complex process of tilling: the native in one go breaks the sod, loosens the ground, plants the seed taytu and covers it up, making a small hillock of soil over it. In spite of the great development of gardening in the Trobriands, in spite of the relative efficiency of agricultural work, and of the amount of time and labour devoted to it, this central act of tilling is remarkably simple.

The implements used are primitive in the extreme; they have been already enumerated (Ch. I, Sec. 4): the digging-stick (*dayma*), the axe (*kema*), the adze (*ligogu*), and the hand.

The stick varies in size according to the person who uses it. A strong man will use a heavy stout pole, about 1·60 to 2 metres long. A woman will usually be satisfied with a slender *dayma* of not more then 1·50 metres; and a child would use a short stick. A *dayma* is always pointed at the end and, should the point break, the stick is resharpened easily.

The axe (*kema*) would at the present day be an ordinary trade tomahawk. In olden days it would have been a light *kema*, that is, an axe with the blade lying in the plane of the blow. The *kema* is used to cut the roots found in the soil. The adze, *ligogu*, with the blade at right angles to the plane of the blow, is used to sharpen the digging-stick, *dayma*. With the modern steel implements now available, both axe and adze are sometimes replaced by a large trade knife, such as can be seen lying on the ground in the foreground of Plate 43. The trade axe can be used instead of the adze for sharpening the digging-stick, which was impossible with the old stone blade. At present, therefore, the gardener's outfit varies, and is apt to be somewhat simplified.

Day after day, over a couple of years, I have seen people going out into the gardens to plant, and I have often examined the implements they took with them. They always carried a cutting instrument or two, but the *dayma* was never taken from the village to the garden. Any straight, strong and light stick found on the field would do, and after use it would be left in the garden, because another could easily be made were it lost or stolen. Thus the main agricultural implement of these enthusiastic and very efficient gardeners has not even risen to the rank of a permanent possession. It can be picked up anywhere, and it has not enough value to be kept from one day to another.[1]

When the gardener proceeds to planting, he must first look for a suitable patch of soil. As mentioned above, there are more or less fertile patches distributed over each plot, as the soil lies in a stratum of varying thickness over a basis of coral. Then the man takes his *dayma* in both hands and, drawing himself up, breaks up the sod with a few blows to a depth of about half a metre within a radius of 25 to 30 cm. (see Pl. 43). If there are any sizeable stones at that spot, the *dayma* strikes against them. They are removed, using the *dayma* as a lever, and put on a *tuwaga* (stone heap) or thrown on a *tukulumwala* (boundary).

The man then squats down, takes the *dayma* in his right hand near the point, and breaks the lumps of soil into loose earth (see Pl. 44). Then with his left hand he feels through the earth, while the point of the *dayma* moves through it at the same time, breaking up the lumps and helping the hand to search out small stones, roots and unbroken sods. This is perhaps the most difficult operation in all the gardening, and the quickness with which the natives will break up and thoroughly clean a planting hole is astonishing. The hand of a master-gardener seems to go through the loosened soil with a swift, caressing movement, quickly throwing out stones and bits of root, pushing the unbroken lumps of soil over to the *dayma*, to be broken up. The soil is thrown out round the hole as the work proceeds downwards.

I made several attempts at planting a taytu and I had the "theory" of it carefully explained and practically demonstrated. But I found it really difficult to co-ordinate the movements of the *dayma* with those of my fingers, and I was always afraid of driving its sharp point into my hand, so that the speed of the natives received my full admiration.

Another difficulty at this stage is the removal of the small roots. They have to be disentangled and laid bare and then cut with the

[1] From enquiries made as to the efficiency of the old stone implements, I am fully satisfied that in the days before steel axes or adzes were known, the *dayma* was as ephemeral a possession as it is now.

axe. Cutting off things which are half in the ground without breaking the blade against stones is not an easy task, especially if you have to do it in a squatting position.

After the soil has been prepared, a whole taytu is placed horizontally in the hole. It is not planted very deep, so that there is plenty of loose soil under it. The seed taytu consists of small, yet healthy tubers. It is called *yagogu*, and both at harvest and in the storehouses it enjoys a special treatment, though it is not given the place of honour. As mentioned in the previous chapter there is private magic to make the *yagogu* good, which is, as a rule, recited in the field, just before planting.

The men have by far the greater share of work in planting, and when this is done by communal labour—and communal labour is very often used at this stage of gardening—only men would be employed on it (see Ch. IV, Sec. 5). Communal planting is enlivened by competitive challenges and has a considerable amount of attraction for the workers. Also there are melodic cries peculiar to planting. The most typical of these, the "*kabwaku*" cry, is to be heard whenever it is being done whether individually or communally, whereas other more elaborate ditties are sung only when it is carried out communally. We have already mentioned a cry uttered during the *takaywa* (cutting of the scrub). The planting cries are different and we shall meet yet other cries at harvest. The natives discriminate between songs properly speaking—that is the dancing songs which are called *wosi*, a term which means indiscriminately dance and song—and all other melodic cries, which have no generic name to denote them. The planting melodies are simply described as *tadodo'usi taytu*—"we call out at the taytu".

The most important is the *kabwaku* cry:—

> "*Kabwaku E-E-E-E-E-E!*
> *Ula'i taytu wakoya*
> *Wawawawawawa* . . ."

The *kabwaku* is a bird with an extremely melodious cry, and in the first line of this formula, the natives exactly imitate it in a descending passage on the vowel E executed in a high-pitched falsetto. The cry: "Kabwaku E . . ." resounds far and wide while planting is going on, and some of the natives are able so marvellously to imitate this bird that I often wondered whether I were listening to art or nature. The rest of the formula "*ula'i taytu wakoya*" means: "Thou, taytu, sprout in the mountains (of the d'Entrecasteaux islands)." The yams of certain districts of the *koya* are renowned (see Ch. I, Sec. 7, the myth about the overturned canoe), but when the Trobriander speaks about the crops of the *koya*, he usually means the *Koyatabu*, the tall slender mountain on Fergusson Island, whose well-watered slopes are apparently really very fertile.

The natives affirm that there is no *megwa* (magic) in this formula; yet it undoubtedly has a magical "look" about it, and it is not only allowed at planting; it is, to a certain extent, regarded as indispensable to the welfare of the planted tubers. It "makes the taytu grow well". Again, these cries are taboo before the planting has begun and they would not be uttered during any other activity.

When planting is done communally, the men utter a kind of antiphonic chant after the work is finished—one man giving out the words and the others answering in a chorus. This is done, as they say, so that "the others in the village may know that the work is over". Here is the text:—

Solo	Chorus
"*Kalupegovaya!*"	"*Yohohohoho*"
"*Tapuropuro, tavayavayo,*	
Tabesabeyso; ya, beyso, ya."	"*Yohohohoho*"
"*Nuwam, porobuyo; buyo, ya.*"	"*Yohohohoho*"
"*Taytu gayewo.*"	"*Yohohohoho*"

The words I was unable to translate. They look as if they belonged to some other language, perhaps a dialect of the d'Entrecasteaux Archipelago, except *taytu gayewo*, which means "taytu (white like) the flower petals of the pandanus".

When properly executed, the last vowels of each word are modified into a long-drawn O, and the choral answer sounds like the neighing of a team of horses.

Another such formula is the following:—

Solo: "*Tamtala kwanada sapusi otuwaga; tapuropuro, tavayavayo, tabesabeso, beso, ya.*"
Chorus: "*Yohohohoho.*"

In this formula the first part has a meaning: "One *kwanada* (species of yam) has been planted on a *tuwaga* (heap of stones)." The rest is identical with the above formula.

Another formula is somewhat obscene:—

Solo: "*Kwaywa'u kibariri Bogina'i.*"
Chorus: "*Yohohoho.*"
Solo: "*Mitaga kwaybogwa bulubolela wim Bomigawaga owokulu.*"
Chorus: "*Yohohoho.*"

This in free translation means: "Bogina'i (name of a woman, about whom nothing else is known) is recently deflowered"; then the next stanza retorts: "But your vulva, Bomigawaga (an equally obscure female personality), over there at the corner of the fence, has for long time had a considerable circumference." The obscene allusions in this spell are connected with the planting; the deeper the soil is broken up at the planting-spot, and the more thoroughly it is worked, the better will grow the taytu. Hence the parallel of the sexual act. This was the interpretation given to me by my native informants. Thus, although the natives do not regard these chants (*vinavina*) as *megwa*—magic in the strict sense of the word—there is a connexion between what they deem good for the fields and the text of the formulae.

As the planting on the *leywota* is usually done by means of communal labour, these plots are quickly finished, but it is a considerable time before all are ready. This work falls in the moons of *Yavatakulu* and *Toliyavata* (cf. Chart of Time-reckoning, Fig. 3). The *gabu* normally takes place some time towards the end of *Milamala*. Then follows about a moon's interval taken up by the *koumwala* (cleaning and preliminary planting) and then at the beginning of *Yavatakulu* the main planting is inaugurated by the *kamkokola* ceremony.

CHAPTER IV

THE GARDENS OF OMARAKANA: THE MAGIC OF GROWTH

We enter now upon an entirely new phase of garden work and begin a new cycle of garden magic.

The gardener has now done most of what lies within his power. The bush has been cut and the ground cleared from extraneous growth and rubbish; the soil has been dressed with ashes, tilled and loosened; the crops have been planted; the surface of the garden has been garnished with a superstructure of geometrical figures and the four corner prisms; it has been fenced round and it forms now a self-contained unit. Within this enclosure only the *leywota*, the standard plots, are, of course, certain to be completely up to date; many of the other gardens lag behind, some of them badly, but by the magical convention to which allusion has been made several times, this does not disturb the ritual and ceremonial progression of things.

1.—THE TURNING-POINT OF GARDEN WORK

Nature now has to do most of the remaining work. The crops have to sprout, to grow, to throw out new roots and tubers, to develop their foliage, and finally to mature. In this natural growth the attention of the natives is concentrated on the taytu plant. True we shall find references to taro and *kuvi* (large yams) in one or two of the formulae which follow, but the taytu will be the principal object of magical endeavour. It is their staple crop; it requires the greatest care, and is most susceptible to blights and to destructive agencies. It is over the taytu that they will presently perform the magic of growth; it is the foliage of the taytu which fills them with delight. The richer the foliage, the fuller the tubers, they say. The garlands of taytu vine must be dark green, the leaves must be luxuriant and there must be no blight on them. The training of this foliage over the supports, which is done by the men, is perhaps the most difficult practical task that still remains.

To start with they have, as we know, a number of supports called *kamtuya*, consisting of the stems left rooted after the cutting (cf. Pls. 20, 23, 25 and 26). When the new yams come up, they fix in the ground slender sticks called *kaygum*, or when these are very slender, *yeye'i*, around which the new shoots will twine; or if there is a sizable tree near-by left over from the cutting, they lean a small stick called *kaybudi* against it to act as a preliminary support from which the vine will jump over on to the big tree. When the vine has developed, however, it must be trained on a real *kavatam*, a pole which, in its largest form, is called a *kayvaliluwa*—that is, unless there is a big tree, *kaysalu*, on whose several branches the taytu can climb. I find the following native statements in my notes, indicating the function of the main categories of support (Texts 31 and 32, Part V, Div. VI, §§ 42 and 44): "The *kamtuya* supports are there already from the cutting; afterwards we plant taytu and it climbs up these. We plant taytu where there is no *kamtuya*; then we cut a small tree named *kaygum*, we make the taytu coil round it. Later on we cut a long tree named *kavatam*; we put it up, we make the taytu coil round it." "The vine climbs up first on a small stick, the *kaygum*. Then we put up the real yam pole, *kavatam*, and the taytu jumps over on to the yam pole." The actual work of training consists in keeping an eye on every vine, twisting the tendril round the pole when it has grown long enough, and tying it in place with a piece of bast (Pl. 45).

I have given here these details, linguistic and descriptive, in order to show, on the one hand, how developed the interest of the native is in the technicalities of this phase in garden work even as his vocabulary is developed; and on the other, to give an idea of the work still left to be done.[1]

The men also have to protect the gardens from pests, to repair the fence, to put up rattles and scarecrows, and to destroy any beetles and other insects. No form of chemical warfare against the winged pests is known as far as I was able to ascertain, not even smoking. One more task remains to the men. When the roots burst into tubers—a process described by the word *puri*, which means to break out into a multitude of things, to erupt, to form clusters—it is necessary to thin out the roots and remove superfluous tubers some time before the real harvest, so that the better ones can develop.

In the meantime, the women have the less arduous but steady work of weeding. Weeding is most important at the time when the crops begin to sprout, but it has to be continued well after the foliage of the taytu vine has been formed.

After the crops have matured comes the harvest, and this is complicated in that the several crops have to be taken out successively, taro first, *kuvi* (large yams) later, taytu, small staple yams, last of all. Each harvest has a special name, each has its specific ceremonies and customary usages.

[1] Cf. Part V, Div. II, § 5, where the correlation between native interest and the degree of terminological discrimination is further illustrated.

Thus we see that the gardeners are kept busy even after the really arduous work comes to an end with the planting of the taytu. But they are not absorbed by their work as they were before. In districts where long overseas expeditions are undertaken, most of the men can go away for weeks at a time, leaving only a few of their fellows to look after the gardens. In other villages this is the time for long fishing expeditions, or for festivities and competitive sports (cf. Chart of Time-reckoning, Fig. 3).

There is one department of human endeavour still to be mentioned, however, and that is magic. The magician has in a way to work hard. In a rapid succession of rites, he has to anticipate each stage in the growth of the gardens, and stimulate the various crucial phases in the development of the plant, thus assisting nature by what we might describe as a special department of garden magic, the magic of growth. He "awakens the sprout"; he "drives up the shoot over ground"; he "throws the headgear" of the taytu; he "makes several branches" of the taytu; he "closes up the canopies" of the taytu. Then descending to the roots again, he "breaks forth the multitude of clusters"; he "pushes the taytu tubers into the soil".[1] Thus, following closely the known development of a plant, he spurs it on its course and adds the favouring power of magic to the natural power of the soil.

But side by side with these rites of assistance, he also performs several inaugurative rites which are more in the nature of those with which we have become acquainted in previous chapters. There is an inaugurative ceremony to the weeding, and one to the thinning out of the unripe tubers. There is at least one inaugurative ceremony marking the change from the small to the large *kavatam* (yam poles), and finally there are the inaugurative rites associated with the several harvests.

2. THE MAGIC OF GROWTH—THE STIMULATION OF SHOOTS AND LEAVES

Let us now concentrate on what is happening underground, where the taytu tuber is coming to life again to begin its new cycle. We shall have to watch its progress through native eyes; for, on the one hand, I am not sufficiently well acquainted with the botany of the Trobriand crops to give objective facts, and, on the other, it is the native point of view which really matters to us.[2] Therefore I shall reproduce here, in a condensed form, the statements of my informants,

[1] Cf. Doc. VI, Growth Magic in Vakuta.
[2] See also Note 16 in App. II, Sec. 4

observations made under their guidance and the result of many an analytic discussion of the magical texts.[1]

Looking at the diagram (Fig. 6), we can see the native picture of what happens to the planted seed yam, *yagogu*. It is always put in the ground horizontally. The blunt, rounded-off end is called by the natives *matala* (eye); the body of the yam *tapwala* (trunk). It is at the *matala* end that the new plant begins to sprout. The sprout (*sobula*) gradually works its way up from underground and emerges (*isakapu*), then, twining round the support, grows into the main

GROWTH OF TAYTU: THE SPROUTING OF THE OLD TUBER

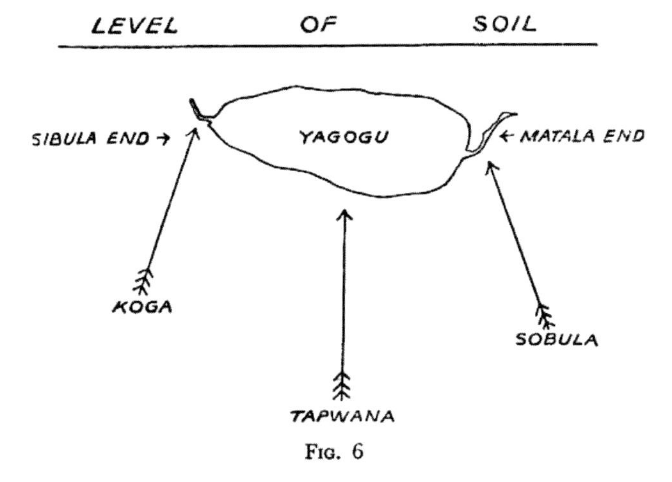

Fig. 6

stalk (*tamna*) and spreads into the various lateral shoots (*yosila*, *kari salala* and *yawila*). The stalk underground—which the natives call, when it is just a tendril, *sobula*, and later when it is young and flexible, *silisilata*—has now grown into the strong main root, the *gedena*. From the *gedena* new roots spring, called again *silisilata*, and on these young tubers (*bwanawa*) are formed (see Fig. 7).

Now the natives know perfectly well that all these processes are worked by nature in the garden very much as they are worked for the wild plant in the bush; but, when they happen in the garden, the magician has to take his share in them.[2] Here is a statement given to me by Towese'i, the younger brother of Bagido'u (Text 98, Part V, Div. XII, § 44):—

[1] Cf. also Part V, Div. III, §§ 2–8. Note, please, that in the frequent possessive suffix *-la* the *l* is interchangeable with *n*. Thus *mata-la* = *mata-na*. The *l* is predominant in the north, the *n* in the south.

[2] See also Note 17 in App. II, Sec. 4.

"(i) We plant taytu, already it lies (in the ground). (ii) Later on, it hears magic above; already it sprouts. (iii) We go to the garden, we walk round, we recite magic—(we charm over) all the garden. (iv) One day only (it lasts) the garden magician goes alone and

GROWTH OF TAYTU. THE NEW VINE

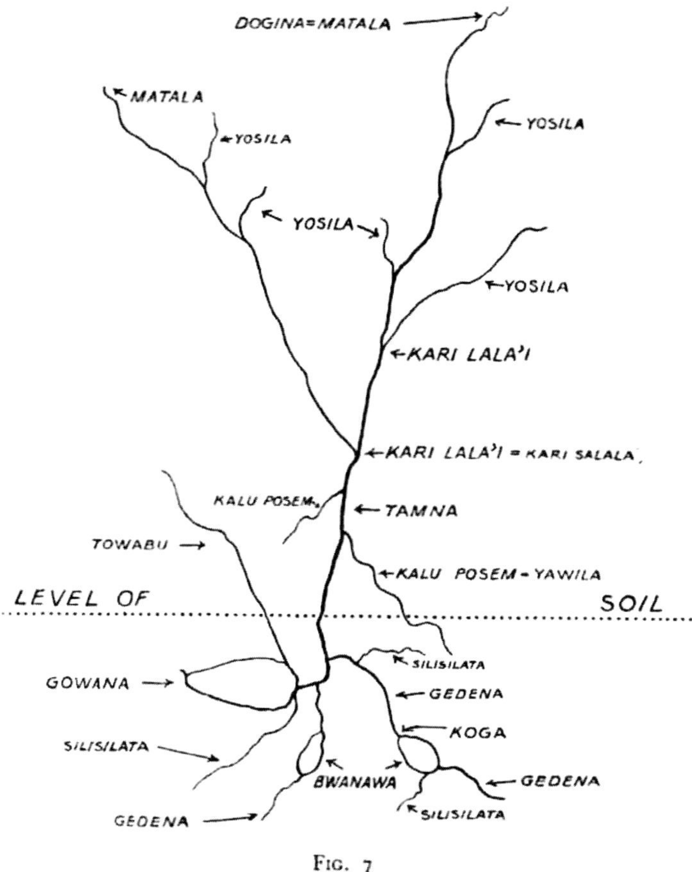

Fig. 7

charms. (v) He (then) remains, he reposes, on the third day he will go and recite magic, he will cause the sprout to come forth out of the ground. (vi) He walks round, he recites magic."

This statement expresses in a concise manner the native view about the magic of growth. The very title of the first spell, the awakening of the sprout (*vaguri sobula*), or its alternative name, the cutting through of the sprout (*tavisi sobula*), indicates the magical function of the charm and the wording throughout is in full harmony with this aim. In the exordium (the *u'ula*) of this spell the word "cut through" (*tavisi*) is constantly repeated. In the main part (the *tapwala*) there is the magical refrain or key-word: *dadeda tavisima*. *Dadeda* is the name of a plant which unfortunately I could not identify botanically, though it was shown to me. It is a very strong creeper which has rich foliage, a rank growth and sprouts readily from shoots; also it will grow again however much it is cut about. As a native described it to me: "*Dadeda* is a plant of extremely rank growth (*kaysususine*); we cut it, already it has sprouted. We cut it and it sprouts again."[1] It is very interesting that in the main part of the spell the magician invokes only the weaker kinds of taytu, those which are slow sprouting, affected by disease and so on. As I was told by one of my informants: "The bad taytu hears the magic, the good taytu hears it also; the good taytu knows that the bad one will grow with the magic, it grows more quickly so as not to be ashamed."[2] Thus the competitive principle enters not only into human activities in the Trobriands, but is carried right down into the soil among the crops.

This is the magic of *vaguri sobula*, the awakening of the sprout, *vavisi* or *tavisi sobula*, the cutting through of the sprout:—

FORMULA 13

I. "Cut through, cut through, cut through, cut through.
Cut through anew, cut through, cut through.
Cut through of old, cut through, cut through.
Cut through in the evening, cut through, cut through.
Cut through at noon, cut through, cut through.
Cut through at daybreak, cut through, cut through.
Cut through in the morning, cut through, cut through.
O *dadeda* tree that sprouts again and again.

II. "O *dadeda* tree that sprouts again and again.
O old skin, cut through, O *dadeda* tree that sprouts again and again.
O slow-sprouting *imkwitala* taytu, cut through, O *dadeda* tree that sprouts again and again.
O slow-sprouting *katumyogila* taytu, cut through, O *dadeda* tree that sprouts again and again.

---

[1] This is a free translation of the native text which will be found in comm. D. to M.F. 13 (Part VII), where also the verbatim translation is given.

[2] Cf. footnote, p. 88.

O *taboula* taytu with the rotten patch on it, cut through, O *dadeda* tree that sprouts again and again.
O rotten taytu, cut through, O *dadeda* tree that sprouts again and again.
O blighted taytu, cut through, O *dadeda* tree that sprouts again and again.
O *tirimwamwa'u* taytu of the heavy growth, cut through, O *dadeda* tree that sprouts again and again.

III. "Thy shoots are as quick as the eyes of the *kapapita*, the quick bird.
Thy shoots are as quick as the *kababasi'a*, the quick black ants.
Thy shoots are as quick as *ginausi*, the quick thing.
Shoot up, shoot up, shoot up, O taytu."

Going slowly across the gardens, the magician turns to one *baleko* after another and recites this spell. With a strong voice, clear and resounding, he sweeps the whole garden, letting the words of the virtue penetrate into the soil on every plot.

While the magician is engaged in reciting this spell and the following one, the gardeners prepare the small supports which presently they will have to add to the stems left rooted after the cutting (*kamtuya*) so that the shoots as they come out of the ground can be supplied with a support each. But first of all it is necessary to help the young sprout to emerge. This is done by the charm *katusakapu* or *vasakapu sobula*, which means literally, 'to make come out the sprout', *sakapu* being almost adequately translated by, "to emerge", "to come out". This spell is entirely symmetrical with the previous one, corresponding to it almost word for word; save that for the word "to cut through", the word "to come out" or "emerge" is substituted. The same plant, *dadeda*, is invoked here also as the sympathetic symbol. Thus in free translation this charm can be rendered:—

FORMULA 14

I. "Come out, come out, come out, come out.
Come out, come out anew, come out, come out.
Come out of old, come out, come out.
Come out in the evening, come out, come out.
Come out at noon, come out, come out.
Come out at daybreak, come out, come out.
Come out in the morning, come out, come out.
O *dadeda* tree that comes out, comes through.

II. "O *dadeda* tree that comes out, comes through.
O old skin, come out, O *dadeda* tree that comes out, comes through.
O slow-sprouting *imkwitala* taytu, come out, O *dadeda* tree that comes out, comes through.
O slow-sprouting *katumyogila* taytu, come out, O *dadeda* tree that comes out, comes through.
O *taboulo* taytu with the rotten patch on it, come out, O *dadeda* tree that comes out, comes through.
O rotten taytu, come out, O *dadeda* tree that comes out, comes through.
O blighted taytu, come out, O *dadeda* tree that comes out, comes through.
O *tirimwamwa'u* taytu of the heavy growth, come out, O *dadeda* tree that comes out, comes through.

III. "Thy shoots are as quick as the eyes of the *kapapita*, the quick bird.
Thy shoots are as quick as the *kababasi'a*, the quick black ants.
Thy shoots are as quick as the *ginausi*, the quick thing.
Shoot up, shoot up, shoot up, O taytu."

This charm also is just "mouth magic", *megwa wala o wadola* (magic just of mouth), that is, a charm with no accompanying rite; and after it the young shoots begin to come out of the ground and twine round the slender supports, the *yeye'i*, *kamtuya* and *kaytusobula*. It is at this stage that the crops are most in danger from weeds and the women have to work energetically. The weeds are either simply torn out with the hand or, when they have grown too rank through neglect, uprooted with a small digging-stick. At this time, with the regular rains and the heavy sultry heat, everything grows at an incredible speed on such fertile soil, and the work of weeding is not easy. Women often carry it out by means of communal labour, accompanied by the usual competitive challenges, joint eating and, above all, joint gossip. It is during such communal weeding that women enjoy special privileges. Men are not supposed to come anywhere near a women's weeding party and, in a few villages in the south of the island, the custom of *yausa* allows women to seize and mishandle any man who comes within sight. A man from their own community they would merely insult verbally; a stranger might be really ill-treated in a sexually degrading manner.[1]

Weeding can also be done by men in the case of a bachelor making his own gardens. There is no shame attached to a male

[1] Cf. *The Sexual Life of Savages*, Chapter IX, Section 8.

PLATE 43

COMMUNAL PLANTING

"The man takes his *dayma* in both hands and, drawing himself up, breaks up the sod with a few blows" (Ch. III, Sec. 5). In the foreground a trade knife is seen on the right: several positions and activities can be observed and the size and depth of planting holes estimated

PLATE 44

MAN IN THE ACT OF TILLING

"With his left hand he feels through the earth, while the point of the *dayma* moves through it . . . breaking up the lumps" (Ch. III, Sec. 5)

PLATE 45

TRAINING OF VINES

This "consists in keeping an eye on every vine, twisting the tendril round the pole when it grows long enough and tying it in place with a piece of bast" (Ch. IV, Sec. 1)

PLATE 46

THE TAYTU VINE

"Shoot up, O head of my taytu,
Heap up, O head of my taytu,
Make mop upon mop of leaves, O head of my taytu" (Ch. IV, Sec. 2)

having to do his own *pwakova*, though whenever there is a woman to do it, it is her work.

The weeding (*pwakova*), or sweeping clean (*sapi*) as it is also called, is inaugurated by a very simple magical rite, the *kariyayeli sapi*. It consists of a conventionalised mimic weeding act: with a symbolic *dayma* the magician scratches the ground, uprooting perhaps a few weeds and utters the following words broadcast across the gardens:—

FORMULA 15

"I sweep, I sweep away.
I sweep, I sweep away.

"Cut thy top, O taro root; split thy leaf, O taro.
Thy stalk bows over.

"O taro of the *kalakayguya* kind; O taro red like blood: O taro of the *kalipadaka* kind, O taro of the *namtamata* kind."

Remarkably enough, the taro, which we have seen figuring very prominently in some of the previous inaugurative rites of Omarakana garden magic, here comes to the fore again. I was not able to obtain any explanation of this from my informants. Perhaps it is that taro is more directly affected by weeds, or that this spell came from some other system of magic in which taro was of greater importance than it is now in Kiriwinian gardening. I have not sufficient data to answer this question.

The weeding ceremony, with its short spell and simple rite, is performed on every garden plot. Afterwards weeding can be carried on throughout the gardens.

Let us return to the cycle of growth magic, of which we have described so far two ceremonies. The third one is a little more elaborate. It involves a small preliminary rite and imposes a taboo, while its spell accompanies the main rite. This spell does not refer to the roots; it is called *kaydabala*, or the head-stick, that is, it is a charm referring to the foliage. "It produces many, many leaves", as the natives put it. The spell is also called *siribwobwa'u*, after the bamboo plant which is invoked in the opening sentences. Though it is a spell chanted over taytu gardens, in its text the *kuvi*, the large yam, figures most prominently. Since the large yam has a specially rich foliage, this might be regarded as sympathetic magic rather than as a direct invocation to benefit the yams; but this is merely surmise, and the explanation that it is a *kuvi* charm introduced into the magic of taytu gardens by some historical process would be, I think, just as plausible. The natives definitely say that this rite is meant both for taytu and *kuvi*.

When the vines have grown and are just budding into leaf, the magician makes a tour of the gardens and on each plot twines some lalang grass (*gipware'i*) round the *kamkokola* or round a *kavatam*. Next morning he makes another tour and, again on each plot, utters a charm and immediately afterwards puts up the head-stick, *kaydabala*—a small stick laid horizontally between two vertical poles, usually somewhere on or near the *kamkokola*. This is the sign for the men to prepare the large supports, the *kayvaliluwa*, which will have to be put up side by side with the small ones (*kaytusobula*), for the growing vines. There is, therefore, an inaugurative element in this magic, though it is primarily a magic of growth, its aim being to make the leaves develop luxuriantly. This is the *kaydabala* spell:—

Formula 16

I. "O many-leaved bamboo, O bamboo,
O many-leaved mangrove, O mangrove,
O taytu of the hillock, O taytu of the mound.
Raise thy stalk, O taytu. Make it flare up, make it lie across.

II. "The yam rises, the yam rises, the yam rises.
The yam rises and swells like a bush-hen's nest.
The yam rises and swells like a baking-mound.
The yam rises and swells like the mound round an uprooted tree.
The yam becomes like an ant-hill.
The yam becomes like a cave.
The yam becomes like a coral boulder.
Thy growth be as the flight of the green parrot,
The boring of thy roots as the nibbling of rats,
Thy reach be as the reach of a thief.

III. "For these are my yams, and my kinsmen will eat them up. My mother will die of surfeit. I myself will die of repletion. The man carrying the yams swears by the outrage of his mother.
"He complains, 'The yams weigh down my head, they crush my shoulders.' He groans: 'Yakakakaka . . .' "

The wording of the spell shows it as a general fertility magic. It is obviously partly a direct stimulus to growth and partly an optimistic forecast of the harvest that is to be.

The aim of the next spell is very much the same as that of the previous one. It has "to produce still more plentiful leaves". "It makes many underground shoots (*silisilata*)." "Where one branch grew, several branches now come; where there was one root, many roots come" (see Pl. 45). This charm is spoken after the large supports have been put in, which may be a week or ten days after the previous performance. There is no rite connected with it; it is merely chanted over every plot.

This is the *kaylavala dabana taytu* spell—the producing of the mop (head) of the taytu:—

FORMULA 17

I. "Millipede here now, millipede here ever!
Millipede of the promontory of Kabulukwaywaya, shoot along, shoot along, shoot along to Kabulukwaywaya, shoot along as far as Dulata. Millipede of Dulata, shoot along, shoot along, shoot along to Dulata, shoot along and shoot along back to Kabulukwaywaya. Millipede shoot along.

II. "The millipede shoots along, shoots along.
Thy head, O taytu, shoots along as the millipede shoots along.
Thy leaves, O taytu, shoot along as the millipede shoots along.
Thy forks, O taytu, shoot along as the millipede shoots along.
Thy secondary stalks, O taytu, shoot along as the millipede shoots along.
Thy shoots, O taytu, shoot along as the millipede shoots along.
Thy overground roots, O taytu, shoot along as the millipede shoots along.
Thy aerial root, O taytu, shoots along as the milipede shoots along.

III. "Shoot up, O head of my taytu,
Heap up, O head of my taytu.
Make mop upon mop of leaves, O head of my taytu.
Heap together, O head of my taytu.
Gather up, O head of my taytu.
Make thyself thick as the *yokulukwala* creeper, O head of my taytu."

This clear and beautiful spell takes the millipede as its leading word because of its rapidity of movement. The millipede, is also associated with the magical and mythological cycle of ideas concerning rain and cloud, since it is a prognostic of downpour. Hence the millipede is also a symbol of fertility.

In the middle part, the direct aim of this magic—the development of the taytu plant above ground and in the soil—is clearly expressed. In the last part the stress is placed on the branches.

The following spell is closely allied to the previous one and pushes the progress of growth one step further. The leaves have formed already, many branches have sprouted. Now it is time to make the whole mop of the vine overflow and spread, leaping from one support to another, until it forms a canopy over the garden. To this end the *sayboda* (closing up) charm is uttered. "The taytu closes up, it makes one branch, another branch, a third branch.

it falls over, the garden becomes dark" (see Pl. 46). Here again the charm is merely "mouth magic" (*o wadola*), with no rite and no manipulation.

Kasayboda—the closing up, covering up of the taytu:—

FORMULA 18

I. "Spider, here now! Spider here ever!

"O spider of Kabulukwaywaya, cover up Kabulukwaywaya, close up the sacred grove of Lu'ebila, cover up the sacred grove of Lu'ebila, close up Kabulukwaywaya. Spider, cover up.

II. "The spider covers up, the spider covers up.
The spider covers up the taytu.
Thy open space, the open space between thy branches, O taytu, the spider covers up.
Thy soil, O taytu, the soil between thy vines, the spider covers up.
Thy dry branches, O taytu, the dry branches on the trees left over from cutting, the spider covers up.
Thy *kamkokola*, O taytu, thy magical prism, the spider covers up.
Thy *kavatam*, O taytu, thy strong yam pole, the spider covers up.
Thy *kaysalu*, O taytu, thy branching yam pole, the spider covers up.
Thy *kamtuya*, O taytu, thy stem saved from the cutting, the spider covers up.
Thy *kaybudi*, O taytu, thy training-stick that leans against the great yam pole, the spider covers up.
Thy *kaynutatala*, O taytu, thy uncharmed prisms, the spider covers up.
Thy *tula*, O taytu, thy partition-stick, the spider covers up.
Thy *yeye'i*, O taytu, thy small slender support, the spider covers up.
Thy *tukulumwala*, O taytu, thy boundary line, the spider covers up.
Thy *karivisi*, O taytu, thy boundary triangle, the spider covers up.
Thy *tamkwaluma*, O taytu, thy light yam pole, the spider covers up.
Thy *kayluvalova*, O taytu, thy tabooing-stick, the spider covers up.
Thy *kayvaliluwa*, O taytu, thy great yam pole, the spider covers up.

III. "Shoot up, O head of my taytu.
Heap up, O head of my taytu.
Make mop upon mop of leaves, O head of my taytu.
Heap together, O head of my taytu.
Make yourself thick as the *yokulukwala* creeper, O head of my taytu."

In commenting on this charm, which as can be seen is symmetrical with the previous one, the natives told me that as the spider spins his web, so should the taytu plant produce many branches. It should cover the spaces and make a green roof, it should fill the spaces between the vine stems, it should cover all the dead wood with which the live plants are supported.

We have now followed the first five rites of the magic of growth—not including, of course, in this cycle, the inaugurative rite of weeding. These first five rites of growth are called *megwa geguda*, the magic of the unripe crops. The next growth magic cycle is called *megwa matuwo*, the magic of the ripe crops, though as the crops are not really ripe for harvest—they have not started forming yet—it would be more correctly described as the magic of ripening crops, and its purpose is to stimulate the growth and formation of the tubers.

3. THE MAGIC OF GROWTH—THE STIMULATION OF ROOTS AND TUBERS

Let us place the events described in the last two sections within the native calendar, which we have brought up to the ceremony of *kamkokola*. This we have placed at the beginning of the moon of *Yavatakulu*, and the planting, which begins immediately, is finished on the standard plots in a week or so. Thus, by the moons of *Toliyavata* and *Yavatam*, the crops are sprouting, the taytu vines begin to climb and it is some time in the second of these two moons, roughly corresponding to our December, that the growth magic of the "unripe crops" (*megwa geguda*) takes place. Near the beginning of this period there is the inaugurative ceremony of weeding; and the weeding, though it is most important at that time, continues off and on well-nigh till harvest.

This brings us to the moons of *Gelivilavi* and *Bulumaduku*, about January and February. Looking at our Chart of Time-reckoning (Fig. 3) we see that *isunapulo* falls in the former and growth magic of the "ripe crops" into the latter. As we have said, the three rites which follow refer to the roots and tubers. The first of these consists of a spell chanted over the fields, *vapuri*. *Va-* is a causative prefix, and the verb *puri* here means to break forth into clusters. As the natives put it, the *vapuri* spell "makes the *bwanawa* (young tubers) break forth in clusters". Here again the rite consists in the utterance of the magical words over the garden plot. This is the text:—

FORMULA 19

I. "One canoe load of plenty,
Another canoe load of plenty,
A third canoe load of plenty,
A fourth (up till ten. At times the magician goes as far as fifty, enumerating after the tenth, the twentieth, thirtieth, fortieth, fiftieth.).

"Break forth, O taytu, in daytime.
Break forth, O taytu, at night,
Break forth, break forth, till you have broken forth your fill.

"Thy return, O taytu, O return to us.
Thy hastening, O taytu, O hasten to us.
Thy quick breaking forth into a multitude of bunches, O break forth.
Taytu breaking forth again and again,
Taytu breaking forth again and again.

II. "O *nakoya* taytu, breaking forth again and again.
O *sakaya* taytu, breaking forth again and again.
O *nabugwa* taytu, breaking forth again and again.
O *kwoyma* taytu, breaking forth again and again.

III. "My grandfather, Tokuwabu, will embrace thee, O taytu,
He will dance with thee on the cross-roads."

The varieties of taytu here enumerated, *nakoya*, *sakaya*, etc., are all good and palatable, but very hard-growing tubers. Here again my informants, asked why the garden magician only mentions those varieties which grow with difficulty, gave me the answer:

"Later on when the *lupilakum* (the best and most easily growing taytu) hear that the garden magician appeals to the wild (not really domesticated, hence difficult of growth) kinds; later on they rush ahead, the *lupilakum*, and they leave the other kinds behind." This answer is similar to the one given above in Section 2.

Tokuwabu, the name mentioned in the last strophe, is one of the ancestors or, perhaps, a magical predecessor of Bagido'u. His spirit will hug the *bwanawa*, that is, will rejoice in the plenty of the new crops, and will dance at the cross-roads.

The next rite, which once more is a simple spell chanted directly over the *baleko* (garden plot), is nothing but a reinforcement of the one just described. It is called *kammamala*, which means the fetching back, the bringing back. Whether this word, which also appears in the text of the formula, carries with it any belief in a return of crops, in a resurrection of the new taytu from the old, I was not able to ascertain. Here is the spell:—

FORMULA 20

"Taytu return, O return for certain.
O taytu of the *nakoya* kind, return, return for certain.
O taytu of the *nabugwa* kind, return, return for certain.
O taytu of the *sakaya* kind, return, return for certain.
O taytu of the *nonoma* kind, return, return for certain.
O taytu of the *kwoyma* kind, return, return for certain."

This spell, as the natives put it, makes the tubers break forth anew. The third and last spell of growth magic directed at the new roots is called *kasaylola* (the spell of anchoring), or *talola silisilata* (the anchoring of the roots).

FORMULA 21

I. "Bush-hen here now! Bush-hen here ever!
Bush-hen from the north-east, anchor on the north-east.
Anchor there on the south-west.
Bush-hen here now! Bush-hen here ever!
Bush-hen from the south-west, anchor on the south-west, anchor there on the north-east.
The bush-hen anchors!

II. "The bush-hen anchors!
My taytu, the bush-hen anchors.
Thy shoots, O taytu, the bush-hen anchors.
Thy new rootlet, O taytu, the bush-hen anchors.
Thy main root, O taytu, the bush-hen anchors.
Thy roots, O taytu, the bush-hen anchors.
Thy black blight, O taytu, the bush-hen anchors.
Thy wounded sides, O taytu, the bush-hen anchors.

III. "The taytu is anchored, it is anchored, it is anchored for good and all."

This formula ends not merely the spells directed towards the development of tubers but all the growth magic spells; those, that is, which aim directly at the fostering of the growth and development of the plant.

Immediately connected with this last formula there is another rite which has an inaugurative function. Now that, owing to the joint workings of nature and of magic, the new tubers have burst forth into rich clusters, another gardening operation becomes necessary. Let us once more turn our attention to what has happened under the soil. Looking at Figures 6 and 7, we see the seed-tuber reclining horizontally in the soil. This tuber has since sent out a sprout which has developed above- and below-ground: it has grown into the vine and at the same time sent out several new shoots underground on which new tubers (*bwanawa*) have formed. The old seed tuber is now spent and decayed, and this, as well as some of the new ones (*bwanawa*), has to be taken out. When the clusters are too thick none of the tubers can develop properly, and a thinning out (*basi*) is, as all the natives have assured me, an indispensable proceeding (cf. Part V, Div. III, §§ 2–8). All those which show any signs of imperfection, blight or disease, or which are too thickly crowded in one spot, or which are too small and do not promise good development, are removed. The bad taytu is described as "black" (*bwabwa'u*); the good is called "white" (*pupwaka'u*). The work is done by men almost invariably. The soil is broken up at the roots of the taytu vine by means of a digging-stick (*dayma*), the earth scraped away, the roots inspected and the bad ones, together with the spent seed tuber cut away with an axe. The new imperfect tubers, unless they are completely diseased, are taken to the village and eaten. As one of the natives told me (Text 30, Part V, Div. VI, § 22): "We dig till the roots are exposed; then we lift and pluck them. The unripe tubers we eat. We bring them to the village and eat them. The good taytu we leave so that it may mature." These tubers are tabooed to the garden magician. They are not regarded as really harvested yet, there is no first-fruit ceremony connected with them, they are never stored in the open ceremonial yam-houses (*bwayma*) but eaten as they are dug up.

There is, however, a ceremony called *momla*, which inaugurates the *basi*, the thinning out of the tubers. Once more the hard-worked *vatuvi* spell (M.F. 2), which, as we remember, has already functioned in the first inaugural rite of striking the soil, at the burning and again at the *kamkokola* ceremony, has to be chanted. It is chanted over the axes and the digging-sticks; and the standard mixture, which is always used with the *vatuvi* formula, is fastened to these. So Bagido'u, or his acolytes, sets out again one morning to collect the ingredients and in the afternoon he charms the mixture between two mats and attaches it to the ends of the digging-sticks and axes brought to him by the men of the village. The next morning all these men assemble before his house, each receives his digging-stick and axe and they go to the garden. There the magician strikes the *kamkokola* on his own garden with the medicated axe, opens the soil at one of the vines with the medicated *dayma*, and performs a representative *basi* on one plant. The men then do the same each in his own garden. The real work begins on the following day.

This ceremony really concludes the main series of garden rites preceding harvest, and since harvest is a complicated and many-sided event, it will be better to deal with it in a special chapter. Before we pass to it, however, it is necessary to say a few words about private garden magic, which is performed independently of the main public ceremonies at certain stages.

4. PRIVATE GARDEN MAGIC

Besides the official magic performed by the *towosi* on behalf of the whole community, which inevitably accompanies the various stages

PLATE 47

CORNER OF FULLY GROWN GARDEN

"The spider covers up the taytu,
Thy open space, the open space between thy branches, O taytu, the spider covers up.
Thy soil, O taytu, the soil between thy vines, the spider covers up.
Thy dry branches, O taytu, the dry branches on the trees left over from cutting, the spider covers up" (Ch. IV, Sec. 2)

PLATE 48

THE SHOW PORTION OF THE HARVEST

Harvesting, the end of agriculture, is in the Trobriands made prominent and important. "The native likes to handle the tubers he has grown, to count them, to arrange them deliberately in conspicuous, well-shaped heaps" (Ch. V, Intro.)

PLATE 49

DISPLAY OF FIRST FRUITS: TARO HARVESTED AFTER THE INAUGURAL RITE

The central place in the village of Tukwa'ukwa: the storehouses can be seen quite empty; the pig, still alive and trussed on a pole, is seen on the left hanging under the foundation-beams of the storehouse; the first fruits are displayed in the foreground (Ch. V, Sec. 2)

PLATE 50

DISPLAY OF FIRST FRUITS: YAMS HARVESTED AFTER THE INAUGURAL RITE

The central place of the village of Teyava: *kuvi* displayed in the foreground. Note the digging-stick held by boy (Ch V, Sec. 2)

of gardening, benefiting all and integrating the work, there also exists private garden magic. There are certain formulae individually owned, which are either used over his own garden plot by their owner, or by an expert on payment of a fee. These establish a differential expectation of fertility. It often happens that the official garden magician of a village is also the practitioner who, for a fee, is prepared to add to a garden some special benefits over and above those which he conveys to everybody, either by using the owner's formula, or by drawing on his own store of private magic or by giving a private performance of a public rite. This private magic is singularly inconspicuous in the Trobriands.[1] Perhaps this is due to the fact that gardening is such an extremely important activity, so strongly fraught with envies and jealousies that it would be dangerous for anyone to claim that his private magic could give him a great deal of individual excellence over and above the communal level. To this must be added the fact that the chief or the headman must invariably be able to claim the best gardens. This he does, or rather can do, by choosing the best soil, but he also usually claims to have the best private magic, and others will beware of countering his claims or boasting of having a better magic than he himself.

There seems also to be a general feeling that the communal magic does all that can be done and that it is not the right thing to carry out private magic. Only a few individuals ever boasted to me, and that in strict confidence, that they had powerful private magic and that the quality of their gardens was due to the superiority of their own formulae over those of their neighbours. Thus Namwana Guya'u, the rebellious, independent, and unscrupulous favourite son of To'uluwa, assured me in 1915 that his private magic was more efficacious than the public magic of his cousin, Bagido'u. He specifically told me that he medicated his own large yams (*kuvi*) with a magic called *bisikola* (cf. also Ch. III, Sec. 1, and Ch. IX, Sec. 2). He also charmed his own *kema*, *dayma* and *yagogu*. The last rite he described as *ayuvi kakavala*, "I breathe on my small yams." The word *kakavala* is a synonym for *yagogu* (seed yams).

I obtained a few formulae from Bagido'u and owe to him most of my information about this department of magic—by no means so complete as that of the official kind. The private magic is performed over the seed yams; over the *tula* (cross sticks dividing the plots into small squares); the digging-stick (*dayma*) and the axe (*kema*). I only once saw such a private ceremony performed. I was walking at the time with Nasibowa'i, the garden magician of Kurokaywa,

[1] See also Note 19 in App. II, Sec. 4.

with whom we shall become more intimately acquainted. He was engaged in carrying out an official ceremony, I think one of the *kamkokola* rites. As he was passing from one plot to another, a man brought him a basket of seed yams (*yagogu*). Nasibowa'i broke a few branches from a neighbouring mimosa tree and, beating the seed yams with the branches, recited a spell. I was not able to record the words. Here, however, is a corresponding spell given to me by Bagido'u from his own repertoire.

FORMULA 22

"Tudava, O Tudava.
Malita, O Malita.
The sun rises over Muyuwa.

"I sit, I turn thee round.
I sit, I sweep thee clean,
I am calling up the taro of my garden; I am calling up the taytu of my garden.

"O fish-hawk hover over my garden."

This spell I was told is recited directly into a basketful of seed yams (*yagogu*).

Another spell which also belongs to a private rite of planting, but which, as Bagido'u told me, must be performed in the garden over seed yams lying ready in one of the squares, is called the magic of the *tula*, or boundary sticks.

FORMULA 23

I. "O crops breaking through the soil—break through!
O crops returning from the soil, return!
O seed yams breaking through,
O taytu returning!

II. "The taytu rises, the taytu turns round in the soil, the taytu swells the ground.
Thy stalk rises, thy root lifts up the soil.
Raise thy young tuber in the belly of my garden.

"The belly of my garden becomes smooth like a pounding-board.
The belly of my garden becomes smooth like a trimming-board.
The holes in the belly of my garden are as the holes the mangrove mollusc bores in mud.

III. "I shall go to the village laden with taytu."

This is obviously a spell which spurs the taytu to grow. The pounding-board is the board on which taro is pounded when it is made into the *mona*, the favourite taro pudding of festive occasions. The trimming-board is the board on which native women make their grass skirts. Both these boards are perfectly smooth. The mollusc (*ginuvavarya*) is one of the marine shell-fishes which makes many holes in the mud of the mangrove swamp at low tide.

As to the private magic chanted over the axe to give it a special efficacy during planting and later on during harvest, Bagido'u said that, if he were specially paid by anyone of the village, he would charm this over with the *vatuvi* spell, the spell which plays such an important part in his official garden magic, medicating it with the same ingredients which he always uses with this magic.

Another private formula owned by Bagido'u belongs to the magic performed over the digging-stick which is afterwards used during the thinning out of the crops. Bagido'u told me that sometimes this magic might also be chanted over a digging-stick used in the planting.

FORMULA 24

I. "O taytu to be plucked out,
O taytu to be pulled away,
Swollen as a mound, as the *sasoka* fruit.

II. "The trunk of the ficus tree is not large—it is the body of my taytu which is large.
The trunk of a big mangrove is not large—it is the body of my taytu which is large.
The trunk of the *bwabwaga* tree is not large—it is the body of my taytu which is large.
The trunk of the great pandanus is not large—it is the body of my taytu which is large.
The coral boulder is not large—it is the body of my taytu which is large.
The trunk of the acacia is not large—it is the body of my taytu which is large.
This is not thine eye, thine eye is the morning star.
This is not thine eye, thine eye is the white flower of the *youla'ula* creeper.
Thy stalk rises, thy root lifts up with the soil.
Raise thy young tuber in the belly of my garden.
The belly of my garden becomes smooth like a pounding-board.
The belly of my garden becomes smooth like a trimming-board.
The holes in the belly of my garden are as the holes the mangrove mollusc bores in mud.

III. "I shall go to the village laden with taytu."

I have kept here the literal translation 'body of my taytu' instead of reducing the figure of speech to its precise meaning "taytu tuber", partly because it keeps closer to the native expression, partly to emphasise the fact that to the Trobrianders the edible roots are the main part or the body of their staple crops.

The following short formula is spoken over a boundary-stick and is accompanied by the simple rite of rubbing the stick into the ground. During its utterance the voice must be directed towards the *tula*, so that the magical virtue may penetrate the mould left by this action.

FORMULA 24*a*

"Move that way, move that way; come this way, come this way;
I shall take hold, I shall place firmly.
It digs in, the mould of my boundary-stick;
It digs in and the mould of my taytu.
The belly of my garden shall recline."

I should like to repeat, however, that my information on private garden magic is not of the same quality as that on the official system. I was so engrossed in the study of the public system of charming the gardens that I failed to enquire at first whether there was private magic also, and since it is by no means prominent, it was only by accident that I stumbled on it when I saw Nasibowa'i beating the seed yams in the basket.

A vague influence of a magical character is that of the recital of fairy-tales, *kukwanebu*. The season for these is when the north-westerly monsoon prevails from December to March. At this time, as can be seen from the Chart of Time-reckoning, the soil has been cleared, the crops planted, the fences put up and the magical wall erected; and now the forces of fertility must produce the crops. At this time, also, the natives are often kept indoors or near the houses by bad weather. Then they tell each other the well-known, interminable stories, mainly ribald, and every speaker must wind up his tale with the following four-line, standardised formula (Text 82, Part V, Div. XI, § 11):

"The *kasiyena* yams are breaking forth in clusters; this is the season when crops cut through, when they grow round. I am cooking taro pudding; So-and-so (some important person present is named in a jocular tone) will eat it. I shall break off betel-nut; So-and-so (another notable is named here) will eat it. Thy return payment, So-and-so (and the man to recite next is named)."

This ditty is intoned with a definite rhythm. It is called *katulogusa*, and I was told: "The telling of fairy-tales exercises a magical influence on new crops. It makes the *kasiyena* yam break forth into clusters. As this yam breaks forth into clusters, so staple food matures" (Text 83, Part V, Div. XI, § 13).

5. COMMUNAL LABOUR

That agriculture is an organised economic activity must be clear now to every reader. The people who make their gardens within the same enclosure are not independent of one another. They obey the initiative, the decisions and the rulings of the chief. They have to follow the rhythm given to them by the sequence of magical rites. They are under the supervision of the same magician. Within limits which are fairly wide, but none the less definite, every gardener has to keep up to time and maintain a certain standard of work. Notably in the making of the fence and in weeding, they actually depend upon one another's thoroughness.

But we find another form of co-operation in communal labour, and it is on this subject that a few words must be added here. Communal labour becomes indispensable in the case of a chief who, as we know already (Ch. I, Sec. 8), could not supply all the male labour for his extensive garden area. Communal work is also favoured when the villagers decide to extend the dancing or festive season and therefore start their gardens late and have to get through the early stages rapidly by intensified effort. Whether there is any technical advantage in communal work at cutting or planting, or whether this method is more effective for psychological reasons, I cannot tell for certain. But the natives are deeply convinced that by doing the work communally, they get over the preliminary stages much more quickly than by individual work. It is never, however, absolutely indispensable in gardening as it is at certain stages of canoe-making or the building of a large storehouse or the sewing up of a canoe sail.

With all this, communal labour is more used in gardening than in any other activity, and the natives distinguish as many as five types of it, each called by a definite name and each having a distinct sociological character.[1]

When the gardens are made privately, each working on his own plot, the natives use the word *tavile'i*, a term which would thus correspond to the expressions "individual" or "non-communal" labour.

When the members of a village are summoned by a chief or headman to do the gardens communally, this is called *tamgogula*. A *tamgogula* is a general arrangement, instituted by the chief on

[1] I am giving here the substance of the account of communal labour to be found in the chapter on Tribal Economics in the Trobriands, in *Argonauts*, pp. 160 ff.

behalf of the whole community, to work jointly on those tasks which lend themselves to communal labour. If such an arrangement is proposed, when the time comes to cut the scrub all the men would be summoned by the chief to a festive eating on the central place (cf. Part I, Sec. 7), after which they would cut the scrub on the chief's plot. When this is done, they all cut each plot in turn and are on that day provided with food by the owner. The same procedure is adopted at each successive stage: at fencing, planting, putting up of supports, and finally at the weeding which is done by the women. Each of these activities is performed by all the men for each individual cultivator. Clearing after the scrub has been burnt, thinning out of tubers and harvesting are always done individually. Several communal feasts are given during, and one at the end of the *tamgogula* period.

If instead of a general agreement to work communally adopted by the whole village under the leadership of the chief, a limited number of gardeners arrange to do their work in common, all working for each, we have a new type of communal labour and new names. When such an arrangement covers the whole cycle (except clearing, thinning and harvesting) it is called *kari'ula*; when it is for one stage only, it is called *ta'ula*. As far as I know, there would be little communal feasting in such cases, perhaps none at all. Each would be rewarded by reciprocal service received. When several villages come to an agreement to make their gardens together by communal labour, the natives call it *lubalabisa*. The principles and organisation of this are similar to the *tamgogula*; *lubalabisa* is directed by one chief or headman, and is only instituted in the case of village clusters (cf. Ch. XII, Sec. 2) or of villages situated in close proximity. The arrangement would not imply that these villages make a common garden within one enclosure; only that all work on each other's fields in turn.

When a chief or headman, or man of wealth and influence, summons his dependants or relatives-in-law to work for him and for him alone, this is called *kabutu*. The owner has to provide food for the workers (cf. Part I, Sec. 3). This procedure may be adopted for one stage of gardening, a headman inviting his villagers to do his cutting or his planting or his fencing for him; or else it may extend over the whole cycle. Incidentally, the expression *kabutu* applies not to gardening only but is used whenever a number of men are needed by another to help him in some undertaking, such as the construction of a canoe, a storehouse or his dwelling.

CHAPTER V

HARVEST

In Kiriwina the harvest is the most joyous and pictureque stage of garden-making. The actual digging out of the roots fascinates the natives in itself; and round this technical activity there cluster a number of enlivening customs and ceremonies which, while they take even more time perhaps and require more work than the mere lifting of the tubers out of the soil, add to the joy of the season. They give it the character of a delightful pastime, and thus greatly help the work. After all, harvesting is the end of all agriculture and, in the Trobriands as elsewhere, this aim is made prominent and important, and is drawn out by its festive setting.

The additional activities consist in the cleaning of the taytu, in its display in the gardens, in its public and ostentatious conveyance to the storehouses and in its ceremonial stacking there. The native likes to handle the tubers he has grown, to count them, to arrange them carefully and deliberately into conspicuous, well-shaped heaps (Pl. 48). He likes others to admire his produce and compare it with that of the rest. He likes to talk about it and to hear others talk. He feels, in a word, the craftsman's or artist's joy in his accomplished work.

Naturally, therefore, certain stages of harvesting are framed into tribal activities and social occasions. The taytu is first heaped up and displayed in the gardens, and then the members of the community and of neighbouring ones walk through the gardens and admire the crops. Later on a part of these are transported to a village, sometimes lying at a considerable distance from the gardens, and offered by the gardener who has grown them to a specific relative, usually the husband of his sister.

A party of young men and women, as many as two score perhaps, are summoned to transport the crops; for it is important to the natives that the whole offering should arrive at once, to make the quantity look imposing, and to lend the act of giving a festive character. The carriers sometimes adorn themselves with facial paint, leaves, feathers and flowers. Accompanied by the owners of the crops and a few older men and women, they stack the taytu into baskets and transport it to the village. At this season the gardens are alive with laughing, chattering men and women. Carrying parties are to be seen in every direction, moving between villages, or sitting by the wayside to rest and gossip and refresh themselves their laden baskets beside them; or running into a village with loud characteristic cries, depositing their offering in the central place and then sitting down to chat with their hosts. Garden and village, roads and groves present at this time a lively and festive appearance.

1. FAMINE AND PLENTY

The harvest thus rewards the industry of the gardener and gladdens his heart with the perennial discovery of his living treasure underground. "The belly of my garden"—*lopoula ula buyagu* (M.F. 2)—as Bagido'u in his spells calls the soil he has charmed—has at last brought forth its fruits and the fruits of man's labour.

But there is a great difference in the social character of the proceedings and in the mental outlook of the husbandmen according to whether it is a year of hunger (*molu*) or of plenty (*malia*). At times of hunger there is a keener interest in reaping, but there may be no joy whatever. When there has been a drought, the natives begin to feel the pinch about the fourth moon of their year, *Toliyavata*, which begins the moons of scarcity (*tubukona molu*).[1] Then women would be seen scouring the grove and jungle for leaves, roots and wild fruits with which to supply their households. If any of the desirable fruits (*kavaylu'a*) of the grove and jungle, such as mango, Malay apple, *menoni* fruit, *gwadila* or *kum* (bread-fruit) have survived the drought, they would gather these; if not, they would have to fall back on the despised fruit of the *noku* tree, which is hardly edible but hardly ever fails. But if a drought lasts for two years, then the natives may find themselves face to face with real famine.

In order to understand the incidence of hunger, it might be well to consult the Chart of Time-reckoning. Imagine that after a good, or at least a normal year, there follows a bad one. Let us label them '1914' and '1915' as an example, though really the harvests of these two seasons were both more or less normal. In '1914' the average harvest allowed the natives to fill their yam-houses with a full year's sufficiency, for taytu will last a whole year if properly stored and healthy. After that time some tubers would go bad and only one in several remain edible; but the question of preserving them beyond a year rarely arises, as even a good harvest does not provide for much longer than that.

We start then with normally filled storehouses in 'July 1914' (roughly the thirteenth moon of our chart), the middle of the dry

[1] On our Chart of Time-reckoning the "hungry" moons are estimated at five. This is, of course, a rough average computation, and indicates rather the season within which hunger usually falls, than any fixed period.

PLATE 51

AWAITING THE MAGICIAN FOR CEREMONIAL HARVESTING OF TAYTU

"Small boys often run ahead to locate and then show to the magician the exact spot where the plants to be cut and harvested grow" (Ch. V, Sec. 3)

PLATE 52

INAUGURAL HARVEST CEREMONY FOR TAYTU IN TEYAVA

In some villages the first tuber taken out has to be ritually split against the pole of the *kamkokola* (Ch. V, Sec. 3). (Owing to a defect in the film, the magician's head is invisible.)

PLATE 53

SHAVING THE TAYTU TUBERS

"The stringy hair has to be plucked or shaved off with a mussel shell" (Ch. V, Sec. 4)

PLATE 54

VERY SMALL HARVEST ARBOUR

No. 1 on Fig. 8 was drawn from this arbour (Ch. V, Sec. 4)

season. Supposing that the following October, November and December, which in average years should have a good rainfall, are dry, the supply of wild fruits and the resources of the jungle will be insufficient. The new taro, yams and taytu, which as can be seen from our chart (Cols. 7, 8 and 9) should begin to ripen as early as the sixth moon, *Gelivilavi*, now fail completely. The natives are driven back on the stored taytu. This, even if economically handled, would not last much longer than 'July 1915'. But it would keep the natives alive. One bad year, as far as I could reconstruct conditions, would not produce real famine. But in 'July 1915', when the stored supplies are exhausted; when during the whole latter part of '1914' and the beginning of '1915' there has been hardly anything to draw upon except the stored taytu; when the new crops fall far short of what is necessary—hunger instead of plenty sets in. If the drought has been such that practically nothing is forthcoming from the gardens, real hunger may be felt immediately. This, however, would be rare, as some remains of the previous harvest would probably have been saved. But if the crops are merely scarce, sufficient for some two or three months, everything would depend upon the next few months. If rains set in during September, October, November, fruit-trees would keep people going until about the fourth or fifth moon, when new yams from the earlier gardens would be available. If, on the contrary, drought sets in again, then with empty store-houses and fruit-trees barren, with the jungle parched and dry, the swamps (*dumya*) 'hard as a rock, broken with crevices', with the lalang grass on them brown and dead and even the *noku* fruit insufficient—the dreaded calamity comes upon the natives.

Up to 1918 no such cases of great *molu* have been recorded for twenty-five years or so. Since white men came, as traders and whalers first, then as pearl buyers and planters, their supply of rice would probably have staved off the worst miseries of famine. Furthermore, the natives have now learned to cultivate the sweet potato, which is a hardier and more prolific plant than their own crops.[1] Growing all round the villages, also, they now have the pawpaw, *carica papaia*, which seems able to withstand drought much better than most of the native fruit trees. Thus nowadays whatever other "blessings" European occupation has brought them, it has averted really bad outbreaks of *molu*.

I have, however, accounts of what used to happen at times of really bad *molu*. The famine would naturally hit the agricultural

[1] My knowledge of botanical geography is defective, but I was assured time after time by my native informants that sweet potatoes were not known in the island before the white man's advent (see below, Sec. 5).

communities of the central and eastern district hardest. These communities had no economic access to the lagoon; in normal years they would get as much fish as they liked by means of standardised 'exchanges' (*wasi* and *vava*: cf. Part I, Sec. 10). On the eastern shore, moreover, it is impossible to fish during the dry season of trade-winds, March to October, since this coast is fully exposed to the continuous blasts of the south-east gale. At this dry season, when wild fruits and the fruits of the grove are scarce, hunger would be felt most painfully.

Starving, the owners of the usually fertile but now parched plains of central Kiriwina would steal to the western shore in hopes of poaching on the lagoon. Hiding as best they could in small encampments among the most remote patches of jungle, they would creep out at night to do some surreptitious fishing. The local men from the villages of the lagoon, anxious lest their fish supply, which in years of famine is barely sufficient for themselves, should give out, would scour the jungle for the encampments, attack the thin, hunted and exhausted inlanders, and kill them by the score. There are some caves near Oburaku on the south and others on the north-west coast which are full of bones; and legend tells that these are the relics of such wholesale slaughter. Some of my informants actually told me stories, which they said they had received from their grandfathers, as to how and when such slaughters happened: how a large party of men from Tilataula (the district round Kabwaku) tried to find refuge on the outskirts of the mangrove swamps north of Oburaku, how they tried to fish, how a fight was fought on the swampy road between Oburaku and Kwabulo, and how all the men from Tilataula were killed, their bodies eaten and their bones thrown into a cave.

There might be some elements of truth in such stories. That a sort of endo-cannibalism was practised in times of real famine is not unlikely. The Trobrianders live on the outskirts of cannibalism and, though in normal times they regard it with scorn and moral disapproval, in times of hunger they might be tempted (as some people in post-War Europe were tempted) to help themselves to an easily available meal. But the bones in the caves which I inspected were certainly not relics of such endo-cannibalistic feasts; neither could they have dated from a few generations back, since most of them were covered with soil and some of them were even overgrown with stalagmites. But that bloody battles were fought is certain; and also that these were outside the normal hostilities of the island, for no vendetta or warfare would follow them.

The following narrative, in which the memories of my friend

Molubabeba are reproduced in the words of his son, Tokulubakiki, illustrates some aspects of *molu* (Text 24, Part V, Div. V, § 6):—

"(i) Molubabeba in his childhood witnessed a famine. (ii) At that time the people first became ill with a skin disease. (iii) Some people died in the bush; some in the swamps; some in the *rayboag*; some round the water-holes. (iv) They went to the water-holes so as to moisten their hands, their feet, and then they died. (v) All this was because of hunger. There was no food to be eaten.

"(vi) Later on when the famine was over, they charmed over wild ginger, so that they might bespit the village. (vii) Then they made rain magic and rain fell down.

"(viii) Such a size, as long as a forearm, would be the valuable with which they would barter seed yams: (ix) a fine valuable for ten basketsful of seed yams; a small valuable for five baskets. (x) (with such seed yams) They would plant and plant:—one single plot, twenty men; one square each. (xi) After that when the seed yams became a little more plentiful; one plot would be cultivated by two men. (xii) Later on still when the seed yams would be plentiful again, one man would make one plot, another man another plot.

"(xiii) In Kulumata, in the western district, there people would disappear, perish. No canoe could go there out on the sea so that we might fish. (xiv) Were a canoe to sail out, they would see us, they would kill us directly. (xv) They should kill us—our kinsmen would not be angry, because it was a time of famine. (xvi) We hide in the bush, we find a canoe, we do not dare to go and fish.

"(xvii) The cause of all this is the witchcraft of drought; the country would be already bewitched by the chiefs because we had killed by sorcery their kindred. (xviii) When Mwakenuva, when Purayasi died, Numakala cast his evil spell. (xix) Whenever a chief died, the country would be bewitched."

Here we have a description of a great famine given in the vivid concrete manner characteristic of all native narratives: the skin disease, with which famine illness is alleged to begin; the starving people besieging the water-holes and dying there; the performance of *vilamalia* magic, bespitting the village with wild ginger (cf. Ch. VII, Secs. 6 and 3); the size of the valuable, in this case a stone axe given in exchange for ten baskets of seed yams, when in ordinary times a hundred baskets would be paid for such a one; the number of men working on a single *baleko*; the abatement of the vendetta; and, finally, the deeply rooted belief that in some way famine was always the expression of the chief's displeasure with his subjects (cf. Ch. I, Sec. 8).

Similar stories from the old days tell how whole inland villages moved to the seashore and encamped there in the dense scrub and jungle.[1] The invaders scoured the mangrove swamps, creeping out surreptitiously at night to hunt for shellfish in the shallows, and all the time they had to fight against each other and against those who tried to prevent them from poaching in the lagoon—such stories I heard many a time in the evening at the fireside without noting them all down verbally. The natives would express the hardships of hunger by the exceedingly small amount of taytu with which they gradually became satisfied; and then, as in the story just told, they would measure the dearth of seed yams by the price which they had to pay for them. At times of real famine, the chiefs would, surreptitiously at least, forgo their taboos and eat bush-pig and wallaby, the despised fruit of the *noku*, and other abominations.

The paramount chief would not himself suffer as badly as lesser men. He usually took credit for both drought and famine as being a manifestation of his supernatural powers, roused to malevolent activity by some misdemeanour of his subjects. Neither he nor his kinsmen suffered so much as lowlier folk, because they claimed and received their tribute in fish and would be given such vegetable food as could be found. But the people died and illness swept over them; and even now, when a native tells you about the great *molu*, you feel the shadow of calamity in his voice and countenance.

Even the ordinary *molu*, insufficiency of food, is bad enough. Before it strikes a native directly in his organic needs, it affects his pride and makes him feel disappointed and dissatisfied with his work. On such occasions the Trobrianders fall back on the unripe crops, they start their thinning out (*basi*) very early, they eat "black and white" tubers alike, and incidentally they expose themselves to epidemics of dysentery, for unripe yams are apparently bad for the digestion. In such years they make bigger taro gardens on the swampy soil, they plant many more of the big yams (*kuvi*) in the *rayboag*, and they systematically exploit the village groves and the jungle—all of which entails much harder work and work which they do not like. After all, they are gardeners and not wild-fruit-gatherers. In such years also the taboo on new garden produce may become somewhat onerous to the garden magician, who has to wait for each crop till it is properly ripe and ceremoniously harvested. In addition to these ethnographic data, interesting light is thrown on the concepts, *molu* and *malia*, by the linguistic analysis of these terms in Part V (Div. V, §§ 3–5).

[1] See also Note 20 in App. II, Sec. 4.

2. THE PRELIMINARY GARNERING

The big ceremonial main harvest of taytu called *tayoyuwa* is set apart from all other forms of gathering crops. The *basi* is, as we may remember from Chapter IV (Sec. 3), not a harvest in any sense of the word, but merely a thinning out of tubers, which may or may not be eaten. Such tubers are never called taytu, but described by the term *bwanawa*; they may not be roasted but must be either boiled or baked in an earth oven. They are taboo to the magician and would not be eaten by people of rank, except under the pinch of hunger. The garnering of taro and *kuvi* (large yams) takes place earlier and is preceded by a special magical ceremony; it entails a magician's taboo of its own and an incantation ritually lifting this taboo.

The magic of this early harvest, as well as the activity itself, is called *isunapulo*. Bagido'u charms over a pearl shell in his house with the following words:—

FORMULA 25

I. "Full moon here! Full moon then, full moon here ever.
Round off in the north, round off here in the south.
Round off in the south, round off here in the north.
Taro round off.

II. "Taro round off, taro round off. . . .
The belly of my taro—taro round off, taro round off—
The base of my taro stalk—taro round off, taro round off—
The top of my taro—taro round off, taro round off—
The foundation of my taro tuber—taro round off, taro round off—
The leaves of my taro—taro round off, taro round off.

III. "They eat the taro.
They vomit taro.
They are disgusted with taro.
Their eyes are burning from taro surfeit.
Their mind is turned from taro.
The *tuvata'u* weed grows out of the taro rotting in the garden.
The *puputuma* weed fruits on it.
The belly of my garden becomes smooth like a trimming-board.
The belly of my garden becomes smooth like a pounding-board.
The holes in the belly of my garden are as the holes the mangrove mollusc bores in the mud.

"I shall go to the village laden with taytu."

After the shell has been charmed over with this formula, the magician wraps it up in dried banana leaf to preserve the magical virtue. The next morning he goes into the garden and with the shell cuts off the top of a taro plant in each *baleko*. The taro top from the main *leywota* he brings home and puts the leaves on one of the rafters of his house. This is an offering to the ancestral spirits, who, on this as on other occasions, are supposed to be hovering over Bagido'u's hearth.[1] While in the garden Bagido'u also takes out a large yam (*kuvi*), which later on is brought by one of his acolytes into the house and placed on the top shelf. This also pleases the ancestral spirits.

Next day a small split stick and some leaves of the *sasari* plant are placed on the *kamkokola* in each *baleko* as a sign of taboo on work.

On the third day after all this the men repair to the garden and each one pulls up a few taro plants and digs a few yams from his plot. These first-fruits of the gardens are brought to the village, and a part of them is displayed on the *baku*, the central place (Pls. 49, 50). Another part is put as an offering on the graves of the newly dead by their relatives.[2] In the old days the dead were buried at one end of the central place while the other end was used for ceremonies, distributions and, when there was no mourning, for dances. The placing of the taro and *kuvi* by each bereaved family at one end, while those who have suffered no recent deaths in the family displayed theirs at the other, was one concerted action. Since Government has ordered the burying of the dead outside the village, the first-fruits have to be displayed in two different places, on the *baku* and, by those recently bereaved, on the graves outside the village. Such publicly displayed food is not eaten by the owners, but given to some friend or relative, preferably to those relatives-in-law who normally receive harvest tribute. The kinsmen of the recently dead invariably share their offering with the widow or widower and those relatives by marriage of the deceased who have taken part in the grave-digging, burial and mortuary rites.[3] On such occasions a pig is often killed and distributed for a festive meal. It would then be displayed on the *baku* beside the first-fruits (Pl. 49; and cf. also Part I, Sec. 7). In some villages taro and *kuvi* are taken out on the same day. In others the first-fruit gathering, display and offering is spread over two consecutive days. This is the case in the district of Kulumata, where Plates 49 and 50 were taken.

A special act connected with this first harvest is the lifting of the

[1] Cf. also Note 8 in App. II, Sec. 4.

[2] Needless to say, taro and *kuvi* can only be harvested on those plots on which it is ripe. Cf. what has been said about keeping up to time and lagging behind in Chapter II (Sec. 5).

[3] See also Note 21, in App. II, Sec. 4.

taboo on the garden magician. Some of Bagido'u's relatives-in-law —that is, some of his wife's matrilineal kinsmen, as a rule her brother, or brothers if they are alive—offer him a bundle of taro plants and two or three yams. A few of these his wife cooks and brings into the house. He breaks off a piece of yam, cuts up some of the taro with a shell, and lays it in offering on the hearthstones. Addressing the ancestral spirits he utters this short incantation:—

FORMULA 26

> "Let us give up last year's food, O old men; let us eat the new food instead."

With this act, called *vakam kuvi, vakam uri*, literally "to make eat yams, to make eat taro", the taboo on new crops is lifted as regards taro and yams, and henceforth the magician must not eat any of these crops from the old gardens. The only hardship connected with this taboo consists, as we remember, in the fact that the other members of the community are allowed in years of scarcity to take not quite ripe yams or taro out of the ground even before the magic of the *isunapulo* has been performed. The magician has to wait till the crops are ripe, perform the magic, and then ceremonially break his taboo.

As regards the purely technical side of harvesting taro and yams, the former are merely pulled out of the ground, the earth shaken off from the root and the whole plant carried home. The digging of *kuvi*, large yams, on the other hand, is the most elaborate form of harvesting, since the roots are large and much ramified, or else very long (see Pls. 19, 63–65 and 68–71), and the earth has to be loosened with the digging-stick over a considerable area. The implements used are the digging-stick (*dayma*) to loosen the soil, the axe (*kema*), adze or trade-knife, to cut off shoots and rootlets, and the mother-of-pearl shell (*kayeki*) to scrape the soil and hair from the tubers. The *kayeki* is also used to cut off the taro top, which is preserved for replanting, from the edible roots.

3. THE RITUAL OF REAPING THE MAIN CROPS

To repeat, the reaping of taytu, the big harvest, *tayoyuwa*, is an activity which stands apart in native ceremonial and in native psychology. It is opened by the last series of magical acts, consisting really of two rites, and it ends the cycle of garden work. When the tubers are ripe for harvesting, that is, when the vine grows sere and begins to droop, the *towosi* performs a ceremony called *okwala*. This will open a short period of taboo lasting some two to four days, at the end of which another ceremony, that of *tum*, will inaugurate the actual harvesting. The magical aim of both these ceremonies is directed to the tubers underground. As the natives put it, "the *okwala* is made so that the taytu may truly, really grow, so that it may ripen." "The *tum* is made, so that the surface of the taytu may darken, so that the taytu may blacken all round."[1] It is clear therefore that both these ceremonies belong to the magic of growth, more precisely to the magic of growth as directed towards the maturing of the tubers.

On the other hand, they are also inaugural ceremonies in that they are an indispensable condition of, as well as the signal for, the opening of the main harvest. I should like once more to make it clear that the concept of a "magic of growth" is not based on a native definition, but is an ethnographic distinction drawn by myself on the basis of the sociological differentiation between one series of rites and another. Those rites that possess a definite inaugural function, which usher in a new type of human activity, are, in so far as they function inaugurally, part of an economic or sociological series which organises human work in the gardens. A few rites referring only to natural processes and carried out by the magician alone do not show this organising function at all, and these I have defined for purposes of convenient classification as the *magic of growth.*[2] It is quite clear, however, that most of the rites show both aspects. In their magical aim and in native psychology they refer to the crops and foster their growth. In their sociological function and to the anthropologist they organise and co-ordinate human activities. The *okwala* and the *tum* show both aspects very clearly.

The rite of *okwala* is simple: the *towosi*, helped by his acolytes and perhaps by some other men of the village, strews the ground with leaves which differ according to the village. In Omarakana the leaves of the *noku* plant, which as we know (see Sec. 1 of this chapter) is specially hardy, are used. The *leywota* (standard plots) are bestrewn pretty thickly, and on them *noku* leaves are also put in the forks of the *kamkokola* (vertical of the prism in the magical corner) and of the *kaybaba* (the sticks slanting from this) and they are tied to the *kavatam* (vine supports). The strong green of the *noku* leaves has a decorative effect against the yellowed and bronzed foliage of the *taytu* vine. Though the *noku* leaves are not previously medicated, they are thought to make the leaves of the taytu droop, as they should about the time of harvest if the roots are to be properly

[1] Cf. Texts 66 and 68, Part V, Div. X, §§ 4 and 6.

[2] Cf. also App. I, and *Argonauts*, Ch. XVII.

PLATE 55

MEDIUM-SIZED HARVEST ARBOUR

This picture was taken when the heap was being dismantled and the taytu placed in baskets to be taken to its destination. In No. 2, Fig. 8, the disposition of the several heaps before they were removed has been represented (Ch. V, Sec. 4)

PLATE 56

A LARGE HARVEST ARBOUR

A complete view of this arbour and heap can be found on the frontispiece. The details of the construction of the arbour and the manner in which the heap is arranged can here be seen. No. 3 of Fig. 8 shows the disposal of the main and lateral heaps (Ch. V, Sec. 4)

PLATE 57

DETAILS OF LARGE ARBOUR

The interior of the arbour and the lateral heaps can be well seen here. Also the arrangement of the taytu in the central heap (Ch. V, Sec. 4)

PLATE 58

HARVESTING PARTY ON THE WAY TO THE GARDEN

"They all set out from the village together", and repair to their host's *kalimomyo*. The party is seen traversing a last year's garden; the man on the right is carrying a couple of empty baskets on his shoulders, the others have poles in their hands, with baskets slung on them (Ch. V, Sec. 5)

ripe. The *noku* leaves are also a sign that the gardens are tabooed, and no work is to be done. The chief aim of the whole rite is, as we have seen, to give a final impetus to the ripening of the tubers.

In Omarakana, Bagido'u invariably sent his acolytes, usually his younger brothers and their friends, early to the garden to garnish it with the *noku*. Just before midday, he takes his *kaytukwa* (an ornamental staff as distinguished from the *kaylepa*, his magical wand), and follows them to the garden where, starting with the *leywota*, he charms over every *baleko* (plot) with this *okwala* spell:—

FORMULA 27

I. "Dolphin here now, dolphin here ever!
Dolphin here now, dolphin here ever!
Dolphin of the south-east, dolphin of the north-west.
Play on the south-east, play there on the north-west, the dolphin plays,
The dolphin plays!

II. "The dolphin plays!
About my *kaysalu*, my branching support, the dolphin plays.
About my *kaybudi*, my training-stick that leans, the dolphin plays.
About my *kamtuya*, my stem saved from the cutting, the dolphin plays.
About my *tula*, my partition-stick, the dolphin plays.
About my *yeye'i*, my small slender support, the dolphin plays.
About my *tamkwaluma*, my light yam pole, the dolphin plays.
About my *kavatam*, my strong yam pole, the dolphin plays.
About my *kayvaliluwa*, my great yam pole, the dolphin plays.
About my *tukulumwala*, my boundary line, the dolphin plays.
About my *karivisi*, my boundary triangle, the dolphin plays.
About my *kamkokola*, my magical prism, the dolphin plays,
About my *kaynutatala*, my uncharmed prisms, the dolphin plays.

III. "The belly of my garden leavens,
The belly of my garden rises,
The belly of my garden reclines,
The belly of my garden grows to the size of a bush-hen's nest,
The belly of my garden grows like an ant-hill,
The belly of my garden rises and is bowed down,
The belly of my garden rises like the iron-wood palm,
The belly of my garden lies down,
The belly of my garden swells,
The belly of my garden swells as with a child."

In this translation I have preferred to keep as near to the native text as is compatible with a readable version, in order to preserve, as far as may be, the characteristic flavour of a Trobriand spell.

The significance of the simile repeated in the key phrase, "the dolphin plays" might, however, have been clearer to the European reader had I developed the mystical associations between the undulating movements of the dolphin and the winding and weavings of the vine, which Bagido'u brought out in his commentary and which are present in the minds of all native hearers. Such a fuller rendering might run: "About my branching support, the taytu winds, the dolphin plays. About my training-stick that leans, the taytu winds, the dolphin plays," and so forth.

After the *towosi* has thus chanted over all the plots, the *okwala* ceremony is over. The gardens lie quiet, undisturbed by human hand or digging-stick, covered with the leaves strewn over them in sign of taboo, while under the influence of magic the tubers mature rapidly underground. After a few days—the length of time depends on the weather and the state of mind and body of the magician—the next ceremony, that of *tum*, is enacted. Once more Bagido'u in his hut chants the standard formula of Omarakana garden magic, the *vatuvi* spell, over an adze (*ligogu*), round the cutting edge of which some aromatic leaves of the *lileykoya* plant are fastened with a wrapping of toughened banana leaf gathered from the *wakaya*, a banana tree with a tall big trunk, bulging out towards the base. The *lileykoya* leaves are used because of their aromatic quality: "they will make the taytu smell sweet". The *wakaya* leaves are used here, as in most ceremonies, because of the size and shape of the trunk. Nowadays an ordinary adze with a steel blade is taken. In olden days the rite was performed over a stone blade of the working type (*kasivi*) mounted in an ordinary adze handle, not over a ceremonial blade.

This chanting of the *vatuvi* spell, important as it is in being the main act of the *tum* or harvesting inaugural ceremony, is even more remarkable in that on this occasion the *kaykapola*, the coconut torches which will be used for the first burning of the new gardens, are charmed side by side with the adze. Each torch has been prepared in the manner described in Chapter III (Sec. 2). Carefully wrapped up in several mats, so that the magical virtue is preserved, the torches are laid on the top shelf of the magician's house, where they will remain till they are lit four moons later at the *vakavayla'u*, the first big burning of the gardens. On the morrow the magician goes to the main standard plot of the garden, carrying the adze charmed on the previous day. On this occasion he is accompanied by a small crowd composed of men, women and boys. Small boys often run ahead to locate the plants to be harvested and show them to the magician and his acolytes (see Pl. 51, taken in Teyava)—

a service satisfactory to the boys' ambition rather than useful to the magician, since the plants on the *leywota* and other plots grow against the *kamkokola* and are easily distinguished. Arrived at the *leywota*, the magician cuts the stalk of the *kwanada* yam, ceremonially planted at the *kalimamata* ceremony (Ch. III, Sec. 1), and with a digging-stick breaks up the sod and takes out the roots. He then cuts the stalk of an ordinary taytu plant, preferably one which twines round the *kamkokola*. The lower part of the stalk flops down on the ground. The magician squats over it, pulls out a handful of weeds, places it over the stalk lying on the ground and presses the whole heap down with a stone. This act, the *tum* (pressing, weighing down), gives the name to the whole ceremony.[1] Thus the first vine has to be ceremonially harvested, and this inaugurates the main gathering of the staple crops, the *tayoyuwa*. In some villages the first tuber taken out has to be ritually split against the pole of the *kamkokola* (cf. Doc. VII).

And here again, as at the harvest of taro and *kuvi*, the magician must have his taboo ceremonially lifted. Some of his wife's matrilineal clansmen who live in his village bring him an offering of tubers which they have harvested. These are roasted by his wife and he again offers a portion to the ancestral spirits, addressing them with the words, "Let us give up last year's food, O old men, let us eat the new food instead," and then partakes himself of the new taytu. This ceremony is called *vakam taytu*, literally 'to make eat taytu'.

The taytu taken out of the soil is called *taytuva'u* or *kalava'u*, and it is of *kalava'u* that the magician partakes. The taytu put away in the storehouses is called *taytuwala*.

We have now arrived at the last turning-point, though not at the end of the harvesting cycle, and we are somewhere in the moons of *Utokakana* or more likely *Ilaybisila*, roughly April or May.

4. WORK AND PLEASURE AT HARVEST

Work now begins. It is invariably done in family groups. Sometimes a man with his wife and children will picnic in the garden for the whole day, sometimes they return to the village for a meal and a short siesta and begin work again in the afternoon. The stem of the vine has to be severed just above the ground, the roots taken out, cleaned and stacked. The plant used to be cut with an adze, holding the flexible vine stem against the *kavatam* pole and then striking; nowadays the trade-knife is used quite as often and the work is

[1] See also Note 22 in App. II, Sec. 4.

lighter. The soil is loosened with a digging-stick, the roots cleared and pulled out. The tubers are still clogged with the earth clinging to them and they have to be rubbed clean with leaves. To make this easier they are usually left for some time in the sun. They are still covered with stringy hair called *unu'unu*, which is also the name for body hair of men and women. The *unu'unu* has to be plucked or shaved off with a mussel shell (*kaniku*, cf. Pl. 53, where the woman in the middle is shaving off the *unu'unu* with a mussel shell, the man on the right is plucking it off with his fingers, and the girl on the left is using a trade-knife for cleaning the tuber). The shaving is done, as is the scraping and cleaning in general, by a movement of the hand away from and not towards the worker. The shell is held between the index and the thumb of the right hand, the other fingers work forward along the surface of the tuber and the shell follows, shaving off the hair. This part of the work, which has mainly, in fact I think exclusively, an aesthetic value, takes longer than most of the really utilitarian activities. Usually it is the man's duty to dig out the roots and the woman's to clean the tubers and carry them to an arbour (*kalimomyo*); but the division of labour is by no means strict.

With the loosening of the soil round the vine roots, the garden naturally becomes dismantled. The yam poles fall down or are taken out and thrown on the ground, garlands of foliage, already sere and brown, litter the soil. The *kamkokola*, no longer of any use, are often taken to pieces and, with the discarded poles of the *kamkokola* and the *kavatam*, a small arbour is now constructed (cf. Frontispiece, and Pls. 48 and 54–56). A number of stout poles are planted vertically, a few horizontals are lashed to the forks of the uprights, and a roofing of slender sticks is laid on the top. On and around this framework garlands of taytu vine are fastened. The number of these arbours that each man constructs, depends on the number of *urigubu* gifts for which he is responsible (see below, Ch. VI). For every *gugula* (*urigubu* heap) a *kalimomyo* is made.

Through a drapery of vine or under the green roof of coconut leaves the gathered crops can be inspected and admired by the natives walking through the harvested garden. Friends and acquaintances from the gardener's own village or from neighbouring communities visit him, sit down, watch the work and show a distinct and laudatory interest in the crops—since this is required by good form.

The crops are classified before they leave the field. The best yield of the whole garden is set aside for the *urigubu*, which later on will be ceremonially presented to another household or households. The *urigubu* heap is always constructed with some care in a conical mound in the centre of the arbour; the best tubers are put on the surface, a good shape is kept, and the heap is built so that it cannot collapse. When it is very large, it is necessary to encircle the base with a framework like a diminutive fence, called *lolewo*.

The good but small taytu, which will be used for the next year's seed and have, as we know, the special name *yagogu*, have the next place of honour. They are usually piled into a square or triangular heap in one of the corners. Damaged or inferior taytu, which will be eaten without having been ceremonially stored are called *unasu*, and are usually thrown in smaller heaps round the central or *urigubu* heap (cf. Frontispiece and Pl. 48). Besides these we sometimes find one or two heaps of good taytu destined for the gardener's own use or for the small gifts presented at harvest. There is an inclusive though rarely used term, *taytukulu*, for all taytu displayed round the central heap (cf. Ch. VI), that is, seed yams, inferior yams and the good crops not placed on the *urigubu*.

To conclude the enumeration of the various forms of taytu: such tubers as are found in the garden after the main crop has been removed from the arbours are called *ulumdala*. Some of the *ulumdala* are good tubers which have ripened very late in the season, others are inferior tubers or the tubers of forgotten or missed plants.

The disposal of the taytu in the *kalimomyo* will be made clearer by a comparison of the diagrams (Fig. 8) of the three typical arbours shown on Plates 54–57; Plates 56 and 57 give two aspects of the same *kalimomyo*. No. 1 on the diagram shows a very small *kalimomyo*, also reproduced on Plate 54, which was made on the garden plot of a commoner, Gumigawaya. It was about two by three metres, and the crops were destined for To'uluwa, the great chief of Omarakana. No. 2 shows the larger arbour photographed on Plate 55; and No. 3 the really big *kalimomyo* of Plates 56, 57 and Frontispiece.

It takes some time to build up the heaps of taytu in the arbours and the work proceeds side by side with the harvesting of the garden. Once in the arbour the crops remain there for a few days or even for a couple of weeks, and during this time the arbour continues to be a centre of social life. As we have heard, the people who visit the gardens are bound by custom to flatter the owner and should never criticise him. It would be a very serious breach of good manners, one might almost say here of morals, to make any allusion to a gardener's laziness, incompetence or bad luck. If you ask a native what would happen if a man were publicly criticised, taunted or laughed at on account of his poor crops, you receive the stereotyped answer, "He would climb a tree and commit *lo'u*

DIAGRAM OF GARDEN ARBOURS

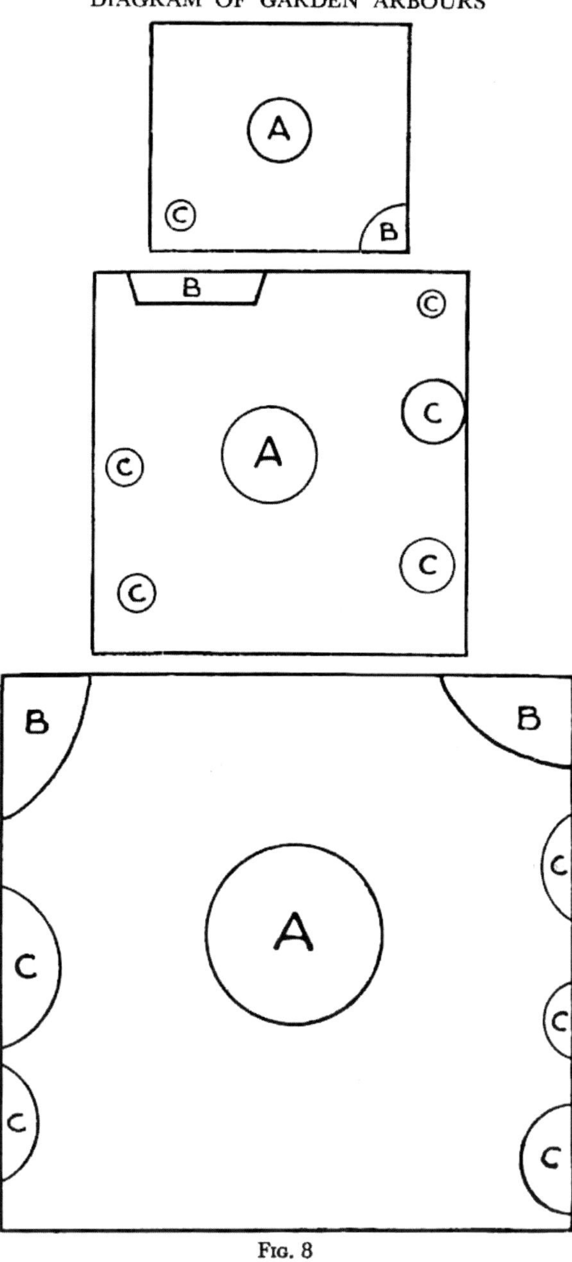

Fig. 8

(suicide by jumping)." Such an answer must not be taken literally, but it shows how deeply the natives feel about their gardens.

At the same time there is always an undercurrent of jealousies, envies and ill-will at this stage of gardening, and much gossip and calumny going on behind the backs of the people concerned. Such ill-natured criticism is of two kinds: some people are censured for the laziness and lack of interest which produces their bad gardens, and since good gardens are a virtue in themselves, a duty towards relatives-in-law and a duty towards the chief, the criticism is not without its venom. A man who acquires the reputation of a bad gardener sinks distinctly in public estimation and his standing in the community may be seriously impaired. On the other hand, no malicious tongues can affect the reputation of a really effective and industrious gardener.

In the second place a man may be attacked because his gardens are too good. He can then be accused of emulating his betters, of not giving a fair proportion of his crops to the chief and to his relatives by marriage, and thus of acting disloyally and pandering to his own vanity and greed. This criticism is also not without its point, and when it comes to death, which is always attributed to sorcery, many a man is believed to have died because of his good gardens. Among the *wabu*, the marks which are found on the corpse at the exhumation which takes place soon after the first burial and which reveal the cause of death, there are not a few which are believed to mean that death was due to an excessive ambition or success in gardening. Taro-shaped tumours or an inordinate craving for this vegetable shortly before death signify that the deceased had too splendid taro-gardens or did not pay sufficient tribute of this commodity to the chief. Bananas, coconuts, sugar-cane produce *mutatis mutandis* similar symptoms, while betel-nut stains the mouth of the corpse red. If the body is found foaming at the mouth, it shows that the man was too much addicted to opulent and ostentatious eating or to bragging about food.

Two men at least had been thus "killed" immediately before my arrival. One of them was Yogaru, the husband of Ibo'una, the grand-niece on the distaff side of To'uluwa the chief. Yogaru had beautiful big gardens, and since the chief and his wife's kinsmen had also to provide him with taytu as he was their sister's[1] husband, he accumulated too much food. He died, and though nobody would openly say so, I was informed by several men privately that he died of sorcery by the orders of To'uluwa. Again a man who had been

[1] The term *sister* must be taken here in the extended or "classificatory" sense—"Kinswoman on the distaff side."

married to the sister of Mitakata—not the Mitakata of Omarakana, but his kinsman, the chief of Gumilababa—and who at that time was a native policeman attached to the white magistrate was "killed by sorcery" because he was getting too influential through his wealth in garden produce. A whole string of names would be forthcoming if you questioned any native on this delicate subject. Many people have met their death through being too effective as gardeners and not dutiful enough in fulfilling their garden obligations. The sociological function of this we shall understand better after the principle of the *urigubu* has been explained.

The self-depreciation which is compulsory when a man mentions his *urigubu* gift in the presence of his betters is well illustrated by the following incident, which is also noted in Document IV. I asked Tovakakita, a commoner of Yourawotu, in the presence of a number of other people how much he received in *urigubu*. He replied (Text 93, Part V, Div. XII, § 34): "No, I have no yam-food; nobody makes gardens for me. They are all already engaged in producing food for To'uluwa." 'They' here refers, of course, to the maternal kinsmen of his wife. I knew him personally, and he was anxious to give me the information I wanted, but before doing so, found it necessary to declare that his harvest gift really amounted to nothing. It is an offence for a commoner to boast that he is rich in the presence of the chief or in his village, or in the presence of persons of higher rank than himself.

When an open quarrel breaks out in the gardens the natives may resort to the hostile and competitive exchange of food called *buritila'ulo*, which often leads to serious consequences, in old times even to war (see below, Sec. 6).

5. THE BRINGING IN OF THE CROPS

After the taytu has been harvested and gathered in the arbours on the majority of the plots and kept exposed to the satisfaction of their owners' pride, the carrying of the crops to the villages for which they are destined begins. This is another festivity. The gardener summons a party composed of his kinsfolk and of his wife's kinsmen, who are in duty bound to assist him (cf. Ch. VI, Sec. 2). It is usually the younger people who do the carrying, but many older ones take part by supervising the proceedings and escorting the party. The carriers receive a small payment called *vakapula* (his cooked food payment) or *vakapwasi* (their cooked food payment). This consists usually of fish or fruit, which is prepared and given to them in the village before they start. Also betel-nut, tobacco or

PLATE 59

HARVEST PARTY ON THE WAY TO THE VILLAGE

"They all keep together because they have to enter the village in a body" (Ch. V, Sec. 5). Note the distinction: those empty-handed are the recipient's party. The recipient's representative, second from the left, holds the tally leaf in his left hand

PLATE 60

LARGE PARTY OF CARRIERS ENTERING THE VILLAGE

"As the carriers approach the village . . . they break into a run, vigorously shouting a sort of litany" (Ch. V, Sec. 5)

PLATE 61

HARVEST PARTY RESTING

"After they have deposited their baskets on the ground (in the recipient's village) the carriers sit down for a short rest" (Ch. V, Sec. 5)

PLATE 62

CONSTRUCTION OF SMALL HEAP IN FRONT OF THE NEW OWNER'S STOREHOUSE

The structure of the heap can be seen; in front are the "male" baskets strung on the carrying poles (Ch. V, Sec. 5)

sugar-cane are distributed among them in the garden while they are loading the baskets. Such half-time refreshments are called *puwaya*. On very great occasions a pig may be killed, and this too would be called *puwaya*. It would usually be killed, carved, cooked and eaten after the donor's party returned to their village. I made notes of the actual payments received on various occasions and find that once two bunches of bananas, a bunch of betel-nuts and five sticks of tobacco were distributed among some fifteen men. On another occasion a taro pudding was made and given to them before they started work, and on yet another, the owner had some fish brought from a coastal village and this was roasted and distributed among a party of twenty. In the next chapter we shall follow one or two of such parties at a *kayasa* (Ch. VI, Sec. 3).

At times the carriers decorate themselves with leaves, scented herbs and facial paint. They all set out from the village together and repair to the garden; there they join their host in one of his *kalimomyo* and sit down to talk and chew betel-nut. Often the party will first scatter over the garden to assess the other crops once more, discuss them and extol the value of their host's *kalimomyo*. I remember once accompanying such a party from the village of Liluta. When we got to the garden the natives paused to collect fruit from a large tree—Malay apple, I think—and then, since it was hot, we went on to a grotto, the mythological grotto of Dokonikan, where we rested and ate the fruit. Parties from Omarakana will often stop at one of the water-holes to bathe, or go across the *rayboag* and bathe in the sea. This is all in the day's work, for the carrying of taytu from garden to village should have a playful and festive character.

At last they all forgather at the *kalimomyo*, where the owner and some of the elders of the village, who came earlier to make arrangements, have been waiting more or less patiently. An elderly man, representing not the owner of the garden but the man to whom the crops are to be carried, takes a circular measuring-basket (such as can be seen on Pls. 18 and 39, and the taytu is first put into this and then emptied into the carrying-baskets. For each standard measuring-basket taken from the main heap, one leaflet is torn off a large cycas leaf (cf. Pl. 65). Every tenth leaflet has only its tip torn off, so that one glance at such a leaf shows how many decimal figures and units have been measured out. This measuring is called *kalawa*. The leaf or leaves are then collected, carried to the recipient village and placed in front of the new taytu heap after this has been erected there. After the taytu has been counted and distributed among the carriers, the whole party proceed to the village of the man to whom the yield of the *baleko* is due. They all keep together as they have to enter the village in a body (Pl. 59). Usually a few men from the recipient village are of the party, but these are not supposed to carry anything.

As they approach the village, the carriers drop into single file or, if they are very numerous, into a column, and, accelerating their pace more and more, they enter at a run, vigorously shouting a sort of litany (Pl. 60). One man calls out a series of words and the rest answer in chorus with a strident high-pitched "*Wi-* . . .". This is the text of the litany:—

Precentor: *Osibwani—bwaniyoyo!*
Sidagu—dagurina!
Yakikoi!
Chorus: *Wi!*
Precentor: *Yakikoi!* (Repeated 4–6 times.)
Chorus: *Wi!*
Precentor: *Siyaloi!* (Repeated as above.)
Chorus: *Wi!*
Precentor: *Iyonoi!* (Repeated.)
Chorus: *Wi!*
Precentor: *Sayseloi!* (Repeated.)
Chorus: *Wi!*
Precentor: *Bom'goi!* (Repeated.)
Chorus: *Wi!*
Precentor: *Yonakoi!* (Repeated.)
Chorus: *Wi!*
Precentor: *Woekayoysa taytu!*
Chorus: *Yuhuhuhuhu . . . !*

I could not ascertain the meaning of the words used in this text. The choral answer, the *wi*, is uttered in an energetic explosive manner, and from a distance sounds like the cracking of some enormous whip. These harvesting screams are called *sawili*; they are meant to give an impressive and joyful tone to the bringing in of the crops and to attract the villagers' attention to the approaching harvesting party. When a really big competitive harvest is being celebrated, the natives will, in addition to the *sawili* cry, blow the trumpet shell and enter the village slowly, usually singing a song. But I shall be describing an actual instance of such a ceremonial harvest which I witnessed in Omarakana in 1918 (cf. Ch. VI, Sec. 3). Even at ordinary harvests there is a great deal of expectation and tension during the busy few days. Early in the morning, sitting in one of the villages, I would hear the *sawili* faintly in the distance. Then as the carriers drew nearer they would again break into the litany, and now the responses would ring out strong and rhythmic at equal intervals. Finally the *sawili* would sound vigorously close at hand, and immediately the carriers, panting and flushed under their dark skins, would rush into the village.

After they have deposited their baskets on the ground the carriers sit down for a short rest (Pl. 61). Then the men of the party proceed to pile up the taytu in front of the recipient's yam-house. They build a heap exactly the same size and shape as that which was made in the arbour. The construction of such heaps is clearly shown on Plates 62–65. A small heap is simply stacked on the ground (Pl. 62); bigger ones have small circular framework round them. You can also see, especially on Plate 63 but also on Plate 64, how the large yams are placed on the outside and the smaller ones in the middle, "topping" being the accepted procedure on such occasions. When there is a great deal of taytu it has to be brought in several instalments. The natives call the final bringing in *yaya'i*. On important occasions, such as a chief or headman's *kayasa*, the completion of a heap is accompanied by a small ceremony. When the last touches have been given—the arbour over the heap erected, one or two specially attractive yams attached to the framework—a shell trumpet is blown and the donor squats in front of the heap exhibiting the tally leaves. Then he approaches the recipient and ceremonially hands over the heap with some such words as: "Thy heap, O So-and-so. It is the *urugubu* gift of So-and-so." Here the name of the wife for whom the *urigubu* is given is mentioned. This constitutes the legal transfer. On ordinary occasions there is no blowing of a shell trumpet, no display of tally leaves; the transfer takes place automatically when the heap is finished.

During the busy days of an ordinary harvest a party has hardly finished the work before the *sawili* again resounds in the distance and another party comes in with the usual hurry—the men with their carrying poles (*katekewa*) on their arms, the women with the circular baskets (*peta*) on their heads.

The rapidity with which the parties follow one another and the number that arrive in a day depends on the size of the harvest and the importance of the village. In a small village of commoners, where there may be some six to ten heaps altogether in the central place and where the bringing in does not extend over more than two days, five parties on an average would enter daily. In a bigger village such as Yalumugwa, consisting of several component hamlets some of them presided over by a *gumguya'u* (chief of lesser rank), about a score of heaps would be brought in every day in an average year. As we can see from Document I, there were as many as thirty-two in one of the component hamlets alone in the harvest of 1915. I cannot say exactly how long it took to bring in this amount, but probably about three days, giving a daily average of ten parties. The other hamlets taken together would receive about the same amount, so that about twenty parties would be entering the village as a whole. In Omarakana during the big competitive harvest of 1918 (Doc. II), there were seventy-six heaps in the village, some of which were of enormous dimensions, and the bringing in lasted about ten days. But then it took a very long time to erect such huge heaps. The central place was covered with them and crowded with men and women. The heaps are always placed immediately in front of the yam-house of the recipient of the crops.

Immediately after its erection each heap is covered with coconut leaves, so as not to be exposed to the sun. If the heaps are very large, or if, for some special occasion, they are to remain on exhibition for some time before being stored away in the yam-house, arbours, similar to those we have seen in the gardens, are erected over them (cf. Pls. 64 and 65).

For after a few days, in the case of a small and insignificant village, or after a week or two, in the case of an important capital such as Omarakana, the donors and their assistants, in fact most of the original carrying parties, will return to the recipient village, and the same hands that constructed the *urigubu* heap in the garden arbour, took it to pieces and reconstructed it in the village, will now store away the yams in the *bwayma* (storehouse) of their final owner. But to this proceeding we shall have to return in Chapter VII.

The harvesting of the taytu is to a certain extent the final event in the cycle of gardening. If we were to return, however, to the now dismantled, dishevelled and untidy-looking garden site, we would not find it by any means completely deserted and useless. In the first place, as has been already stated, the "gleaning" of the *ulumdala*, the tubers left over, has to be carried out. And this at times represents quite a substantial contribution to the household economy, and may require some considerable amount of work. Besides this, however, there is at least one more very important crop which will be harvested later—and these are the sweet potatoes (*simsimwaya*). Some of these were planted with the earlier crops, but they take a longer time to grow, and only mature a month or two after the taytu. I believe also that at the main harvest some more seeds are put in. The sweet potato is a hardy crop. It is not easily choked off by weeds, and thus can grow in an old garden site without the need of weeding or any other agricultural cares. A garden on which the old main crops have been harvested, on which there are only some tubers to be gleaned and sweet potatoes growing, is called *ligabe* (Pls. 58 and 59). No work is done on the *ligabe* except for an occasional visit to take out such sweet potatoes as may be required. The weeds soon begin to grow—some of them develop into the saplings of the low jungle, and two or three years after the harvest it is very difficult to distinguish a *ligabe* from the *odila*.

At the present time sweet potatoes are important, I think, from the economic point of view. I was told by the natives, however, that they are an agricultural innovation, brought to the islands by the Europeans. They play a remarkably insignificant part in the ceremonial side of native economics. They never appear at public distributions, are never given in any of the ritual exchanges of presents. They are never as much as referred to in magic or in native comments on magic. When there is plenty in the village, they are mainly used as fodder for pigs. For the natives do not like them. In times of scarcity, however, they form an important food reserve.

A few banana trees are planted from time to time in the *ligabe*, i.e. when the garden has been made on moist and really fertile soil (cf. Ch. X, Sec. 5). In this case, the plants would be kept clear of weeds for the two or three years necessary for the maturing of the bunch.

6. *BURITILA'ULO*—THE COMPETITIVE CONTEST IN HARVESTED WEALTH

The under-current of malice, suspicion and envy which accompanies the display of food and the show of praise and admiration, may lead to bitter personal animosity, which in the Trobriands usually ends in attempts to kill by witchcraft. When this happens between people belonging to two different communities, the quarrel may be taken up by their fellow-kinsmen and fellow-villagers. Then there comes into play the contest of wealth, the comparison of the respective harvest yields, which is called *buritila'ulo.*[1]

The *buritila'ulo* is one of the most characteristic examples of the double-edged nature of gift among the Trobrianders. On the one hand it is a present given with the grandiloquent yet calculating generosity which the natives affect on such an occasion, and received with the vigilant and grudging scrutiny which is always ready to perceive meanness. Thus marked as a gift, the *buritila'ulo* is a mutual pitting of economic resources in which each of the opposing sides means to score, to show that it is the richer, the superior and the more powerful. For the present will have to be returned immediately in exactly the same quantity and quality. If the repayment

[1] See also Note 23 in App. II, Sec. 4.

is too small, its inadequacy will be thrown into the face of the givers. If the return be too generous, this will be taken as an insult to the recipients.

The *buritila'ulo* occurs only at harvest and only in connexion with quarrels about food. Were a man, contrary to the established code of manners, to criticise the quality of another man's harvest yield, the latter would naturally reply with invective. It is characteristic of a Trobriander, if told that he has got bad yams, to reply almost automatically: "It is you who have bad yams." When such an argument starts, it is quite obvious where it will lead. The insult: *gala kam* 'no food thine', 'thou hast no food', is inevitable sooner or later, and this leads to a stream of mutual abuse of that type which cannot pass without serious consequences. Usually the quarrel degenerates into an immediate fight on the spot. This, however, may not go very far if there are persons of authority, a headman or a man of rank, present. These might intervene and then the matter would be sooner or later put to the arbitrament of competitive food exchange, *buritila'ulo*.

As an illustration I can quote what took place in June, 1918. A commoner of the village of Kabwaku, by name Kalaviya Kalasia, quarrelled in the garden with Mweyoyu, a commoner of Wakayse. Both villages, which lie near to each other, belong to the district of Tilataula, which is ruled by the chief of Kabwaku, Moliasi. War could never occur between the two communities since they owe allegiance to the same chief, but quarrels are frequent, and small fights (*pulukuvalu*) are not unknown. The two men quarrelled as usual about the quantity and quality of their harvest products. The Kabwaku man in the course of the quarrel destroyed the garden arbour of Mweyoyu. A fight took place on the spot, but was stopped. Later on, however, the headman of Wakayse went to the chief of Kabwaku and remonstrated with him about the destruction of the garden. The chief of Kabwaku, Moliasi, backed up his subject, and said something to the effect that since the people of Wakayse had no decent gardens and could not give proper tribute, they should not boast about their food. In reply to this challenge the headman of Wakayse offered to present the village of Kabwaku with all the yams produced by the people of his village. This was the announcement of the *buritila'ulo*, which was accepted by Moliasi, the chief of Kabwaku, and immediately put into action.

Now the principles underlying a *buritila'ulo* are in brief the following: Community A, which is either worsted in the quarrel as Wakayse had been, or which received an injury, or which is first severely taunted, issues the challenge. This community then has to muster all the yams possible, for the *buritila'ulo* is invariably carried out in terms of *kuvi*, large yams, and never in taytu. All the yams which community A can muster will be accumulated, carried over to community B, displayed there, ceremonially given, and then community B will make a return gift. If the return is made in exactly the same quantity, all comes to a happy ending; otherwise, as said already, further trouble will arise.

The following text, extracted from my conversation with several informants and noted down while the *buritila'ulo* was in progress, documents some of the points laid down (Text 88, Part V, Div. XII, § 24): "(i) A quarrel might arise in the garden. Our companion would say: 'You, is there any food with you? You have no food!' (ii) He says: 'Come now. Let us make *buritila'ulo* (competitive food display).' (iii) (The narrative passes here to the concrete events of the present.) It was the men of Kabwaku who started the dispute, saying: 'There is no food with you.' (iv) The people of Wakayse then spoke: 'Wait a little. Let us bring the yams for the display. Or will you bring first?' (v) Then spoke the people of Kabwaku: 'Good!', upon which the people of Wakayse at once fetched yams. (vi) Yesterday the people of Wakayse brought the yams. To-day the people of Kabwaku repaid to the full, and over and above they brought the excess contribution and presented it to the people of Wakayse. (vii) In olden days, whenever the excess payment was given, the people of Wakayse would at once be angered and war would arise."

Two or three points of interest emerge from this text. The initiative in the quarrel was taken by the people of Kabwaku, hence the challenge to the *buritila'ulo* came from the people of Wakayse who were insulted. The assertion of the last verse that war would arise, is so far correct that probably in old days there would have been some fighting between Kabwaku and Wakayse. But not real war, only the *pulukuvalu*; that is, an encounter between two normally friendly villages in which blood might be spilled, but usually no deaths occurred. This text gives a good idea of the type of information which a native will give spontaneously. Had I not been on the spot, observed the details of the quarrel and of the transaction, and elicited concrete facts by direct questions, it would have taken a very long time for me to get to the real inwardness of this custom.

Returning to the facts observed, let us start with the preparations. All the large yams have to be taken out of the *bwayma* (storehouses) and displayed in heaps in the village. The long yams called *kwibanena* are then sandwiched between two sticks and ornamented with pandanus streamers and dabs of white paint. These contraptions, which the natives call *kaydavi*, can be seen laid on top of the crate in Plates 68, 69 and 70 and less distinctly in the foreground of Plate 71. The two largest of the yams, each from one of the competing villages, appear on Plate 70. Then as much sugar-cane and betel-nut as possible is accumulated. Only these two products may be used besides the large yams. The contributions in yams have to be made exclusively from the villagers' own produce. No one from the outside is allowed to contribute to the joint store. There is no *dodige bwala*, that is helping out by relatives-in-law, as at ordinary displays or distributions. On the other hand every man has to give all the yams he possesses. Each man carefully counts his contribution and keeps a rough tally of the size of each tuber. The long yams are measured by sticks of equivalent length, one stick for each yam. The round yams are measured by means of string, knots being made to indicate their size. Each owner keeps this private tally of yams so that he can claim back his share and neither more nor less from the common pool of the return gift.

Then the natives have to make an approximate computation of the cubic capacity necessary to contain their accumulated yams. They go to the bush and collect a few stout poles and some sticks. With these they roughly construct a crate (called *liku*) and fill it with the yams to test its capacity. Then they take it to pieces again, and the whole village (A) starts on the work of transporting the yams and the component parts of the crate to the challenged village (B). Here the yams are deposited on the *baku* while the crate is reconstructed, this time more solidly, because it will have to be carried bodily by the men of village B back to village A.

When the crate is finished, each man places his contribution into it. The long tubers, each tied between two sticks, as well as pieces of sugar-cane and bunches of betel-nut are put on top. The filled and decorated *liku*, standing ready as the final challenge, can be seen on Plates 68 and 69.[1] Sometimes, the natives told me, a number of *pwata'i*, prism-shaped receptacles, are also erected, and filled with smaller *kuvi* and topped with betel-nut and sugar-cane. When there is a great wealth of produce, vertical frames (*lalogwa*) are set up, decorated with yams, bananas and betel-nut.

Then comes the actual transaction. First of all the exact measurements of the crate are taken. Community B will have to return the same crate, in no way changed, to community A, and fill it exactly to the same height. In order to ensure against any fraud

[1] These photographs were not taken in community B of our theoretical exposition (these plates being unfortunately under-exposed) but represented the return gift of the Kabwaku men, displayed in Wakayse.

PLATE 63

CONSTRUCTION OF LARGE HEAP

First layer placed within the circular framework, showing small tubers in the middle, and large tubers, some of which are decorated with white markings and pandanus streamers, on the outside (Ch. V, Sec. 5)

PLATE 64

VERY LARGE HEAP IN CONSTRUCTION

Several heaps are already finished: on the left smaller ones, covered with coconut leaves, on the right one under a small arbour. Within the miniature fence the building of a heap has just been started. The large tubers in the foreground are decorated with markings and pandanus streamers. Photographed at chief's *kayasa* (Ch. VI, Sec. 3 and Ch. V, Sec. 5)

PLATE 65

LARGE DECORATED HEAP

This photograph was taken at the chief's *kayasa* (Ch. VI, Sec. 3), at the ceremonial moment when the shell trumpet was blown and the tally leaves exhibited. The white markings on the large tubers, and the large yam hanging from the framework are characteristic decorations (Ch. V, Sec. 5) and have a clear standard of measurement, a number of sticks are cut on which the length, width and height of the *liku* are recorded. Then the size and quality of the most important of the gifts, the long yams, is measured. Community B have by this time prepared their *kaydavi*, that is their yams tied between sticks, and for each *kaydavi* brought to them by A, they check off a corresponding *kaydavi*, which, however, they do not yet present (see Pl. 70). Then the number of bunches of betel-nut is ascertained and their size roughly estimated and recorded. The contents of the *liku* are now distributed; each man in community B receives his share in exchange for an exactly corresponding contribution to the return gift from his own storehouse. Next day the *liku* is transported bodily to village A where it was first built. Some twenty men were necessary to lift and transport it on the occasion when I was present. The rest of the villagers, men, women and children, were busy carrying the yams. Arrived at village A, the proceedings of the previous day are exactly repeated, only now the transfer is from community B to community A.

And now comes the dramatic moment. Community B have been straining all their resources not only to repay the full quantity of yams but to provide a surplus. The strict return measure is called *kalamelu*, which might perhaps be translated 'its equivalent', 'the equivalent of the gift received'. If they can offer an extra quantity, this will be put on the ground and declared to be *kalamata* 'its eye'. The word 'eye' is here used in the figurative sense of something which is ahead of, which overtakes, goes beyond.

Now such a surplus gift would not be offered in a very friendly spirit. Community B would boast of having given it. They would also immediately clamour for a repayment of it. But since community A have strained all their resources for their original gift, they cannot repay. They would have recourse to argument, they would say that the surplus was not a real surplus but due to the fact that the *kalamelu* was not honestly and fully meted out. A quarrel will break out again and another fight arise from the *buritila'ulo*.

Since, however, community B, by supposition the richer one, would also be stronger, the people of A would obviously be beaten on every point. But two communities practising a *buritila'ulo* against each other are not essentially hostile, so the fight would probably have no very serious consequences. I was told, however, that in old days, especially when the *buritila'ulo* was not between two adjoining communities normally friendly, but between two communities who, though not on terms of recurrent warfare might yet fight if occasion arose, a serious regulated combat might follow.

But I will exemplify one or two points in this general account by what occurred between Wakayse, which corresponds here to community A, and Kabwaku, corresponding to community B. Here the Wakayse men were obviously the weaker; they had been insulted, they had been told that they had no food, and they issued the first challenge. On Plates 66 and 67 we see them constructing the *liku* on the central place of Kabwaku. Feeling ran high. In spite of the fact that Moliasi was the acknowledged chief of the whole district, there were several quarrels between him and Kulubwaga, the head-man of Wakayse and between their people. Both of them can be seen on Plate 70. Moliasi is seen standing in the middle, a turban on his head; Kulubwaga, decorated with a belt of cowrie shells, is on the right behind the enormous yam about five feet long, which is held by two of his henchmen.

Both headmen from time to time made impassioned speeches, ostensibly addressing their own subjects, but really directing their adverse comments at the other side. Moliasi, for instance, while the Wakayse men were rushing into Kabwaku with poles and yams and erecting the *liku* rather effectively and quickly, commented on the slowness with which everything was being done. Remarking on the betel-nut which was offered, he directly taunted the Wakayse people with having no betel-nut of their own and having to get it from other villages (Text 89, Part V, Div. XII, § 26).

"Why do you bring betel-nut from Kaybola, from Kwaybwaga? Take back this betel-nut of other villages, of Kwaybwaga and Kaybola. I do not want it. Bring us your own betel-nut from Wakayse."

These insults were not replied to directly because the native is always subdued in another man's village, but on the following day in Wakayse I overheard a great many insulting remarks levelled at the Kabwaku men.

For when, breathless from their hard work of carrying the crate and the yams, these were moving as in a trance, excited and absorbed in their task, they were taunted with slowness, with the *prima facie* inadequacy of their return gift and the distortion of the *liku*'s shape in process of carrying.

Discussing what happened that day with some Omarakana men who had been present there with me, an informant thus reproduced the boasting of the Kabwaku men (Text 90, Part V, Div. XII, § 28):—

"Some of them said: 'Let us throw away this crate. Let us take a new one. Let us exceed the people of Wakayse.'"

Such words were obviously boasting because it would be very

incorrect in a *buritila'ulo* to construct a larger crate. In fact, this is never done. Any excess should be presented by laying down the *kalamata* on the ground beside the crate.

To this the Wakayse men retorted that the *liku* had been too small from the outset, and that they really wanted to build a much bigger one to accommodate their gift.

The comparison of the long yams was by no means as peaceful and pleasant as Plate 70 would suggest, because there the performers paused for a moment to pose for the camera. On the whole it was riotous, full of quarrelling and threatening.

However, nothing serious occurred, and the people of Kabwaku refrained from adding the insulting surplus to their return gift. The natives are now afraid of fighting and try to avoid such situations as would almost inevitably tempt them to use spears and throwing-sticks. I was told that Moliasi, as a matter of fact, went as far as to ask the Assistant Resident Magistrate whether he could give a *kalamata* (surplus gift). Since at that time the official was entirely ignorant of native custom, he gave the permission not knowing where it might lead. But Moliasi when it came to the point was afraid of the possible consequences.

I measured the *liku* which functioned in this transaction. It was 4·6 metres long, 1·85 wide and 1·7 high. The longest *kaydavi* was about 2·4 metres, of which the tuber itself measured about 1·8. Towards the end of the performance, when all the measuring, discussion and quarrelling was over, a small distribution of food was made for the onlookers from other villages. It consisted of a piece of sugar-cane and some betel-nut, put in small heaps on the ground and assigned to the men of the neighbouring villages. Such a distribution is called *kokouyo*.

Even about the reapportionment of yams within the village there is always some dissatisfaction and quarrelling, but on the whole this arises more from personal ambition and vanity than from actual greed, and springs from a desire to prove that one has given more than one receives.

CHAPTER VI

THE CUSTOMARY LAW OF HARVEST GIFTS

So far we have merely watched throngs of people forgathering, carrying the fruits of their labour, not to their own village but to quite a different one, displaying them in this latter, and offering them on the part of someone to someone else. We have vaguely divined that there is some system in this apparent confusion, that there is a giver and a recipient; that among the helpers and carriers, the measurers and the admirers, there obtain certain sociological relations as between one group and another. But what are these relations, what forces move these people, what incentives make them toil and pant and find satisfaction in their work? Above all, what motive can make one man offer the best part of his harvest to another? These questions still remain unanswered. And indeed the answer is neither simple nor obvious. Traditional decrees are framed into a complex system of economic, legal and sociological rules, which at first appear to us almost perverse in their complexity and obliqueness. So difficult are they to grasp that most of the long-time white residents in the Trobriands, some of whom are married to native women and benefit under the Trobriand harvesting system, are incapable of understanding, still less of explaining them. It will therefore be necessary for us to go into these rules carefully and minutely. It will also be necessary to see whether we can reduce them to the mainsprings of human action, to hunger, love and vanity.

1. DUTIES OF FILLING THE STOREHOUSE

The economic side is really startling and paradoxical. For, baldly speaking, it comes to this: each man works for another and each man receives from another a large share of his household maintenance. Perhaps it would be more correct to say that no household is fully maintained by its head—the husband and father, the man whom we would deem the natural bread-winner—but that someone else, who hardly ever partakes of a meal in the house and is as a rule rarely a guest there, also substantially contributes to it. At the same time, the household thus supported is working and watching the crops in order to offer the best tubers harvested to a third somebody who equally does not live in the same village and who is hardly ever seen at the house.

The main sociological principle in this transaction between the harvest giver and the recipient can be stated in a few words. I, a male Trobriander, have always to work for my sister's household, supplying it with the best taytu I myself produce, in a quantity sufficient to form roughly 50 per cent of its total consumption for the year.[1] The proportion of each harvest allotted by the wife's brother to the sister's husband is called *urigubu.* My wife's brother, on the other hand, has to work for my household. The rule, though it does not appear reasonable to us, is simple enough. But in actual practice it is complicated almost indefinitely by corollary rules and by the realities of human life.

I, the giver, may have one sister and several brothers. In this case my younger brothers have to help me though I remain the titular donor. Or I may have several sisters, some of them older and some younger. The older ones may be married and may have children. In that case some of my (matrilineal) nephews will work with me and help me, not only with the maintenance of their mother's household but also with that of her younger married sisters. If I am the eldest of the family and have several younger sisters, I may have to work very hard and distribute my produce so that every one of my sisters has some provision. It is obvious that in the Trobriands the ratio of brothers to sisters is of importance; the more brothers the merrier for each sister, the more sisters the less endowment for them.

Another complication is introduced into the system by the fact that the recipient, after he has been given his harvest offering, has to redistribute some of it among his kinsfolk and nearest relatives. So that the taytu must be received first as an *urigubu* gift and then some five to twenty baskets of it are handed on under the name of *kovisi.*[2] Not only that, but the giver after he has presented the best part of his harvest, the *urigubu* allotment, to his sister or other maternal kinswoman, has still a number of minor gifts to distribute, which are called *taytupeta.*

Even more complicated is the custom called *likula bwayma*, "the untying of the yam-house." When a man has his storehouse very fully loaded by his wife's kinsmen, his sister and her husband may decide to claim some of it, for themselves and for their household. In this case the husband of the sister would give her a "valuable", say an axe-blade, telling her: "Take this valuable and untie your brother's yam-house." When this is given, the owner of the very full yam-

[1] See also Note 24 in App. II, Sec. 4.

[2] This word carries with it an automatic complication in that it is used also to denote a gift economically analogous, almost identical with *urigubu*, but sociologically different (see below).

house has to unload one of the compartments, *kabisitala*, and give the contents over to his sister's husband. Concerning this custom I was given the following statement by a native of Vakuta and resident in Sinaketa; a missionary teacher, known to me under the papal name of Leo, who was very well up in native custom—better, I am afraid, than in ecclesiastical matters (Text 92, Part V, Div. XII, § 32).

"(i) Already my yam-house has been filled to overflowing by my relatives-in-law (my wife's kinsmen). (ii) Later on, my sister would come and see my *bwayma* filled to overflowing. (iii) Then she would go to her husband and speak to him about it, and he would say: (iv) 'Take a valuable and untie your brother's yam-house.' (v) This woman would then bring the valuable and give it to us; whereupon we would empty one compartment of our yam-house. (vi) Were she to give us two valuables, we would empty a second compartment. (vii) We call this 'the untying of the yam-house' or 'the snapping of the rope'—the rope would be snapped, they would untie the yam-house."

This—one of the best definition statements I ever obtained—clearly indicates the conditions under which the custom would become operative.[1] Also the economics of the transaction are given: for each valuable one compartment of the yam-house would be emptied. The husband's remark is a semi-legal, semi-ceremonial phrase, which would probably always be uttered on such occasions. The expressions 'untying of the yam-house', 'snapping of the rope' are figurative. No rope enters into the construction of the yam-house, nor is there any 'tying' (cf. Ch. VIII). They merely denote the opening of the storehouse.

A subtle source of confusion for the European arises from the fact that it is not easy to state simply and exactly whether the gift at harvest is offered to the male head of the household or to his wife, who is usually the donor's sister or other matrilineal kinswoman. In name and in legal principle, the gift is handed over to the husband: he is the owner of the yam-house which is going to be filled; the gift is transported to his village community, and he becomes the owner of the crops. It is also he who has at intervals to repay the harvest gifts by occasional presents of valuables called *youlo* and *takola*. But in reality the *urigubu* is only handed over formally to the husband in order to finance the household of the wife. After her death it is discontinued.[2] It is because of her, for her and for her children's maintenance that the annual gift is given.

[1] What I mean by definition texts will be explained in Part IV, Div. IV.

[2] See Note 25 in App. II, Sec. 4.

Not less puzzling to the European is the rôle of the children or more correctly of the sons in this transaction. As a rule you would find that the sons fill their father's yam-house, but these sons after reaching maturity have, in obedience to tribal law, to join their maternal uncle in his village. After this they will not fill their father's storehouse directly, but give the taytu to their maternal uncle, who then conveys it together with his own produce to his sister's husband. Nominally they work for their maternal uncle. To a European enquirer, this might appear a roundabout and incorrect way of putting it. To the Trobriander it describes exactly what happens in sociological reality.

The result is that often a man will tell you one day that he works a certain garden plot for his father and next day that he works it for his maternal uncle. Both statements are correct, but each of them stresses one aspect of the transaction and it takes some time before you unravel the actual relations.

As if all this were not complicated enough, another element enters to introduce an additional confusion into the terminology and into the facts of the case: while the harvest gift received by a poor man is small and simple, in the case of a chief it is large and the sources of it are very complex on the sociological and economic side. Terminologically the ethnographer becomes easily confused, in that the word *urigubu* is often used to designate the chief's tribute as well as his harvest gifts. As a matter of fact most of the chief's tribute, though not all, is given as a glorified contribution from wife's brother to sister's husband, only, in the case of a woman married to a chief, her whole sub-clan would labour for her and not merely one of her male relatives. Hence a chief's wife brings her husband a much bigger quantity of food than she would have received had she been married to a commoner (about five times as much, I estimated), her gift consists of finer taytu and is offered with greater display. Now the chief has the privilege of polygamy. Before the decline of their power, the chiefs profited by this to the extent of some fourscore wives, and To'uluwa had a couple of dozen at the beginning of his reign. Even in 1918, when he had only a dozen, his *urigubu* was, on a rough estimate, sixty times that received by a commoner and, at the beginning of his reign, when he had twice as many wives, it must have been at least double this. In fact, I think it was much more than that, since at that time the power of the chief was unquestionably greater (cf. Doc. III). In the old days when a paramount chief had some sixty wives or more he must have received something like four hundred times as much as commoners. The brother of a chief's wife and the maternal kinsmen of this man would also have to render a much wider range of services than is normally due from brother to sister's husband.

Thus the chief's *urigubu* became a tribute levied on a number of village communities, but always levied in virtue of his position as a glorified brother-in-law of the whole community. Some of the chief's *urigubu*, however, although it went by this name, was not given by people related to him by marriage, but by vassals, who paid it on account of their residence in his village, and still other of the tribute he received could not be brought within the term *urigubu*, but was named *pokala* and *tabubula*.

So far, by the word *chief* I have meant, primarily, the paramount chief of Omarakana, but between his *urigubu* and the amount received by an ordinary tribesman or by a poor commoner with some fifty baskets a year, there was a whole range of gradations. There were the sub-chiefs, *gumguya'u*, with about half-a-dozen wives each; and of lower rank but even greater power was the *toliwaga* of Kabwaku.[1] Next came the lesser chiefs with a couple of wives each, and after them the headmen of villages, who were better endowed than the ordinary citizen, though as a rule monogamous. This again introduces some diversity into the picture.[2]

One more complication has to be mentioned here: the harvest gift presented by relatives-in-law, the *urigubu*, cannot be received by someone whose wife has died without leaving him male progeny.[3] If the wife has left him sons, these, being of her kin and at the same time his sons, give him an annual harvest gift which to the native is regular *urigubu*. By the anthropologist this type of *urigubu* has to be distinguished from the ordinary gift which is given for the wife to the husband.

On the other hand, a mature man, who for one reason or another has not remarried, has yet to have his storehouse filled by someone else. For a grown-up man of rank and position to fill his own *bwayma* is a disgrace. A man thus placed will receive an annual present at harvest—not any more from his deceased wife's brother—but from his own matrilineal relatives, in the first place from his younger brothers or maternal nephews. Such a gift might be called *kovisi*, but this term is very rarely used in this sense (see above). Such a gift

[1] Cf. Part I, Sec. 9; also Ch. XII, Sec. 3.

[2] Cf. Docs. I–IV which give examples of actual transactions, and see also references given in the previous footnote.

[3] Comparing Doc. II with this, it will be seen that the general rule is always stretched in the case of people of rank. The paramount chief receives one or two harvest gifts (notably numbers 2, 36 and 37) which are attributed to dead wives although they are actually presented to a living wife and placed in the latter's storehouse. See also Note 25 in App. II, Sec. 4.

PLATE 66

LIKU IN PROCESS OF CONSTRUCTION

The component parts of the crate are transported to the village challenged, where it is constructed. The yams to be loaded into the finished crate can be seen left centre. Some men are at work on the crate, and members of the two communities are sitting round watching the proceedings (Ch. V, Sec. 6)

PLATE 67

CLOSE-UP OF LIKU SEEN ON PLATE 66

The men are seen lashing the poles together (Ch. V, Sec. 6)

PLATE 68

THE BURITILA'ULO CHALLENGE

The filled *liku* can here be seen with the largest yams displayed on the outside, the long yams tied between sticks laid on top, bunches of bananas to the left and of betel nut to the right (Ch. V, Sec. 6)

PLATE 70

COMPETING VILLAGES DISPLAYING THEIR LARGEST YAMS

"Moliasi is seen standing in the centre, a turban on his head; Kulubwaga, decorated with a belt of cowrie shells, is on the right behind the enormous yam, about five feet long, held by two of his henchmen" (Ch. V, Sec. 6) would be more usually described by the phrase: such and such a man is filling my storehouse; thus *bwadagu idodige ulu bwayma*, "my younger brother fills my storehouse."[1]

Again if a woman is married to a chief and has no kinsmen near enough or important and efficient enough to fill her yam-house, her father may assist in this. Examples of this will be found in Document II, where a number of cases are quoted, especially of wives of the paramount chief, where the yam-house is filled by the wife's father.

Another complication of the system is that the harvest gift, the *dodige bwayma*, the filling of the storehouse, is only a part of the duties of relatives-in-law, but all these duties are called by the generic term, *urigubu*. When a man brings taytu or yams to his sister's husband before one of the big ceremonial distributions (*sagali*) it is called *dodige bwala*, the filling of the house. And this again is part of the *urigubu* duties, though to call the gift itself *urigubu* would not conform to native usage.

Again we have mentioned that it is the custom for the husband to present his wife's brother with a *youlo*, gift of a pig or some object of value, in acknowledgment of the *urigubu*. In repayment the wife's brother will offer some twenty baskets of taytu as a *vewoulo*; this, in legal theory, is part of the *urigubu* and will be admitted as such, though it is not usually styled *urigubu*.

Matters are not made simpler by the fact—and it is a fundamental fact of native harvesting—that each man counts on producing a really substantial part of his taytu for himself, that is for his own household. This part, consumed by the same household that produces it, is to a certain extent minimised or passed over in silence and stowed away *sub rosa*. It is the less showy and less honourable result of the harvest; no one will boast of it and, as a rule, the *taytumwala*, as this part is called, the own taytu, is stored away in covered houses or in covered parts of an open *bwayma*.

The *taytumwala*, the self-produced, self-consumed share of the harvest, consists of *yagogu*, next year's seed which will be "consumed" by planting; the *unasu*, inferior taytu; the *ulumdala*, the gleanings after the harvest is taken in; and also of a reasonable share of good taytu, as much as can be decently kept after the *urigubu* duties have been honourably satisfied.

It will be perhaps well to add that all the produce which we have described above, in our account of gardening, as the secondary crops, go invariably to a man's own household—that is, the large yams (*kuvi*) which can be stored, taro, sugar-cane, peas and pumpkins.

[1] For the sake of brevity I am, however, using the term *kovisi* in the sociological analysis in Documents I–IV.

It is not called *taytumwala*, but economically it is equivalent to that part of the crops, since it is consumed by the household that produces it. Most of the produce of the early gardens, the *kaymugwa*, thus belong to the category of crops consumed within the household though the best taytu might be kept for the *urigubu* gift.

The distinction between *taytumwala* and *urigubu* is marked in the gardens where some plots are called *gubakayeki* and others *urigubu*. All the produce of the first goes to the own household; most of the produce of the latter is given away. In the harvesting arbours the division is continued: the big central heap is the *urigubu* taytu and around it, in the corners, are heaps of seed taytu, *yagogu*, of inferior taytu, *unasu*, and also of some good taytu reserved for a man's own use and for *taytupeta* gifts (cf. Ch. V, Sec. 4, and Fig. 8).[1] These smaller heaps represent the less showy part of the harvest, the one about which there is no boasting and which is hardly ever spoken of. It is described sometimes by the generic term *taytukulu*. I was under the impression that since it is the right thing to display a very small amount of *taytukulu* and make the *urigubu* look 90 per cent, rather than the 50 per cent or so that it really is, of the whole yield, some men take the harvested taytu straight home and put it away unostentatiously into the covered part of their main storehouse or into a small completely enclosed storehouse. And again, in harvesting that they would "overlook" a good many tubers which would be collected later on as *ulumdala* for their own use.[2]

The *urigubu* are the choice crops. They are the only crops which are piled into conical heaps in the arbour and in the village and which are ceremonially stored in the open well of the yam-house. They are the only crops over which the magic of plenty and permanence, the *vilamalia*, is spoken. They are the only crops which are called *taytuwala* (real taytu—the words *taytumwala* and *taytuwala* are not to be confused), and they are the crops of which the garden magician is allowed to partake when the taboo on new crops is lifted for him. Hence the taytu which he eats then must be given to him by his wife's brother (cf. Sec. 2 of Ch. V).

I have so far presented the harvest duties in all their complexity, showing how they appear to the amazed ethnographer; how their ever-increasing intricacy seems to entangle him in a continuous web of terminological, sociological and economic problems, even of apparent contradictions. For any reader will easily assess how difficult it is to unravel the strands of the whole system, and how easy it

[1] Cf. also Note 26 in App. II, Sec. 4.

[2] For the difficulties of ascertaining the proportion of harvest that a man retains for his own use, especially in the case of commoners, cf. Ch. VII, Sec. 5.

PLATE 69

LATERAL VIEW OF THE CRATE SHOWN ON PLATE 68
(Ch. V, Sec. 6)

PLATE 71

FILLING THE LIKU IN WAKAYSE

"The Kabwaku men, breathless from their hard work of carrying the crate and yams, moved as in a trance, excited and absorbed in their task" (Ch. V, Sec. 6)

PLATE 72

THE MAIN STOREHOUSE OF THE TROBRIANDS

To'uluwa's *bwayma*, after its rebuilding in 1918. It stands in the middle of the central place in Omarakana. To the left centre the chief's personal dwelling; to the left of this, half hidden by the two palms, the new residence of Bagido'u, who is seen standing in front of the storehouse with his wife. The scaffolding ladder used in thatching the roof is still *in situ*. The tall slender platform made for the spirits can be seen to the left of the storehouse; the picture was taken during the *Milamala* season in 1918 (Ch. VII, Intro.)

is to become enmeshed in almost insoluble antinomies, especially considering the terminological looseness and multiplicity of meanings, and the fact that we have met hardly any rule without exceptions.

Let me, therefore, restate as simply as is possible the main rules in a positive manner. The average man in the Trobriands makes a garden. He owns also a main storehouse and one or two smaller storehouses. The produce of his gardens he has to divide into two parts. The best tubers and those which he will most carefully clean, most conspicuously display, and ceremonially transport—this part of the produce he gives to his sister's husband or to the husband of some other kinswoman. The other portion of his harvest he keeps for himself, and puts away in the smaller storehouse or in the less conspicuous portion of his main storehouse.

The open well of his main storehouse must, however, be filled by a kinsman or kinsmen of his own wife. This is the *urigubu* gift which he receives every year. His wife's nearest relative in the mother line, her brother, her maternal uncle, later on her son or her sister's son, will be the giver of the *urigubu*. If his wife dies leaving no children, his own kinsmen will fill his *bwayma*.[1] If he is a chief, his wife's father may help to fill his *bwayma*, though since this is not a father's duty, he would not do it if his daughter were married to a commoner. Or again in the case of a chief, a fictitious relationship may be established between his wife and certain other people who, though really filling his *bwayma* as his subjects, perform this duty, by a legal fiction, as his relatives-in-law.[2]

In any case a man must have his storehouse filled every year by someone, otherwise his social status is adversely affected. The man can and must provide himself with next year's seed. He usually provides himself and his family with most of the food for daily consumption, but the food which is to be displayed as show food and will be reserved as long as possible for festive occasions, gifts and exchanges,[3] must be given him at the *dodige bwayma*, either as *urigubu* from his wife's brother, or from his junior kinsman as a harvest gift.

It will be well now briefly to summarise the native distinctions with regard to the distribution of crops at harvest. The most generic concept and the widest term is that of *dodige bwayma*—"filling the storehouse". Whether the gift be given by the wife's brother or by the husband's own kinsman, or by some of the more remote relatives,

[1] See also Note 25 in App. II, Sec. 4.

[2] Cf. Doc. II, and also Note 27 in App. II, Sec. 4.

[3] For the different use of the *taytumwala* (own taytu) and the *urigubu*, see Ch. VII, Sec. 5.

it will be said that "so-and-so fills the storehouse" of the recipient. *Urigubu* in its strict sense denotes this duty when it is carried out by the wife's brother. As can be seen from Documents I–IV, *urigubu* constitutes almost one hundred per cent of harvest gifts when it comes to commoners, while, in its genuine form, it constitutes only about half of the gifts presented to the paramount chief.

The word *urigubu*, however, has two extensions which are derived from its primary and essential meaning. On the one hand it is made to cover various tributes to the chief over and above his genuine *urigubu* (cf. Doc. II), on the other hand it designates also the other duties incidental to relationship by marriage (cf. below Sec. 2). The word *urigubu* may be applied to garden plots cultivated for a sister's husband (in the south, indeed, it applies only to these, cf. Ch. X, Sec. 2), to the crops destined for him, to the gift and to the legal principle. Only in loose usage can the term *urigubu* be applied to the gifts given by kinsmen or dependants. Such loose usages are very frequent in native speech.

I have made this digression on terminology, not so much to push all these distinctions to the utmost limits of pedantry, but rather to illustrate, on the one hand, the difficulties of an ethnographer and the only way of surmounting them, that is, by the study of concrete instances and of actual verbal usages; and on the other hand, as an illustration of native legal usage. Of course there are regional and dialectical variants. In the south of the island, for instance, the principal harvest gift of taytu is called *taytumwaydona*, the "full taytu" or "taytu altogether". There, the term *urigubu* is not used primarily with regard to taytu gifts, but to gifts of taro which, as we know, is the staple crop in the south. In the south, moreover, the conditions of chieftainship are very different to those in the north, except at Olivilevi, where the Kiriwinian terminology has been introduced since the establishment of the chiefs—chiefs, that is, who belong to the ruling family of Omarakana—a generation ago. But I will not complicate matters by going further into such variants here.

The little table appended may be useful as a diagrammatic presentation of the conditions in Kiriwina, and of the terminology used there.

2. HUNGER, LOVE AND VANITY AS DRIVING FORCES IN THE TROBRIAND HARVEST GIFT

We are now in possession of most of the facts concerning the disposition of harvest in the Trobriands. But out of these facts several puzzling questions have arisen, which we have not been able to

THE PROGRESS OF THE *URIGUBU*

N.B.—The terms *bomaliku* and *sokwaypa* will be explained in Chapter VII.

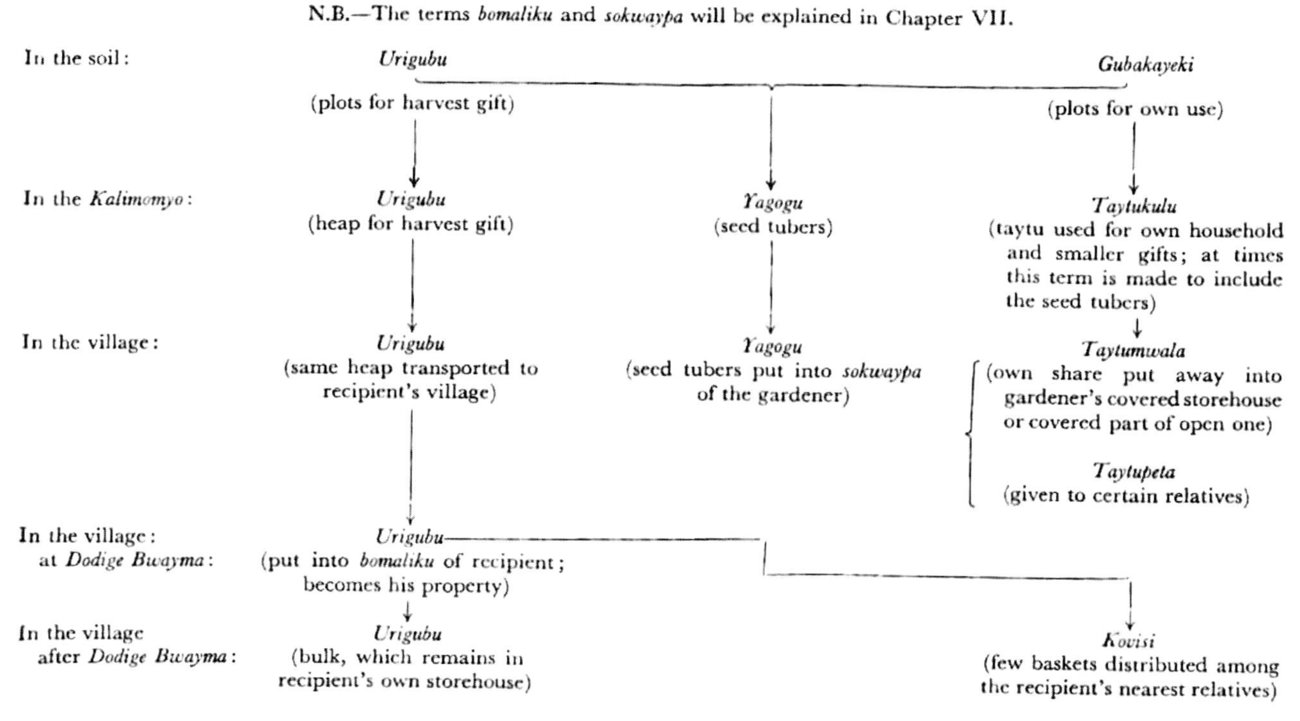

answer yet. It is still not clear who is the real recipient of the gift; is it the husband or the wife? Or is it the household as a whole? What is the rôle of the wife's brother in this household? Why is it that he who is practically debarred from any intimacies with his sister and her family, who cannot have a free personal conversation with her, has yet to support her household? What is the reason for the ambiguous and complicated part played by sons, and the no less intricate and roundabout way in which daughters have to be treated before and after marriage?

To the European reader the whole arrangement of harvesting must appear absurd. Economically it is cumbersome. A great deal of time and labour has to be wasted on transport and also on display, on cleaning and on other aesthetic trappings which are obviously connected with the fact that the *dodige bwayma* is a ceremonial act. The transport is necessary because marriage is patrilocal, that is, the wife lives in the husband's community; while household economics are matrilocal, in that the brother has to produce the harvest in his own village and carry it thence to his sister's husband. This of course could be simplified if it were possible to regard the *urigubu* as a commercial transaction; that is, if the wife's brother could arrange for someone in his sister's community to provide for her while he in turn provided for someone in his own community. But this is not allowed by the customary law of the Trobriands. Each man must offer to his sister taytu grown by himself on his own land and carried to her by his own kinsfolk and relatives-in-law.

It is only if we understand how much personal pride as well as moral duty is involved in a man's producing a generous share for his sister's household; how much a gardener identifies himself with his crops, and especially that part of the crops offered as the *urigubu* gift, that we can realise why a simple commercial exchange is not possible here, though it obtains in so many other transactions.

The scheme is not only economically clumsy to the European, it is also morally unfair. If I am industrious, if I am a good gardener, if I strain all my forces, somebody else derives the benefit from it. If my wife's brother is lazy, incompetent, or in bad health, I suffer for it. This is in a way true, but with the Trobriander the argument does not hold, because, as we shall see, a Trobriander has, sociologically speaking, a split identity: on the one hand his interests and his heart are in his own household, in the household, that is, of his wife and children; on the other hand his pride and his moral duties are in the household of his sister.

What puzzles the European most, what puzzled me for a long time very badly, is the question of motive. Why does the Trobriander spend an enormous amount of time and energy in work for someone else? What spurs him to such endeavour? Here again the split or double set of interests, ambitions and emotional incentives will supply the answer.

I am pointing out these difficulties in order to justify the somewhat lengthy sociological digression on which we have now to embark. The ethnographer's duty is not merely to lay down the overt, external facts; his duty is also to interpret the ideas, the motives, the feelings, even the emotional reactions and suppressed wishes of the natives. And this world of ideas can only be understood when we realise those differences in social structure, in legal principle, and in moral compulsion which make a Trobriander feel, think and aspire somewhat differently from a European.

The *urigubu* principle is the very core of both social and economic life. To understand it on the economic side it is necessary to grasp the law of marriage, native ideas about procreation and the native system of kinship. Some of these problems have also been treated in other of my publications, but their restatement in connection with harvesting is indispensable. To begin with the law of marriage. In the Trobriands marriage puts the wife's matrilineal kinsmen under a permanent tributary obligation to the husband, to whom they have to pay yearly gifts of *urigubu* for as long as marriage lasts. The marriage contract is effected by the exchange of reciprocated gifts the balance of which is on the whole favourable to the husband, or rather to the new household, mainly owing to the large gift of taytu presented at the harvest immediately following the union. This is packed into decorative, oblong, prism-shaped receptacles which are erected by the brother in front of the freshly constructed yam-house of the new household. This gift, called *vilakuria*, is the first instalment of the *urigubu* contributions, which year after year will be placed on the same spot as that where the prism-shaped receptacle now stands, though later contributions will on the whole be less substantial and will be arranged simply in heaps. These gifts are every few years repaid by the husband, with a gift of valuables to his wife's brother. But such gifts never total to the economic equivalent of the *urigubu*.

Kinship is counted only in the maternal line. The children of a union are not in any way regarded as kin (*veyola* 'kindred his') to their father or to his lineage. They are of the same body as their mother: she has made them with her own blood; they belong to her clan and sub-clan; they are members of her village community. They are successors to her brother. That is, her male children have the right to the offices, positions, and social status which is now occupied by her brothers. Her daughters will continue the lineage and bear children who will belong to the same kindred group. The unit of filiation, the group, that is, which embraces successive generations bound together by identity of flesh, is then not husband, wife and children, but brother, sister and her offspring.[1] I say advisedly the brother and his sister, for wherever there are several brothers and sisters, one brother is specially paired off with one sister.

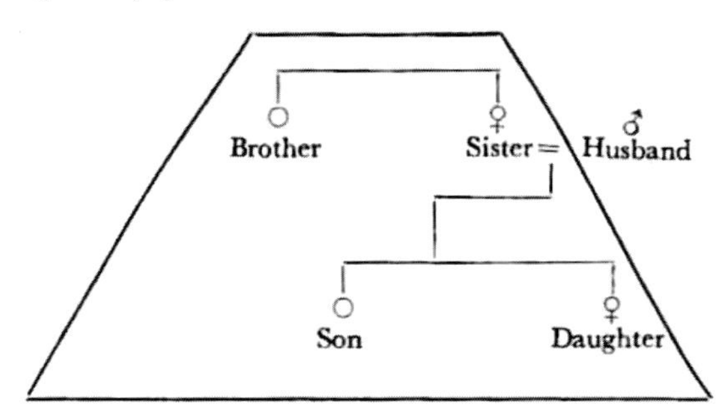

The above diagram, superfluous for the sociologically trained reader, will yet bring these conditions clearly to the eyes of one not so familiar with primitive kinship. The husband, though head of his own household, is outside the group of kindred, in the legal sense of the Trobriand word. Brother, sister and her children are the real genealogical unit of descent. In the Trobriands, as the diagram shows, there are therefore two units which correspond to our term 'family'. One of the units is the physiological procreative group, the husband, wife and her offspring. This group, as among ourselves, is united by the intimacy of daily life, by the economic interests of direct housekeeping, by the sentimental ties which spring out of daily contacts on the basis of innate emotional tendencies. This group is very clearly defined by the Trobriand law of marriage. But it must be carefully noted that in speaking about the physiological foundation of the group, about the procreative unit and about the bonds of blood which unite this group, I have been speaking as a sociologist and not reproducing the native point of view. For the natives have a procreative theory which, by ignoring the father's share in pregnancy and ascribing it to the influence of the wife's dead ancestors, supplies the dogmatic foundation for the matrilineal doctrine of bodily identity only and exclusively in the female line. The husband presides over the household as a member extraneous in all matters of legal kinship. He is styled *tomakava* (stranger, outsider). He is united only to his wife by the contract of marriage and only

[1] For the principles of native filiation cf. Part I, Sec. 9.

through her and indirectly is he related to her children and to her brother.

The other unit which, in one way and in one way only, corresponds to our term 'family' is the filiational or genealogical group formed by brother, sister and her children. It is this group which comprises the males of two generations, who replace each other, who are united by pride of lineage, by ambition, by that sense of continuity which comes from being successive links in the chain of descent. This group also, the matrilineal group, comprises the women who can continue the lineage. It is a strange fact to a European or to any human being brought up in a patrilineal culture, but it must be clearly realised : in the Trobriands, as in any other matrilineal society, a man is a barren shoot on the genealogical tree. In the Trobriands, moreover, and in those communities which like the Trobrianders or the Central Australians do not admit physiological paternity—a man can have no bodily issue in his own right by virtue of his own physiological contributions. The only children to whom he is related by blood are his sister's children. They are his legal issue, as they are in any matrilineal society, even though, in such a society, it were known that he had physiologically procreated the children of his wife.

It is best, therefore, in the Trobriands to apply the term 'family' to the paternal household, to the group of husband, wife and children. The other group I have termed here 'genealogical unit', because all relevant genealogies are to the natives in the mother line, but perhaps 'unit of filiation' would be the most suitable expression. The two groups, the family and the unit of filiation, are held together by the law of marriage and by the marriage contract.

One question arises out of this duality of grouping. If the Trobrianders do not admit physiological paternity, if a woman conceives and produces children by the agency of her own dead kinswomen and bears a child for her own kinsfolk, particularly for one of her brothers, why is the husband necessary at all? What is his place in the household? Is he merely a drone who lives on his wife and benefits from her brothers? The answer to this is two-fold. In biological and psychological reality, the husband lives with his wife because he is attached to her and is fond of her, because he was at one time in love with her and married her very largely for that reason. He also has received by the fact of marriage a much higher social status, and a comfortable home to live in. He looks after the children because he is fond of them, because he has desired to have them, even as his wife has. This is the reality of living men and women in the Trobriands, as elsewhere—a reality which, however, can only be ascertained and formulated by an outside observer and of which the natives themselves are unaware and could not put into words.

If you were to enquire into native theory and collect native opinions, you would receive an answer in a way less true but more interesting and sociologically more relevant. The husband, you would be told, has married the woman because a man without a household is not a real adult man. He has married her also because a grown-up man must have a woman who lives sexually with him and with him only. It is good for young people to have all kinds of intrigues, but a mature man must have a wife. In the Trobriands, however, a wife marries her husband as much as she is married by him. And if you enquired into her motives the answers would be even more illuminating. A woman does not need a husband as a legal guardian: her legal guardian is her brother, who remains in that relation to her after she marries, quite as much as before. But her brother is absolutely debarred from one whole side of her life. There is an extremely rigid and all-embracing taboo forbidding any friendly intimacies between brother and sister, and forbidding a man to take any interest whatever in his sister's sexual affairs or reproductive processes. For that purpose another man is necessary. Not only that, but this other man, the husband, is indispensable in order to render her reproductive functions honourable, in order to give her offspring full tribal status within the community. Although the sexual life of the natives is very lax and unmarried girls can have as many lovers as they like, a girl is not supposed to become pregnant before she is married.[1]

Marriage is thus an absolute prerequisite of maternity, of the legal status of the offspring, and hence of the honourable continuity of lineage. A brother who is directly interested in having his lineage honourably continued, of having his nephews real young men (*to'ulatile*) and not bastards (*tokubukwabuya*) is, therefore, directly interested in the sister becoming married. A husband is indispensable.

The brother has therefore at a certain time of his sister's life to waive his rights of guardianship over her and allow her to go under the direct and intimate control of another man. The brother never does exercise any intimate control over the sister's affairs. Up to her marriage such little control of a girl's private life as is exercised, together with all arrangements connected with her marriage, have been in the hands of her father, who here again acts as the delegate of the girl's matrilineal kindred; and after marriage the real super-

[1] Cf. for the full substantiation of this and the following statement my *Sexual Life of Savages in N.W. Melanesia*, especially Chs. III–VII and X.

vision of his wife's sexual and reproductive processes is vested in the husband.

This is connected with the principle of patrilocal marriage. The girl joins her husband in his village and in his house, and henceforth, having a joint household to run, the two must have a number of interests in common. They have joint economic duties and obligations; the husband watches over his wife's interests in many matters, and, when the children come, he undertakes a number of educational duties, especially with the sons, and advises and protects his daughters in their pre-matrimonial affairs. When the daughter becomes pregnant, it is his sisters who will take her pregnancy concerns and ceremonies in hand.

We see, therefore, how out of the native concept of marriage the idea of fatherhood gradually emerges, and that the rôle of the father and his relation to the children are derived from and based upon his personal relation to his wife. Let us gain a clearer insight into this relation.

Marriage is, in the Trobriands, the outcome of a series of intrigues. It is the intrigue which has endured, the intrigue which has been accepted by the girl's family as a permanent arrangement, and been legalised by them in a series of gifts, that finally, by public eating in common and the open cohabitation of the two, becomes legal marriage. Now since we are studying here the economic side of legal relations, what is the economic aspect of a sexual liaison? All sexual services in the Trobriands are repayable by gifts which have a special name, *buwana* or *sebuwana*. Not that the natives regard the pleasure derived from intercourse as one-sided, or that women in the Trobriands take a passive or retiring part in it; on the contrary, both in behaviour and in theory, female initiative and the female share in a liaison is equivalent to that of the man. But with the characteristic inconsistency of traditional decree, custom has it that for almost every sexual act, and certainly for continued sexual relationship, the man should pay and the woman should receive gifts. It is the payment "for her scarlet", by which generic name natives describe the female genitals.[1]

Marriage differs from a sexual liaison in two respects. In the first place marriage is a contract which leads inevitably to the establishment of a new household, it is the foundation of a new type of social grouping, the family. This means that, in its personal aspect, marriage is a much fuller and more complicated relationship between man and woman than that which obtains between two lovers. In the second

[1] Cf. *The Sexual Life of Savages*, Chapter X, Section 7, "The Commercial Aspect of Love."

place, it is a publicly acknowledged relation, approved and accepted by the girl's family, and binding them to definite economic prestation.[1] Consequently the economic side of the relationship is complicated and two-fold. No longer free to give her body to anyone she wants to, the woman is now under severe sanctions bound to be for her husband's exclusive sexual pleasure. He, on the other hand, is expected still to go on repaying the sexual gifts which he receives from his wife. This he does, not by occasional gifts, but by continuous services rendered to her, and above all by services rendered to her children. The children are regarded legally as hers and not his, the love and care and the material benefits which he bestows on them are, by the traditional convention, regarded as a repayment: "the payment for sleeping with the mother", "the payment for the mother's scarlet," such and similar phrases are the stock answer to the question, Why does the father care for his children?

The wife, besides the erotic services which she renders to her husband, has also to cook for him, prepare his food, bring his water in the water-bottles, look after him when he is sick, and mourn for him when he is dead. Her brothers have to assist her by supplying the *urigubu* gift, in watching over him in illness, in defending his interests, and by being ready to do work for him whenever communal labour is necessary.

The *urigubu*, as we know already, designates not merely the annual gift at harvest, but comprises also the other services due from the wife's family to the husband.[2] Naturally, therefore, it is one of the central features of the marriage contract, which again is the result of native ideas about procreation, their doctrines of kinship and relationship, and of the moral duty of the brother to look after his sister and her offspring; and the equally important and more arresting moral duty of brother and sister to avoid each other.

With all this it must be repeated once more that the father loves his children, that he needs them emotionally, that sociologically his status is enhanced by them. He is their assistant nurse when they are young, he is their companion when they are growing up, he educates his sons and looks after his daughters. Paternity devoid of the rigid legal guardianship which devolves on the mother's brother

[1] Since the English language has a really unaccountable and intolerable gap, I am deliberately introducing here the word "prestation" in the French sense, that is, of legally defined services to be rendered by one individual or group to another.

[2] Communal labour (cf. Ch. IV, Sec. 5) is used in a far wider range of pursuits than gardening. The communal labour supplied by wife's brothers and kinsmen refers more to such matters as carrying heavy weights, fishing, construction of canoe, storehouse, or dwelling, etc.

becomes a lively companionship and real friendship. The father, in actual fact, always tries to give as much as he can to his own sons at the expense of those of his sister, who are his legal heirs. His natural inclinations are seconded by customary usage which almost defies and certainly circumvents the rigid matrilineal law, by giving the father a number of opportunities to favour his sons and to curtail the rights of his matrilineal nephews.

Thus the sons receive a great many benefits, especially in the form of certain kinds of magic, as a free gift from their father, while the nephew has to buy such benefits from his maternal uncle. The father usually tries to arrange for a cross-cousin marriage between his son and his sister's daughter, whereby he places his son in the position of an intermediate heir. Cross-cousin marriages, moreover, in opposition to ordinary ones, are matrilocal; that is, the son remains in his father's community, which is also the community of the father's matrilineal niece.[1]

Thus the attitude of the father towards his children is built up very much on the same pattern as the patriarchal attitude, with a few important differences. The father has no legal power over the children; he is therefore much nearer to them emotionally, he tries to win their affection and usually succeeds in doing so. The father has no right officially to control the destinies of his sons, nor of his daughters after marriage. Both of them have to leave his house and his community when they reach the age of full citizenship, that is some time between eighteen and twenty-two. The sons, according to the strict law, should go to the village community of their maternal uncle.

The father, however, has definite claims, reciprocal claims, that is, on the services of his children, and above all of his sons. When they are young they both work on their father's plots and in his village, and with their maternal uncle for their mother's *urigubu*. In the latter case they would have to repair to their maternal uncle's community. Either way they contribute yams for use in their own household. Also the sons as successors and helpmates of their mother's brother are at their father's service. Thus, though in theory the fact that they are legally beholden to their maternal uncle fundamentally alters the normal son-to-father relationship as

[1] Cf. *The Sexual Life of Savages*, Ch. IV, Sec. 4. The reader acquainted through this work with Trobriand society will realise that matrilocal marriage in the Trobriands often presents the paradox of the husband remaining in the village of his birth and adolescence, and the girl joining him there. As will be made clear in Chapters XI and XII, citizenship in a community and early residence there hardly ever coincide.

familiar to us in patrilineal societies, in practice it seldom makes much apparent difference. It does, however, mean that before acceding to the wishes of their father in any important matter they would have first to consult their uncle.

Later on they either remain with their father—through cross-cousin marriage, or, when he is a chief or a man of power, through his arbitrary decision—and then help him in his own community; or else they return to the maternal uncle and from there work for their father's and mother's *urigubu* each year in the gardens of their own village.

At the father's death it is the sons who carry out most of the services. They have to wash and adorn the corpse, bury it and carry out some revolting mortuary duties, such as tasting the dead man's flesh, using certain pieces of his bone as lime spatulae and his skull as a lime-pot—revolting, that is, not only to the missionary or ethnographer but to the natives themselves and regarded as such, or rather as onerous and exacting duties. Here once more we see the play of give and take, of repayment for services rendered. If you ask why a son should render these mortuary services to the father, reference is again made to the father's relation to the mother or else to his nursing and educating of the children. "It is the repayment for the father's hand that was soiled with the child's excrement, for the skin which was wetted with the child's urine." "The father has received the child in his arms and nursed it, now the child nurses the father's body."

Thus the relationship between father and children is founded on natural affection, the sentiment grown out of intimacy and love. It has as its legal framework an artificial but neatly balanced system of reciprocal duties, the centre of these remaining always the annual harvest gift, the *urigubu*.

Let us in turn focus our attention on the group consisting of mother and children and the mother's brother. This group, as we know, forms the real unit of kinship in native theory. The females continue the line, the males represent it in each generation. The males inherit the property, the females through the *urigubu* gift are made to benefit from a large part of the males' stewardship. The males of several generations work for the females; the females, under the guardianship of their husbands or fathers, lead their sexual and reproductive lives, continuing the lineage of the kindred.

It must be emphatically stated here that it is not the "clan" which acts as a unit in this complementary division of functions, but definitely the group consisting of brother, sister and her offspring.

The woman's brother then is her natural legal guardian. But he is not enough. He cannot enter in any way into his sister's sexual life, not even indirectly, not even from a great distance. He leases his sister's procreative life as well as her offspring to her husband, as his maternal uncle has leased his mother's reproductive life to his father. Since he has to keep at a distance, physically as well as morally, the marriage must be patrilocal. But with all this the fundamental principles of matrilineal kinship oppose the development of any strong patrilocal or patrilineal family life. In the first place, the matrilineal, that is to the Trobriander the real, kindred of the husband, do not want him to bestow too much attention, privileges and gifts on his children; for a man's possessions and advantages should be kept within his own matrilineal kindred. On the other hand the real, that is matrilineal, kindred of the children, want them back in their community, want them as helpmates and legal successors. The patrilocal household thus is usually broken up, the girls leave the paternal house to marry, the boys return to their maternal uncle. Husband and wife, both old, decrepit and dependent now, remain alone and are still maintained by the wife's family, including her sons.

We are now, after this long but necessary digression on kinship, in a position to understand the nature of the *urigubu* gift. We can also answer our previous questions, first as to the motives which make this gift be given and given generously; second, as to the reason for the elaborate and competitive display incidental to the giving; and third, as to the function of the obliqueness in the fulfilment of economic obligations.

The *urigubu* is the endowment by its real head of the matrilineal unit of filiation, and this real head is neither within the household, nor even, as a rule, in the same village. The patrilocal household, on the other hand, which benefits by this endowment has its real head within it. This head also contributes economically to it, but is regarded as a stranger in matters of kinship and his position is legal only in virtue of a series of reciprocities which bind him to his wife, to his wife's brother, and to her offspring.

The *urigubu* is, therefore, the expression of the real constitution of Trobriand kinship grouping. This grouping is not simple as with us, where it consists of one household, one family, and one unit of filiation. The kinship grouping in the Trobriands embraces on the one hand the unit of filiation—brother, sister and offspring; and on the other hand, the household—husband, wife and children, including sometimes old parents and more distant relatives. The core of the household is always the family; that is, to the sociologist, the reproductive unit, husband, wife and offspring; to the Trobriander, the matrimonial grouping, founded on the contract of marriage, which makes the husband the deputy guardian of his wife and her immature children. The *urigubu*, therefore, is the outcome and economic expression of a compromise or adjustment between the principles of the patrilocal household and matrilineal filiation. Into the composition of the *urigubu*, however, there also enters the brother-sister taboo, the principle of legitimacy which decrees that all children must have a father, and the customary rule that adult man and adult woman must enter into permanent and full sexual partnership.

We can now understand not merely the legal sanctions, but also the personal motives for the *urigubu*. If the *urigubu* is regarded as a gift from an outsider to a household which he is scarcely even permitted to visit, it appears absurd, unfair and cumbersome. But when the *urigubu* is understood as the endowment of his own kindred group by its head, it becomes natural, almost obvious. Exactly as in a patriarchal society the *pater familias* feeds his household and works for its firm economic foundation to endure after his death, so in the Trobriand matrilineal community the maternal uncle provides for his descendants freely, generously and with a will. In the process he also feeds his sister's husband, but the latter repays by reciprocal gifts that share of the *urigubu* which falls to him and furthermore himself contributes to the feeding of his wife and her children, that is of the filiational group, a group which is not his kindred. All in all, the economic arrangements of a household reflect very neatly the various sociological, legal, moral and personal inclinations, interests and reciprocal duties of the native.

We can also perceive now why it is that a man will not procure taytu locally at so much a basket for his sister's household and pay for it by some service rendered in his own community. According to native law, custom and morals, his real duty lies with his sister's household. It is consequently for this household that he has to raise and harvest the *taytuwala* (real taytu), the *taytumwaydona* (the full taytu). It is on the size and quality of this part of the result of his labours that his reputation as a gardener depends. Moreover, in the giving of it he satisfies both his vanity and his sentiment. For on the one hand he is giving to a stranger—a man who by definition belongs to another clan—his sister's husband, and of such a gift he is allowed to brag. But on the other hand he is providing for his sister, for his own descendants, for those who will in future work for him and for the glory of his lineage, and therefore his heart is in the work.

The *urigubu* gift therefore lends itself to that boasting, display, comparison—to the running mythological elaboration, which is so dear to the Melanesian's heart.[1]

The whole ceremonial side of the transactions which we followed in the previous chapter is not merely an expression of a strong individual inclination; it contains also an effective sanction. The display, the measurement in public, the taking and recording of tally, provide both a psychological spur to the giver and a handle to the community, whether it be for praise or blame. When his work has been successful as well as efficient, the generous gift is appreciated, the glory of the giver and his lineage extolled, and the moral approbation of the community bestowed, and this to a Trobriander is a great satisfaction and a real reward.

We see, therefore, that the idea of satisfying *urigubu* obligations by a simple commercial transaction would be grotesque as well as repellent to the natives. Whenever I propounded such an idea to my informants, or enquired why they did not simply *gimwali* (barter) the harvest gift, they took it as a joke in bad taste, and expressed their scorn clearly. We also see how a better understanding of native psychology and of the native social framework enables a European to appreciate the function of certain customs and their effective value for a community. Such white residents in the Trobriands as had even an inkling of what the *urigubu* was were extremely contemptuous of its useless waste of energy, and were keen on having it stopped by Government order. To us, however, it will be obvious that half the incentive of the Trobriander to work would be destroyed and more than half of his moral sense and sense of responsibility blighted if the *urigubu* were in any way interfered with.

We can also understand why a man may be a good gardener as long as his object is to satisfy the fundamental obligations of the *urigubu*, but that it is dangerous to be a good gardener on his own behalf. To have large gardens and to use them for his own food runs obviously counter to one of the fundamental principles of Trobriand sociology, and the community, there as elsewhere, reacts with a sound conservatism and a deep hostility to anything which offends its established conceptions.

We can now quite briefly assess the functions of the *urigubu*, the influence, that is to say, which it exerts on Trobriand social life. From all that has been said above it can be seen that it is one of the main elements in the stability of marriage. It is closely connected with the position of the father and his relations to the offspring, as

[1] By 'running mythological elaboration' I mean the stories which gather round the current events of tribal life, in time grow into legends and become the established verdict of tribal opinion.

well as with the latter's relations to the maternal uncle. But one of the most important elements in the *urigubu* is that it is the channel through which every headman of a village community, every chief of subsidiary rank in his district, as well as the paramount chief of Omarakana, levies his tribute.

3. THE THEORY AND PRACTICE OF THE HARVEST GIFT

It will be best to illustrate the principles of *urigubu* by concrete examples. Let us visit at harvest-time a component village of Yalumugwa, a typical medium-rank community, now controlled by a sub-clan of lesser chiefs (*gumguya'u*). As can be seen from the data contained in Document I, the eleven storehouses which have to be filled ceremonially belong to members of three sub-clans (cf. Ch. XII, Sec. 3). The people who fill these storehouses are all closely related, and they all reside either in the same village cluster or near at hand. In such a village, the fillers or donors can be easily reached and the date arranged on which they will start bringing in the crops. On such a day, crowds of carriers will arrive at short intervals in the ceremonial manner described (Ch. V. Sec. 5) and erect the heaps in their proper places. Some donors may be behind-hand, and then the bringing in will extend over two or three days perhaps. In 1915, as far as I remember, all the heaps were erected within three days in the component hamlet of Yalumugwa described in Document I.[1] The heaps will be displayed in such a village for some three to five days, or, at the outside, for a week. Then a day is appointed for the return of donor and carriers and all the *bwayma* are filled at the same time, the proceedings taking an hour or so (cf. Pls. 75–77).

When I witnessed and photographed the filling of the *bwayma*, Yalumugwa was animated but not crowded. Carriers and villagers were in semi-festive attire, their bodies were well oiled and all were wearing new pubic leaves and fibre skirts. The procedure in Yalumugwa is typical of a small *gumguya'u* village and the result of checking my records of figures, relationships and distribution against such villages as Kwaybwaga, Liluta, M'tawa and Kabwaku shows that the number of heaps ranged between twenty-five and fifty, and that the conditions of relationship and legal principle were very much the same in all of them.

In much smaller villages such as Tilakayva, Wakaysi or Yourawotu, the proceedings would be the same only on a smaller scale. Here the number of heaps would be about a third or even a fourth of

[1] See also Note 28 in App. II, Sec. 4.

what we have seen in Yalumugwa. At one harvest I counted five heaps in Kupwakopula, seven in Tilakayva and six in Yourawotu.

As regards the commoners who reside in a high rank village, they receive one, two or three heaps each. Many commoners who live in a small village will just get one heap. The data presented in Document IV show us a few concrete examples. Such examples must be taken as illustrations only. In my walks through the villages I have time and again enquired at harvest seasons into the *urigubu* conditions. In every single case I found that it completely conformed to the pattern established in the illustrations which were recorded by me. The inductive generalisation is, therefore, based on much wider material than that presented in the documents.

In smaller villages also the donors would live near-by and the whole ceremony would be over in a shorter time than in a *gumguya'u* village. The heaps would be brought in in one day, left for twenty-four hours, and then stored away.

In villages of chiefs of high rank or of great wealth, that is first and foremost in Omarakana, but also in Kasana'i, Gumilababa, Kavataria and Olivilevi, or further south in Vakuta, the arrangements would be on a much larger scale. Perhaps it will be best if I briefly describe what happened in Omarakana during the exceptionally large *urigubu* of 1918, of which I have recorded some numerical data in Document II.

In that year it became necessary to rebuild the main storehouse of To'uluwa which was falling to pieces, and to repair a number of his smaller storehouses, each of which was assigned to a special wife. Also in that season (1917–18) the natives of Kiriwina wanted to express their loyalty to the chief, very largely, I think, to remove the tension and hostility which had crept into the relations between the villages after the expulsion of the chief's eldest son in 1915. At the same time the villages of Kwaybwaga on the one hand and of Liluta and M'tawa on the other, had been carrying on a private feud of their own. This was also connected, I think, with the quarrels in the chief's family, but the last straw, or perhaps the ostensible cause, was a difference of opinion over some scores at cricket, a then newly introduced boon of civilisation and Christianity.

I will quote the account which was given to me of this quarrel because it contains some interesting statements as to one of the reasons for holding a *kayasa*,[1] and it is also a good illustration of how difficult it is for real diffusion to take place. Cricket, which to an Englishman has become a synonym for honour and sportsmanlike

[1] Cf. *Argonauts*, Ch. VIII, where a type of *kayasa* called *uvalaku* is described; and also *Sexual Life of Savages* s.v. *kayasa*.

behaviour, is, to a Kiriwinian, a cause for violent quarrelling and strong passion, as well as a newly invented system of gambling; while to another type of savage, a Pole, it remains pointless—a tedious manner of time-wasting.

"(i) When you lived here before, there was no *kayasa* because there was no quarrelling. (ii) At the present time there is a *kayasa* because there is a reason for it in a quarrel. (iii) The Kwaybwaga people on the one side and the people of Liluta and M'tawa on the other quarrelled. (iv) When the quarrel was over they said: 'Good, let us wait and see. We will make a competitive display of crops at harvest, and we will make an inaugural feast at Omarakana to open the *kayasa*, (v) Then when the distribution of To'uluwa was over, they went to their villages. They held another inaugural feast and another distribution in their own villages. (vi) The Liluta people held one, and the people of Kwaybwaga another. They ate their fill until it was finished. (vii) The reason for the custom of *kayasa* is that we should see that one man is the more powerful, his magic sharp. (viii) For when the bringing in of the crops is over, we would know (i.e. learn by enquiring): 'Thou, how far did thy tally go?' and at the counting of the tally it would become clear to us: 'Their tally stands there.' (ix) In the time ahead, when the bringing in of the taytu is over, when the *kayasa* is over, their quarrel would be over also. (x) Supposing another quarrel broke out, another *kayasa* would be arranged.

(xi) These people quarrelled because of the cricket. (xii) The people of Kwaybwaga went to M'tawa and cricketed. (xiii) They cricketed, they finished, they counted; they counted and they said: 'Who has won?' (xiv) The people of Kwaybwaga spoke and addressed the people of M'tawa: 'You lie, we others have won'. The M'tawa people answered: 'No, you have not really won.' (xv) They quarrelled: 'Good, we shall beat you.' They hit one another with throwing-sticks. The people of M'tawa drove off the people of Kwaybwaga, and these latter departed to their village, saying: (xvi) 'Good, you have driven us off; but to-morrow come you to Omarakana. We shall beat you.' (xvii) Later on they came to Omarakana, the people of Kwaybwaga stood up against the people of M'tawa, they took their revenge, fighting with spear and shield. (xviii) The people of M'tawa and Liluta ran away. They went to their village and said: 'We have quarrelled, but let us make a *kayasa* and see who is more efficient in gardening.' (xix) The master of the *kayasa* is Kwoyavila of Liluta. (xx) In old days it was like that: they quarrelled, they fought, and then they arranged for a *kayasa*. (xxi) The quarrelling was because of women, garden plots, or food."

Parts of this account may be somewhat obscure to the reader, but I must not dwell further on it here. It is fully commented upon in (Text 85, Div. XII, §§ 18–20).

At any rate, owing to a combination of reasons it was decided to organise a competitive and ceremonial harvest on the *kayasa* principle. *Kayasa* is a generic name for any period of competitive, obligatory activity. It is always organised on a definite pattern, with a ceremonial according to its kind and it has, in some aspects, the binding force of law. The activity engaged in may be of a purely festive character (dancing, games and so forth) or it may, as in this case, be economic and concerned with gardening, fishing or the production of shell ornaments. The organiser is always a man of rank, and the *kayasa* principle enables him to bind to himself a number of people for the execution of the enterprise in question. He makes an initial distribution of food and this imposes on the recipients the duty of carrying out the enterprise in the teeth of any obstacles or difficulties that may arise. Thus in 1917 when a harvest *kayasa* was decided upon, the natives were expecting a good gardening year with plenty of rain at the right season. Conditions turned out quite the contrary and the crops over the district were very bad. In spite of this all the participants in the *kayasa* strained their forces to the utmost, and the yield given to the chief was enormous. The rest of the island was distinctly under-stocked with taytu, but this did not matter very much, since fruit and secondary crops were excellent. But it illustrates the principle that once a *kayasa* is started, strong ambition as well as a sense of duty makes the natives remain true to their aim.

I was in residence at Omarakana when the Kiriwinian harvest began. The arbours (*kalimomyo*) constructed in the plots were of exceptional size (see Frontispiece and Pls. 48, 56 and 57). Many of them I think were not filled from one plot, nor even from the plots of one man, but the heap was the result of many men's labour. Some of the heaps, as can be seen from Document II, reached the exceptional size of nearly two thousand baskets. The visiting, admiring, boasting, criticising and gossip also reached to quite unusual dimensions in the villages. Associated with it incidentally was a *buritila'ulo* (competitive display of food) between two villages of the Tilataula district. In each community the headman was the leader and organiser of the harvesting. Most conspicuous among these were Kwoyavila of Liluta, Simdarise Wawa of M'tawa, Kumatala of Kwaybwaga, Kaniyu of Liluta and Tokunasa'i of Kaytagava (main hamlet of Kwaybwaga). Document II shows that these men contributed most to the harvest. They were among the leaders of the two contending factions in the quarrel over cricket.

After the crops had remained for some time in the gardens the bringing in began. It lasted not a day or two, as in minor villages, nor even a week or a fortnight, as usually happened in Omarakana, but extended over a full month. Each time a *gugula* (heap) was transported from one of the more or less distant plots to Omarakana, the two villages, that of the donor and the capital, were thronged with people. Hundreds helped in the carrying, and the entry of the parties was attended with full ceremonial: shell trumpets blown, songs and ditties recited, carrying screams uttered. The men tricked themselves out for the occasion, not of course in the full dress reserved for dancing, but with all sorts of paint and fantastic adornment. Palm leaves held in the hand, paddles used for carrying poles, ornamental staffs, drapery of leaves, feathers stuck in the hair, big hornets' nests used as casques—such were the decorations on that occasion. In the unconventional, freakish features of this "fancy dress" there was expressed the rough sense of fun which is permissible during a *kayasa*. Day after day I used to hear, from a great distance at first, then gradually approaching, the mournful and usually discordant blowing of trumpet shells. By degrees voices could be heard singing some slow marching song. From time to time the explosive *sawili* interrupted the singing. Then again, as the party neared the village, the trumpet shells were blown with renewed strength and the *sawili* uttered, and to a slow song the party would enter, not as usually at a run, but walking ceremoniously, several abreast, laden with the over-filled baskets.

In the central place each man laid down his burden and then, standing round, they gave voice to a final *sawili*. Panting, eyes glittering, faces eager and excited, they sat down to rest and discuss the erection of a heap and the arbour over it. A few men, mostly the commoners of the chief's village, would busy themselves in bringing round refreshments, speaking to people and carrying out practical arrangements. The *tolitaytu*, "master of the taytu"—in this context the title invariably designates the donor—remained sitting in a dignified manner. He neither worked nor directed the proceedings, but from time to time, when things did not go well or when something roused his temper, he would jump up and in jerky, explosive sentences harangue his people or those of another village, never, of course, any belonging to the chief's community.

To'uluwa on these occasions is seated on top of his *kubudoga* (high platform, cf. Pl. 27); never active or apparently interested in what is passing, he sits there throughout the proceedings surrounded by the equally calm and indifferent local aristocrats and by his retinue, placidly chewing his betel-nut. It is the inferior men who work, who move, who talk and joke, who in short inspire the proceedings with life.

At a given signal the carriers jump up and start building the heap and preparing the little fence, the *lolewo*. They place the taytu carefully and cunningly, so that the best tubers remain on the outside, and then erect an arbour very much on the pattern of the *kalimomyo* (garden arbour, cf. Pls. 64 and 65).

After all sorts of decorations have been attached to the arbour and the *kalawa* (tally leaves) tied to it, the party takes a short rest and then, with a farewell blast on the shell trumpet, moves homeward.

I will now describe an incident which occurred during this *kayasa*. I witnessed it myself and it was afterwards put into words for me by another witness of it, Tokulubakiki (Text 86, Part V, Div. XII, § 21). It was a quarrel on the eternal theme: the respective generosity of gifts, both sides boasting of being able to give generously, which always implies the capacity for giving, the lavishness which springs not from good will but from affected nonchalance about possession. The quarrel actually arose out of some remark passed by the people of Kwaybwaga on the crops which were being brought in by the people of Liluta. At the same time, they reminded the Liluta men that these latter had promised to give them yams, areca-nut, and a pig, and had not kept the promise. Kaniyu, headman of Liluta, took up the challenge. Let Tokulubakiki take the word:—

"(i) At first Kaniyu held forth: he spoke and told the people of Kwaybwaga: (ii) 'I would give you yams and areca-nut and pig, but you would not repay. (iii) That being so, don't come to Liluta too often.' (iv) Tokunasa'i got angry, he came up, he answered the words of Kaniyu: 'Kaniyu, you speak about areca-nut, pig and yams. (v) To-morrow fill two log cases with yams; bring one pig and one bunch of areca-nut. Bring these, and we shall see.'"

Tokunasa'i was really angry, and he barked out his challenge in short, clipped sentences. The two headmen, divided by a sufficient distance, walked up and down in front of their respective groups, who sat listening, ready to back up their leaders. In the old days, I was told, there would unquestionably have been a fight. Kaniyu having said in effect: "I'd give you lots of stuff if I didn't know you wouldn't repay it," asks them, for this reason, to be sparing of their visits—a request which carried the insulting implication that visiting on their part means begging. Tokunasa'i then challenges them to bring along their gifts and see whether they are repaid or not.

The words quoted (which I think were actually uttered) are only the gist of the dispute. More words passed, but the important thing was that the opponents refrained from uttering any of the really insulting challenges, such as *gala kam* 'thou hast no food', etc. The quarrel was cut short by the intervention of some people of Omarakana.

At this same *kayasa* I overheard and noted down the following conversation between Kumatala, headman of Kwaybwaga, and his assistants (Text 87, loc. cit., § 23): "(i) Is all the taytu in already?" "No, some still remains." (ii) "Get back quickly then, lift it and bring it, so that it may be finished. There is a pig waiting for your refreshment at work."

The initial question and answer were, of course, rhetorical. Kumatala knew that there was still lots of taytu to be brought. He wanted to emphasise its abundance.

After we had gone back to his village, the distribution of *puwaya* (refreshment at work) took place with the following *kolova* (loud crying of names, Text 87, loc. cit., § 24):

"(i) People of Omarakana, coconut for your refreshment at work! (ii) People of Tilakayva, areca-nut for your refreshment at work! (iii) All of you, a pig for killing for your refreshment at work!"

As was usual on such occasions, the refreshments were distributed according to localities. All the people present had assisted the master of the proceedings—in this case Kumatala—in carrying his taytu to the chief.

Several times I have accompanied such parties from the gardens to Omarakana and back. On the return journey they would pause in the jungle to eat the refreshments which had been given them. Then they would go on to the village from which they started; they would eat and drink and chew and rejoice. Sometimes they would be given a pig, which would be presented ceremonially with the words: *mi bulamata* ("your pig to be killed and eaten"). If so, it would be roasted alive, carved, and eaten directly on their return to the village. It was on such a day, when I had returned with a party to Kwaybwaga, that my young friend, Toyagwa, epitomised—gorging himself with pig and taro—all that is desirable in life: "We shall chew, we shall eat: we shall vomit during the day; we shall vomit at night—so much fat all round us."[1] This was while the remainder of the pig was being cut up and its horribly fat entrails were scattered about the ground.

After everyone had brought their share of the harvest, the heaps remained for a few days on exhibition, and it was then that I was able to draw up my Table of To'uluwa's Harvest Gift (Doc. II), inspecting each heap, collecting numerical details, pedigrees and

[1] This idea of surfeit as the symbol of great plenty is found in M.Ff. 16 and 25.

traditions. On that occasion better than on any other was I made to understand how inextricably the economic side of chieftainship was bound up with their political power, how their tributes were sanctioned, and how they were later used (cf. Part I, Sec. 9).

It was a piece of very bad fortune for myself, and perhaps for this book, that owing to a spell of ill-health I had to leave Omarakana for a few days and move into the verandah of my friend, Billy Hancock, on the shores of the lagoon, while the new yam-house was being built and the taytu stored. The glowing accounts which I afterwards received—how the people could hardly find standing-room on the *baku*, how several people almost speared one another, how quarrels broke out which in old days would have led to a war, did not compensate me for my lost opportunities in the way of photography and observation of ceremonial technology and economic acts.

I was able, however, to ascertain for a definite fact that there is no special magic connected with the building of a new *bwayma* except the magic of *vilamalia* described in the next chapter. I was also able to inspect the yam-house after its erection and to discuss most of the technological problems, which, moreover, I have several times seen practically solved on minor storehouses. But the *bwayma* is an object of special interest and will be discussed in Chapters VII and VIII.[1]

[1] See also Note 29 in App. II, Sec. 4.

CHAPTER VII

THE WORK AND MAGIC OF PROSPERITY

Now that we have grasped the sociological principles of the filling of the storehouse, the *dodige bwayma*, we must resume the trend of our narrative. We left the crops arranged in heaps in front of the storehouses, to which they had been carried with much pomp and circumstance. Now the taytu has to be stowed away in the storehouse, a process quickly finished but of momentous importance. Taytu, the pre-eminent, the staple food, is the basis of tribal enterprise; it can be transformed into objects of permanent wealth, by the simplest form of capitalising, that of feeding the workers; it can be kept and paid out for services, thus giving power to those who possess it (cf. Part I, Sec. 10). Therefore it is the foundation of native expansion, wealth and power that is being stored away in the act of filling the *bwayma* (storehouse); it is the *bwayma* that makes the accumulation and preservation of wealth possible. Hence the *bwayma* is a permanent centre of interest as well as the centre of the activities we are about to witness.

Every visitor to the Trobriands will be impressed by the prominent position which the *bwayma* occupies: higher and more imposing than the living-house; more lavishly decorated, more scrupulously kept in repair; surrounded with many more taboos and rules of conduct (cf. Pls. 72, 73, 75 and 49). Even as the Irish or Polish peasant takes more count of his pigs than of his children, and looks better after his cattle than after his wife, so the Trobriander is more interested in the dwelling of his yams than in the dwelling of his family. He is obliged to look after his storehouse, because if it falls to pieces, his fortunes will fall with it; but he tends it with a care which passes quite beyond what is required by practical necessity. The storehouse is more to him than a mere mechanical contrivance for preserving the taytu. We shall follow the magic of prosperity performed over it; we shall watch the ceremony of filling and the care with which this is done; we shall see how the *bwayma* is adorned with food; and how even the emptying of it is subject to strict rules. The taytu is, as we know already, not merely an article of food to the native: it is a means of carrying out many enterprises; it is a symbol and vehicle of value and it is an object of aesthetic satisfaction (cf. Part I, Sec. 10). To the different functions of the taytu there correspond differences in the *bwayma*. The large open *bwayma*, with its partially exposed interior, provides for the display of taytu as well as for its accumulation; whereas the small enclosed storehouses subserve only the latter end. To these functions there corresponds a differentiation in structure, place in the village, sociological rôle and magical treatment, as well as in the ideas, beliefs and sentiments as the native.

It is the correlation of all these elements which gives the real significance to the object which we are to study, that is, to the Trobriand storehouse as a centre of native interest and as the foundation of native economic life.

In the magic of *vilamalia*, I think I was fortunate in being able to observe a most interesting form of magical induction referring to nutritive processes and the appetite. The Trobriander's misapprehension of the fundamentals of human procreation[1] is here matched by his misunderstanding of the processes of nutrition and metabolism.

1. THE MAGICAL CONSECRATION OF THE STOREHOUSE

Returning now to the taytu, we remember that in a big village as well as in a small one, the date for the filling of the *bwayma* once fixed, the work is rapidly carried out in one morning. On that day all the donors, together with their retinue of helpers and the recipients, are present. No owner may fill his own *bwayma*, it has to be done by those who give and by those who help the giver.

But before the work begins, the *towosi*, the garden magician, has to perform an act of supreme importance: the last act but one of the magic connected with gardens and food. For it he assumes a different title. The magic which he is to perform now is called *vilamalia*, and in his new capacity he is called *tovilamalia*. But both types of magic, the *towosi* and the *vilamalia*, are invariably done by the same man. In Omarakana it is Bagido'u again who has to act. Since it is his spells that I have recorded and his system of magic which we have followed most closely, it will be best to return once more to the capital village and to watch him at his work among the storehouses, even as we have watched him in the gardens.

The *vilamalia* is a magical frame to the filling of the *bwayma*: the first ceremony inaugurates the filling, and the second winds it up. This is the case not only in Omarakana, but in all other villages. In fact the main outlines of the magic and the ideas underlying it are the same throughout the region, and everywhere we find a curious discrepancy between the symbolism of rite and spell, and

[1] Cf. *Sexual Life of Savages*, Ch. VII, on "Procreation and Pregnancy, in Native Belief and Custom".

the native theory of this magic. But whatever the discrepancy, both ritual and comment agree on one point: the magic is to make the taytu last, remain; it is to make the village full of *malia*, prosperity. Hence its name: *vilamalia; vila-* prefix corresponding to *valu*, village; *malia*, prosperity.

We have spoken before about the contrast between *molu* (hunger) and *malia* (plenty, prosperity, and satiety). *Malia* has also the wider meaning of 'wealth' and of 'absence of disease, dangerous influences and disaster'.[1]

On the eve of the day agreed upon for the filling of the storehouse, Bagido'u goes into the bush and collects three bunches of magical herbs, *setagava*, *kakema* and *kayaulo*. *Setagava* is a tough weed with roots which are very strong and difficult to pull out of the ground. The *kakema* is a dwarf tree also with powerful roots and immovable save with great effort. The *kayaulo*, the totemic tree of the Malasi clan, is extremely tough; the wood can be cut with an axe or knife, but it is impossible to break it. Thus all the magical substances of this ritual are associated with tenacity, toughness and compactness.

Next morning before sunrise, at the time when the first *saka'u* bird utters its melodious wail, the magician repairs to the storehouses. He begins with the large *bwayma* of the chief, the *bwayma* which stands in the middle of the village and which has a personal name, *Dudubile Kwaya'i*. This name "Darkness of the Evening" is associated with the impression of darkness produced by the wealth of the stores which it contains. In order to understand the native distinction between light and darkness, compare Plates 74, 81, 87 and 91 where the empty stores let the light through, giving an effect of transparency, with Plates 72, 79, 82 and 83, where the full storehouses give an impression of darkness and solidity. Before refilling, the storehouses are completely cleared, the useless tubers thrown away and the sound ones taken into the dwelling-house. On Plate 81 this process can be seen in execution: the right-hand *bwayma* has been completely emptied; in the left-hand one a few tubers still remain in the interstices between the beams. It is characteristic of the natives' love of display that this outer layer—this shadow of wealth—is left to produce an impression of fullness, when the cabin behind is empty. Bagido'u climbs up the wall of logs, using the interstices as a ladder, and descends to the floor (*bubukwa*) of the well. He squats on the floor, lays down a bunch of the leaves,

[1] Whether the word *malia* corresponds to the Polynesian and Melanesian word *mana* (power of magic) I cannot decide, though there are certain etymological indications pointing in that direction (cf. Part V, Div. V, §§ 4 and 5; also Note 30 in App. II, Sec. 4).

and takes up a stone which has been there since the building of the *bwayma*. Holding the stone close to his mouth he charms it over with the following spell:—

FORMULA 28

I. "Rattan here now, rattan here ever, O rattan from the north-east!
Come, anchor thyself in the north-east.
I shall go, I shall fasten in the south-west.
Come, anchor thyself in the south-west.
I shall go, I shall fasten in the north-east.
My bottom is as a *binabina* stone, as the old dust, as the blackened powder.

II. "My yam-house is anchored; my yam-house is as the immovable rock; my yam-house is as the bedrock; my yam-house is darkened; my yam-house is dusky; my yam-house blackens; my yam-house is firmly anchored. . . .
It goes, it is anchored for good and all.

III. "Tudududu. . . .
The magical portent of my yam-house rumbles over the north-east."

This spell and rite, performed on the floor of the storehouse, is named "the pressing of the floor", *tum bubukwa* or *kaytumla bubukwa*. In it we have first of all the symbolism of the "lawyer cane" (rattan), to the native associated with an obstinate toughness and tenacity, surpassing that of all other vegetable growths. The lawyer cane is invited to encompass about from the north-east to the south-west and to be firm. Then the bottom of the taytu is identified with the *binabina* stone, the stone over which the magic is spoken. The taytu is also identified with ancient dust and with blackened dust such as is found on the *kuroroba*, the shelf in the house where the clay pots (*kuria*) are put. The idea underlying this is that the taytu should remain so long in the storehouse that the floor of the latter should become covered with black dust. Then in a direct statement the storehouse is said to be anchored, to be like a coral outcrop still joined to the bedrock; to be like the bedrock itself. And towards the end, the idea of darkness, which here means fullness, plenty of taytu, is again developed.

In this spell therefore we have definitely an impression of the desire to give stability to the accumulated crops: the same idea as we saw embodied in the magical herbs used in the rite and also in the stone which the magician picked up on the floor of the *bwayma*. Such stones are called *binabina*, which is a generic name applied to all the volcanic and basaltic rocks and stones found in the d'Entrecasteaux archipelago. Coral stones, the only kind found in the Trobriands, the natives call *dakuna*. *Binabina* stones are imported from the south, and in the principal storehouses of the Trobriands there are always one or two such stones on the floor. In this capacity they are called *kaytumla bubukwa*, "the pressers of the floor". Their function is to impart their qualities to the stored food; they are heavier, hardier and less brittle than the local dead coral.[1]

After the magician has chanted over the stone, pressed the magical herbs and finished with the main storehouse he visits the other open *bwayma*. Where there are no stones the magician takes a sound taytu left over from the old crops for this purpose and uses it in the same fashion.

2. THE FILLING OF THE *BWAYMA*

Although the magician started at daybreak, the filling of the storehouses has to begin almost immediately after he has finished. For his work is by no means easy or short: he has to make the round of a large village, climbing into one storehouse after the other. The filling of the *bwayma*, which has to be completed in one morning, is usually begun fairly early, about eight o'clock. I have seen the filling of the *bwayma* in Omarakana only once, and as it was raining heavily I could take no pictures. The photographs (Pls. 75–77) of the *dodige bwayma* in the village of Yalumugwa, however, illustrate well the method of filling. A boy or young man stands in the log cabin of each storehouse and another man hands him the yams through the interstices, and these are disposed in layers on the floor. This process is well illustrated in the plates just mentioned. If there are several donors, each of them has his own delegate and his own compartment inside. One donor may have more than one heap outside, but he never has more than one compartment within (cf. Ch. VIII, Sec. 4).

The same principle is followed in filling the *bwayma* as in stacking up the heaps, that is, the best yams are displayed on the outside. First a few of the finest tubers are handed up and these are arranged in the lowest interstices between the beams of the cabin. Some of the poorer ones from the interior of the heap follow, and are placed on the floor. Then again some better tubers are disposed in the higher interstices, and so on. Since most of the big tribal ceremonies at which good taytu is used, follow almost immediately after harvest, there is a tendency to store the better taytu at the top of the *bwayma*, whence it can be easily removed.

[1] See also Note 31 in App. II, Sec. 4.

When there is a good crop there is enough taytu to fill the log cabin and to overflow into the space under the thatch. The vertical sticks which divide the well into compartments run close up to the roof, so that the allocation of space to various donors is continued from top to bottom of the storehouse (cf. Ch. VIII, Sec. 4).

All this refers only to the open, decorated storehouses in which the show crops, the taytu whose main function is display and the representation of wealth, are stored.

After they have finished their task the fillers sit down in groups and partake of refreshments. Since by the very principle of *urigubu* each donor has some near kinswoman in the village, he and his party will settle down near her house, open and drink the green coconuts provided for them, crack and chew betel-nuts, eat cooked yams, taro and bananas. In a chief's village a pig or two would often be killed and distributed among those who have filled the yam-house.

After this the strangers retire, and within the village a new distribution takes place: taytu is taken out of the *bwayma* almost immediately after it has been stored there and *kovisi* and *taytupeta* gifts are distributed. The *kovisi* is taken out of the *urigubu* just received, the small gifts are taken from the own harvest crop.

3. THE SECOND ACT OF *VILAMALIA* MAGIC

The day after the filling, or perhaps two or three days later if the weather is bad, the *tovilamalia* performs his second and concluding rite—the *basi valu*, "the piercing of the village". In the morning he goes to the bush to collect leaves from the tree *lewo* which grows on the belt near the seashore, and from the *kayaulo* tree; also wild ginger root, *leya*. The *lewo* is a hardy but stunted tree which is said to live for a very long time. The *kayaulo*, as we know, is a tree with tough, unbreakable wood, and the *leya* is in magic always associated with fierceness and toughness. In his house, about noon, the magician chants over the *lewo* leaves, the *kayaulo* and *leya*. This is the spell:—

Formula 29

I. "Anchoring, anchoring of my village,
Taking deep root, taking deep root of my village,
Anchoring in the name of Tudava,
Taking deep root in the name of Malita.
Tudava will climb up, he will seat himself on the high platform.
What shall I strike?
I shall strike the firmly moored bottom of my taytu.
It shall be anchored.

II. "It shall be anchored, it shall be anchored . . .
My soil shall be anchored.
My *ulilaguva*, my corner-stone, shall be anchored.
My *bubukwa*, my floor, shall be anchored.
My *liku*, my log house, shall be anchored.
My *kabisivisi*, my compartments, shall be anchored.
My *sobula*, the young sprout of my taytu, shall be anchored.
My *teta*, the sticks that divide my log cabin, shall be anchored.
My *bisiya'i*, my decorated front board, shall be anchored.
My *kavalapu*, my gable-boards, shall be anchored.
My *kiluma*, the supports of my thatch, shall be anchored.
My *kavala*, my roof batten, shall be anchored.
My *kaliguvasi*, my rafters, shall be anchored.
My *kivi*, my thatch battens, shall be anchored.
My *katuva*, my thatch, shall be anchored.
My *kakulumwala*, my lower ridge pole, shall be anchored.
My *vataulo*, my upper ridge pole, shall be anchored.
My *mwamwala*, the ornamented end of my ridge pole, shall be anchored.

III. "It shall be anchored.
My village is anchored.
Like an immovable stone is my village.
Like the bedrock is my village.
Like a deep-rooted stone is my village.
My village is anchored, it is anchored for good and all.
Tudududu . . . .
The magical portent of my village rumbles over the north-east."

With the utterance of this formula, the magician has come to the end of all the spells connected with gardening, harvest and crops. A comparison of this spell with the main formula spoken in the *kamkokola* magic (M.F. 10, Ch. III, Sec. 4) will show that the two are identical in pattern, and differ only in the substitution of the various parts of the *bwayma* for the parts of the *kamkokola* and different kinds of vine support. By comparing the native text of this spell (see M.F. 29, Part VII) with the structural description of the *bwayma*, it is possible to assess the extraordinary accuracy and minuteness with which the natives reproduce technical details in their magic.

Comparing both this and the first *vilamalia* spell with the various magical formulae used during the gardening cycle, it is clear also that the *vilamalia* is part of the magic of gardening. Both spells definitely express the desire to make the food strong, tough, resisting all forces of decay and consumption.

PLATE 74

EMPTY STORES READY FOR FILLING

Immediately before filling all storehouses are completely cleared of old tubers. The sun shining through the *liku* shows the effect of transparency and lightness produced by emptiness. The new heaps covered with palm leaves are stacked in front, and a large *kuvi* decorates one of the storehouses (Ch. VII, Sec. 1)

PLATE 73

STOREHOUSES IN OMARAKANA DURING A TEMPORARY INUNDATION

A somewhat irregular section of the inner ring: on the right front two show storehouses belonging to men of rank; in the background one or two enclosed *bwayma*, and in the centre background two living-houses. The small boys are playing with toy canoes, and Tokulubakiki has joined them (Ch. VII, Intro.)

PLATE 75

THE CEREMONIAL FILLING OF BWAYMA IN YALUMUGWA

The heaps are carried to the storehouse. A man can be seen standing on the platform, handing the tubers to those inside (Ch. VII, Sec. 2). Note that there are no heaps in front of the enclosed store in the centre background. Through the coconut matting which has slipped it can be seen that it is still empty, but it will not be filled simultaneously with the show *bwayma*

After the herbs have been charmed they remain between two mats until near sunset when the magician starts to make his last round of the village. He begins naturally with the main storehouse. In front of it he makes a hole in the ground with a small stick called *dimkubukubu* or *katakudu*, made of *kayaulo* wood and charmed with the other substances. Into this hole he puts some *lewo* leaves and a twig of the *kayaulo* tree, squats down and chants into it the charm just quoted. Then, through one of the interstices of the *liku* (log cabin), he thrusts some of the *lewo* leaves among the tubers. He chews some wild ginger root and ritually spits right in amongst the taytu in the *bwayma*. Subsequently he makes the round of the other show storehouses and repeats the same proceedings at all of them, but without chanting the spell over the hole. For the minor storehouses, those standing outside the inner ring, no hole is made and only some leaves of the *lewo* are inserted and the yams are spat upon ritually.

In passing over the *kadumilagala valu*, the points where a road strikes the village (cf. Pl. 78), at Omarakana seven in number, the magician chews ginger and spits on the ground. On the broad entrance which opens from the twin village of Kasana'i, Bagido'u makes one more hole and chants the spell into it. He also spits with *leya* over the spot.

I saw the *basi valu* (piercing of the village) performed on a wet, sultry afternoon in the southern mid-winter of 1915 (June or July). This was the first time that I was allowed to witness any important magical ceremony in the Trobriands. Bagido'u was accompanied only by his younger brother Towese'i, who held a mat spread over the magician and his ritual paraphernalia. They went to the large *bwayma* without any pomp or ostentation. From a distance the two might have been taken for men repairing or adjusting something about the storehouses. There was no unction, solemnity, or display of any transcendent or esoteric qualities in the proceedings. The whole action was businesslike—everything was done quietly and deftly. The complete absence of any crowding spectators or even of any interest or curiosity on the part of the villagers, seated under the eaves of their houses or *bwayma*, also contributed to divest the performance of any solemn, ceremonial character. Although there is no definite and explicit taboo, it would not be considered the proper thing to crowd round the magician or to gaze at him or show any undue interest, and children would be kept back on such an occasion. This principle accounts for the general lack of ceremonialism in Trobriand magic, with a few exceptions, such as the garden rite described in Chapter IX (Sec. 2) where a larger number of people have to take part in the ceremony.

4. THE OBJECT AND FUNCTION OF THE *VILAMALIA*

As I have mentioned, perhaps the most remarkable feature of this magic is the discrepancy between its meaning as revealed in an objective analysis of spell, rite and context, and its aim as laid down in the comments of everybody concerned, including the officiating magician himself.

Both of course agree that the magic is a magic of plenty, that it is a magic meant to prevent hunger. But whereas the objective facts reveal to us that the whole performance is directed at the yam-house, at the food accumulated there, the comments of the natives make the human organism the real subject-matter of magic influence.

Let me once more survey the magical facts. Both rites are performed over the storehouses. In the first rite the magician presses the floor, even as the title of the ceremony indicates, *tum bubukwa*. The substances used in this ceremony are all symbolical of tenacity and strength. It is these substances which are pressed on the floor, and they are pressed by means of a stone which symbolises stability. The words of the first spell—with its metaphors of anchoring, of the lawyer cane encompassing the storehouse; with its comparisons, and direct comparisons at that, to bedrock, to a coral outcrop; with the invocations of darkness and plenty—leave not the slightest doubt that it is directed at the yam-house and is meant to make the yam-house—that is, its contents—resistant, tenacious and enduring. The second rite is also performed over the storehouses; the hole is made in front of a storehouse and resistant substances are inserted into it. The same substances are put among the stored food, and this food is spat upon ritually. And here again, the spell tells the same tale. The anchoring and heaping up refer to stability and to plenty. Every detail of the storehouse is enumerated; the whole village, which means really all the food stored in the village, is made immovable and unshrinkable by verbal imprecation. The piercing of the ground of the village would obviously mean a magical isolation and entrenchment. Looking at Plates 72 and 79, where the largest *bwayma* and one of the largest are shown, each standing in the middle of its village like a solid block of prosperity and plenty, we can appreciate the correspondence between the visual impression produced by the aesthetic manipulation of food and the aim of *vilamalia* magic.

What, on the other hand, is the gist of native comments? They have not the slightest doubt that the magic does not act directly on the substance of the food but on the human organism, more specifically on the human belly; or, to use a non-native word, on the appetite. It is not the food to be eaten up which is made resistant to nutritive destruction, it is the mouth which eats and the oesophagus which swallows that are made sluggish and disinclined. "Supposing the *vilamalia* were not made," I was told by Bagido'u, "men and women would want to eat all the time, morning, noon and evening. Their bellies would grow big, they would swell—all the time they would want more and more food. I make the magic, the belly is satisfied, it is rounded up. A man takes half a taytu and leaves the other half. A woman cooks the food; she calls her husband and her children—they do not come. They want to eat pig, they want to eat food from the bush, and the fruits of trees. *Kaulo* (yam-food) they do not want. The food in the *bwayma* rots in the *liku* till next harvest. Nothing is eaten."

I have put together here a number of statements received on various occasions from Bagido'u which I noted down at the time in English. Time and again I discussed the same subject with other men and I found everyone confirming his view. For instance, the same point was put more briefly thus by another informant (Text 72, Part V, Div. X, § 12): "When we do not make the magic of prosperity, the belly is like a very big hole—it constantly demands food. After we pierce the village the belly is already satisfied."

The theory is not astonishing in itself. The natives are not aware of the need for supplying the organism with new material, and their ideas about the digestion and the physiology of nutrition are rudimentary. They believe that food is transformed in the stomach (*lulo*) into excrement (*popu*). Eating in their opinion is mainly done to satisfy appetite, and because it is pleasant. They have a dim glimmering of the connexion between food and life. They know that famine produces all sorts of illnesses and can ultimately kill a man. Also, in speaking of old age, they will say that the stomach becomes closed up when a man is very old and then he dies. None the less abstention from food is to them a virtue and to be hungry, or even to have a sound appetite, is shameful. Hence you must not speak about being hungry, especially in a strange village. Therefore to reduce the desire for food to its minimum, to make man dispense with food as much as possible, must appear to them an excellent device. And again *kaulo*, the farinaceous staple food, is to the native his daily bread and not a great delicacy.[1] To eat somewhat less *kaulo* every day would not be even a serious hardship to him. I was puzzled as well as amused by this point of view, and I often discussed it with various informants. They would tell me in praise of Bagido'u's magic of Omarakana that his was the best *vilamalia*.

[1] Cf. Part V, Div. II, § 12.

He himself boasted of it often, telling me that many can do good *towosi* (garden magic) but nobody can match him at *vilamalia*. "And what would be the good of harvesting splendid crops if they were eaten in a hurry because people had too much desire for food?"

In 1915, when I was in Omarakana for the first time, I had not yet acquired much taste for taytu, though later on I grew really to like it. Bagido'u told me as a proof of the efficiency of his magic that I myself preferred mango, bread-fruit, or bananas; that I preferred even pineapple to taytu; that I ate a great deal of taro and things out of tins, instead of eating taytu. This he said was the result of his *vilamalia*. He also pointed out to me, which was quite true, that in spite of a very meagre taytu harvest in 1915, the *bwayma* remained full for a long time. He forgot to mention or to notice that, following the poor harvest, there was an exceedingly good season for fruits as well as for taro.

Thus the natives believe that the magic acts on the human organism while the magic itself tells quite as clearly and consistently that it aims at the storehouses. The discrepancy must remain, as this is not the place to discuss possible explanations of it.[1]

5. THE FUNCTION OF THE STOREHOUSE[2]

Our account of the ceremonial filling of the storehouses and the magic performed over them refers only to the open show storehouses which are, strictly speaking, the privilege of rank. These are large rather than numerous, as size in itself has an aesthetic value to the Trobriander; they are built with an open log cabin to permit of an ostentatious display of their contents, and they are conspicuously placed, usually in the first or inner ring round the *baku*. A few, indeed, stand right in the centre of the *baku*—those of Omarakana, Kasana'i, Olivilevi and Kabwaku (Pls. 72 and 79). They can be, and even nowadays (1918) often are, decorated with carving and with white, black and red paint; and, especially if they have been recently repaired, hung with an array of pandanus streamers, shells, maize cobs, large painted yams, coconuts and pigs' jaws.[3] Thus, in those villages which consist in a double ring

[1] See also Note 32 in App. II, Sec. 4.

[2] The function and structure of the storehouse are fully dealt with in Chapter VIII, where, for methodological reasons there set out, technology has not been treated apart from function. Therefore I shall only touch here on such aspects of the *bwayma* as may help to the understanding of Chapters V and VI and the three first sections of the present chapter.

[3] Pandanus streamers can be seen on Plate 72; painted boards on Plates 75 and 76; a large yam on Plate 86; and coconut trimming on Plate 79.

THE FILLING OF THE STORES IN PROGRESS

This picture, taken a little later than the previous ones, shows a more active phase in the filling. Two men are putting tubers into baskets: four are at work on the large *bwayma* on the left, two on the next, and three on the one on the extreme right (Ch. VII, Sec. 2)

PLATE 77

DETAILS OF FILLING A STOREHOUSE

A man is rearranging tubers in a relatively small storehouse. There are two boys inside, the toes of one show just above the hand of the man, the hand of the other to the left. A man and woman are transporting the taytu from a dismantled heap. In the background a hut is seen, flanked by two enclosed stores (Ch. VII, Sec. 2)

PLATE 78

WHERE THE ROAD FROM THE SEA STRIKES THE OUTSKIRTS OF OMARAKANA

(Ch. VII, Sec. 3)

of houses, the *baku*, the centre of village life, the place of dancing, festivals and rejoicing, is surrounded by an imposing circle of *bwayma*, through the interstices of which the Trobriander can see and gloat over his accumulated wealth of taytu.

A stranger crossing the district and passing through a number of villages is puzzled and impressed by the fact that the highest and best buildings are not the habitations but the storehouses. Even in villages of rank the chief's storehouse is bigger and better built than his personal hut, as can well be seen on Plate 72. This is because, as we already know, the storehouse is much more important as a source of power and a symbol of it. Add to this that dwellings must be built flush with the ground and huddled together, for fear of the sorcery that might otherwise creep underneath or between them, and we understand how it is that these are always overtopped and outshone by the *bwayma*.

But besides the show *bwayma* there are small enclosed storehouses, modestly situated in the outer ring among the dwelling-houses; yet so placed as to be well in view of the community and the owner in order to minimise the danger of pilfering (cf. Pl. 77). They may at times present a rather ramshackle appearance (cf. Pl. 80, which shows an extreme example of this). No taboo restricts their use and they are easily accessible, as the husband or wife have constantly to repair to them for the daily yam.

The chief has no inferior, domestic storehouse; *urigubu* and *taytumwala* (own taytu) are alike arrogantly displayed. The commoner, on the other hand, has no show storehouse; and none of his stored food is displayed. His *urigubu*, though stacked apart and reserved for ceremonial occasions and exchanges, is stored in the enclosed *bwayma* with his *taytumwala*.

These two extremes, the chief and the poor commoner, shade gradually into each other through the various degrees of rank. The more aristocratic and richer citizens would have bigger show *bwayma* and would store more *taytumwala* in them; those not so well endowed and of lower rank less, and so on.

Though less showy, the inferior storehouses are more important than the show *bwayma*, for they contain the food used for daily consumption, and also the seed yams for next year's planting.

The filling of these *bwayma* I cannot, unfortunately, describe with the same fullness of detail and documentation as the ceremonial ones, though I have seen it done countless times and have even taken part in it. I am not even able to state with any degree of accuracy what proportion of the aggregate crops harvested is stored in them. This is due to that defect in my method which led me to pay more attention to the ceremonial and dramatic than to everyday events (cf. introductory remarks to next chapter).[1]

Even had I not made this mistake, however, an accurate computation of *taytumwala* as compared with *urigubu* would have been very difficult. The *urigubu* is counted and the count recorded with *kalawa* leaves; it is displayed, boasted about and well remembered. Quite the opposite is the case with the taytu kept for a man's own use. The transport and storing of it goes on unostentatiously from day to day. Far from being a subject for boasting, the quantity of food reserved by a man for his own use may even be concealed. If a man for any reason receives a small *urigubu* so that he has to keep a large proportion of his harvest to supply the needs of his own household, this is considered shameful. It would be concealed by the man himself and it would be extremely bad manners in anyone else to talk about it. Add to this that such a misfortune would be most likely to befall commoners who do not display their *urigubu* to any extent, and the difficulties of getting accurate information are obvious.

Speaking, however, from innumerable though undocumented observations, I should say that about half of the taytu harvested is show taytu, the other half being used for daily consumption and for next year's seed.[2]

Food is collected from the time of the *basi*, the preliminary thinning out of tubers. I am not quite sure whether the *bwanawa*, the thinned-out tubers, can be stored or whether they must be eaten immediately, but I think the latter is true. Certainly they are never put in the show *bwayma*, and if kept would be kept in the dwelling or hidden away in an inferior storehouse.

At the regular harvest, the *tayoyuwa*, most of the inferior taytu, called generically *unasu*, is brought straight to the domestic storehouse, though some of it is displayed in the *kalimomyo* (arbour). The seed taytu, of which a man is proud, is always first exhibited in the *kalimomyo*, but afterwards it is transferred without ostentation by the owner and his family to the storehouse. The *ulumdala*, the gleanings after the main harvest, are also put in the inferior *bwayma*.

As in the filling, so also in the removal of crops, a distinction is made between the show garners and the enclosed ones. The latter have to be visited daily and their structure is adapted to their use. The man or woman has only to enter them and reach up his hand to secure the necessary yam. The show storehouses, on the other hand, are only emptied occasionally and then much larger quantities

[1] See also Note 33 in App. II, Sec. 4.

[2] See also Note 24 in App. II, Sec. 4.

PLATE 79

THE MAIN BWAYMA OF KASANA'I FILLED

This photograph, taken in 1915, which was a lean year, shows this storehouse filled almost to the top, bordered with coconuts and with a shell trumpet lying on the platform (Ch. VII, Sec. 4)

PLATE 80

A RAMSHACKLE STOREHOUSE

The dwelling-hut and the *bwayma* of a fisherman of a lagoon village (Ch. VII, Sec. 5)

PLATE 81

STOREHOUSES WHICH HAVE BEEN ALMOST EMPTIED

Stores being emptied for refilling, and the taytu stacked in baskets. Whether in this case the taytu was removed and stacked in the baskets to be used on some ceremonial occasion, or merely for clearing, I am not quite certain. The vertical division into compartments can be seen; one such compartment is not quite cleared, at least the outer layer still remains in the interstices. Note the long sprouting stalks of the tubers, both in the baskets and in the store (Ch. VII, Sec. 5)

are taken out. At a ceremonial distribution (*sagali*), or on occasions when large presents such as *vewoulo*, *dodige bwala* or *yaulo* are given; or again when considerable quantities of food are sent to the coastal villages at a *vava* or *wasi*, the exchange of vegetable food for fish, then only will men and women climb the *liku* and take out several basketfuls of taytu, which are usually left for a few hours on show in front of the storehouses before they are carried to their destination. Such a procedure is shown on Plate 81. On such occasions the crops are graded. Some of the really perfect tubers which line the interstices and are disposed on top of the compartment are selected and the baskets topped with them; and at the subsequent distribution or exchange they are always kept in this position. In the case of the exchange of taytu for fish, the commoners would contribute their quotum from their *urigubu.*[1]

A characteristic of the show storehouse is that tubers once taken out of it are never returned to it and its contents are never added to until it is filled again at the next harvest. The only exception to this rule is found in coastal villages where it is customary to refill storehouses with the taytu received in exchange for fish.

In years of plenty it sometimes happens that the best tubers, those placed in the interstices of the *liku*, are never eaten. Exposed to sun and rain, they sprout and send out long shoots, and become less palatable for eating (cf. Pl. 81, where the shoots can be seen creeping out of the *bwayma* and sticking out of the baskets). Because this is a sign of *malia*, it is not a matter for regret but for congratulation.

Both types of storehouse have their associated or extraneous uses. The big *bwayma* furnish the decorative setting to the central place of the village. Dancing, games, athletic competitions, social and official gatherings all take place on the *baku*, and the surrounding *bwayma*, filled with taytu to the brim, speak to the villagers of prosperity, feasting and satiety (cf. Pl. 82). To outsiders they speak of the welfare of the village, they advertise its wealth, they provide the necessary *butura* (renown). In villages where there are no special shelters or sitting platforms provided for the comfort of guests and villagers, such as we find in the capital (Pl. 83) and in the pariah village of Bwoytalu (cf. Pls. 84 and 98), the front of the show *bwayma* may be used instead (cf. Pls. 85 and 86). But it is not strictly proper to use a ceremonial storehouse in this familiar fashion. Where other platforms are provided, there would be none under the overhanging gables of the *bwayma*, and only the owner and such few privileged persons as he might choose to invite

[1] See also Note 34 in App. II, Sec. 4.

would venture to sit on the projecting ends of the foundation-beams.

The smaller storehouses, on the other hand, play a prominent part in the day to day social life of the village. On the open platforms, elevated and dry, men will sit (cf. Pls. 95–97). Within the store itself, when not too full, boys and girls, taking advantage of the perfect privacy of such retreats, will meet for love trysts and amorous dalliance. The domestic storehouse does afford privacy, for good taste forbids any to enter it but the owner and his relatives, that there may be no suspicion of *vayla'u* (theft of food) which is regarded as specially despicable. Also their discreet position in the outer ring among the dwellings and the inconspicuous entrance from behind, make them very suitable for private meetings.

Associated with their rank and the position of storehouses in the village is the idea of the sensibility of taytu to the smell of cooking. It must not be pervaded by the steam of cooking-pots nor by the smoke of baking tubers. The yam-houses of the inner ring, which are open and therefore specially accessible to smell and smoke, are protected by a taboo on cooking imposed on any dwellings that may stand in the inner ring. Only the chief's personal hut (*lisiga*) (cf. Pl. 72) or a bachelors' house (*bukumatula*)[1] is ever erected there, and in these cooking is forbidden. A conventional or symbolic protection is sufficient for the small storehouses, which stand in the same ring as the dwelling-houses (*bulaviyaka*) where the meals are prepared: their covering of plaited coconut leaves screens the taytu from the sight and at the same time protects it from the obnoxious smells of cooking.

A storehouse, whether big or small, always belongs nominally to the husband. He is, as a rule, a citizen of the community in which the household is situated, but even if he is an outsider and lives there through cross-cousin marriage or because he is a chief's son, he still owns the storehouse. At the same time in a polygamous, and therefore aristocratic, household, each wife has an open *bwayma* specially allotted to her and filled by her kinsmen. The smaller enclosed *bwayma*, owned by the man and used by the woman, really belongs to the household.

An attractive feature of Trobriand villages is the diminutive storehouses on high stakes owned by small boys (cf. Pl. 75, in distant centre background, and 99). The taytu from them is used by the mother, but the ownership, with all that it implies of pride and vanity, is vested in the boy.

[1] Cf. Pls. 20 and 21 in *Sexual Life of Savages*, where the institution of the bachelors' house is described in Sec. 4 of Ch. III.

PLATE 82

FILLED STOREHOUSES

"The surrounding *bwayma*, filled to the brim with taytu, speak to the villagers of prosperity, feasting, and satiety." These are the same stores which were seen empty on the previous plate. Now they produce the solid, heavy, dark impression appreciated by the natives (Ch. VII, Sec. 5; cf. also Sec. 1)

PLATE 83

THE CHIEF'S YAM-HOUSE AND SHELTER IN OMARAKANA

On the right-hand side, in front of the store, we see the sacred clump of aromatic and decorative bushes, near to it two men displaying arm-shells (Ch. VII, Sec. 5) Note the shell trumpets on the shelf.

PLATE 84

LARGE COVERED PLATFORM IN BWOYTALU

The industrious carvers of the village sit on such a platform and do their work. A number of stores are seen, all of them carefully covered (Ch. VII, Sec. 5)

6. THE MAGIC OF HEALTH, WEALTH AND PROSPERITY IN OBURAKU

I have described the *vilamalia* as it is practised in Omarakana. I shall now add a brief account of another system which I recorded in the southern parts of the district, in the village of Oburaku, and then compare the two and discuss once more the native theory of the effects and *modus agendi* of this magic.

In Oburaku, as in Omarakana, the *vilamalia* frames the harvest, but it is not as closely connected with the filling of the *bwayma*; it is in fact much less the magic of the yam-house and the yam crops and more the magic of the village, its welfare and that of the community. Thus the first ceremony does not take place immediately before the *dodige bwayma*, the filling of the garners. Nor is it called by any name associated with the structure of the storehouse, as is the case in Omarakana, where the first rite is "the pressing of the floor".

In Oburaka also there are two acts of *vilamalia* and the first one is carried out at the new moon preceding the *tayoyuwa* (the main taytu harvest). The magician, who is as ever both *towosi* and *tovilamalia*, repairs to his hut, taking with him a shell trumpet (*ta'uya*), devoted especially to this magic from year to year, some dry banana leaves, and some wild ginger root. First of all he takes the shell trumpet, which is a *cassis cornuta*. He stuffs the shell with dry banana leaves and into its open lips he utters the following charm:—

FORMULA 30

I. "Restore, restore . . .
Restore this way, restore that way.
Trumpet shell, restore, restore.

II. "Trumpet shell, restore, restore.
The hunger-swollen belly, trumpet shell, restore, restore.
The hunger exhaustion, trumpet shell, restore, restore.
The hunger faintness, trumpet shell, restore, restore.
The hunger prostration, trumpet shell, restore, restore.
The hunger depression, trumpet shell, restore, restore.
The hunger drooping, trumpet shell, restore, restore.
The throbbing famine, trumpet shell, restore, restore.
The utter famine, trumpet shell, restore, restore.
The drooping famine, trumpet shell, restore, restore.
Round the *tatum* (house), trumpet shell, restore, restore.
Round the *kaykatiga* (house), trumpet shell, restore, restore.
Round the earth oven, trumpet shell, restore, restore.
Round the hearth-stones, trumpet shell, restore, restore.

Round the foundation-beams, trumpet shell, restore, restore.
Round the rafters, trumpet shell, restore, restore.
Round the ridge pole, trumpet shell, restore, restore.
Round the front frame of my thatch, trumpet shell, restore, restore.
Round the shelves of my house, trumpet shell, restore, restore.
Round the threshold boards of my house, trumpet shell, restore, restore.
Round the threshold of my house, trumpet shell, restore, restore.
Round the ground fronting my house, trumpet shell, restore, restore.
Round the central place, trumpet shell, restore, restore.
Round the beaten soil, trumpet shell, restore, restore.
Round the *yagesi*, trumpet shell, restore, restore.
Round where the road starts, trumpet shell, restore, restore.
Round the roads themselves, trumpet shell, restore, restore.
Round the seashore, trumpet shell, restore, restore.
Round the low-water mark, trumpet shell, restore, restore.
Round the shallow water, trumpet shell, restore, restore.
Restore this way, restore that way.

III. "This is not thy wind, O hunger, thy wind is from the north-west.
This is not thy sea-passage, the sea-passage of Kadinaka is thy sea-passage.
This is not thy mountain, the hill in Wawela is thy mountain.
This is not thy promontory, the promontory of Silawotu is thy promontory.
This is not thy channel, the channel in Kalubaku is thy channel.
This is not thy sea-arm, the passage of Kaulokoki is thy sea-arm.
Get thee to the sea-passage between Tuma and Buriwada.
Get thee to Tuma.
Disperse, begone.
Get old, begone.
Disappear, begone.
Die away, begone.
Die for good and all, begone.
I sweep thee, O belly of my village.
The belly of my village boils up.
The belly of my village is darkened with plenty.
The belly of my village is full of strong beams.
The belly of my village streams with sweat.
The belly of my village is drenched with sweat."

After the spell has been recited, the magician breathes a strong guttural aspirated "Ha" into the open lips of the trumpet shell. Then he places it with its mouth downwards on a mat, so that the virtue shall not evaporate. Presently he takes a piece of dry banana leaf, folds it over so as to form a conical bag, and placing some wild ginger root in it, charms it over, again reciting the formula just given.

Late in March 1918 I witnessed this performance in the house of Navavile, the *tovilamalia* of Oburaku. Directly after the magic had been recited he and two or three of his nearest kinsmen went to the northern end of the village, where the road from Kiriwina strikes it. There one long blast was blown on the trumpet shell, and then Navavile chewed the ginger and ritually spat it several times towards west, north and east. Then we walked across the village to the south, where exactly the same ceremony was repeated. Then the trumpet shell was carried to the shore, and a man waded into the lagoon and dropped it into the water at a well-marked spot. A few months hence it will be fished out and used again.

Whether this last act, the submerging of the trumpet shell, means that prosperity should be anchored at the mooring-place of Oburaku; or whether it means that the trumpet shell, carrying with it the various evils and disabilities of hunger exorcised in the spell, should be drowned—I failed to ascertain.

All this happens before the beginning of the harvest. After the harvest is over and when all the storehouses have been filled, the magician has to perform the second ceremony. He repairs to the jungle and there he makes an extensive collection of leaves. Most of them come from the eastern seashore, the *momola*. Among them are the leaves of the *lewo*, the same tree that figures in the *vilamalia* system of Omarakana. This tree grows on the coral ridge and down towards the open sea. It is a very long-lived tree, with a strong stout stem, never growing very tall, but sending down its roots to the coral outcrops or the bedrock and gripping this very tightly. The natives told me that, as this tree lives long and is difficult to uproot and grows right in to the bedrock, so the taytu should remain long in the storehouses.

Another small but stout tree, the *bulabula*, which grows to a great age and has deeply penetrating roots, is used for the same reason. The *kavega'i* again is a tree of very stunted growth which spreads its branches very wide and low on the ground. The *kayaulo*, which we also know from the Omarakana magic of *vilamalia*, has a very hard wood. The leaves of the croton tree are also used in Oburaku, since the croton is associated with the sacred stones standing on the central place, which symbolise the stability and permanence of the village. The leaves of the casuarina tree are used because of the density and darkness of its foliage, while the flowers of the hibiscus symbolise to the natives the joy and festivity which always go with plenty. Besides these, Navavile has to collect the leaves of certain fruit trees, the bread-fruit, the *menoni*, the *gwadila* (a tree with nuts), the *saysuya*, the *sayda* (nut tree), the *gegeku*, and the leaves of the Malay apple. Curiously enough, the leaves of the pawpaw, an importation of the white man, are also used.

The magician tears up the leaves and makes a hash of the green stuff which he places between the folds of a large mat. Meanwhile the men of the village have assembled in front of his house, each man bringing with him a small *vataga* (oblong basket) and a digging-stick. The latter is handed over to the magician, who places the folded mat and the digging-sticks on the platform in front of his house, even as is depicted in Plate 109, though this refers to a different form of magic. Then in the presence of all those assembled, the magician chants the following spell :—

FORMULA 31

I. "Padudu, Pawoya,
Thy mother is Botagara'i,
Thy father is Tomgwara'i.
. . . .
. . . .[1]

II. "I exorcise, I exorcise, I exorcise.
I exorcise his illness.
I exorcise his weakness,
I exorcise his black magic.
I exorcise the foundation-stones of my village.
I exorcise the foundation-beam of my house.
I exorcise the big logs of my yam-house,
I exorcise the rafters,
I exorcise the ridge pole,
I exorcise the floor of my yam-house,
I exorcise the sticks that divide my log cabin,
I exorcise the rafters of my gable,
I exorcise the sprouting of my taytu,
I exorcise my gable wall,
I exorcise the beaten soil,
I exorcise the belly of my village.

III. "I sweep the belly of my village.
The belly of my village boils up.
The belly of my village is darkened with plenty.
The belly of my village is full of strong beams.
The belly of my village streams with sweat.
The belly of my village is drenched with sweat."

[1] Here several lines are omitted because they cannot be translated satisfactorily.

After the chanting of the spell, the magician distributes the magical mixture. Each man puts his portion into the *vataga* (oblong basket) and departs to that point on the outskirts of the village nearest to his house, where, making a small hole with the medicated digging-stick, he buries some of the leaves. Then he returns to his house, inserts part of the remaining mixture in the foodhouse among the taytu and thrusts the rest between the *urinagula* stones, the three stones forming the domestic hearth. This rite is called "piercing the village", *basi valu*, the same name that is given to the second rite in Omarakana.

So far there is very close parallelism between the magic of Omarakana and that of Oburaku. In both villages we have a magic connected with harvest and with the storing up of the crops. In both villages the magic is in some way associated with the storehouse, though this association is less pronounced in Oburaku than in Omarakana. The magical substances in both systems are partly identical and certainly of exactly the same type of magical symbolism: we have everywhere the symbolism of stunted growth above-ground and strong gripping roots, of longevity and endurance.

There is, however, one radical point of dissimilarity: the magic of Omarakana is carried out only at the harvest of taytu, the magic of Oburaku is carried out also at the harvest of the large yams, *kuvi*, the harvest which, as we know, has a special name and special inaugurative ritual everywhere. The *vilamalia*, moreover, in Oburaku is also a magic which can be performed at times of hunger, sickness or disaster.

At *isunapulo*, the harvest of the *kuvi* and taro, the *tovilamalia* of Oburaku collects leaves from the aromatic mint (*sulumwoya*) and from the *tuvata'u* (a native plant which looks like a marigold). He medicates them, reciting the first spell of "restoring" (M.F. 30), and then distributes them among the men. Each of them puts some of the herbs under the front platform of his house and some at the entrance of the road to the village nearest to him. This is called the closing up of the front part (*vaboda kaukweda*), and the closing up of the roads (*vaboda kadumalaga*). This magic bars the way of hunger and ill-luck into house and village.

Even more interesting is the use of *vilamalia* which is not connected either with harvest or with the filling of the storehouse. When there is sickness in the village or when hunger threatens the community, or even, I was told, when a falling star drops in or near the village, the magician may be called upon to perform the *vilamalia*. He will then carry out both rites exactly as they have been described here. Especially in times of hunger the *vilamalia* will be performed, and these occasions in Oburaku are not rare. Sometimes in the month of October or November, when the taytu from the foodhouses is finished, and the new gardens have just been planted so that many yams have been used as *yagogu*, the pinch of *molu* is often felt in this village, which, though subsisting largely on fish, always needs vegetable food. When hunger is so bad that the women have to go out to the jungle in search of food, the people will say to the magician: *Kuyovilaki m'malu*; *boge ilousi vivila, ikalipoulasi o la odila*, "turn your village (make the luck of your village turn); already the women have gone, they are scouring the bush". Then the magician will again medicate the trumpet shell and the ginger root, and perform the first ceremony. Again he will collect roots and make the second magic over them.

An interesting topographical comment was given to me by Navavile, the head garden magician of Oburaku, and some of his associates. When the trade wind starts blowing, at the end of the calms in April or May, it blows the magic over from the island of Kitava, where the magicians have been performing their *vilamalia* and exorcising all the evil, bidding it depart with the wind. The evil influences driven out of Kitava will strike the village of Oburaku, bringing sickness, hunger and death. As the wind continues blowing, more and more of these evil influences will come, till towards the end of the trade wind season, hunger may set in seriously, and sickness and death. About November comes the change of the seasons, when the north-west wind begins to blow. Now is the time to retaliate. Navavile sends scarcity and misfortune away from his island and they are carried by the drift of the wind towards the island of Kitava. Thus the tide of wind and the tide of human fortunes turn together.

It is an interesting piece of magical interpretation of natural events, and as usual the logic of wizardry is elastic and can be used both ways. The turn of the seasons, the time of calms, is I think on the whole the period when illness appears, and that for several reasons. In Oburaku, where the natives rely a great deal on fishing, there is no direct incidence of hunger during the calms; on the contrary, the natives often over-eat at this time and become ill on that account. Also with the calms, in April, they have just begun harvesting the new yams and early taytu, which often gives them diarrhoea or even dysentery. This is also put to the account of evil magic, of which adverse *vilamalia* is but one instalment.[1]

[1] It is also at this season that the malignant spirits, *tauva'u*, are supposed to visit the district, coming from the south; while flying witches often choose the calms for their aerial journeys. Cf. *Argonauts of the Western Pacific*, Ch. X, and *Sexual Life of Savages*, pp. 39–40, 128, 360 and 369.

During the following months, they may or may not suffer from food scarcity. When the crops are good and they need not fish, they are quite happy and bad magic is forgotten; but in a bad agricultural year, when their gardens, small and poor in any case, become quite inadequate and fishing is still impossible, they feel the pinch of hunger acutely towards October. The abundance of fish, which follows with the calms, may again be a doubtful blessing, leading to surfeit and illness. With the calms, however, in a village which relies mainly on fish, hunger would cease, so that the *vilamalia* then performed might really appear beneficent. In Kitava on the other hand, where the large yams are the staple crop and fishing is of very little importance, hunger would be more likely to set in at the normal time indicated on our Chart of Time-reckoning, that is from the beginning of November on. And this again roughly corresponds to the magical theory.

It is, however, certain that the *vilamalia* in the south of the Trobriand Islands is definitely a magic of dearth and sickness; on its positive side it is an exorcism of evil influences from the own village, and on its negative side it is an evil magic directed at an outside community.

At harvest the magic in Oburaku is performed exactly for the same reason as in Omarakana; it is meant to make the vegetable food remain in the yam-houses. As in Omarakana, it acts, not on the food primarily, but on the digestive system and on the appetite. "Their minds are nauseated (*iminayne ninasi*); they decline food, so that it remains in the storehouses" (*ipakayse kaulo, bisisu wabwayma*). "Fish they like, fruits from the bush they like, but not the yam-food" (*magisi yena, magisi kavaylua, kaulo gala*).

CHAPTER VIII

STRUCTURE AND CONSTRUCTION OF THE *BWAYMA*

PERSONALLY I am interested in technology only in so far as it reveals the traditional ways and means by which knowledge and industry solve certain problems presented by a given culture. The Trobriand storehouse, as we already know, enables the natives to realise a number of economic ends indispensable to their tribal order. It makes possible the accumulation of food—in some cases also the display of it, in others its modest concealment. It is the knowledge of how the structural technique of the *bwayma* satisfies the wants created by the demands of Trobriand culture that appears to me of real value.

In this chapter I have, therefore, made an attempt, not so much to give a bare account of technology, as to correlate this with social and economic requirements, with native ideas of value and of magical influence, to the limited extent to which my material makes it possible.

At the same time I have felt obliged to incorporate in my account all the technical details which I recorded. However one may scorn the purely technological interest which ignores the cultural conditions subserved, yet details of construction and of manual procedure remain cultural facts which must not be neglected.

Therefore the following account may irritate the sociologist because it is over-technical and upset the technologist because it is too much larded with social, economic and magical digressions. Yet both would be stultifying their own interests were they to keep these two aspects of the subject in watertight compartments.

With all this I wish emphatically to state that my method of presentation is rather an expression of good intention and a clearly conceived methodological aim than a finished product. It was my good fortune to work in conditions where the correlations between sociology, economics, magic and technique were so conspicuous that I was driven, even in the field, into adopting the functional approach as regards the present problem. But I started my work insufficiently prepared. Almost to the very end of my sojourn in the Trobriands, I was keeping the several aspects of culture each in its own pigeon-hole, not merely in my own mind, but also in the procedure of my note-taking. Thus in recording technology, my diagrammatic sketches and structural comments were in one

PLATE 85

PLATFORM IN FRONT OF SHOW BWAYMA

The ethnographer with some informants on the main *bwayma* of Tukwa'ukwa (Ch. VII, Sec. 5)

PLATE 86

PEOPLE RESTING ON BWAYMA PLATFORMS

This photograph, taken during a ceremonial dance at the *yalaka*, shows groups of guests resting under the eaves of *bwayma*. Note the decorative yam hung in front of the store in the centre. Also the difference between the enclosed *bwayma* extreme right, and the open show *bwayma* in front (Ch. VII, Sec. 5)

PLATE 87

THE MAIN STOREHOUSE OF THE TROBRIANDS, EMPTY

To'uluwa's *bwayma* before its rebuilding in 1918 (cf. Plate 72). The light shining through the empty log cabin shows the details of construction. The picture was taken during the harvest of 1915 (Ch. VIII, Intro.)

set of notebooks and my sociological observations in quite another. Having conceived the structure of the *bwayma* as a task by itself and not as a means to an end, I failed, as will be seen from the following pages, to investigate and record the natives' theory of why certain proportions are kept; to study sufficiently their ideas of ventilation, though I remember perfectly well that they discussed these with each other. Even as regards purely technological observations I missed one important point. I know, because I heard frequent references to it, that the Trobrianders have a certain theory as to the need for solid foundations and that they can estimate the size of foundation-stone necessary for a given size of *bwayma*, but I failed to record this in sufficient detail and cannot document my knowledge.

Otherwise, by a strange irony, my material is quite full on that very point which I now regard as irrelevant unless correlated with its context—I mean the technology and structural details of the storehouse. Since, however, I studied the uses of the storehouse pretty completely and recorded the ideas associated with them, I am able, to a limited extent, to supplement the deficiencies of my method in the field. But just because I regard the functional method as so important, I want there to be no doubt that it is much more practicable than the data of this chapter would warrant us in assuming. Anthropology and its servant the field worker have still to learn that the relations between the various aspects of culture are quite as important as these aspects themselves.

Another way in which I have been badly hampered by a wrong theoretical approach is in the greater emphasis which I gave throughout my field-work to sensational large-scale events. Thus we have several well-drafted documents with reference to the filling of the big *bwayma* (Docs. I–IV). The filling of the small *bwayma*, which I have seen much more often and in which I have taken part, I failed to record with the same concrete detail; hence I have to trust to my memory. Thus there is a considerable lack of balance and "functional thoroughness" in my documentation, because I did not treat the drab, everyday, minor events with the same love and interest as sensational, ceremonial, large-scale happenings; and I did not maintain a constant interest in the use of an object as correlated with its structure and form. The reader will find the same lack of balance in this chapter in that the show *bwayma* are treated in considerable detail, while Section 6, which is devoted to the small *bwayma*, is by no means so complete and reliable.[1]

[1] See also Note 35 in App. II, Sec. 4.

1. THE FORM OF THE *BWAYMA* AS CONDITIONED BY ITS FUNCTION

Let us once more state the structural problem presented by Trobriand sociology and economics to the Trobriand artisan. Trobriand culture, from the economic point of view, is conditioned by the natives' capacity for accumulating, controlling and distributing large supplies of food. This again is made possible by the existence of the storehouse, *bwayma*. A Trobriand storehouse must provide a dry, well-ventilated interior, protected alike from rain and from sun, and lifted sufficiently above the ground to exclude certain obvious pests. Since wealth, especially accumulated vegetable wealth, serves not only as a practical means of sustenance, but also as an index and symbol of power, the show storehouses must express this in their lofty and imposing dimensions, their decoration, their elegant shape and their conspicuous position. Also since the direct impression of food has a combined aesthetic and economic fascination to the eye of the Trobriander, their contents must be visible, to a certain extent at least. Commoners' *bwayma* and the inferior *bwayma* designed to preserve the modest everyday yam in good condition must be easy of access to the owner and sufficiently in view both of the community and the owner to safeguard it from pilfering.

Hence all storehouses are raised, carefully roofed and ventilated; some are large, prominently placed and allow of a partial display of their contents, while some are small, low, enclosed and lie in a retired position near the owner's house.

Native villages are generally built in a circle round an open central place; hence prominence can be given to the show *bwayma* by disposing them round the inner ring. In some villages the central place is surrounded by *bwayma* and in a few capitals the chief's yam-house stands right in the centre (Pls. 72 and 87; cf. also Plan of Omarakana, Fig. 2). Yam-houses of commoners are as a rule placed inconspicuously next to the dwellings in the outer circle (cf., e.g., Pls. 73, 75, 77, 80, 90 and 93), and the food contained in them is not visible, their walls being covered. The strict distinction, however, between covered and uncovered *bwayma* obtains only in villages of rank or in villages which immediately surround one of the capitals. In distant villages, even in old days, there were relaxations of the rule (cf. Pl. 62). Nowadays upstart villages have been built on the high-rank pattern, notably the village of Teyava (Pl. 50), where a solid inner ring of open *bwayma* surrounds a central place, though sociologically this village should have its stores all closed up and discreetly tucked away amongst the dwellings (cf. below, Sec. 6).

The materials at the natives' disposal are stones, used as foundations for the larger garners; and wood, for the heavy logs, poles, stakes, rods, sticks and boards necessary for the structure. The joining of the various parts is generally done by vegetable bast and tough, flexible creeper, and the thatching with lalang grass, sago or coconut leaf.

As will be seen from the following detailed account of structure, the material used for the storehouse is much more solid and elaborate than that used for the dwelling-house, its foundations are stronger and more permanent, and require more labour. The difficulty of building is greater and the amount of care needed for repairs more considerable. The houses are usually squeezed against each other so as to prevent sorcerers from prowling along the lateral walls. For the same reason they are never raised, be it ever so little, above the ground. They have no log structure in them nor stone foundations (cf., e.g., Pl. 89).

The fact that the yams are better housed than the human beings, which might, at first sight, seem paradoxical, is thus correlated with one or two conspicuous features of Trobriand belief and custom: the fear of sorcery and the desire for display of wealth; as well as with the practical requirements noted above.

Before passing to our account of how the yam-house is constructed, I wish briefly to indicate the method adopted for illustrating and documenting my technological descriptions. I have tried to make these latter neither too concise nor yet too verbose, and to describe rather the activities connected with the building of the yam-house than the parts of the building one after the other. For the details of structure the reader has to a large extent to rely on the diagrams and plates. The Diagrams I to XII might be alone sufficient to the technologist. I have indicated each structural element by the same numeral throughout the set. Thus (3) represents in every diagram the floor (*bubukwa*). In order to facilitate cross-referencing and to enable the reader to find the meaning of each structural item, I have listed these at the end of the chapter in a table which gives their native names, with their current numbers and with references to all the diagrams on which each can be seen.

2. THE PREPARATION OF MATERIALS

As in most other tasks of construction, be it of a building or a canoe, a Trobriander has to prepare the various structural elements before he puts them together. With the type of labour used and the implements available, work goes slowly. The implements employed in producing the logs, rods, stakes, poles and sticks are the axe, *kema*, the adze, *ligogu*, and nowadays the trade-knife; also to a slight extent the polishing implement, *kisi*, made of the skin of the shark or stingaree. A mounted shark's tooth, a wallaby bone and a hammer are used for the carving of the decorated boards. The trees have to be felled, lopped and trimmed, and this is done with axe and adze in the jungle. They have to be pulled into the village, and there the craftsman with his adze works on them round and round, planing them into an almost perfectly cylindrical log. At times the cylindrical logs of a *liku* are carved (see Pl. 87).

The parts that are really carefully planed and polished and occasionally carved are the *liku* (5), the logs of the cabin; the *po'u* (4), logs which hardly differ from the former in their shape and function except that they are the lowest and highest beams of the log cabin; the *kavalapu* (12), the decorative though not necessarily decorated front and back boards framing the gable wall and the *bisiya'i* (20), board. This last frames the gable at its base, also one is sometimes put across the gable about midway.

The powerful longitudinal beams which rest immediately on the stone foundation, the *kaytaulo* (2), are on the other hand only very roughly trimmed as a rule (cf. Pls. 87, 88, 90 and 91). This is also true of the clumsy boards, sometimes nothing more than stout poles, called *bubukwa* (3), which form the flooring (also called *bubukwa*) of the log cabin.

Some of the component parts of the roof have to be bent into the proper arch. It is not easy to find wood which lends itself to the elegant, Gothic-shaped arch required; and usually the necessary curve has to be obtained by heating the board over a fire, holding one end between two logs and bending the remaining part gradually. The *kavalapu* (12), the front and back gable boards, shown on Plates 90 and 91, have to be treated in this manner; so have also the *kavilaga* (11), which run parallel to them within the roof and have to have exactly the same curve and a considerable supporting power. In old days I understand that all the gable boards, *kavalapu* (12), the front board *bisiya'i*, (20), and most of the logs, *liku* (5), were carved. This was done by the master carver, the *tokabitam*, who would sit over them for days and weeks with his wooden hammer and chisel made of wallaby bone. He would finish them off with his mounted shark's tooth, and finally paint them in black, red and white, with red ochre, charcoal and calcareous earth. Even nowadays, as can be seen from Plates 75 and 76, a great many of these structural parts are carved and painted, though some of the principal *bwayma*—even the principal *bwayma* of the island itself—have but few carved parts on them, and no paint at all (cf. Pls. 72 and 87).

There is less work in the preparation of the strong ridge poles (9 and 10) of the roof; of the two longitudinal foundation poles of the roof (8); of the scaffolding poles (23) which are necessary to support the roof in very large storehouses and of the other parts of the scaffolding. Also a number of long rods for the longitudinals in the framework of the roof, the rods to be bent for the laterals of the roof and a quantity of short slender sticks for the battens, are quickly prepared. Towards the end lalang grass is collected for the thatch. Coconut is used for covering up holes and rents in old thatch. It also supplies the plaited mats for covering the floor and at times the gable walls, and most important of all, it serves to enclose the log houses of commoners' *bwayma*. Sago leaves are seldom used except right in the south, where they are imported sometimes from the Koya (d'Entrecasteaux group). On Plate 86, the store shown on the left has a gable wall made of sago, while that on the right has one of coconut matting. Pandanus leaves supply the gable wall in more elegantly finished garners.

3. THE CONSTRUCTION OF THE STOREHOUSE

I think that the component parts of the storehouse are prepared by relatives-in-law at their own villages, or, in the case of the chief, in tributary villages. In 1918 only the roof of the chief's *bwayma* was rebuilt (Pl. 88), and unfortunately I fell ill some time before the putting together of it and my record, made a fortnight later, was imperfect. I know that there was no great accumulation of building material in Omarakana itself and I know also that some of the bigger component parts were being prepared in Kasana'i, Kwaybwaga and Tilakayva.

The *liku*, the large beams which take a great deal of time and care to work up, used in old days, as I have just said, to be made in the subject villages, each village producing one or two *liku*. The decorative carved boards were made at the home of a master carver. The master carver might be the chief himself. Such was the case with the predecessor of To'uluwa, Numakala, who did all the decoration of his houses and storehouses himself. These were burnt during the last war, when Omarakana was destroyed by Kabwaku.

When a chief or headman had to construct a large storehouse and was not a *tokabitam* (carver and carpenter) himself, he would have to apply to a specialist. This transaction is based on the same principle as the hire of any other specialist's services.[1]

The organisation of work, then, in the construction of *bwayma* would be based on the following principles: the owner or chief would summon his relatives-in-law, those who fill the storehouse with *urigubu*, to work for him. They would prepare in their villages the material for the storehouse. I was told that each village would produce one log of the *liku* and a number of less important structural parts. The chief would at the same time choose a specialist whom he would finance by the regular series of food gifts. After the yam-house had been constructed, a big *sagali* or distribution of food would take place at which, besides pigs and *kaulo* (vegetable food), some valuables also would probably be distributed to the most important contributors to the work.

Looking at Diagrams I, II and III, and most of the plates referred to in this and the previous chapter, we see that the structural elements of the *bwayma* naturally group themselves into three main parts: the foundations, the log cabin and the roof. Accordingly there are three big structural tasks before the natives, which are to a certain extent independent of each other: the laying of the foundations, which takes place when a village is built; the construction of the log cabin, which I think, barring war or accidents, need not be done more than twice or three times in a century; and the building of the roof. Thus the present foundations of the principal *bwayma*, as well as of the lesser show ones, in Omarakana survived even the last burning. The log cabin, which was erected in 1899, was not rebuilt in 1918, nor was it even retouched, and I was told by the natives that it was good for another twenty or thirty years. The roof, on the other hand, has to be reconstructed every ten or perhaps fifteen years.

The same people who fill the storehouse have to do the repairing when necessary, and also to help the owner to rebuild his small garner. A show *bwayma* is invariably repaired between the bringing in of the taytu and the filling. The *dodige* (filling) takes place, in fact,

[1] Cf. Part I, Sec. 10. I have also dealt with this question briefly in *Argonauts of the Western Pacific*, p. 183: "The gifts to the specialists are called *vewoulo*—the initial gift; *yomelu*—a gift of food given after the object has been ceremonially handed over to the owner; *karibudaboda*—a substantial gift of yams given at the next harvest. The gifts of food, made while the work is in progress are called *vakapula*; but this latter term has a much wider application, as it covers all the presents of cooked food given to the workers by the man for whom they work."

immediately after the store has been repaired and after the first *vilamalia* magic has been performed. In the case of a small *bwayma* a few men finish it within a day or two. For a big *bwayma*, which would always belong to a chief or headman, three or four days are necessary and a whole community or several communities are employed on it.

There is no magic connected with the building or repairing of storehouses, nor, so far as I was able to ascertain, with the laying of the foundation-stones. I was told that the first act of *vilamalia* is the only ritual which accompanies this work. Ordinarily the architect and workmen find the foundations of a show *bwayma in situ*, and these foundations determine the length of the *kaytaulo* (2), foundation-beams, the length and width of the log cabin, and consequently, since certain proportions must be kept, the height of the *bwayma*.

The laying of the foundations is apparently a laborious task. It is necessary to go down to bedrock, place large stones there and on top of these the overground foundation-stones *ulilaguva* (1). The size and shape of the latter is roughly pyramidal, some of them have a remarkably narrow top, others are almost cubic.

On these stones rest the two foundation-beams, the *kaytaulo* (2). The *kaytaulo* and the roof are both considerably longer than the distance between the corner-stones, and the protruding ends of the former are sometimes covered with boards crosswise, thus affording a covered platform which plays an important part in the social life of the village. Remarkably enough there is no word for such a platform, and the natives speak about people "sitting in the garner", *isisusi wa bwayma*. On Plates 85, 86, 95, 96 and 97 people can be seen using the platform; while on Plate 94 there are people sitting simply on the protruding beam ends. It is characteristic that the biggest and best *bwayma* have no elaborate platforms; it would not be a proper thing to use such storehouses in this way, and only specially privileged people are allowed to do so. The owner himself would as a rule be the only person to associate himself so intimately with his storehouse (cf. Pl. 28, where we see Bagido'u sitting in front of his *bwayma*, and Pl. 94). Often when a *bwayma* has been abandoned and its upper structure burnt or used for other purposes, the two long beams are left and covered with a few boards that serve as a platform (see extreme right forefront of Pl. 79).

The foundation-stones, *ulilaguva* (1), can be seen on Diagram I, where there are four of them, on Diagram II, from the front, and Diagrams III and IV, where they appear in the unusual number of six. They are also to be seen on Plates 72, 73, 75, 76, 77, 79, 81,

82 and 87).[1] They are all made of coral stone and are about 50 cm. high. The height does not greatly vary with the size of the *bwayma*, To'uluwa's big *bwayma* being relatively lower down on its foundations than the small *bwayma*. This can be seen on Plate 75, where the large *bwayma* on the left appears sunk in comparison to the small *bwayma* on the right. Compare also the large storehouses on Plates 72, 74 and 87 squatting on the ground, with the smaller ones on Plates 28, 73 and 77 well lifted above it.

The foundation-beams, *kaytaulo* (2), are made of hard wood and are about 20 to 30 cm. in diameter or less in very small *bwayma*. They can be seen on Diagrams I–IV and on Plates 28, 72 and 76 and on almost every subsequent plate which shows a *bwayma*.

4. THE LOG CABIN

The position of the foundation-stones determines the dimensions of the *liku* (5). The four sides of the rectangle formed by the four corner-stones are measured and recorded on two standard lengths of creeper, and the longitudinal and transversal logs are cut to correspond. The four angles of the log cabin have to come immediately over the centre of gravity of the foundation-stones. This can be seen in the diagrams and on every plate, but best on Plate 90, which shows the stones and log cabin.

Immediately on the foundation-logs, *kaytaulo* (2), are put the two transversal logs of the cabin. These, although they do not differ in form or function from the other logs, have a special name, *po'u* (4). Sometimes the *po'u* are much longer than the other logs, as can be seen on Plates 75, 77, 79, 81, 82 and many others. They may be specially ornamented or left in the rough. The same name is given to the top pair of transversals in the cabin (4*b*). On the bottom *po'u* (4*a*) are placed the first longitudinal pair of cabin logs, the *liku*. All the longitudinal logs are called *kaybudaka* (5*b*), and the transversal *kaylagim* (5*a*).

This terminology introduces an analogy between the storehouse and the canoe; for in the canoe the boards running alongside are called *budaka* and the two decorated transversal boards, which enclose the well of the canoe at both ends, are called *lagim*. This analogy we also found running through some of the magical

[1] Only a few *bwayma*, and not those necessarily the biggest, are founded on six stones. These can be clearly seen on Plate 28 and less clearly, owing to the accidents of photography, on Plates 60, 72 and 87. The large *bwayma* of Kasana'i also had six foundation-stones though these do not show on the photograph (Pl. 79).

PLATE 88

THE CHIEF'S BWAYMA JUST BEFORE REBUILDING

The roof is off, and the structure of the log cabin is plainly visible. One of the gable or barge boards has veered round, and is seen resting on the foundation of the roof. The poles forming the inner compartments are sticking out of the log cabin. The central supporting pole of the roof still stands upright, giving the measure of the height (Ch. VIII, Sec. 3)

PLATE 89

THE FRAMEWORK OF A ROOF

This photograph represents a native dwelling in process of construction. The roof of a *bwayma* is built up in an identical manner (Ch. VIII, Sec. 5)

PLATE 90

CONSTRUCTION OF A SMALL BWAYMA

"The roof is made straight on top of the cabin without any additional scaffolding" (Ch. VIII, Sec. 5)

PLATE 91

THREE SMALL BWAYMA, ONE IN PROCESS OF CONSTRUCTION

This photograph gives another view of the storehouse in Teyava in process of construction. In front two empty stores are seen, showing structural details (Ch. VIII, Sec. 5)

formulae. Thus in Formula 19 (*vapuri*) a prolific development of the tubers was directly invoked by the simile of a richly laden canoe. Again in the magic aimed directly at the *bwayma*, translated and analysed in the previous chapter, we find that the leading word of the second spell of *vilamalia* (M.F. 29) is *kaylola lola*, 'mooring stake', and the key word is *bilalola*, derived from 'to moor', 'to anchor', both borrowed from the vocabulary of seafaring. In the first spell uttered over the floor of the *bwayma* (M.F. 28) we have again the simile of anchoring, applied to the storehouse through the medium of rattan.

Even in a wider sense we find this analogy running through the magical and mythical ideas of the natives. Thus in various spells, notably Formula 2, we find that the evil influences, the pests and blights, are magically laden into an imaginary canoe and sent away. What remains will be laden into the "firmly anchored canoe" of the storehouse in the village. Whether this simile enters consciously into the ideas of the natives I cannot say. Once you suggest such an idea to your informants, they readily accept it, but it was never volunteered to me. A study of the spells, however, especially in Part VII where they are given in literal translation and with commentaries, will convince any reader, I think, that my suggestion is not far-fetched.

Returning now to the construction of the cabin, the logs are simply laid one on top of the other, each pair resting in a broad, flat scooping made near the end of the pair below (cf. diagrams and plates). There is, I think, always some additional cutting and planing done during the construction. I have only seen the log cabin of a small *bwayma*, such as that shown on Plates 90 and 91 in the building, and here the work went very easily and smoothly, the whole construction being finished in an hour or so. With big heavy logs the work is apparently much more difficult. The low cabins of the small storehouses, such as those on Plate 73, and of the storehouses of medium height, such as Bagido'u's *bwayma* (Pl. 28), and the somewhat bigger garners of Yalumugwa (Pls. 75-77) can be constructed without any scaffolding, the men lifting and placing the logs from the ground. But when a big *bwayma*, such as those seen on Plates 72, 79 and 81, is constructed and long heavy, unwieldy logs have to be lifted, often to a height of three or more metres, special corner platforms have to be erected at about shoulder or head height. A very big storehouse, such as that which stands in the centre of Omarakana, may require platforms which run all along its sides as well (cf. Diag. XI, showing a small side ladder used at the constructing of *liku*). Ordinarily, men standing on the two corner platforms would take the log from those below and place it in position.

There are ten rows of longitudinals (5*b*) in the *liku* of To'uluwa's big storehouse; eight or nine was the average number I found in the large garners surrounding the *baku* of the capital, while the smaller houses have some five or six. The number of *kaylagim* (transversal *liku*, 5*a*) is less by one, or more by one if we include the *po'u*.

The upper *po'u* or framing logs (4*b*) are really the topmost of the transversal *liku* and are held in place in the same way by grooves. Parallel to these outer or wall *po'u*, a row of *po'u* are laid on top of the topmost *kaybudaka* (5*b*) which form a sort of roof to the *liku* and are the foundation of the roof itself (see Diags. I, III, IV, V, VIII and X).

I have followed the usual order of construction, in which the walls of the cabin and its upper covering are finished before the floor is made. The floor, *bubukwa* (3), of the log cabin is often made by inserting a series of boards or poles into the opening between the foundation-beams, *kaytaulo* (2), and the lowest two *kaybudaka* logs (5*b*). The boards or poles of the floor rest, therefore, side by side with the lower *po'u* (4*a*) on the *kaytaulo* (2). The floor, *bubukwa* (3), is perhaps the most sacred part of the *bwayma*, in that on it are placed the *binabina* stones, over which the *vilamalia* magic is performed. The *bubukwa* gives the title to one of the rites of the prosperity magic: the first rite of *vilamalia* is called *tum bubukwa*, or *kaytumla bubukwa*, "the pressing of the storehouse floor". In the spell (M.F. 28) corresponding to this rite, reference is made to the *binabina* stones; to the dust and the black powder which should accumulate on the *bubukwa*. On the *bubukwa* year after year the first layer of taytu is carefully placed, after it has been meticulously swept and cleaned so that no decaying remains of the old year's harvest may be mixed with the new crops.

Yet paradoxically enough the floor is one of the least carefully constructed parts of the *bwayma*. Made of any material which will serve: of long clumsy boards, badly cut stakes, broken poles, the floor is the most slapdash, jerry-built part of the storehouse. Looking at Plate 90, where the floor of the newly constructed small store of Teyava can be seen, we find there about eight poles, some of which are not even trimmed, their broken ends sticking out in slovenly fashion. Or again, glance at the photographs of the biggest *bwayma*, especially Plate 87, which shows how rough and ill-assorted in size and shape are the beams used. These incomplete foundations.

which are freshly made each year. Whether it is the need of ventilation, or the idea that the floor or bottom does not matter, it still remains one of the vagaries of Trobriand culture that this one detail which plays such an important part in magic, belief and the technical handling of the crops, should remain so slovenly and ill-finished. The *bubukwa* (3) is shown on Diagrams I, III and IV in profile, and on Diagrams V and VI in longitudinal and transversal section respectively.

With this we have finished the description of the outer frame of the *liku* (5), the principal part of the large *bwayma*, the pride of the owner, and that part which, through its construction, allows the stores to be exhibited and surveyed. It is characteristic that, in Formula 29 of the *vilamalia* magic, the *liku* is the only part of the *bwayma* which is given the prefix of personal possession, *agu liku*, so that the word does not refer so much to the log cabin as to its contents.

Let us now enter the interior. As can be seen from Diagrams IV and IX, it does not consist of one chamber but contains several compartments, *kabisitala* (28) or *kalikutala*. From our sociological analysis we know already the importance of these, for each compartment is filled every year by one of the *urigubu* contributors (cf. Ch. VII, Sec. 2). They are constructed as follows: two long poles, the *teta* or *katuveyteta* (6), are placed on top of the *po'u* (4*b*), the top transverse log of the *liku* wall, so that they divide the cabin longitudinally into two equal parts and might be called the median dividers. They are lashed to the *po'u*, though these lashings have not been indicated in the figures so as not to confuse the outline. Between these two *teta* a series of long rods, *kabisivisi* (7), are put, their lower ends being inserted between the interstices of two boards of the *bubukwa* (3), and thus kept in place, while their upper ends reach high above the level of the *po'u* (4*b*) and almost to the lower ridge pole of the roof, *kakulumwala* (10). Similar rods, also *kabisivisi* (7), are placed along the transversal *po'u* (4*b*) and lashed to them. These are cut to follow roughly—very roughly indeed—the slant of the roof from ridge pole to *liku* wall. In this way the well of the cabin is divided into six, eight, ten or, in the main *bwayma* of the chief, as many as sixteen compartments. In the chief's main *bwayma* thirteen of these are nowadays filled by outside contributors and three by *taytumwala* crops, produced in the chief's own gardens by his wives and sons. In old days the chief's *bwayma* had a few more compartments but never as many as there were wives; only headmen of the most important tributary communities from which the chief took his wives would be allowed to fill the principal storehouse. On some of the plates the inner divisions can be seen; either the partition-sticks show through the interstices of the logs, as in Plate 81, or, as in Plates 60 and 88, they protrude above the roofless *bwayma.*

The *kabisivisi* (7) are illustrated on Diagrams IV to IX; since they are interior arrangements and cannot be quite so well seen on the photographs, I have devoted more diagrams to them. On Diagram III the *teta* (6) can be seen, as well as the *po'u* (4*b*), which supply the upper framework of the compartments. Diagram IV shows the *kabisivisi* of one or two compartments in position. Diagram V shows the section about one-third from the lateral wall, illustrating the relative position of floor, front and rear wall, *kabisivisi* (7) rods and upper *po'u* (4*b*). Diagram VI shows by a transversal section of the house the construction of the roof as well as the median division. Diagram VII, also a transversal section, in the plane of one of the *kabisivisi* partitions, illustrates the position of the sticks. Diagram VIII is a section made between the two *teta* (6) and showing the median plane of the compartments. Diagram IX gives us a glance from above into the log cabin, *liku* (5). In this case an average size *bwayma* with eight compartments is illustrated. This was the number of compartments in the *bwayma* of which the measurements were recorded in detail and which can be seen on Plates 81 (left centre) and 82 (centre).

5. THE ROOF

We have now climbed above the log cabin (*liku*) into the roof. No floor divides this part from the lower compartment, no material surface is placed there, and when the yams overflow from the cabin they naturally enter the upper part, since the *kabisivisi* (7) or dividing rods usually stretch right up into the roof. The ideal surface which divides the lower from the upper compartment of the *bwayma* is the surface running along the top of the *po'u* (4*b*). Two poles (*kiluma*, 8), resting on the *po'u*, form the foundation of the roof framework, the other main constituents of which consist of a system of frame-boards and one or two ridge poles. The big arched frame-boards rest on the *kiluma* (8); their lower end is either tied to the *kiluma* or the latter is pushed through holes or notches in the frame-boards. The other ends of the arch-boards are fastened at the top to the ridge pole (10), forming thus a roughly triangular prism, if the curvature of the arch-boards be disregarded. This framework, however, is completed by a double layer of horizontal and longitudinal rods (13 and 15) interspersed with bent rods (14), parallel to the arch-boards. The whole structure from the technical point of view will be quite clear by a glance at Diagram X and Plate 89.

A few words must, however, be said about the technical problem which the natives have to overcome. There are three ways in which the roof can be placed in position.

(1) It can be made on the ground, a provisional small scaffolding being constructed of four pillars, *kokola* (27), on which the *kiluma* (horizontal support of roof, 8) are placed. The ridge pole (10) is then laid on two forked scaffolding-sticks (*tutuya*, 23), and on this frame the roof is constructed exactly in the same manner as that in which the ordinary house is built on the ground (Pl. 89). After the framework and thatch have been made the whole structure is lifted and placed on the top of the *bwayma*. As a matter of fact since the foundation poles, *kiluma* (8), are usually laid free on top of the *po'u* (4*b*), fitting into the grooves of the outer *po'u* and into the notches made in the other transversal *po'u*, the roof of any ordinary small *bwayma* can be lifted off with small effort, put on the ground as a temporary shelter, and then lifted up again on to its permanent foundations.

(2) Another way of making the roof is illustrated on Plates 90 and 91 (*Teyava*). The roof is made straight on the top of the cabin but without any additional scaffolding except the two *tutuya* poles (23), on which the lower ridge pole (10) is placed in position and to which the frame-boards (11) are then fastened. On these plates we see the two *kavalapu* (12), gable-boards of the ridge pole, set up. The rest of the structure and the manner in which it is done in such a case would be quite plain to the technologist.

(3) In the case of very large *bwayma*, where it would be impossible to lift the roof bodily, and where it is equally impossible to construct it aloft, standing on the ground or even standing on the top of the *liku*, an additional scaffolding is necessary. Since I have not seen the construction of such a *bwayma*, I have to rely on native statements which, in technological matters, are always somewhat unsatisfactory. In the construction of such a big roof overground a couple of stout scaffolding supports, *tutuya* (23), are put up. Such supports have been left in position in the case of the large *bwayma* of Omarakana, and they can be well seen on Plate 88, with the roof off, and Plate 72 after the roof has been finished. On these two supports the ridge pole—I am not quite certain whether it is the lower or upper, but I think the lower—is laid. This is done by placing one end of the ridge pole against the forked top of the *tutuya* and pulling it by means of ropes which are passed through the forked end of the second *tutuya*, so that it passes between the other fork and rests on the two *tutuya*. The platforms already constructed for the making of the higher parts of the log cabin, also serve for the

placing of the *kiluma* (8), and the handling of the lower ends of the *kavilaga* (11). Since the upper ends of the arched frame-boards have, however, to be fastened to the ridge pole, more scaffolding is needed: a rough ladder, *daga* (24), is put against the upper rim of the *liku* (5), log cabin, and a few rungs are made above the *liku* so that men can stand on them and handle the upper ends of the arched frame-boards. After this is done, the ladder is removed and the lower framework of the roof is fixed in place at both sides from the scaffolding platforms. Then a ladder, seen on Plate 72 standing where it was left after the work had been finished, is placed against this lower framework, the upper parts of the framework are placed in position and thatching can begin.

Storehouses are thatched in the same manner as dwellings, that is, the broad end of a lalang wisp or of a palm leaf is pulled through the interstice between two *kivi* (15). It is then folded back and pulled through the next interstice below. Thus it remains fast. The process is clearly shown on Plate 92. It is started from the lowest rungs of the framework so that each successive layer presses on the one below and keeps it in position. Thus a completely waterproof and relatively smooth surface is formed which, after several heavy rains, becomes well flattened and compact.

Although the roof part of the *bwayma* is not floored throughout, the space between the protruding gable-ends has a floor called *bomakayva* (17), a word used also to designate the sticks of which it is made. These sticks run between the two upper *po'u* (4*b*) of the *liku* (5) and a front *po'u* (19, Diag. I), which in large *bwayma* rests on special supports, *kaynubilum* (18, Diag. I); or in smaller *bwayma* is merely lashed to the *kiluma* (8), the horizontal supports of the roof. The gables are filled in with an open lattice work, consisting of verticals, *bisiboda* or *kavituvatu* (21), and horizontals, *yobilabala* (22). At times the lattice is covered with coconut matting or broad pandanus leaves, presenting the neat elegant appearance seen on Plates 75 and 86, but in the "best" *bwayma* it is left open, allowing the accumulated food to be seen (Pls. 79 and 82).

Glancing at the magical formulae of *vilamalia*, it is interesting to compare these with the terminology of the component parts of the *bwayma* with which we have now become acquainted from the technical side. We see that most of the fundamental parts of the structure are mentioned in all the formulae: *ulilaguva* (1), the foundation-stones, called in Oburaku by the generic term *kaylagila*, which means 'foundation', *bubukwa* (3) and *liku* (5). In my version of the Omarakana *kaylola* spell (M.F. 29), I do not find the word *kaytaulo* (2). It must, however, be remembered that in the *tapwana*, that is in the litany of the spell, the magician does not always need to enumerate the full list of inventory words, although he seldom omits very important ones.

The *tapwana* of no spell, therefore, is absolutely complete, and the absence of a word does not signify much more than a lapse of memory or attention on the part of the magician during the one or two times he recited it to the ethnographer. We find also, in one of the spells at least, such words as *kiluma, kakulumwala, vataulo, kavala, kaliguvase, kivi, katuva.* It is characteristic that, in all the spells, the ornamental elements, *kavalapu, bisiya'i* and *mwamwala* are mentioned. Also the words *kabisivisi* and *teta* referring to the internal compartments occur in Bagido'u's spell.

6. THE STRUCTURE OF THE SMALLER *BWAYMA*

The main difference between the show *bwayma* and the inferior ones, which are usually also smaller, is to be found in the part lying between the foundations and the roof. Here also they show a considerable range of structural variation among themselves.

As regards the foundations, the inferior garner of average size also rests on stones. Only the really small ones, and especially such as have no middle part, are erected on the short, upright wooden pillars (*kokola,* 27) seen on Plate 93. Such pillars invariably have a fork at the upper end, on which the two longitudinal foundation-beams rest. Since these *bwayma* differ, it will be convenient to divide them into types.

(*a*) We have first of all the small storehouses in imitation of the larger ones, that is comprising an open cabin constructed of more or less well-finished logs. Such is the small *bwayma* seen in process of erection on Plate 90 and again with two others on Plate 91; such are a great many of the ordinary small *bwayma* found in other villages. Some of them, when they belong to commoners living within the sphere of influence of a chief, have their log cabins more or less carefully covered with coconut leaves (see Pls. 73, 77, 84 and 86). Others in villages more independent and more distant were left open, even in olden days, especially when they belonged to the headman of the community (see Pl. 97). Nowadays, when the influence of chiefs has been undermined, many people who would not have dared to have open *bwayma* fifty years ago arrogantly exhibit their taytu. But a great many of the *bwayma* built with a regular *liku* exactly on the pattern, inside and out, of a big yam-house are covered up with coconut leaves (see Pl. 98, taken in the village of Bwoytalu, where even now the "inferiority complex" of the natives is so strong that they carefully enclose their stores). The garner seen in the background of Plate 77 (Yalumugwa) is built exactly like the one in front, but it is carefully enclosed.

(*b*) Another type of lesser *bwayma* is produced when the log cabin is of the regular height, stretching from fairly low stone foundations to the junction of roof with walls, but is shorter by half than the *kaytaulo* (2) and the roof, thus leaving a long sheltered platform in front (cf. Pls. 93, 95 and 96). Such a construction is made possible by two strong supporting poles in front, which correspond to the *kaynubilum* (18) of the larger *bwayma*, and I think are called by the same name. Usually the front ends of the *kaytaulo* (2) of such stores are supported by pillars. The log cabin of such a storehouse again can be either enclosed or open, and this depends on the social status of the owner.

(*c*) The *liku*, instead of being shortened and left with its full vertical dimension, may be given its normal front to back length, but only about half its usual depth or less, by raising it high above the ground, as in Plate 94. This type occurs at times in the small stores placed immediately against the dwellings, which might be called domestic *bwayma* (cf. Pl. 94). They are always built on wooden pillars and not on stone foundations. Such *bwayma* may or may not have a platform underneath, and this may be either enclosed or open.

(*d*) At times the *liku* completely disappears and the *bwayma* consists merely of a thatched platform raised on pillars. Most of the small *bwayma* belong to this structural type (cf. Pls. 96 and 97). Sometimes a lower platform is added, especially when the *bwayma* is somewhat bigger, as is the one shown on Plate 97, left foreground. Imagine the lower platform enclosed with coconut matting, and you have a type of construction often found in enclosed stores. When thus enclosed, the platforms are a convenient shelter for people who want to sleep during the daytime or meet for amorous dalliance at night.

(*e*) An inner construction frequently found within the enclosure of coconut leaves is the one represented on Diagram XII. It differs but little from the two previous types and is, in a way, an intermediate form. These *bwayma* have a rudimentary and very shallow log cabin, consisting of about one pair of logs, and also a lower platform half the length or so of the *liku*. The upper compartment of the *liku* would be used for the storing of seed taytu. It permits of better ventilation, I was told, than does a platform flush with the roof. On the lower platform inferior taytu would be stored. I was told that this is the classical pattern for a *sokwaypa* (*bwayma* for seed taytu).

PLATE 92

TECHNIQUE OF THATCHING A ROOF

"The broad end of a lalang wisp, or of a palm leaf, is pulled through the interstice between two frame rods. It is then folded back and pulled through the next interstice below" (Ch. VIII, Sec. 5)

PLATE 93

YAM-HOUSE BUILT ON WOODEN PILES

This illustrates also the alternation of store and dwelling found in a number of villages of low rank, where there is only one ring of buildings. In such villages there are no open show stores. This picture was taken in Bwoytalu (Ch. VIII, Sec. 6)

PLATE 94

TYPICAL VILLAGE STREET

The section shown curves away to the right; on the right are storehouses, on the left dwellings, with one domestic *bwayma* between the second and third huts (Ch. VIII, Sec. 6)

PLATE 95

TWO BWAYMA WITH SPACIOUS FRONT PLATFORMS

They are of the small, light type, built on wooden pillars; the top compartment is fronted with a pandanus screen (Ch. VIII, Sec. 6) (*f*) There is finally the makeshift or jerry-built *bwayma*, not infrequently met in the coastal fishing districts where tubers are not of such great importance. It can be seen on Plate 80 and consists simply of a roughly built roof, covering a raised platform.

(*g*) One more type of *bwayma* should be mentioned here, the toy *bwayma* erected for small boys who have just started making gardens, at first almost pretence gardens, which they will have gradually to treat more seriously. It is usually raised very high above the ground and is an attractive sight (Pl. 99).

Comparing the inferior types of *bwayma* with the large storehouses, one or two general observations may be made. The best *bwayma* in the village of high rank are almost exclusively devoted to the twin purpose of harbouring and displaying the crops. There is hardly any platform accommodation connected with them; the best *bwayma* of the island allowing only one man to sit on each of its front ends. The more inferior the *bwayma*, one might almost say, the more accommodation it affords for use in the daytime and at night.

Another point is that the less pretentious ordinary storehouses are very much more accessible and convenient for daily use. To take a tuber out of the big chief's *bwayma* necessitates a climb of considerable difficulty, and one which the housewife, in this case one of the chief's wives, cannot undertake alone, for reasons of delicacy. For with the chief's *bwayma* one has first of all to climb all the rungs of the *liku*, perch on its top, if the *bwayma* is full, and select a tuber from the interior. To the natives, with bare feet and enormous experience in climbing trees, this is not such a difficult performance as it would be to a European, but it always takes some time.

In the inferior *bwayma*, on the other hand, a man or woman stands on the platform and reaches up into the compartment above. Since the floor of such *bwayma* is usually made of little sticks covered with coconut mats, it is only necessary to push aside a piece of the matting in order to reach a tuber.

7. SUMMARY OF THE STRUCTURAL, SOCIOLOGICAL AND ECONOMIC CHARACTERISTICS OF THE *BWAYMA*: LINGUISTIC TERMINOLOGY

We can now briefly summarise the structural and functional types of storehouse. The *bwayma* properly speaking, called also by the natives *bomaliku*, or, in its less conspicuous form, *bwayma goregore*, is both a receptacle for food and a medium for display. Structurally it consists of strong foundations, a well-built log cabin, and an elegantly shaped, gable-ended thatched roof. Sociologically it can be owned only by a chief, a sub-chief (*gumguya'u* or *toliwaga*), a man of rank or a headman of an important village. Economically, most of the compartments have to be filled by a man other than the owner; and their contents are largely used for gifts, for ceremonial distribution and exchange, and as the staple food for financing enterprises. Aesthetically, it can be and usually is adorned with carved boards, shells and streamers. Topographically it stands in the inner ring, and on that account it is called *bomisisunu*. In a few capitals, such as Omarakana, Kasana'i, Kabwaku and Sinaketa, the chief's *bwayma* stands in the middle of the *baku*, and then enjoys the title of *bomilala*. With the position of the *bwayma* in the inner ring are associated the taboos against cooking in that part of the village. These ceremonial storehouses usually contain *binabina* stones, but in any case the magic of *vilamalia* is always performed over them.

Inferior storehouses are structurally smaller, need less strong foundations, and often lack the log cabin. Sociologically they may be owned by anyone. The sitting platform often constitutes a social centre, especially for the men. Economically they are entirely filled by *taytumwala* and seed yams when owned by men of rank, and contain the *urigubu*, the *taytumwala* and the seed yams of the commoner. Aesthetically they almost strain after modesty and inconspicuousness. Because of the cooking taboo and because they are not meant to impress the onlooker, they are completely enclosed, barring such holes as are due to wear and tear. There is no magic associated with them, no taboo restricts their use, and they can serve for sleeping and fornication.

The classification, therefore, if we study function as well as structure, is clear. The terminology, considered by itself, is confusing, inconsistent and indefinite; but if we study the term not only in its context of speech but in its context of situation, we find a very clear and consistent use. The word *bwayma* is a generic term which means storehouse in general, as well as specifically 'show garner'. Speaking about his yam-houses collectively a man would use the word *bwayma*, but if he wanted to make it clear that some of them were built with an open *liku* and others enclosed, he would use the word *bwayma* for the first and the word *bwaymaya* or *sokwaypa* for the latter. The *sokwaypa* are not, however, invariably enclosed. The two fine show storehouses on Plate 81 are obviously in the highest class of open *bwayma*. Yet the upper part of one of them—the one on the right covered with plaited coco-leaves—is used for storing seed yams and, though structurally identical with the *bwayma*, functionally it would be called *sokwaypa*.

Thus, as regards structure, the words *bwayma*, *bwaymaya* and

PLATE 96

HOUSES AND BWAYMA IN OBURAKU

On the right a typical small storehouse with a large platform. The taytu is mainly stored in the top compartment, which is fronted with pandanus leaf. On the left a small domestic *bwayma* (Ch. VIII, Sec. 6)

PLATE 97

A VILLAGE STREET SHOWING TWO TYPES OF STOREHOUSE

On the left a raised roof store covering a sitting platform; on the right a domestic *bwayma* placed against a house (Ch. VIII, Sec. 6)

PLATE 98

SITTING PLATFORM AND STOREHOUSE IN BWOYTALU

In the centre the covered raised platform shows the structural details of the supporting pillars, longitudinal beams, and floor. On the right, a typical enclosed ordinary store (Ch. VIII, Sec. 6)

PLATE 99

TOY BWAYMA WITH OWNER LEANING AGAINST IT

Such stores are erected for small boys who have just started making gardens (Ch. VIII, Sec. 6)

sokwaypa can be used almost indiscriminately. The terms receive their definite meaning only when it comes to contrasting them, and then *sokwaypa* is functionally a storehouse for putting away seed taytu, *bwayma* specifically a show house, and *bwaymaya* anything which is neither *sokwaypa* nor *bwayma*. On the whole the word *bwayma* is by far the most frequently and widely used, while the word *bwaymaya*, which logically has the widest connotation, is very seldom heard.

8. NOTE ON PROPORTIONS

The proportions of storehouses differ according to their size and character. Glancing at Plates 75, 77, 80 and 93 we get examples of all the main types. We see that some of these are squat, others slender, while yet others are almost shapeless. It is the big *bwayma* which have a more or less well-established pattern of proportion. How far these proportions are determined by practical considerations and how far by considerations of elegance and convention I am not able definitely to say.

If the length of the roof from point A to B (Diag. I) and the length of the foundation logs A′ to B′ (Diag. I), which are approximately the same, be assessed at the figure of 5, the width of the thatch at its widest point, C to D (Diag. II) would be 2·5, which very often is also the width C′ to D′ between the outer edges of the foundation-stones; and the total height from the ground to the upper ridge pole, E to F (Diags. I and II), would be 5·5. Further, the distance from the earth to the top of the log cabin, E to G (Diags. I and II), would be 3, and the height of the log cabin itself (G–H on Diags. I and II) 2·30, the elevation of the floor above the earth (E–H) being 0·7. The length of the log cabin from front to back (I–J, Diag. I) would be 2·8, and its width (K–L, Diag. II) 1·6. The figures here quoted are, as a matter of fact, the measurements of a *bwayma* which stands in the inner ring of Omarakana and can be seen on Plate 82 in the middle and on Plate 81 on the left-hand side. This *bwayma* I measured exactly, but I had before taken a number of measurements which all agreed substantially as regards proportions. The following table can, therefore, be taken as representing the average proportions of a good *bwayma*.

Metres	
5·50	from ground to top of ridge pole.
0·70	from ground to upper surface of foundation-logs (*kaytaulo*); (foundation-stones 0·50; diameter of *kaytaulo* 0·20).
3·00	from ground to top of log cabin, i.e. upper surface of *po'u*, top framing logs.

Metres

2·30	height of *liku* (log cabin).
1·60	width of *liku*.
2·80	length of *liku* (log cabin), i.e. length of each cabin log, front to back.
5·00	length of roof at top; length of *kaytaulo*. Usually 10 to 20 cm. difference between length of roof at top and at base, since base is a little shorter.
2·50	Distance from top of log cabin to top of upper ridge pole, i.e. height of roof.
2·50	width of thatch at widest.
2·30	inner width of roof at basis.
3·90	distance between front and back partitions.

If we wanted to calculate the cubic capacity of the *liku*, we can base ourselves on the following figures: the inner height of the *liku* measured from the floor is 2·30 as given in the table; the inner length of the *liku* as compared with the outer is about 30 cm. less, 2·50, 15 cm. being the average diameter of beams; and on the same principle of subtracting 30 cm. the width is 1·30. Multiplying these three figures, we have therefore the result of 7·475 cubic metres. To get the approximate cubic capacity of the roof, we can assume that the gable is a triangle, and multiply its surface by the length of the roof from front to back. We have therefore 2·30×2·50×3·90×one-half =11·2125 cubic metres.

All the other dimensions of the *bwayma* can be read off directly from the figures by applying a centimetre or inch ruler. They are all drawn to scale, 1 in 50. The various structural parts of the *bwayma* are marked by the same numbers throughout the diagrams. These numbers are listed in the next section, and also briefly explained and described.

9. TECHNICAL TERMINOLOGY OF THE *BWAYMA*

1. *Ulilaguva*—'foundation-stone'; one of the four or six coral stones which support the whole structure of the storehouse. These stones rest as a rule directly on the coral bedrock. (Diags. I to IV, VI.)
2. *Kaytaulo*—'foundation-beam'; the long beams lying horizontally on the foundation-stones and supporting the rest of the storehouse. (Diags. I to IV, VI, VII.)
3. *Bubukwa*—'floor', 'floor-board'; boards, stakes, or poles forming the flooring of the log cabin. (Diags. I, III to VIII.)
4. *Po'u*—'framing log'; the horizontal logs which form the top and bottom pair of the cabin. (Diags. I to X.)

4*a*. *Po'u*—'framing log at the bottom of the log cabin'.

4*b*. *Po'u*—'framing log at the top of the log cabin', that is, at the foundation of the roof.

5. *Liku*—'log cabin'; 'the beams or logs which encase the well of the storehouse'. (Diags. I to IX.)

5*a*. *Kaylagim* (*liku*)—'transversal cabin log'; the shorter logs which form the front and back wall of the cabin. (Diags. I to V, VIII.)

5*b*. *Kaybudaka* (*liku*)—'longitudinal cabin log'; any of the logs which form the longer side walls of the cabin. (Diags. I to IV, VI, VII, IX.)

6. *Teta* (or *katuveyteta*)—'median divider of log cabin'; one of the two longitudinal rods or poles laid horizontally over the upper *po'u* and halving the cabin of the storehouse. (Diags. III, IV, VI to IX.)

7. *Kabisivisi*—'upright divider of cabin'; the vertical rods which are put in rows between the two *teta* and the longer *po'u* and form the compartments of the storehouse. (Diags. IV to IX.)

Kabisivisi means also 'compartment' of the storehouse. Such compartments are counted by means of the formatives *kabisi-* or *kaliku-*. Thus *kabisitala* or *kalikutala* means 'one compartment'. *Kabisiyu* or *kalikayu*—two compartments; *kabisitolu* or *kalikutolu*—three compartments, and so on.

8. *Kiluma*—'horizontal support of the roof'; the two longitudinal poles running on top of the *po'u* and forming the support of the whole roof. (Diags. II to IV, VI, VII, IX.)

9. *Vataulo*—'upper ridge pole'; one of the three essential poles which form the structure of the roof, corresponding to the *kiluma*. (Diags. I, II, VI to VIII, X.)

9*a*. *Mwamwala*—'ornament of ridge pole'; a ring or a carved representation of a bird, attached to the end of the upper ridge pole. (Diags. I, II, VIII.)

10. *Kakulumwala*—'lower ridge pole'. (Diags. II, VI to VIII, X.)

11. *Kavilaga*—'frame-board of roof'; one of the inside boards which form the framework of the roof, together with the supports of the roof and with the ridge pole. (Diags. VII, X.)

12. *Kavalapu*—'gable-board'; the two extreme frame-boards of the roof, lying on the plane of the gable and visible to the eye looking at the storehouse from the front or from behind. The front *kavalapu* on *bwayma* of rank are now often decorated, in olden days invariably so. (Diags. II, X.)

13. *Kavala*—'inner frame-rod'; the horizontal rods which, resting on the frame-boards of the roof and lashed to them, form the inner longitudinal layer of the framework of the roof. (Diags. II, VI, VII, X.)

14. *Kariguvase*—'curved frame-rod'; the bent rods running outside the *kavala*, resting on them and lashed to them, and forming the outer curved surface of the roof, parallel to the surface of

the frame-boards of the roof, the *kavilaga* (11). (Diags. II, VI, VII, X.)

15. *Kivi*—'outer frame-rod'; the longitudinal rods laid horizontally at small intervals against the *kariguvase* (14), and forming the outer surface of the framework of the roof, to which the thatch is directly attached. (Diags. II, VI, VII, X.)
16. *Katuva*—'thatch' of roof. (Diags. I, II, VI, VII.)
17. *Bomakayva*—'gable-end floor'; 'one of the sticks forming the gable-end floor'. (Diags. I and II.)
18. *Kaynubilum*—'support of gable-end'; long poles placed in the case of specially large or somewhat dilapidated storehouses to support the protruding gable-end. (Diag. I.)
19. (*Bomakayva*) *po'u*—'gable-end foundation-pole'; the pole, or in the case of a large *bwayma*, the log, tied to the *kiluma* (8), the supports of the roof, from underneath and used as support for the *bomakayva* stakes. (Diag. I.)
20. *Bisiya'i*—'base-board of gable'; board sometimes decorated and laid at the bottom of the front gable-end. Sometimes there are two such boards one above the other. (Diags. II, VIII.)
21. *Bisiboda* (also called *Kavituvatu*)—'vertical gable-rod'; rods placed upright and forming the vertical part of the framework of the gable. (Diag. II.)
22. *Yobilabala*—'horizontal gable-rod'; placed across the gable and forming the horizontal part of its framework. (Diag. II.)
23. *Tutuya*—'ridge pole support'; tall and strong vertical pole used in the construction of a storehouse and sometimes left as permanent support. (Fig. 11.)
24. *Daga*—'ladder'; made for reaching the upper parts of the storehouse during the construction of the framework and thatch. (Diag. XI.)
25. *Unawana* (also called *Daga* in the stricter sense)—'vertical of ladder'. (Diag. XI.)
26. *Getana* (also called *Yobilabala*)—'rung of ladder'; generic term. (Diag. XI.)
27. *Kokola*—'pillar'; strong wooden post usually forked at upper end used as a pillar to support the foundation-beams, *kaytaulo* (2) of smaller storehouses; any forked pole used to support a platform or a bedstead. (Diag. XII.)
28. *Kabisitala*—'inner compartments of log cabin'. (Diag. IV.)

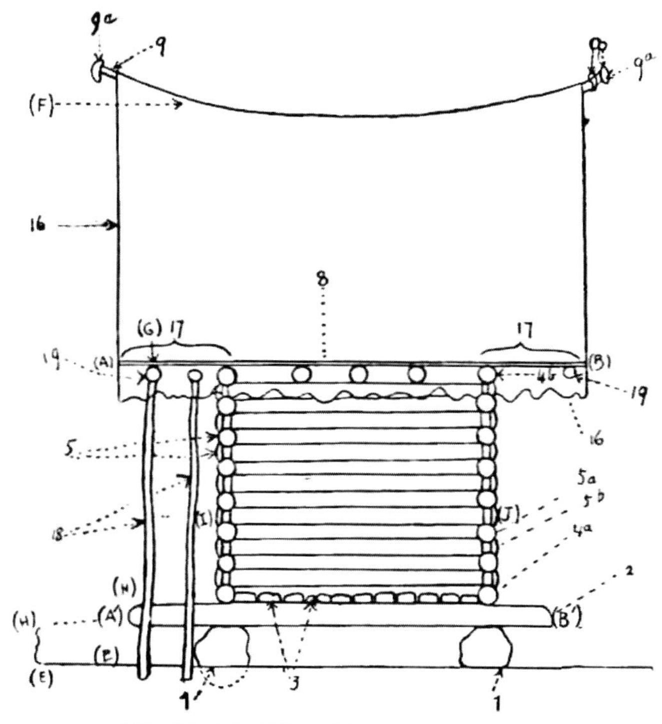

DIAGRAM I. SIDE VIEW OF YAM-HOUSE

1. Foundation-stone.
2. Foundation-beam.
3. Floor-board.
4. Framing log *a* at botom of log cabin.
   b at top of log cabin.
5. Log cabin *a* transversal cabin log.
   b longitudinal cabin log.
8. Horizontal support of roof.
9. Upper ridge pole.
9*a*. Ornament of ridge pole
16. Thatch.
17. Gable-end floor.
18. Support of gable-end.
19. Gable-end foundation-pole.

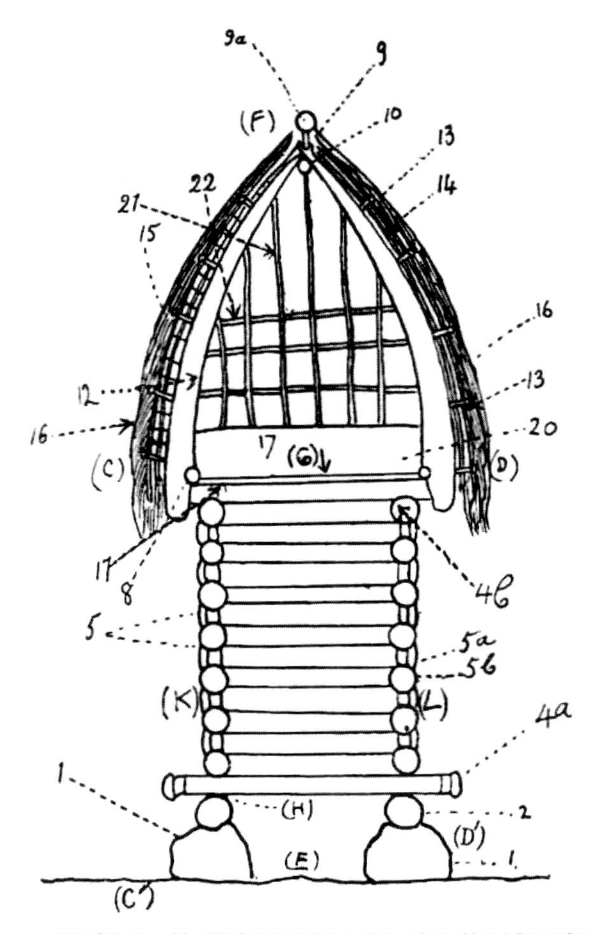

DIAGRAM II. FRONT VIEW OF THE YAM-HOUSE

1. Foundation-stone.
2. Foundation-beam.
4. Framing log *a* at the bottom of the log cabin.
   b at the top of the log cabin.
5. Log cabin *a* transversal cabin log.
   b longitudinal cabin log.
8. Horizontal support of thatch.
9. Upper ridge pole.
10. Lower ridge pole.
12. Gable-board.
13. Inner frame-rod.
14. Curved frame-rod.
15. Outer frame-rod.
16. Thatch.
17. Gable-end floor.
20. Base-board of gable.
21. Vertical gable-rod.
22. Horizontal gable-rod.

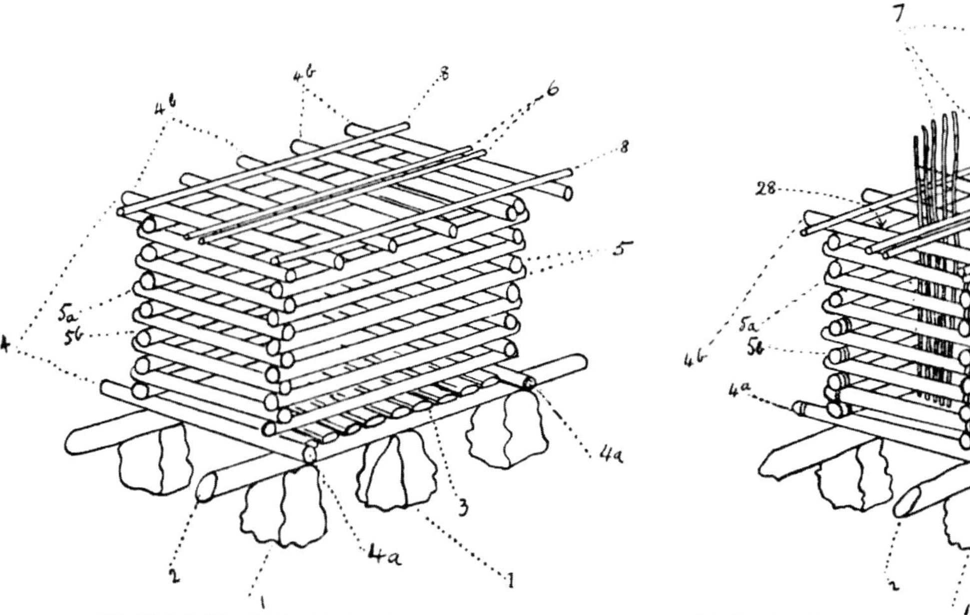

DIAGRAM III. STRUCTURE OF LOG CABIN

1. Foundation-stone.
2. Foundation-beam.
3. Floor-board.
4. Framing log.
5. Logs of cabin *a* transversal. *b* longitudinal.
6. Median divider of log cabin.
8. Horizontal support of thatch.

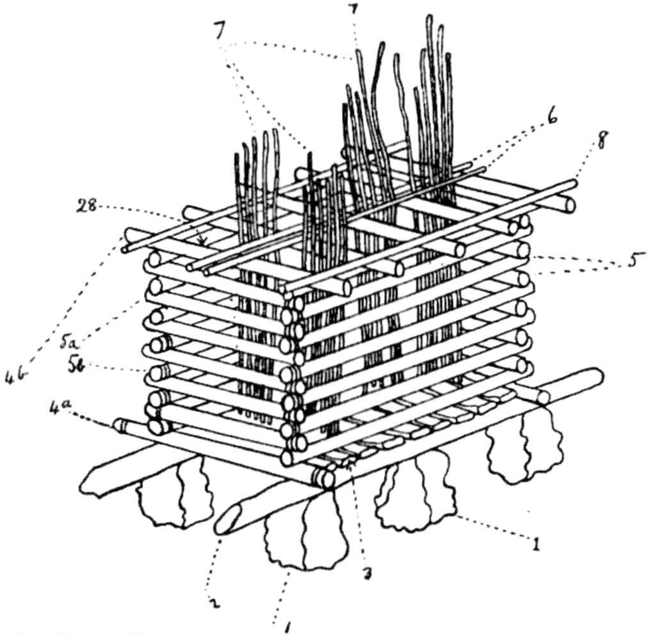

DIAGRAM IV. STRUCTURE AND DIVISIONS OF LOG CABIN

1. Foundation-stone.
2. Foundation-beam.
3. Floor-board.
4. Framing log *a* at bottom of log cabin. *b* at top of log cabin.
5. Logs of cabin *a* transversal. *b* longitudinal.
6. Median divider of log cabin.
7. Upright divider of cabin.
8. Horizontal support of thatch.
28. Inner compartment of log cabin.

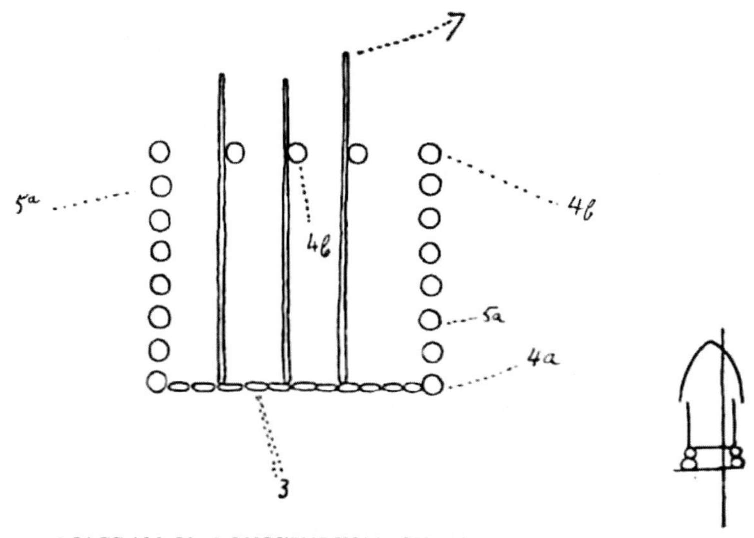

DIAGRAM V. LONGITUDINAL SIDE-SECTION OF LOG CABIN

3. Floor-board.
4. Framing log *a* at the bottom of log cabin.
   b at the top of log cabin.
5. Logs of cabin *a* transversal.
7. Upright divider of cabin.

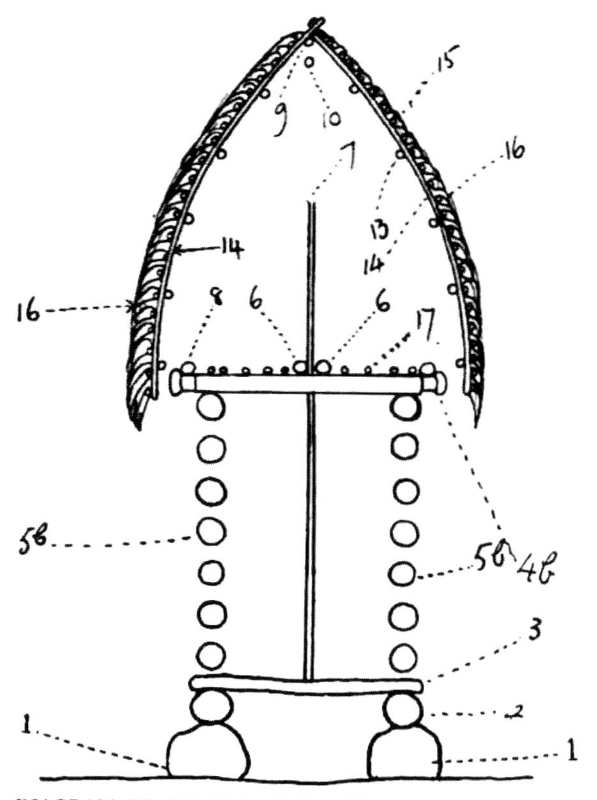

DIAGRAM VI. TRANSVERSAL SECTION OF YAM-HOUSE

1. Foundation-stone.
2. Foundation-beam.
3. Floor-board.
4*b*. Framing log at the top of the log cabin.
5*b*. Longitudinal cabin log.
6. Median divider of log cabin.
7. Upright divider of cabin.
8. Horizontal support of thatch.
9. Upper ridge pole.
10. Lower ridge pole.
13. Inner frame-rod.
14. Curved frame-rod.
15. Outer frame-rod.
16. Thatch.
17. Overhanging gable-end.

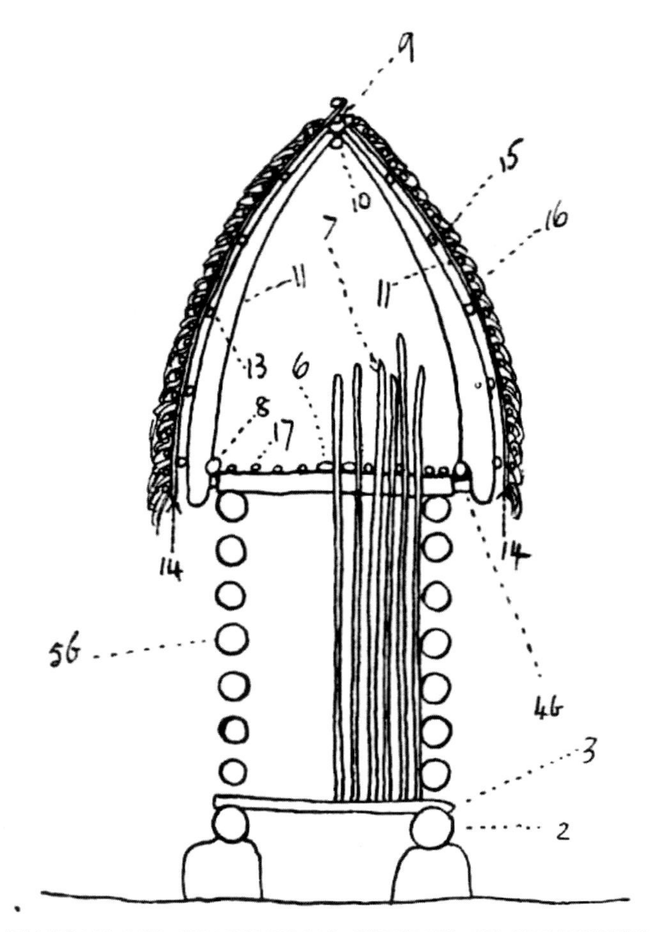

DIAGRAM VII. TRANSVERSAL SECTION OF YAM-HOUSE

(showing structure of the transversal compartment wall)

2. Foundation-beam.
3. Floor-board.
4*b*. Framing log at top of log cabin.
5*b*. Longitudinal cabin log.
6. Median divider of log cabin.
7. Upright divider of cabin.
8. Horizontal support of thatch.
9. Upper ridge pole.
10. Lower ridge pole.
11. Frame-board of roof.
13. Inner frame-rod.
14. Curved frame-rod.
15. Outer frame-rod.
16. Thatch.
17. Overhanging gable-end.

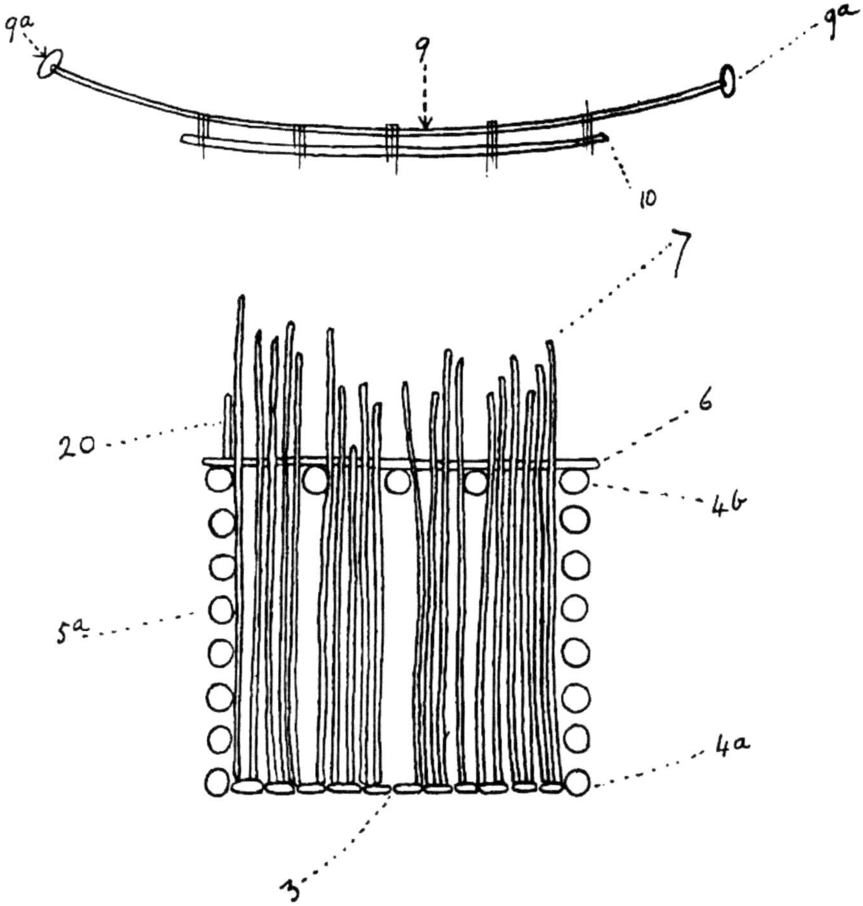

DIAGRAM VIII. LONGITUDINAL SECTION OF YAM-HOUSE

(showing structure of longitudinal compartment wall)

3. Floor-board.
4. Framing log *a* at bottom of log cabin.
   b at top of log cabin.
5*a*. Transversal cabin log.
6. Median divider of log cabin.
7. Upright divider of cabin.
9. Upper ridge pole.
9*a*. Ornament of ridge pole.
10. Lower ridge pole.
20. Base-board of gable.

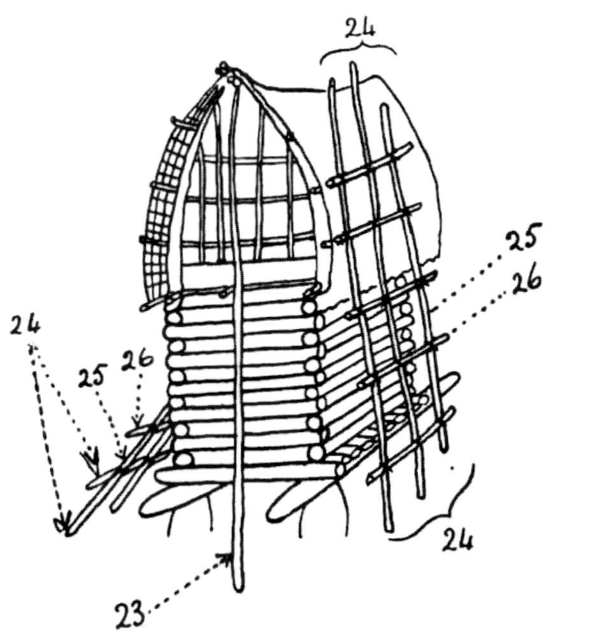

DIAGRAM XI. STRUCTURE OF SCAFFOLDING OF YAM-HOUSE

23. Ridge pole support.
24. Ladder.
25. Vertical of ladder.
26. Rung of ladder.

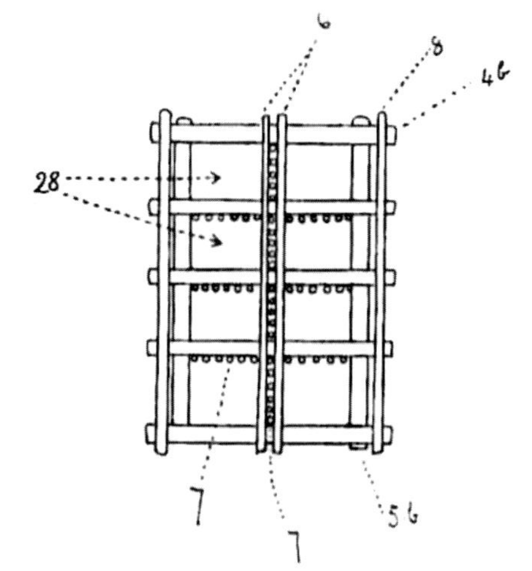

DIAGRAM IX. HORIZONTAL SECTION OF TOP OF LOG CABIN

(showing structure of compartment division)

4*b*. Framing log at top of log cabin.
5*a*. Transversal cabin log.
6. Median divider of log cabin.
7. Upright divider of cabin.
8. Horizontal support of thatch.
28. Inner compartment of log cabin.

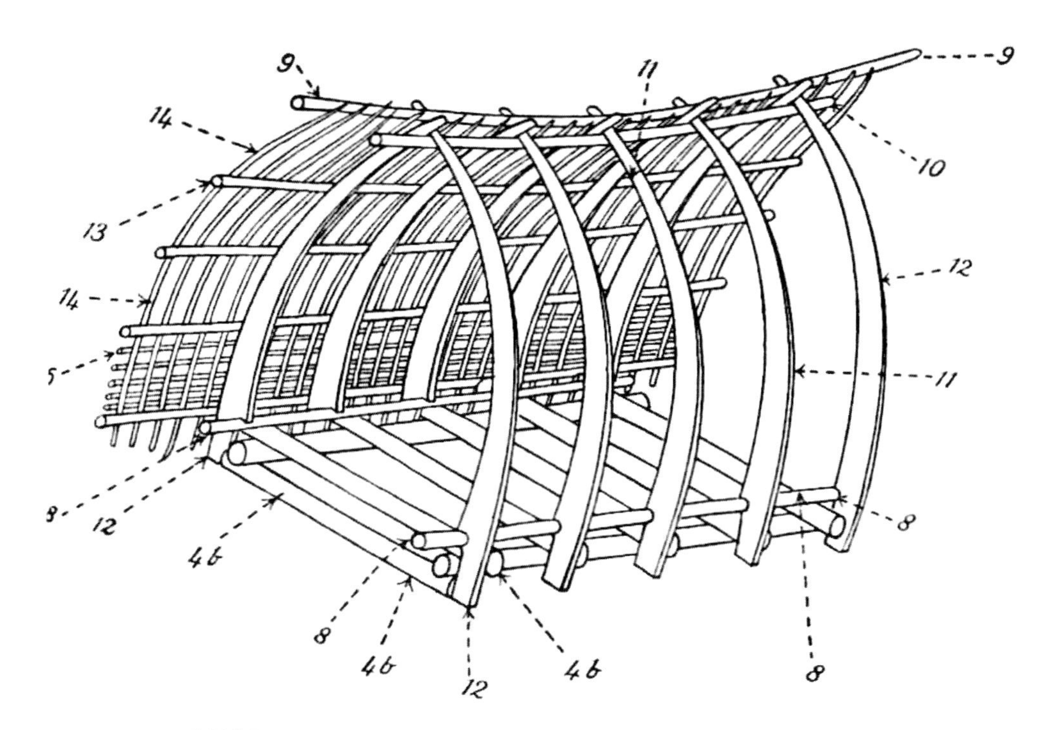

DIAGRAM X. STRUCTURE OF THE ROOF OF YAM-HOUSE

4*b*. Framing log at top of log cabin.
8. Horizontal support of thatch.
9. Upper ridge pole.
10. Lower ridge pole.
11. Frame-board of roof.
12. Gable-board.
13. Inner frame-rod.
14. Curved frame-rod.
15. Outer frame-rod.

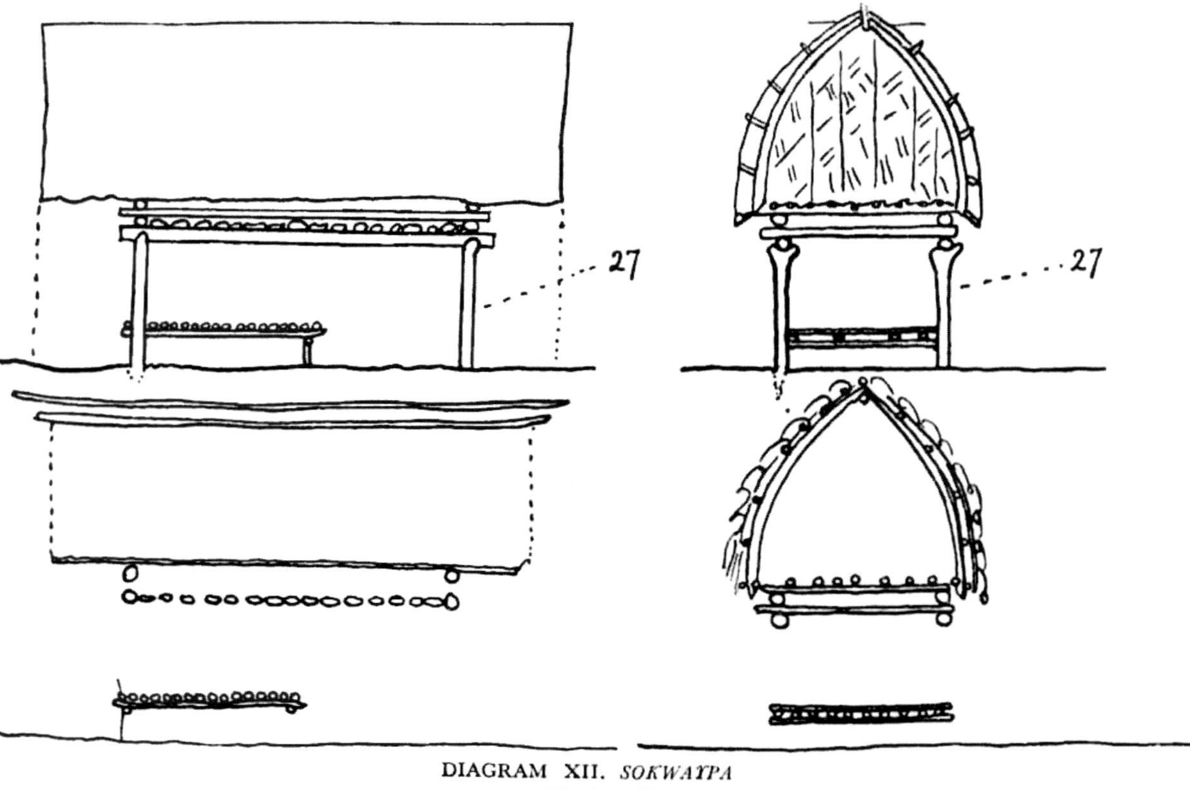

DIAGRAM XII. *SOKWAYPA*
27. Wooden pillar.

CHAPTER IX

A COMPARATIVE GLANCE AT TROBRIAND GARDENING

So far we have described only one system of gardening in full detail. All our general remarks about work and magic, the storage of harvest gifts, and the principles of division of labour were built round a concrete account of the Omarakana gardening system. I have meant it to be taken as a pattern and type of gardening in general, for, as we know, Omarakana gardening stands for the gardening of Kiriwina, and to every Trobriander, indeed to every native of the Northern Massim area, Kiriwina stands for good gardens. It was obviously better to describe one system fully than several superficially, and to give the same detailed account of some score or so of gardens would be impossible, both from the point of view of the field-worker and of the reader. The value of such details lies not so much in the knowledge of them for their own sake as in that they reveal the structure of the Trobriander's gardening customs and beliefs, the length to which he goes in the definition of ritual procedure and the way in which these elements integrate into a consistent system of behaviour.

But it will be necessary to indicate briefly how far the gardening system of Omarakana really is representative of Trobriand gardening in general.[1]

1. THE ESSENTIAL UNITY OF TROBRIAND GARDENING

In the first chapter of this book I have made it clear that there is an essential similarity between the various systems of garden-making. This similarity goes so far into detail that certain information, especially about the magic of burning, which I received in a preliminary way on the west coast when I first arrived, I was able to check in actual performance during my early months at Omarakana in the centre of the island.

At that time, as a matter of fact, having just arrived in the district and having been assured that the culture was identical throughout, I went on the assumption that there would be just one type of gardening and one type of magic. I was not prepared to find garden magic developed with the amazing intricacy and elaboration of detail both in ritual and spell which I later on discovered.

[1] See also Note 36 in App. II, Sec. 4.

For the sake of its methodological interest, I shall reproduce my first notes on garden magic almost word for word as I took them down from a native who spoke pidgin-English.

"Each village has one garden sorcerer, called *towosi*. When the garden is ready for burning, the *towosi* sends a boy to get a 'bud of coconut' (*kaykapola*). The boy is also told to bring a *wakaya*, 'dried banana leaf', the leaves of the plant called *bulabula*, and *yayu* (leaves of the casuarina tree). Then with a *kaybomatu* (large mussel-shell) he scrapes some 'sandstone' (*kaybu'a*).

"Then they take some stuff that has been scraped by the bush-hen—the idea being that the taytu will swell like the bush-hen's mound. The *kaybu'a* stone bites right down into the earth; so the taytu should do. The casuarina grows very quickly and the taytu ought to grow quite as quickly. The banana has a thick trunk and it bulges towards the base like the trunk of a coconut tree, so should the taytu do.

"They take the 'buds' of the coconut, so that the taytu leaves should be dark." Some more plants are used, but my informant does not remember them.

"The sorcerer puts all the ingredients together on a mat (*mo'i*). He puts another mat over his head and sings some spell over the stuff. He is alone in the house. *The words have no sense*. (Obviously incorrect!) The ceremony takes place in the early morning. Then the sorcerer ties the whole stuff between two mats.

"Then he himself must make some fire. He has to take two sticks and rub them against each other. He makes it on his 'verandah' (in front of his house). He then takes two burning sticks and carries them into the garden. Then he takes the two sticks in each hand, his sister or daughter takes the mats, and they go into the garden, heading a procession of other men and women.

"The woman puts down the mats on a spot called *omile'ula* (this is the place where the road strikes the garden and where the stile will be erected). He then gives the two sticks (*sulu'a*) to two men. They go and burn the garden.

"The magical stuff inside the mats is distributed among the individual gardeners. It has been previously wrapped into small parcels by the sorcerer in his house. Each parcel is wrapped up in a piece of banana leaf and attached to a stick. (Here my informant has probably mixed up the two ceremonies of burning.) Each man buries his parcel of magical stuff on his plot in the gardens."

This account is correct except for one or two minor points, the most glaring of which I have rectified by the remarks in brackets. There are, of course, the usual crudities of early wording, borrowed directly from the interpreter's pidgin-English: such words as "sorcerer", "boy", and incorrect interpretations of the words *kaybu'a, kaykapola, wakaya.*[1] But on the whole this account could refer to the ceremony of the first burning in almost any community, and having seen this ceremony performed in several villages and witnessed it twice in Omarakana, we remain impressed by the accuracy of my informant's statement rather than by the mistakes in details.

During the first month of my field-work in Kiriwina I used to collect my information about gardens indiscriminately from the various communities on the west coast and from those round Omarakana as well as in the capital itself. In the course of this investigation I soon became aware that every village has its own system of magic; that is, in every village there is a series of magical formulae handed on in the maternal line, always in possession of the ruling sub-clan of that community, and performed on behalf of the community by the head of this sub-clan, or his delegate. By system I mean a sequence of magical acts correlated with practical work and integrated into a progressive series of activities (cf. App. I). The natives have a very definite realisation of this sociological and cultural fact. They know that every village has its own garden magic. They have names for each individual system. They have a clear conception that one system may be more powerful than another, although I suspect that each community believe their own magic to be the best, at least for their own soil and gardening technique. On the whole, however, it is acknowledged that the system practised in Omarakana is the strongest, and since, roughly speaking, the more fertile the district the more powerful the community who occupy it, there is a clear correlation between the assumed excellency of magic, the real quality of gardens, and the power as well as the rank of the community owning them (cf. Ch. I, Sec. 7).

Thus the general character of garden magic, its correlation with work and the sociology of its enactment—that is, the position of the *towosi*, the payments given to him by the villagers, his character as a leader—are the same throughout the district, as has been briefly indicated in Chapter I.

There also I have stated that the main garden ceremonies are identical in type. With the detailed account of one system behind us, we can now amplify this statement. The *yowota* and the *gabu*, that is the inaugural rites of cutting and burning the scrub; the *kamkokola*, the magic of planting and of the taytu supports; the series of growth magic; the inaugural rites of thinning out and

[1] *Wakaya* is not "dried banana leaf" but the species of banana with the bulging trunk.

weeding, and the inauguration of harvest—these are found in every system. In every system also they are carried out with special fullness and precision on the standard plots, the *leywota*. In every system they stand in the same relation to practical work.

But each system has its own spells, its own magical substances. It presents certain variations in ritual, and the taboos on the magician differ slightly from one system to another. But though the words, the substances, the ritual gestures vary, the type is the same. A study of Documents VI and VII will exemplify the points of divergence as well as the essential sameness.

In Document V I give a somewhat dry enumeration of the names of the various systems in Kiriwina. At first sight this document does not seem very informative, containing as it does only a correlation of names. Its analysis, however, shows that the same system is used in as many as seven communities. The Kaylu'ebila system, which has been recorded in detail in Chapters II—V, is found not only in Omarakana but in six other villages. It was practised in the now extinct village of Omlamwaluva; it is practised in its original home of Lu'ebila; in the village of Laba'i, where the Tabalu sub-clan came out of the ground, and in Kuluvitu, Kapwani and Olivilevi. Now this means, first that information obtained in Omarakana applies more or less completely and without change to all these villages; and second, that, since the system is named after Lu'ebila and practised in two more villages of the extreme north, its probable home was on the northern shores of the island. It most likely moved south with the Tabalu sub-clan, hence its use in Omarakana, Olivilevi, and formerly in Omlamwaluva (cf. Ch. XII, Sec. 3).

It would have been very tempting to check the data obtained in Omarakana in one of the other villages, especially in Lu'ebila. It would have thrown a great deal of light on the interesting problem of how far magic changes in the process of independent use, and how far the belief of the natives that magical ritual and formulae are immutable is borne out by fact. But to everyone who knows how difficult it is to obtain magical texts from the average informant, especially before one has become acquainted with him and gained his confidence, it will be obvious why I did not carry out this tempting experiment. Even with an exceptionally intelligent, well-disposed and patient informant like Bagido'u, it took me many months to write down, check, translate and obtain a full commentary on his magic.

Looking at the further data contained in Document V, we can see that two more systems are practised by five villages each. One of these, the Giyulutu, is practised in Kabwaku, and in the villages of Wakayluva, Kaurikwa'u, Tubowada and Wakayse, which form a definite cultural and even political unit, dominated by the capital (Kabwaku). The Bisalokwa system, on the other hand, is scattered from Kavataria, on the western coast, to Kabululo in the north, and is practised also in Obweria (Tilataula), Kudukwaykela (Kuboma) and Kuluwa on the east coast. Momtilakayva, again, is practised in the Kurokaywa villages (Yourawotu, Tilakayva and Kupwakopula), which are a stone's throw from Omarakana; in Tukwa'ukwa on the coast of the lagoon; and in Suviyagila, in the extreme west. Some systems such as the Gayga'i are equally practised by the people of Bwoytalu, who are the lowest of the low, by the aristocratic village of Yalumugwa, and by the commoner village of Moligilagi.

There is on the whole not very much consistency in the distribution of magical systems, but the very fact that they are thus scattered, that they are not definitely associated with rank nor with any topographical principle, shows the essential unity of garden magic throughout the Trobriands. I observed the system of Silakwa at close quarters through its whole extent, for it was practised in Kasana'i, the sister village of Omarakana. With the help of some informants from Kasana'i I was able to check every detail of Omarakana magic as against their own and found the two agreed so closely that I did not feel it necessary to keep a record of this work. The Giyulutu system I saw practised in Wakayse, a village about ten minutes away from Omarakana. The magic carried on by Bwaydeda in the village of Obowada I photographed (Pl. 100) and observed at several stages though I did not take detailed notes. The Momtilakayva system I observed in part and noted down to the extent shown in the brief account found in the next section.

Over and above all such consecutive observations, it must be remembered that I also had a vast experience through casual and occasional glimpses of gardening and its magic. On my walks through the district I used to come unexpectedly on one ceremony or another, and observe it in actual performance as well as listen to the spells chanted ritually in the gardens. On my visits to other villages I used to discuss, conversationally, the state of the gardens and the work or magic then in progress. The Trobrianders, like all peasants, are always ready to talk about their gardens. They are not so keen to talk about their magic to anyone who is not already thoroughly acquainted with it, but they knew that I was conversant with it, and so I could always obtain an exact description of the minor points in any system. Whenever I discovered anything which seemed to diverge from the norm, I made an attempt afterwards to follow it up.

2. A PUBLIC CEREMONY IN THE GARDENS OF KUROKAYWA

Thus, for example, in the system practised in Kurokaywa, just next door to Omarakana, I found one or two ceremonies divergent in detail from any contained in the Kaylu'ebila system used by Bagido'u. The complex of villages called Kurokaywa consists of three settlements: Kupwakopula, Tilakayva and Yourawotu. The first two make their gardens jointly, the third separately. There are therefore two garden enclosures made each year. The magic is made for both gardens by Nasibowa'i, the headman of Kupwakopula.

The system which he uses is called Momtilakayva. Unlike most of those in Kiriwina, it begins, not with the *kayaku*, but with a ceremony in the sacred grove, Ovavavile. This consists of a large clump of trees which has not been cut for many generations, and it lies about midway between the villages of Omarakana and Tilakayva (see plan of Omarakana garden lands, Fig. 13). It occupies the centre of a field which really belongs to Omarakana, but the tabooed grove mythologically and traditionally plays a rôle only in the Tilakayva magic. It is strictly tabooed to all save the magician, and even he would only enter it for ritual purposes. Anyone who violates this taboo is liable to be stricken by the *pwawa*, a swelling of the sexual organs. The natives are so averse from anyone entering it that I never inspected its interior, though I had to pass within a stone's throw of it almost every day during my long sojourn in Omarakana. In the middle of it, I was told, there is a large stone, and on this the *towosi* of Kurokaywa performs a rite. Just before the *kayaku* is held, he carries a large tuber of a species of yam called *kasiyena* into the *kaboma*, and laying it on the sacred stone as an offering to the ancestral spirits, utters the following spell:—

FORMULA 32

"Who is it that bends down in the grove of Ovavavile?
I, Nasibowa'i, I am bending down in the grove of Ovavavile;
I shall carry this bending down in the grove of Ovavavile;
I, Nasibowa'i, I am bending down in the grove of Ovavavile;
I shall carry this my basket on the head into the heart of Ovavavile;
I shall carry this my (pledge of) new growth into the heart of Ovavavile."

In this rite we have a direct association between a tabooed grove, ancestral spirits, a sacred and tabooed object, the stone and the magician. Although in the Momtilakayva system, as in the

Kaylu'ebila, the offering to the spirits, the *ula'ula* (cf. Ch. II, Sec. 4), is made in the magician's own house, this ceremony is said to bring the whole cycle of gardening under the direct tutelage of the ancestral spirits. In this case the ancestral spirits are those of the predecessors of the magician.

Soon after the inaugural ceremony follows the *kayaku*, which is held by the two villages who make their gardens together in front of the house of Nasibowa'i, and by the people of Yourawotu in front of the house of their headman Giyokaytapa. Subsequently the buying of the fish by the villagers and the collection of the herbs by the magician, the presentation of the ceremonial payments and the *ula'ula* are conducted in the same way as in Omarakana.

The first important ceremony after the *kayaku* is in all its essentials identical with the corresponding one at Omarakana. The *kema* (axe) and some of the magical mixture are charmed by the *towosi* in the village with the following spell:—

Formula 33

"Clustering, clustering. . . .
Coiling, coiling. . . .
Thy clustering, O taytu, thy coiling, O taytu,
Thy fullness is that of the *waybitu* plant;
Thy foliage is that of the *yokwa'oma* creeper."

Next day the men, each with his axe, go to the garden with the *towosi*, who carries his *kaylepa* (garden magician's wand). On the *leywota* the *towosi* cuts the bad stick (the *kaygaga*) and throws it away, uttering this spell:—

Formula 34

"I strike thee, O soil,
Arise, O soil,
Lift and raise thy crops, O soil,
Lift and let thy crops sag, O soil."

In this spell the word soil stands metaphorically for garden and more especially for the produce of the garden. It is thus an invocation for the taytu to rise, that is, to grow; to lift and raise, that is, to develop the plant above the ground; and to lift and sag, that is, to produce an abundant crop of tubers underground. Comparing this spell with the *kaygaga* spell of the Omarakana system, it is clear that it is built on an entirely different pattern. It is really the spell of striking the ground, and will be repeated when the magician performs that rite. But it has not got the same exorcistic function as the *kaygaga* spell of the Omarakana system. Whether this is due to the fact that I obtained insufficient or confused information from Nasibowa'i, or whether in this system the same spell is really used for two performances, I am not able to decide with full conviction.

Then the magician cuts the good stick (*kayowota*), plants it in the soil, squats down, sways the stick to the right and left and utters another spell, which he recites again while he rubs the soil with the magical herbs:—

FORMULA 35

"O bush-hen of my magic,
O small bush-hen of my magic,
O mirthful laughter, O playful playing!
I shall cry out with the throats of my garden companions.
It is not their voice, their voice is the night-jar's.
It is not their throat, their throats are of the *kabwaku* bird.
We cry out at our work,
We brag about our gardens."

Nasibowa'i then strikes the ground with the *kaylepa*, repeating Formula 34, "I strike thee, O soil", etc., and the men disperse to their *baleko*, to cut or mark trees. The magician rubs and strikes the soil on each *baleko*, when he has finished with the *leywota*. As we remember, all details in this ceremony have a definite meaning in Omarakana, and the same meanings are ascribed to them here.

The cutting of the bad stick is to drive away evil influences and to make the earth fertile. The good stick is a symbol of fertility, and the rubbing of the ground with the leaves is to make the soil good. The striking with the magical wand is also to make the ground fruitful. The whole ceremony in this system, as in every other, has a general inaugurative function, and is meant to instil the vital forces of fertility into the earth.

As we see, the spells in this system are different from those used at Omarakana. So also is the magical mixture. Instead of the thirteen ingredients used by Bagido'u, Nasibowa'i uses only four: *ge'u*, the sand or earth taken from a bush-hen's nest, a substance used in Bagido'u's magical recipe; *nunuri*, leaves from a tree which is very prolific and has big fruit; *wokubila*, a plant with large, thick and deeply green foliage; *kaytagem*, also a plant with exuberant foliage. All these are used for their sympathetic qualities.

All these substances are divided into two parts, which are laid between mats in the usual manner and separately medicated in the magician's house. One part is charmed over in a heap, and afterwards portions of the mixture are inserted between the cutting edge of the axe blade and the folded banana leaf, and then the axes are

PLATE 100

BWAYDEDA THE GARDEN MAGICIAN OF OBOWADA

This picture is not posed; it was taken during the actual performance of the *gibuviyaka* rite, and shows the concentrated expression of the magician at work. Behind him is his maternal nephew, holding a few torches in the mat (Ch. IX, Sec. 1)

PLATE 101

NASIBOWA'I WITH THE CEREMONIAL AXE ON HIS SHOULDER

(Ch. IX, Sec. 2)

PLATE 102

CHANTING THE SPELL OVER THE KAMKOKOLA
(Ch. IX, Sec. 2)

PLATE 103

PEOPLE GATHERING AT THE STILE FOR THE KAMKOKOLA CEREMONY
"The women brought the taytu, some leading a child by the hand, others carrying one astride the hip" (Ch. IX, Sec. 2)

charmed (cf. Ch. II, Sec. 4). The loose herbs are taken to the garden and there the magician mixes them with some *sisiye'i* (bracken) which he pulls up on the spot. He rubs the ground, uttering the *yowota* formula.

The next ceremony, the burning of the cut and dried scrub (*gabu*) with its associated rites, does not differ substantially from the proceedings in Omarakana. Here also the magical *kaykapola* torches have been prepared at the previous harvest, when the magician charmed them with the same spell (M.F. 32) which he utters in the grove of Ovavavile. The first burning (*vakavayla'u*) is carried out with as scant ceremony as in Omarakana, even women being allowed to be present.

The next burning, the *gibuviyaka*, is begun as at Omarakana by making a small heap of dried twigs on each *baleko*, inserting some of the magical substance into it, and setting fire to the heap (*lumlum*) with a *kaykapola* charmed with Formula 32 immediately before use.

Then follow the *pelaka'ukwa*, or chanting of the taro, and the *kalimamata*, a rite performed over a *kuvi* or large yam. There is no *kwanada* spell, as in Omarakana.

At the *pelaka'ukwa* this short formula is said over a taro top before it is planted in the magical corner:—

FORMULA 36

"O taro, tenacious as the bracken,
O taro, anchored, anchored firmly,
Blossom on."

Apart from its obvious aim this formula is intended, I was told, to frighten away the bush-pigs, although its text does not express this as does the Omarakana spell.

The *kalimamata* spell spoken over a *kuvi* is a little longer:—

FORMULA 37

"Whose children are crying out for food?
My children, the bush-hen's, are crying out for food.
The *uku'uku* weed pierces, the reed pierces through.
Come out (O young yam tubers) and surround (the old one)."

The *towosi* does not erect the miniature house of twigs, *si bwala baloma*, as is done at Omarakana.

After the *koumwala*, which follows this series of rites, has been finished, the *kamkokola* ceremony takes place. Of this ceremony I shall give a much fuller account than of those which precede and follow it. It is the only ceremony in the Kurokayva system which I witnessed personally and with close attention. I have seen one or two of the other rites, coming upon one by accident and catching a glimpse of another. But the *kamkokola* ceremony I saw under particularly favourable conditions and I was specially interested in it because, according to the preliminary accounts which I obtained, it was a much more developed and ceremonial affair from the sociological point of view than most Trobriand magic. I have mentioned already that, from the first ceremony I witnessed—the second rite of *vilamalia* in Omarakana (Ch. VII, Sec. 3)—I was impressed by the extreme bareness of Kiriwinian magic. All that I had seen afterwards of magical performance in the gardens had the same sober and practical, non-mystical, non-social and non-collective aspect. The fact that in reality even those rites which seemed most unregarded, from which the community seemed to remain completely aloof, were yet very much in the mind of every member of the community—a fact I emphasise in my introduction to the magical formulae (Part VI) I did not then fully appreciate. And since at that time I was suffering from a super-sociological, almost Durkheimian, attitude, I was very much interested to hear that a ceremony was to be performed next door to me in which the whole village community, including women and boys and even the spirits of the departed, would participate. I knew also that the main outline of the *kamkokola* proceedings was more or less the same as in Omarakana. The magician would prepare the *kayluvalova* stick, the symbol of the taboo; then for several days the men would be bringing in the *lapu* (stout poles); and after a few days the magician would charm a magical substance which would be put under the *kamkokola* pole on the following day. But after all this there was an important communal ceremony which has no counterpart in the Omarakana system.

Let us follow Nasibowa'i, the garden magician of Kurokayva, and watch him organising, haranguing, directing and uttering his spells. On a Wednesday afternoon, early in October, 1915, he harangued his community and announced the beginning of the *kamkokola* cycle on the next day. He imposed the usual taboo on the gardens, saying that from the morrow all men must go and collect the heavy *lapu* for the *kamkokola*. Next day he himself went round the *baleko*, planting the *kayluvalova*, the stick which marks the taboo on the gardens, in each one of them.

Some of the more ambitious or careful gardeners would immediately add two slanting horizontals to the *kayluvalova*, making a small *kamkokola* of it (see Pls. 102, 105 and 106). During Thursday, Friday, Saturday and Sunday the men were busy running with the *lapu*—literally running, because when carrying a very heavy load they prefer to run instead of walk, resting for a good while at intervals.

On Monday morning the magician recited Formula 38 (see below) in his house over two kinds of herbs, *nunuri* and *kaluwayala*, together with *ge'u* (earth taken from the bush-hen's nest). The *nunuri* and *ge'u* we have met in a previous rite; *kaluwayala* (hibiscus) is used as a symbol of fertility.

On Monday afternoon, about three o'clock, the *towosi* went into the gardens accompanied by his suite, including the ethnographer with a camera. There were two different acts to be performed: the *kamkokola* had to be erected in the magical corner, which task must be done by the chief magician or his accredited representative. Again, a spell had to be chanted over the erected *kamkokola*, and this also must be done by a *towosi*. We delegated the former task to the junior magician, and reserved the more important one, the chanting of the spell, for ourselves. The nephew of Nasibowa'i went, therefore, ahead of us, preparing the *kamkokola*, and I watched him at work on two or three *baleko* before rejoining Nasibowa'i. He dug a hole with his *dayma* and, taking a handful of the leaves medicated in the morning by Nasibowa'i, rubbed the three poles throughout their length, put the herbs into the hole and, with his own hands, inserted the *kamkokola* into it The *kamkokola* on the remaining three corners were then put up by the owner of the *baleko*, perhaps assisted by some other men. In many places the *karivisi* had already been erected. The fields were lively from one end to the other with men carrying poles, putting them up and adjusting them.[1]

Nasibowa'i followed in the track of his assistant and consecrated the *kamkokola* one after another. On his right shoulder he had a *beku* (large ceremonial axe), in the handle of which, under the blade, some of the mixture charmed over in the morning was inserted (see Pl. 101). It is remarkable that, in one or two of the systems, the old technical implements or processes survive. Thus, as the reader no doubt has noticed in the account given above of the burning at Kavataria, my informant told me explicitly that the garden magician makes fire by friction. Subsequently I found that this is still done at Kavataria, and in those villages where the same magical system, Bisalokwa, is used. In all other villages without exception the ordinary Bryant and May vesta is considered sufficient. Again, in the Momtilakayva magic of Kurokayva, Nasibowa'i still uses the old ceremonial blade. Whenever an axe is required in the garden magic of Omarakana and elsewhere, an ordinary steel blade has superseded the old *beku*. As far as I could ascertain, the other three communities which used Momtilakayva have also gone over to steel.

At each magical corner, the *towosi* first stroked the two *kaybaba*

[1] For technical expressions, such as *kamkokola, karivisi*, see Ch. III, Sec. 4.

(the slanting arms of the *kamkokola*) with his open palms and subsequently uttered a spell, standing in a characteristic attitude near the *kamkokola*. He laid both hands on one of the *kaybaba* and rested the sole of one foot on the inside of the thigh of the other leg (see Pls. 23 and 102, which were taken during the actual chanting of the spell). This position is often assumed by men while resting. He chanted the spell in a loud, solemn voice, with his face turned towards the *baleko* he was medicating:—

FORMULA 38

"Anchor on, anchor on, do anchor, do anchor!
There is one anchoring of my companions,
A weak anchoring, a feeble grip,
My anchoring is firmness.
I am anchoring the belly of my garden.
It rises and stands up, the belly of my garden.
It is anchored firmly, the belly of my garden.
It is anchored in the bedrock, the belly of my garden.
It darkens, the belly of my garden.
It stands up like the interior of iron-wood palm groves, the belly of my garden.
It stands up like an interior of shady leaves, the belly of my garden.
O ancestral spirits Kutorawaya, Torawaya, Wasa'i, Iluvapopula and Tomlawa'i;
O grandfathers of the name of Mukwa'ina, Uluvala'i and Mwoysibiga;
And thou new spirit, my elder brother Mwagwoire.
Anchor on, anchor on, do anchor, do anchor!"

When he had finished the spell, Nasibowa'i lifted the *beku* from his shoulder and struck the upright *kamkokola* pole (see Pl. 40). That finished the rite.

We went all over the fields, doing one *baleko* after another, passing people at work and, from time to time, catching sight of the junior *towosi* busy digging, rubbing over and planting his *kamkokola*. As we passed the Yourawotu gardens, which were in a more advanced stage, we saw people planting *yagogu* (seed taytu). A man brought a basket of *yagogu* to Nasibowa'i who charmed it over with a private magic, beating it with a branch of mimosa (cf. Ch. IV, Sec. 4). On another *baleko* he performed a *bisikola* for a man, this rite being private and not official magic in Tilakayva (cf. Ch. III, Sec. 1). For both perfomances Nasibowa'i subsequently received an official fee. The work in the gardens was very hard—it was a clear, hot day, and we were not nearly finished as the sun was setting. This was

THE FOOD FOR THE SPIRITS DISPLAYED

This is the religious phase of the *kamkokola* ceremony. The food for the spirits "is left for a few minutes on the platter before the women take it away" (Ch. IX, Sec. 2)

PLATE 104

THE MATERIALS FOR MAGICAL RITE AND RELIGIOUS CEREMONY

A man carrying magical herbs is seen "standing beside the magician with two large bunches of leaves on his carrying pole". On the right and left, two women with platters of "food for spirits" on the head (Ch. IX, Sec. 2)

PLATE 106

THE RECITAL OF THE SPELL

Nasibowa'i "squatted on the ground in front of the large bunches of magical herbs; to his right, the special *bwabodila* leaves in oblong baskets" (Ch. IX, Sec. 2)

remedied, however, by delegating the junior *towosi* to complete the work.

The main ceremony, *keliviyaka*, the big burrowing or burying of the *kamkokola*, was to be performed the next day. It corresponds to the charming and inserting of the *kavapatu*, the leaves buried at the foot of the *kamkokola*, as described in the Omarakana system. But in the Momtilakayva system, two kinds of magical mixture, named according to their function, are used: the *kavapatu*, which are buried at the foot of the *kamkokola*, and the *bwabodila*, which are tucked into the fork of the *kaybaba* where it rests against the *kamkokola*. Both the *kavapatu* and *bwabodila* are chanted over in the garden by the magician. He is helped by some acolytes who, though on this occasion they need not be junior magicians, must keep a complete fast on that day till the ceremony is over. This is incumbent on the *towosi* and on all who take an active part in the magical ceremony. Only these helpers may perform certain of the rites, such as the putting up of the *bwabodila* leaves on the *kamkokola*. The *kavapatu* may be handled by any of the men and each does his share by burying them on his own *baleko*. This ceremony, in which practically all men, women and consequently children from Kupwakopula and Tilakayva took part, is the only garden rite known to me which is carried out communally and with a certain amount of public display.

As there is a *sagali* during the ceremony, the women of the village also have their share of work, and on the previous day they are already eagerly employed. On Monday evening when I returned to the village with Nasibowa'i after our first day's round of magic, women were coming in laden with dry branches and firewood, with stones for the *kumkumuli* (baking in earth) and with water. Others were preparing the large cooking-pots (*kuria*) and whatever else was necessary for the morrow's *sagali*. On my arrival the next morning the village was full of bustle and smoke; the *kuria* were steaming on their stone supports and the heaps of *kumkumuli* were being opened to take out the baked yams. Each woman put her contribution of baked taytu on a *kaboma* (wooden dish) or into a *peta* (basket) and brought it into the garden.

In the meantime some of the men had gone into the bush, some right down to the seashore and others to the *rayboag* (coral ridge) in order to get the herbs necessary for the magic. Large quantities of these have to be gathered as, after the ceremony, the *kayvapatu* herbs are distributed among the gardeners, each taking his portion to use on his own plot.

The whole population of the village does not march in a body to the gardens; everybody goes when it suits him. As a matter of fact the *leywota* plots were not ten minutes from the village, so the way was not far. The *towosi*, myself and a few others started at about ten o'clock and we were first at the *leywota*. There we seated ourselves in what shade we could find and patiently waited till everyone had arrived and everything was ready. On the Omarakana side of the fence was the dense, low scrub of the *odila* (uncut bush); on our side the gardens lay stripped and bare, save for the sticks left standing from the *takayva* (cutting of the bush) and the newly erected poles. Across the fields one could see several groves—those of Yourawotu close to and of Tilakayva and Kaulagu beyond; also a few *kapopu* (tabooed groves), and in the distance the wooded ridge of the *rayboag*. The two rows of *karivisi* planted along the road from Yourawotu to Omarakana formed a regular espalier (cf. Pls. 103–108) terminated at the *leywota* end by two specially fine *kamkokola* bridged by a horizontal, so as to form a kind of gate (see Pl. 103). This, however, was not essential to the structure but a fancy touch added to the *leywota kamkokola* where magic was to be performed.

Soon after we arrived the other people began to gather, dropping in by ones and twos. The women brought the taytu, some of them leading a child by the hand, others carrying one astride the hip (see Pl. 104). The people seated themselves on either side of the alley through the field, everyone trying to find some shade. The women with the taytu were soon all present, but we still had to wait for the men who had gone to the seashore in search of magical herbs. In this ceremony four kinds of leaves are used: three for the *kavapatu* and one for *bwabodila*. The herbs used for the former are *nunuri*, *kaluluwa* and *wokubila*. The first grows on the beach and is not too easily found; the second in the *weyka* (village grove) and the third in the *odila* (uncut bush). The magical properties of the first and third have already been mentioned, and I failed to ascertain those of the *kaluluwa*. For the *bwabodila*, *kaytagem* leaves are used, and these have also been mentioned before. The leaves are not torn up before using.

At last, however, the party from the seashore arrived, and we can see one of them, on Plate 104, standing beside the magician with two large bunches of leaves on his carrying pole (*katakewa*). When all was ready the *sagali* took place. The *taytu* was divided into small heaps, each one putting his or her contribution on to one of the *kaboma* (wooden dishes) which were all spread out on the broad path in readiness to receive these. The men went from heap to heap, allotting them to those who played a leading part in the ceremony, the magician and his helpers. There was no *kolova* (loud crying out of names), such as is usual at distributions in the village.

This *sagali* has two aspects: ceremonially it is an offering of food to the *baloma* (spirits). From this point of view it is called "the food of the *baloma*" (*baloma kasi*), and after it has been allotted it is left for a few minutes on the path before the women take it away (see Pl. 105). As the natives say, it is exposed that the spirits of the departed may take their share, and this is supposed to please them. But the natives have no more than a vague idea and a very generalised feeling about this tribute to the spirits. No one was able to explain to me what spirits they were to whom this offering was given, nor even if there were any spirits present (cf. Part VII, M.F. 1, A, and App. II, Sec. 4, note 8). A comparison with kindred customs in native sociology may, however, make this rite more intelligible; such, for instance, as the offerings of food made to the *baloma* during their annual visit to the village at *Milamala*, or the exhibition of valuables round a dead man's body immediately after death.[1]

In its other aspect this particular *sagali* is a payment or gift offered by every household to certain outstanding members of the community, in this case to the garden magician, his helpers and some other notables. I noted the following allotments:—

3 heaps for the *towosi* and headman of Kupwakopula;
1 heap for his son, who was helping him;
1 heap for his *kadala* (maternal nephew), who was helping him;
1 heap for the *tobwabodila* (the acolytes who placed the *bwabodila* leaves);
1 heap for the helpers who fetched the herbs;
1 heap for Giyokaytapa, the headman of Yourawotu;
1 heap for Mwaywaga, a notable of Tilakayva;
1 heap for Toviyama'i, the younger brother of the headman of Tilakayva.

After the food had been allotted the women took the *kaboma* away to the village. They took the smaller children with them, and such boys as remained were dismissed, and vanished after some slight friction. The ethnographer and his servants and the men who came from Omarakana had to step back to the Omarakana side of the fence.

Then Nasibowa'i proceeded to the main part of the ceremony, the chanting over of the magical herbs. He squatted down on the ground in front of the large bunches of magical herbs; to his right, the special *bwabodila* leaves in oblong baskets (*vataga*, see Pl. 106). He chanted Formula 38 for about a quarter of an hour, the men keeping quiet and not moving about during the performance. After he had

[1] Cf. also my *Argonauts of the Western Pacific*, p. 512, and Pl. LXV.

finished he took some of the leaves from the bunches in front of him (the *kavapatu*) and buried them at the foot of the *kamkokola* in the two *leywota baleko*. As soon as this was done the men rushed towards the *kavapatu* leaves, snatched up a handful and ran off, each to his *baleko*, to bury the leaves at the foot of the *kamkokola* (see Pl. 107). Their proceedings were distinctly hurried and I was told that this was the correct thing at this ceremony.

The two or three baskets containing the *bwabodila* (leaves tucked into the fork made by the *kaybaba* and the *kamkokola*) were taken by the two or three helpers (called in this connexion *tobwabodila*) who were to go round all the *baleko* and perform their rite. This has to be done in the following manner (see Pl. 108). The *tobwabodila* approaches the foot of the *kaybaba* to the right, that is, he moves behind the structure from left to right. He does this in a crouching attitude and mimicking a stealthy and careful gait, as if to take something or someone by surprise. Once near the spot where the *kaybaba* touches the ground he springs erect and, abandoning all mimic precaution, runs round to the foot of the other *kaybaba*, the one which can be seen touching the ground in the centre of the picture. There he takes hold of the *kaybaba* and rubs it all over with the *bwabodila*. Then he climbs up the *kamkokola* and tucks the *bwabodila* under the *kaybaba*, the act reproduced on Plate 92. All is done in a rush, from the moment when the *tobwabodila* with a sudden jerk begins to run round from one *kaybaba* to the other. With the conclusion of this rite the *keliviaka* ceremony comes to an end.

My knowledge of what follows the *keliviaka* or second *kamkokola* ceremony in the Momtilakayva system is very fragmentary. Soon after it I left the district, and was not able to continue my researches in 1918, having more pressing observations to make during my relatively short stay in the neighbourhood. I know that it contains a series of spells and rites associated with growth magic, but I cannot even give a list of the ceremonies in succession. They have inaugurative rites connected with *basi*, the thinning of the taytu, with *pwakova*, weeding, and with harvest. Of these I recorded only the spells associated with weeding and with harvesting. This is the weeding spell:—

FORMULA 39

"Weeding, I shall weed.
Weeding, I shall weed.
Weeding villagewards, clearing bushwards.
Weeding bushwards, clearing villagewards.
I shall clear, I shall weed.
I shall clear, I shall weed."

PLATE 107

MEN WITH MAGICAL LEAVES

"The men rushed towards the *kavapatu* leaves, snatched up a handful and ran off, each to his *baleko*, to bury the leaves at the foot of the *kamkokola*." But paused for a moment to be photographed (Ch. IX, Sec. 2)

PLATE 108

RITE ON THE KAMKOKOLA

The performer "climbs up the *kamkokola* and tucks the *bwadodila* leaves under the *kaybaba*" (Ch. IX, Sec. 2)

PLATE 109

NAVAVILE, THE GARDEN MAGICIAN OF OBURAKU

"We see the magician seated beside the magical herbs with the seven digging-sticks for his own and his younger brothers' plots" (Ch. X, Sec. 1)

These words are spoken while the magician carries out a mimic act of weeding with a digging-stick.

The following formula is spoken over the adze with which the magician cuts the first plant at harvest:—

FORMULA 40

"It is going, it might go;
It is dropping, it might drop.
I shall make you come out, I shall help you to come.
One tree, my tree, a magical tree, as the iron-wood palm,
I shall set up on my right side.
It will break through in daytime,
It will cut through at night.
It will break through so as to provide for our ceremonial exchanges."

This evidence, fragmentary though it is, has some value in showing that the formulae, in wording and ideology, are built on the same pattern as those of Omarakana.[1] Also the magical substances, which are partly identical, are certainly selected on the same principle. Of the formulae, I heard only one in actual performance repeated over and over again in the field, the formula given in the *kamkokola* ceremony (M.F. 37). I came across Nasibowa'i once or twice at other times in the gardens, but was not able to check the words spoken by him as minutely as in the case of a spell repeated more than a score of times in my presence. This spell I am confident was given to me in its entirety, but I doubt whether the other spells—which are conspicuously shorter and more fragmentary than those of Omarakana—are complete. None of them really develops the main part, the litany, but this is true also of the *kamkokola* spell. In my opinion there seems to be no doubt that in the course of their history magical formulae undergo processes of attrition and degeneration (cf. Part VI), and probably in Kurokayva they had recently suffered severely in this way.

It will be noted that among the several minor ceremonies following the *gabu*, the rite of *bisikola* does not appear (cf. Ch. III, Sec. 1). I was told that, in Kurokayva, the word *bisikola* is reserved for a private magic performed by each man over his large yams (*kuvi*). Each man has his own ritual and the spells apparently also differ (cf. also Ch. IV, Sec. 4).

[1] See also Note 37 in App. II, Sec. 4.

CHAPTER X

THE CULTIVATION OF TARO, PALMS AND BANANAS

1. THE GARDEN SYSTEMS OF THE SOUTHERN DISTRICT

So far I have been speaking mainly about the north. Here again this is due not altogether to the fact that northern gardening is more important, but to certain defects in my information.[1] Nevertheless the north is the agricultural district of the Trobriands and gardening plays a relatively greater part there than in the south. In the south, on the narrow strip of land running from the isthmus (as we might call it) of Kwabulo down to the strait of Giribwa where, at one point, there is no more than half a mile of dry ground between the eastern shore and the swamps which encroach upon the land from the west, there is relatively little cultivable soil. Whole stretches of it—the wide club-shaped expanse in the south of Kaybwagina, the portion north of Kaulasi and south of Wawela—are completely barren, being covered with swamps and coral rocks.

But even in the best parts of Kaybwagina, between Sinaketa and Okayaulo, there is but a narrow strip of relatively good soil between the brackish mud of the mangrove swamp and the broad back of the *rayboag*, and then it is much stonier than in the northern district.

It is therefore on fishing that these districts rely as much as on gardening; on fishing for their own cooking-pots and to barter fish for tubers (cf. for this and the following paragraphs also Part I, Secs. 4–6 and 10). Some communities, such as Oburaku and Kaulasi and, in the old days, Sinaketa also, used to derive a great deal of their vegetable diet from such exchanges. Nowadays (1918) Sinaketans live very largely on rice and tinned food, for this complex of villages is the main pearl fishery of the district and is well supplied with European goods by the traders.

Moreover, in the old days, the Sinaketans, who specialised in the fishing of spondylus shell and the manufacture of the red shell-discs, used to marry women from Kiriwina. Marriage always entails, as we know (cf. Ch. VI), reciprocal gifts: vegetable food from the wife's kinsman to her household, that is, the annual *urigubu*, and in return valuables—in this case red shell-discs—from the husband's to the wife's kindred. Now people of rank in Kiriwina were able to supply the Sinaketan husbands of their sisters with a large quantity of food.

[1] See also Note 38 in App. II, Sec. 4.

The Sinaketans in return gave to the paramount chief and to other notables what to these was of greater importance, necklaces and belts of shell-discs.

The island of Vakuta used to have better gardens than Sinaketa, although they also produced shell-discs and married Kiriwinian women.

Another difference between the north and the south is that in the latter taro, which can grow on swampy ground, and large yams (*kuvi*), which thrive in the holes of the *rayboag*, play a somewhat greater part in their agriculture than taytu. Taytu is the staple food in both districts and the only crops stored for any length of time, displayed and handled; but the proportion of yams to taytu is somewhat larger in the south.[1]

Magic and the technique of gardening, however, are much the same throughout the Trobriands, and the garden magician has a very great influence even in such villages as Sinaketa and Oburaku. In Oburaku Navavile, the garden magician, is still the first man in the community by virtue of his office, and he exercises that office, in virtue of his personality, for he inherited it from his father and not in the mother line. In Sinaketa, where chieftainship is well established, the garden magic is invariably practised by a member of the junior Tabalu clan which exercise political power there.

I was able to check the list of gardening activities and garden magic in Oburaku, and it agrees pretty closely with the system of Omarakana. There are minor differences, such, for instance, as that in Oburaku only one *kamkokola* is made on each plot, and that the harvest ceremonies of *isunapulo*, that is, the harvesting of large yams and taro, are more developed. As we have said already (Ch. VII, Sec. 6), the magic of *vilamalia* has got a slightly wider scope there than in Kiriwina. Also the *isunapulo* ceremony seems to differ more substantially from that of Kiriwina than do the other rites. I have to speak guardedly, for not counting the *vilamalia*, which is not a garden rite, the *isunapulo* is the only ceremony I have actually witnessed there; so for the others I have to depend on informants. I will here give a brief description of it.

The garden magician, Navavile, collects two big bundles of leaves in the morning; one gathered from the *menoni*, a fruit tree, the other from a tree called *kekewa'i*. These two bunches are laid on the platform before the magician's house and he seats himself beside them. The villagers, in the meanwhile, have been preparing new digging-sticks made of strong red mangrove wood, the best material for the *dayma*, and they now bring these one by one to the magician's hut.

[1] See also Note 39 in App. II, Sec. 4.

Round the sharpened end of each *dayma* a piece of dried banana leaf is folded in the usual way, so that one flap is open. Every man has to prepare as many digging-sticks as he has plots. On Plate 109 we see the magician seated beside the magical herbs with the seven digging-sticks for his own and his younger brothers' plots. I had to take the photograph before the real gathering began, because by the time there is a big heap of digging-sticks the crowd round the magician obscures the platform.

As soon as all the men have brought their sticks, these and the two bunches of leaves are laid between two mats, and the public recital of the spell over them begins. The spell I was unfortunately not able to record. Immediately after this rite each man takes his digging-sticks and, closing the open flaps of the banana leaf, binds them tightly round the pointed ends. No herbs were enclosed within the leaf, so that it was only "the magical substance" of the spell which was thus imprisoned. Then, with the digging-sticks and a few leaves from each bunch of herbs, they go to the garden. There each man plants a *dayma* in his plot at the foot of a support (*kavatam*), usually in the neighbourhood of the *kamkokola*, of which, as I have just said, there is only the one standing at the magical corner.

On this occasion I was able to observe "the magical corners" in the gardens of Oburaku, and I noted there the following traces of previous magical ceremonies: a pointed stake, which is inserted at the first inaugural rite, the *yowota*, and is called *si gado'i baloma*, "the stake of the spirits"; a special taro which is planted at the *pelaka'ukwa* rite; a yam corresponding to the *kwanada* of Omarakana planted during the *kalimamata* rite; a taytu, officially planted during the *kamkokola*, making this ceremony a definitely inaugural planting rite, as it is in Vakuta also and in Sinaketa (see Doc. VI, and further on in this section); a *kaydabala* stick inserted during one of the rites of growth magic, and a tuft of lalang grass (*gipware'i*), which is planted during another rite of growth magic.

These observations bore out the result of the direct enquiries which I had made, i.e. that there is the same series of magical activities in Oburaku as in the communities of Kiriwina, though there are variations in rite as well as in spell.[1]

To return to the *isunapulo* ceremony: after planting his digging-stick near the magical corner, each man tucks some of the medicated leaves under the *kaybaba* where it meets the *kamkokola*, some are inserted among the leaves of the taytu climbing up the *kavatam*, and some are buried at the foot of the *kamkokola*. These last are called *boda*, while those put under the *kaybaba* and among the taytu leaves

[1] See also Note 40 in App. II, Sec. 4.

are called *paku.*[1] No spells had to be recited in the gardens, and the actual placing of the substances was not, as we have seen, done by the magician but by the ordinary gardeners.

The charming of the leaves and *dayma* was performed in public in the presence of all those who were afterwards to carry out the ritual disposition of these, but there were no special formalities and taboos. The women went on cutting their grass petticoats, some of the old men were trimming their nets. Only those who had to go to the gardens crowded round the magician. The ceremony was finished just about noon or a little later, and the last men to come from the gardens arrived in the village at three o'clock. Those who were ill or otherwise occupied were replaced by their kinsmen. At the end of the day some new yams and taro were brought from the gardens and placed in front of the garners, while some leaves were put on the beam ends (*kaytaulo*, cf. Ch. VIII, Sec. 3).

Next morning the villagers started to dig up yams and taro and to cut sugar-cane. Green coconuts were also collected. First-fruit offerings of the new food were laid on the graves of the *karigava'u*, the people who had died during the last year, by their maternal kinsmen, who are not the official mourners. Afterwards there was a small distribution in which the kinsmen of the deceased gave food to the mourners, that is to the children and the relatives by marriage of the deceased.[2]

The heaps did not remain on the graves for much more than half an hour or an hour, and there was no incantation or prayer to the ancestral spirits. The whole day had a certain atmosphere of reverential sanctity; everyone made an effort to remain in or about the village and to play his part in the handling of food; no fishing expeditions would have set out on that day, and no work was allowed in the gardens except the preliminary, half ritual, digging up of the new food (see Pl. 96, which was taken on such an occasion.)[3]

Moving further south to Sinaketa, we would find that there also the general system of gardening and garden magic differs but little from that of Kiriwina. As a matter of fact, during my early work in Omarakana I had attached to me, in the capacity of interpreter, an expert in modern languages, one Gomaya, a notorious scoundrel of the highest Sinaketan rank, a member of the junior Tabalu sub-clan

[1] The reader will have perceived a similarity between this ceremony and the *kamkokola* rite of Kurokaywa described in Chapter IX (Sec. 2), though the rite here described is a part of the *isunapulo*.

[2] For the sociology of mourning and other mortuary services, see *Sexual Life of Savages*, Ch. VI, especially Secs. 2–4. Also Part I, Sec. 7, of this book.

[3] See also Note 21 in App. II, Sec. 4.

of that community. Gomaya was perfectly well acquainted with the gardening of Kiriwina, for he had resided there for several years, and of course he knew his own system quite well. At that time I was trying to get a clear view of the Kiriwinian system and, struggling as I was with the fundamentals, I preferred not to be confused by minor comparative details, and therefore did not note the differences which Gomaya, with his local pride, indicated whenever we were discussing Kiriwinian garden customs. But though I cannot reproduce these, one thing I do remember—that they were all of a subordinate character.

In Sinaketa land tenure, the choice of fields and the apportionment of plots at the *kayaku* are the same as in Kiriwina. It was here indeed that I noted one of my best *kayaku* texts (Text 28*c*, cf. Ch. II, Sec. 3 and Part V, Div. V, § 21). Here also Motago'i, one of the best informants I ever had, gave me some interesting information about garden sorcery and the rôle of bush-pigs (Text 78, cf. Ch. III, Sec. 2 and Part V, Div. XI, § 2). Unfortunately I was so intent on studying the trade and sailing customs of the *kula* that I was not able to discuss gardening customs with Motago'i, whose comments were always worth having. There are minor differences in the rites and, of course, a different set of spells, but these do not affect the fundamental similarity in the structure of magic and its relation to work. In Sinaketa there are the four main types of rite: the *yowota*, the *gabu*, the *kamkokola* and the harvest rite. This last consists of *okwala* and *tum*, preceded by the complicated sequence of growth magic and thinning and weeding magic performed between the *kamkokola* and harvest rites. The *kamkokola* magic is, in Sinaketa, definitely a magic of planting, and of the planting of taytu.

In so far as I can trust my memory, there are no essential differences in the first three ceremonies, that is in the inauguration of cutting, of burning and clearing, and of planting (for a description and analysis of the Sinaketan *pelaka'ukwa* rite, see Ch. III, Sec. 2). In Sinaketa, as in Vakuta (cf. Doc. VI), it is in growth magic that differences occur. There is a magic connected with the putting up of the yam poles, *kavatam*, which does not exist in Omarakana, but which we find in Teyava (see Doc. VII). It is called *talovasi kaykela kavatam*, "we cast over the foot of the *kavatam*". They also have a rite named *ikayosi dabwana taytu*, the getting hold of the head of the taytu, which makes the vine produce more leaves and more roots; and an inaugurative magic of thinning out called *mom'la*, associated with a two days' taboo. They have a rite of *pwakova* which is called *talova kaydabana*, "we throw the head-wood"; another rite, called *tasasali*, is a preparation for the harvest of yams and taro; the *isunapulo*, which is of course the main ceremony inaugurating the harvest of these two crops. Finally they have *okwala*, during which the magician medicates his adze and cuts the stalk of the taytu in a ceremonial manner, and *tum*.

These details and names I have culled out of occasional information and a few notes jotted down while I was walking with my informants over the gardens in Sinaketa (cf. Pl. 110).

As to the differences in Vakuta, where there is more fertile soil and where conditions are rather like those of Kiriwina, it can easily be seen from Document VI that, whatever the minor differences, gardening and garden magic are fundamentally identical with those of Kiriwina. The harvest customs are different in so far as a man will neither harvest, transport nor store away the *urigubu* gift for his sister's husband unless he receives *vaygua*, valuables, as *karibudaboda*, payment. Otherwise the sister's husband has to do all the harvesting himself, that is, dig up the taytu, stack it in the *kalimomyo*, and then transport it to his own yam-house. Only the seed yams, the *yagogu*, does he leave for the man who has gardened the plot. Besides cultivating a whole *baleko* for his sister, a Vakutan usually gives her from two to half a dozen basketsful of taytu, *kala taytupeta*, from the plot or plots which he cultivates for his own household.

As I have mentioned already (Ch. VI, Sec. 1), in the south of the main island the word *urigubu* refers to taro plots or that part of one which is cultivated by a man for his sister, as well as to the crops taken from such a plot. Such *urigubu*, however, obviously cannot be ceremonially harvested and offered in bulk for the simple reason that taro has to be taken out, which is done by the sister's husband, as it ripens and eaten within a short time. The taytu given to the sister's household is called *taytumwaydona*, 'taytu altogether', probably because, in contrast to taro, it is given all at one time. In Vakuta taytu is sociologically treated in the same way as taro, in that it is offered in the ground and not taken out by the recipient save in exchange for payment (see above).

2. *TAPOPU*—TARO GARDENS

Any discussion of the gardens of the south—in Oburaku and Sinaketa and Vakuta—brings us back at every point to the subject of taro. Therefore a few words must be said about the *tapopu*, the gardens made exclusively for taro. We see on the Chart of Time-reckoning (Fig. 3, Col. 9) that the harvest of taro belonging to the first cycle of special gardens takes place in the thirteenth native moon (*Kuluwasasa*). To the southerners the *tapopu* are the gardens which provide the food offering to the spirits, *kasi baloma*. It is out of the taro harvested from the *tapopu* that the pudding and porridge, the *silakutuva* and *uripopula*, are made which will be presented at the *Milamala* festival to the returning spirits.

In order to understand the relation of the *tapopu* to the other gardens we must remember that taro differs considerably from both yams and taytu in the time it takes to mature, the conditions in which it thrives, and in its capacity for being stored. Its period of growth is much shorter. The natives told me that it takes four moons for the taro to ripen, and that in the fifth moon they must begin to harvest it. There can, therefore, be three taro crops in a year, as against one for both yams and taytu.

Again taro is less dependent on the season when it is planted, as long as it gets enough moisture. Unlike yams, it thrives on any damp soil short of a quagmire. Thus, during the wet season it can be planted in the ordinary soil alongside of taytu, *kuvi* and the other garden produce, and during the dry season special taro gardens can be made on the *dumya*, the swamps. If taro were planted in the *dumya* during the wet season, it would be swamped and would rot during the latter stages of its growth, but it will thrive then on dry soil.

Thus taro takes less time to mature and will grow in more varied conditions. Roughly speaking,[1] I think that the taro gardening of the Trobriands is conducted as follows: one crop is planted in the early gardens, the *kaymugwa*, a little before the beginning of the new year, normally in the moon of *Kuluwasasa* (see Chart of Time-reckoning, Col. 8, Row 13). The crops of the first cycle of the *tapopu* (as we have said above) are now maturing and being harvested (Col. 9, Row 13), and the taro tops are being replanted in the *kaymugwa*. As we know, the new plant grows from the top of an old plant put into the ground (cf. Part V, Div. III, §§ 23–25). This crop matures in the moon of *Toliyavata* and is harvested and replanted in the *tapopu* (second cycle, Cols. 8 and 9, Row 4). The planting of taro in the *kaymata* (Col. 7, 2nd moon) is probably fed from the later stages of the first cycle of the *tapopu* gardens (Col. 9). These start in the thirteenth moon, but since harvesting of taro extends over a long period, they last well into the first or even second moon of the new year. When the *kaymata* are very late they might also be fed from the earliest taro harvested on the *kaymugwa*.

So far our data fit reasonably well, but here I come to a slight discrepancy in my material. From Column 7 we see that the taro harvest on the *kaymata* is listed in moon 6, while the planting of the

[1] See also Note 41 in App. II, Sec. 4.

TWO GARDEN MAGICIANS OF SINAKETA ON THEIR ROUNDS

Motago'i and To'udawada at the entrance to the *kaymata*. To'udawada, who is acknowledged as the *primus inter pares* among Sinaketan chiefs, is also a garden magician. Motago'i is seen with his ebony walking-staff. The two separated afterwards to perform their duty in their respective gardens (Ch. X, Sec. 1)

PLATE 111

MOTAGO'I AT A MAGICAL CORNER OF A TARO GARDEN

"My friend Motago'i who gave me the best account of taro gardening and taro magic which I ever obtained" (Ch. X, Sec. 2)

PLATE 112

GLIMPSE OF A NEW TARO GARDEN

The fence, which has to be made early in a taro garden, is seen in the background. The spacing between the plants amounts roughly to half a yard (Ch. X, Sec. 2)

tapopu, which ought to coincide with this, takes place in moon 9, Column 9. Of course this later planting is fed from the harvest of the second cycle of *tapopu*. The harvest of the *kaymata*, therefore, does not appear to be used for replanting. This may be explained by the fact that the main gardens, which are principally devoted to taytu, have but very little taro in them.

The rest of our taro cycle runs smoothly. The second cycle of *tapopu* has its main time of ripening in moon 8 (Col. 9), and from it the first cycle is replanted which, as we know, matures four moons later (in moon 13) and here we have arrived at our starting-point.

Such a clearly defined three-fold garnering and replanting of taro must be regarded as a schematic representation of what actually takes place. In practice the harvesting and replanting of taro is an almost continuous activity; throughout the year the taro plays an auxiliary part to the other gardens and there is almost always a place from which taro can be harvested and where it should be replanted. I think, however, that there may be three or, when the *kaymata* are important for taro, four peaks in this activity.

Unfortunately I realised the entirely different character of these crops only towards the end of my stay in the Trobriands. I think it was first explained to me by my friend Motago'i (Pl. (111, who, during my last visit to Sinaketa, gave me the best account of taro gardening and taro magic which I ever obtained, though it was all too brief. I only spent a short time in Omarakana after that, and could but rapidly check the results obtained from Motago'i. Apparently in Omarakana there is no special magic for the *tapopu*, which play a far less important part there than in the south. Whenever magic is performed over a taro garden, the rites and spells used are taken from the regular system of garden magic. Bagido'u, however, told me that he himself never medicates the *tapopu*, though some men perform a private magic over them, and also a magician would, if paid by the community, perform certain ceremonies from the regular system over these gardens, more especially the *gabu* magic, with the incidental rites of *pelaka'ukwa*, *kalimamata* and *bisikola*. At times a *kamkokola* might be erected on a *tapopu* plot, but it would be a small and merely symbolic one, as the taro does not need a support to twine round (cf. Pls. 111 and 114).[1]

[1] As a matter of fact it was due to the repeated assertion of Bagido'u, during the early days of my stay in Omarakana in 1915, to the effect that there is no special *tapopu* magic and that the ordinary garden magic only is chanted over taro gardens, that I failed to find out the importance of these gardens in the south until it was too late to obtain a knowledge of them and of their magic comparable to that which I obtained about the gardens in the north.

Thus to put it briefly: there is no special official magic of taro gardens in Omarakana, and probably this holds good with regard to most places in the north. Some parts of the official magic of the main gardens can be enacted over the *tapopu*, even as they can be enacted, and usually are enacted, over the *kaymugwa*. There are never any standard plots in the taro garden, the plots are often shared by several men, the *tula* boundaries in this case fulfilling a utilitarian and economic function. There is no *kayaku* for the *tapopu*. The *tapopu*, in fact, unless special reasons connected with the moisture during the year demand otherwise, are usually made directly adjoining the main gardens. As a rule the *yowota*, the main inaugural rite, is carried out on those *baleko* which beforehand have been designated for *tapopu*, at the same time as it is performed over the main gardens.

Let me now state briefly the information which I obtained, mainly from Motago'i, about the taro gardens in the southern part of the island. In Sinaketa and other villages of the south there are no standard plots on the *tapopu*. The *baleko* are shared by several men and the boundaries between the different portions are marked out by the *tula* sticks. Sometimes such dividing boundaries are slightly raised above the ground, the horizontals being placed on small forked sticks. Of these divisions one or more would be cultivated for a man's own household, and these are usually called *kubuna yamada* (the place of our hand?); and one or more for the sister's or maternal kinswoman's household, and these are called *urigubu*.

There is no *yowota* on the *tapopu*; probably here, as in the north, the grand inaugural rite refers to all gardens designate. A special magic is performed at the *gabu*, which closely resembles in its ritual the magic performed over the yam gardens. *Kaykapola*, young shoots of coconut leaves, are charmed over in the magician's house. The *towosi* goes to each *baleko* and sets fire to the dry refuse *o nunula*, at the magical corner. Next day each man obtains a *kaykapola* and some medicated herbs from the magician, makes a *lumlum* on his garden plot, inserts the magical herbs into the heap, and sets fire to it. This corresponds to the *gibuviyaka* and is, I think, called by this name. Then they have a rite called *pelaka'ukwa*, in which they drive away the bush-pigs and attract a mythical pig called *bulukwa gado'i*, the pig of the garden stake (cf. Ch. III, Sec. 2).

It is in connexion with the second burning that the *kamkokola* of the taro garden is erected. The rite is called *kalipwala kamkokola*, the making of the hole for the *kamkokola*. If the *kamkokola* magic is on the whole a magic of planting, it is only natural that it should be done at this stage in the south where taro is the main crop, for it is planted immediately after the *koumwala*. In the *kaymata* the planting of the subsidiary crops, including taro, takes place immediately after the *koumwala*, but the *kamkokola* rites are not performed until later, before the planting of taytu, the *sopu* proper.

I have watched and assisted at the planting of taro several times. On the *sapona*, the clear patches of ground between the stony heaps on the drier soil, or in the holes on the *dumya*, about half a yard or so is left between the plants (see Pl. 112). The soil which has been carefully cleared, does not need be dug into as deeply as for taytu; it is only loosened superficially, the sods broken up, and stones and the roots of weeds taken out. Then a taro top, that is, the upper part of the tuber with a few leaves attached to it, is inserted into the ground (cf. Pl. 113). At times the plant is attached to two sticks which keep it in position and prevent it from toppling over or being uprooted by the wind.[1] Later on there is some weeding to be done. When the new plant begins to sprout out of the old top, the dried-up outer skin of the tuber, which is called *bam*, "the after-birth" of the taro, has to be cleaned away. Also wind-rattles, *kaygogwa'u*, have to be erected, and bird traps, *sikuna*, set, and the plants must be carefully protected by fences and pointed stakes against bush-pigs (cf. Pl. 114). As taro grows on the surface it is more exposed to birds, bush-pigs and insects than taytu.

The wind-rattles were made out of an empty *kwaduya* (*melo diadema*) shell, which is bound to a *kamtuya*, a stem left standing after the cutting. A stone is put in the shell and a pandanus streamer or a broad piece of wood attached to it so that it catches the wind. In the south, where European influence is stronger, the empty kerosene or benzene tin has replaced the shell. In the north, I have seen the old shell rattles in many places.

After this there is little work to perform in the taro garden till it is time to harvest it, but some growth magic is performed. There is a rite called *vapwanini yotata*, "to make the new leaf sprout". This consists in a spell recited over the gardens, accompanied by no special taboos on work, after which a bit of the lower stalk, *ikipwoysi sikwaku*, of one plant on each *baleko* is pinched off (cf. Pl. 115). This makes the taro leaves lie down, droop; for with taro, as the natives say, "a big top makes a small foundation", *dabwana kayviyaka, kaylagila pikekita*. One other rite, the *katutauna*, is performed with the same object. The magician recites a spell over his palm in the corners of the garden, scratches up some soil and throws it over a taro. This makes it lie down on the ground. Finally there is the magic of *isunapulo*, which I think is identical with the same ceremony

[1] Cf. also Part V, Text 15, Div. III, §§ 23–27.

as carried out in the ordinary gardens, but which is performed separately over the taro in the *tapopu*. This ends the cycle of taro magic

3. THE MAGIC OF PALMS

Apart from the crops cultivated in the gardens, the most important vegetable produce for the economics as well as for the pleasure of the natives are the two palms, the coconut and the betel-nut. Both palms are appropriated under a similar system of ownership, both of them are cultivated in a similar manner, both of them are used in a variety of ways and both enjoy a similar magic; while round both there cluster a number of similar ideas and beliefs. But the coconut palm is the more important, since it provides one of the essential elements in native diet—in years of scarcity, indeed, the nut may become the staple food during the months of *molu* (famine) —whereas the betel-nut is merely a stimulant.

Both palms are grown in and round villages, of which they form a characteristic feature. The trees are owned individually, though they are planted on what is regarded as communal soil. The head-man usually enjoys a certain over-right, while the chief is the titular owner of all the palms of the district (cf. Part I, Sec. 10). The coconut palm, its fruit, its leaves and buds are put to various uses. As an article of food the unripe coconut yields the "milk" which is a favourite drink, while its flesh is regarded as a delicacy; cream and oil are prepared from the grated substance of the ripe nut and are used as a seasoning and relish to ordinary vegetable diet. In domestic crafts the leaf plays an important part: it provides the material for inferior thatching; plaited, it forms a covering for the walls of storehouses and makes doors and partitions for dwellings; it supplies the material for basket-work and the rough mats prepared whenever there is need for carpeting the ground with a substance which can be thrown away without regret. We have constantly come across the coconut palm in magical practices; it also plays a part in beliefs about witches and sorcerers, and in aesthetic interests.

The cultivation of the palm is simple. It is grown from a coconut which has been kept until it sprouts. After the sprout has grown to some height, the coconut is carefully opened, and the edible kernel and the spongy substance which forms in the liquid are eaten. A small bit of the nut only is left attached to the sprout, and this is planted. As a native described it to me (Text 19, Part V, Div. IV, § 9):—

"(i) We leave the whole coconut until it ripens, and when we see it is already big we say to our mate: (ii) 'Go fetch the coconut and bring it here. Go plant our coconut.' (iii) The young sprout emerges here (and the speaker demonstrated on an actually sprouting coconut). We tear off this part; we tear off the other part; some husk remains, together with the young shoot and the root. (iv) The coconut meat we eat. The spongy kernel we eat. The husk, with the shoot and the root, we plant. (v) Together with it we plant the *moroba'u* (native lily). (vi) We dig the ground, and when we have finished we put in the *moroba'u* and the coconut and we cover them over. (vii) Later on the roots of the lily give strength to the roots of the coconut, so that the coconut might sprout." I was told by local white planters in New Guinea that this procedure is detrimental to the growth of the tree, and that it is very much better to plant the whole coconut. The young tree is fenced round to preserve it from bush-pigs and wallabies, children and dogs, and the ground about it kept sufficiently clear from weeds to prevent it from being choked; but apart from this no other tending is necessary. The coconuts in the Trobriands are neither very numerous nor very good; they are quite insufficient for any substantial trade in copra.[1]

The betel-palm is owned and cultivated in exactly the same way and has the same aesthetic interest. It is an agreeable stimulant and therefore plays an important part in native sociabilities. This rôle is now shared by imported tobacco, but in the old days the betel-nut was the typical gift offered at all social gatherings. Chiefs were at pains to conceal any betel-nut which they might have about them, and even used three-tiered baskets to shield it from the inquisitive glances of onlookers who, on the principle of *noblesse oblige*, had to be regaled with anything exposed to the eye (cf. Part I, Sec. 10).

The payment given for erotic services is called *buwana*, and this name is, I think, derived from the word for betel-nut, *buwa*.

Both palms are the object of an interesting institution, the *kaytubutabu*, the magic of the taboo imposed on palms in order to make the fruit plentiful.[2] To the natives the *kaytubutabu* magic is

[1] Only recently a considerable amount of planting has been done under Government pressure and very much against the wish of the islanders who, while recognising the material advantages of the plan, cannot be made enthusiastic about it.

[2] To the European reader the word *kaytubutabu* will naturally suggest the idea that this is to the natives the "taboo" *par excellence*. It must, however, be remembered that the Trobrianders have another specific term for taboo, *bomala*, a term which incidentally has also a wide distribution in the Indonesian and Oceanic area. A number of linguistic queries and sidelights can be derived, however, from the survey of this and cognate terms in the Trobriand language, and these have been given in the appropriate place (Part V, Div. IV, § 11).

primarily associated with its symbolic expression—the stick planted in the middle of the village with two girdles of leaves, one of pandanus and one of banana, tied round it, and the band of coconut leaf (*gam*) tied round every palm stem. These are the effective symbols of the taboo. For the duration of the ceremony certain taboos have to be observed to make the magic effective; above all, the coconuts must not be picked. To the ethnographer it is a complex institution with ritual, moral, legal and economic aspects—an institution associated with a number of other concerns and activities. But let me present the facts before these aspects are considered.

The season of the *kaytubutabu* is determined by two considerations; on the one hand the coconut is too important an article of diet to be tabooed during the five or six months when other food may be scarce; on the other hand the period must be over before the ceremonial distribution and other festivities connected with the *Milamala* moon begin; for then also the coconuts could not well be spared. In fact it is in preparation for some one of the more important *sagali* (distributions) or for a dancing period (*usigola*) that the *kaytubutabu* taboo would be imposed. As I was told (Text 21, Part V, Div. IV, § 13): "When we decide on a *sagali* (ceremonial distribution), we might set up the *kaytubutabu*. Then we would charm, the palms would ripen quickly, the nuts would become abundant and we could make a *sagali*." Therefore, when a big distribution or competitive undertaking (*kayasa*) or dancing season was being planned, coconut magic had to be done. As a comparison with the Chart of Time-reckoning (Fig. 3, Col. 10) will show, the coconut taboo cannot be imposed before the moon of *Kuluwotu*, nor extend over the new moon of *Kaluwalasi*; but it can take place any time within the five moons from the beginning of *Kuluwotu* till the end of *Kaluwalasi*, which, as it only lasts about two moons, still leaves a fair latitude of choice.

As a rule it is performed early in this period, that is, it begins at the full moon of *Kuluwotu* or *Utokakana*, and lasts till the full moon of *Ilaybisila* or *Yakoki*. In a place of high rank such as Omarakana, Olivilevi or Gumilababa, it would usually take place earlier than in a commoner village, since the chiefs take the lead in everything and invariably start their ceremonial distributions and festivities first.

If a big *sagali* were pending and there was some concern about the coconuts, the head-man of a community would summon his fellow-villagers to a *kayaku*. It would then be decided whether a *kaytubutabu* was to be performed, when it should start, and who was to perform it. In some villages there are magicians who know the spells and ritual; others have no such specialists and have to invite some outsider to do it. At present (1918) the only man who knows the magic in Omarakana and Kasana'i is Tokulubakiki, who resides in the capital but whose magic comes from his true or own village, Tukwa'ukwa. Only a few people know it in the neighbourhood—a man in Kwaybwaga, one in Obweria, one in Obowada and one in Kabwaku. The following account is based on data supplied to me by Tokulubakiki.

KAYTUBUTABU POLE

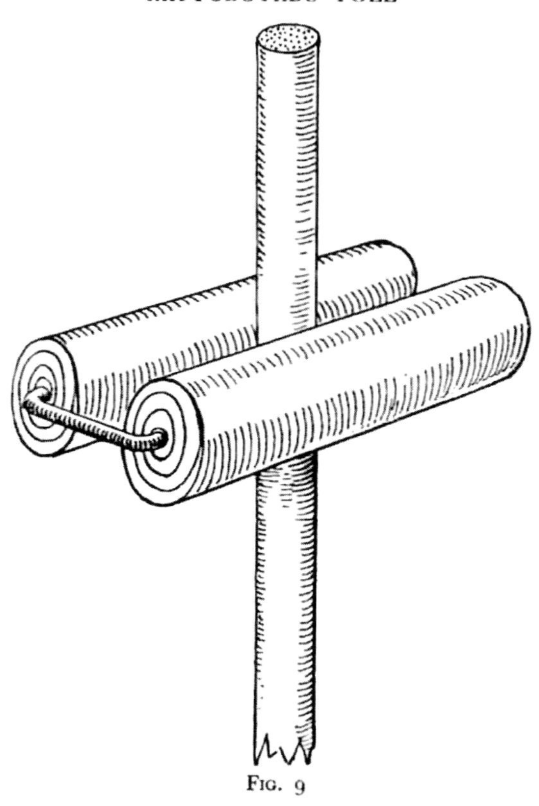

FIG. 9

In any case the magician receives a substantial payment from the headman, a ceremonial stone blade, a necklace or a pair of armshells. The other villagers would bring him contributions of smaller value, a few mats, a basket, spears or a cooking-pot.

When the *kaytubutabu* has been arranged, the palms are trimmed, all the nuts, ripe as well as green, are gathered, and the lower branches, such as might dry and fall down within the next two months, are lopped off, leaving only the tuft of young fronds in the middle. The magician also has his paraphernalia to prepare. He selects a straight pole some two and a half to three metres long, made of light wood from the *busa* or *yonakiu* trees. He also collects some *modigiya*, the dried leaves of the pandanus trees of the species of *kaybwibwi*. He trims and strips the pole, and coils the pandanus leaves into neat rolls, such as we see on the diagram (Fig. 9).

Then the magician collects a number of herbs to be used in the ceremony: leaves of the same tree from which the pole has been prepared, *busa* or *yonakiu*—why these are used my informants were unable to tell me—and leaves from a leguminous tree, the *kwaygagabile*, which grows on the seashore, because this tree is very prolific, producing many large pods, and the coconut palm should do likewise. Also leaves of *yayu*, the casuarina tree, which plays such an important part in all garden magic because its dense foliage conveys to the natives an idea of fertility, are brought from the seashore, together with an aromatic creeper called *makita*. *Makita* has a strong, sweetish smell, and the magician will chew and spit it over the palms during the ritual performance, so that, as the natives say, the coconut may be as sweet and rich in its fragrance as the *makita*. *Ubwana*, a plant with edible roots and a non-edible but prolific fruit which grows in clusters, is used that the coconut may imitate the fruit; and the *noriu*, a plant from the *odila* which is eaten during *molu*, because it bears abundant fruit even in the worst and driest seasons. *Setagava* is a grassy weed which grows round the village and in the grove (*weyka*); it is very tenacious and very difficult to uproot or to destroy, as it will grow even from a leaf or a piece of stalk; *boyeya* is a tiny plant with small leaves like those of thyme but not aromatic, which produces a plentiful crop of small fruit sprouting out of the leaves "like the coconuts in the *kaykapola*", as the natives say. The reason for the use of these herbs in a magic of fertility is obvious.

These leaves are collected on the day before the first ceremony. The magician tears them up and mixes them together. He also prepares the pole, finds some dry banana leaves (*siginubu*), sees that the pandanus leaves are neatly coiled up and, collecting all these things on a mat, leaves them for the night in his hut. Meanwhile the villagers have been providing every coconut and betel-palm in the place with *gam*, that is with a band of coconut consisting of a section of midrib with a few leaflets attached, which is fastened round the trunk by means of one pair of leaflets growing opposite each other (cf. Fig. 10). Both these bands and the *kaytubutabu* pole will remain as visible signs of the taboo during the whole period.

On the day of the ceremony the magician and his helpers for-

PLATE 113

THE PLANTING OF TARO

"A taro top, that is, the upper part of the tuber with a few leaves attached to it, is inserted in the ground" (Ch. X, Sec. 2)

PLATE 114

TARO GARDEN ON STONY SOIL

A few plots near the *rayboag* in Sinaketa. In the left front, a small *kamkokola* is seen. In the centre a native stands against a wind-rattle, consisting of a wide pandanus leaf attached to an empty kerosene tin. In old days a large *melo diadema* shell was used. Note the enormous stone heaps due to this type of soil (Ch. X, Sec. 2)

PLATE 115

A MAGICAL RITE IN THE TARO GARDEN

"A spell is recited over the gardens . . . and a bit of the lower stalk of a taro plant is pinched off on each *baleko*" (Ch. X, Sec. 2)

PLATE 116

A BANANA PATCH ON THE MOMOLA

A woman ready to go home after having done some work on the banana plants (Ch. X, Sec. 2)

gather on the central place of the village. A mat has been spread in the middle of it, and on this the *kaytubutabu* stick, a shell trumpet and the magical herbs have been deposited. An assistant with a digging-stick stands near the magician, and with this a hole in the ground will presently be made for the taboo pole. On such an

TABOO MARKS ON PALM

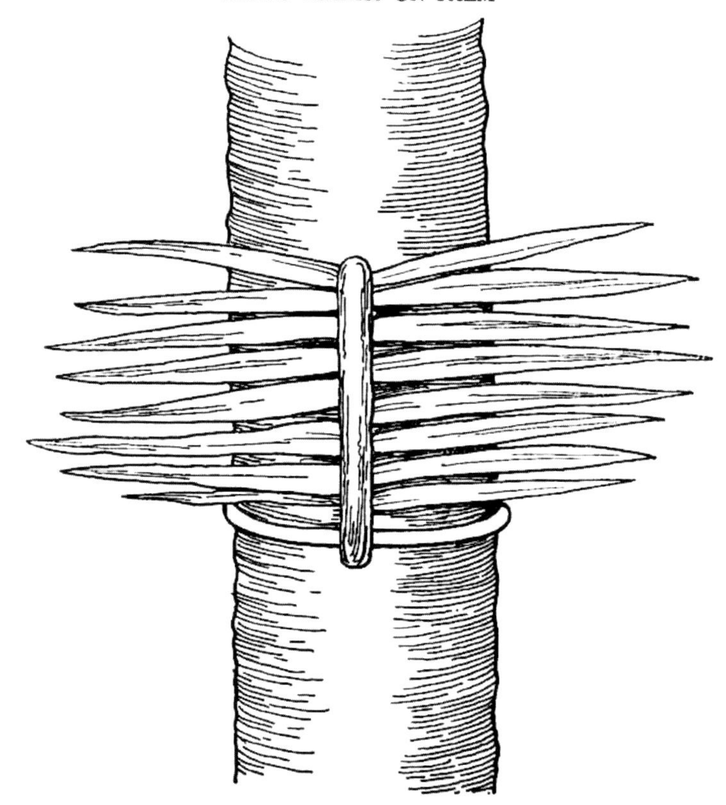

FIG. 10.

occasion the chief sits on his raised platform, surrounded by his wives and children, lesser notables are grouped somewhat more in the background, commoners stand on the outskirts of the central place, and the women and children keep near their houses. When all is ready the magician squats in front of the mat, places another mat on top of it in the usual manner so that the breath carrying his words may be imprisoned between the two of them, and then solemnly and in a loud voice recites the first spell of the *kaytubutabu* :—

FORMULA 41

I. "O Seulo, O Milaga'u, go, load our coconuts!

Throw away those which are blighted by blackened and discoloured leaves; by their tops being eaten away; those which are malformed and those which are grown out of shape.

Load the coconuts fit for us, the coconuts good for the making of water-bottles; fetch them to our village.

II. "I shall cast off the bad coconuts, I shall spill them on the island of Nadili,

Those which are blighted by blackened and discoloured leaves; by their tops being eaten away; those which are malformed and those which are grown out of shape.

I shall launch my canoe (made of) the stump of a coconut leaf.

I shall load coconuts fit for me, refined coconuts, the large coconuts good for ceremonial use, the strong coconuts suitable for water-bottles, the white coconuts.

I shall cast off the bad coconuts, I shall spill them in the village of Koymara'u.

Those which are blighted by blackened and discoloured leaves; by their tops being eaten away; those which are malformed and those which are grown out of shape."

(The magician then repeats the whole strophe beginning with "I shall cast off" till "white coconuts", inserting various spots instead of Nadili and Koymara'u. He gradually approaches Omarakana by way of the villages of Woodlark Island, the islands of the Marshall Bennett archipelago, the villages of the east coast of Boyowa.)

III. "It cuts, it rounds off."

(These words probably refer to the cutting through of the buds and to the rounding off of the fruit of the coco palm. After these words, the magician repeats the opening strophes up to the words "island of Nadili".)

It can well be imagined how long this spell lasts, seeing that the *tapwana*, the chief part, is repeated with a lengthy list of geographical names. The chief object of this formula is to exorcise the bad nuts and to attract the good ones. In common with many other formulae it combines an exorcism, a spell of black magic casting evil things on neighbouring places, with a positive appeal for good luck in the own village. The magical principle that charity begins at home combined with the other principle governing Trobriand magic, that things cannot be equally good everywhere, determines this form of spell.

I should like to add that, according to Tokulubakiki, if he were to recite this magic in his own village, he would not enumerate villages and islands in the east. He would start with Dobu or Tu'utauna or Bwayowa in the d'Entrecasteaux archipelago, and move northwards along Sanaroa, Tewara, Wamea, Gumasila, Domdom, and so on, passing thence through various islands of the Lusançay Archipelago to villages on the western shore.

After the spell has been chanted over the herbs, the pole and the shell trumpet, this latter is blown over the whole village while the pole is being erected. The assistant with the digging-stick makes a hole in the ground and plants the *kaytubutabu* pole, after the coils of pandanus leaves and the band of dried banana leaf (*siginubu*) have been tied round it. Some of the magical herbs are first inserted into the hole, others are tucked into the band of banana leaf. After the pole has been erected, the men utter in chorus the shrill intermittent yell, *kutugogova*, made by clapping the hand against the mouth, while the *tauyo* is blown with renewed vigour.

Then each man takes some of the medicated herbs and goes to his tree or trees, and tucks them under the *gam*. The magician takes some leaves of the *makita*, the aromatic creeper, and after chewing them spits them over the pole and over every palm tree in the village. With this the ceremony is finished.

From this day on an elaborate taboo falls on the village. Most heavily is the magician taxed, for he is absolutely forbidden throughout the whole period to partake of any coconut or betel-nut. He can neither eat the ripe nut nor drink the milk nor use coconut oil in anointing himself or in seasoning his food. Should he break his taboo, at home or abroad, the magic would become ineffective. Nor are the villagers allowed to eat or in any way use coconut in the village itself, but the taboo does not hold outside the village precincts. Should an unripe green coconut fall down from a tree, it may be eaten in the bush, but it must not even be opened in the village.

A further taboo forbids the people to make noises, especially to break or chop wood or to hammer. The reason for this was thus explained to me by the natives: "Supposing a man sits down and looks away. You come, you stick your finger into his ribs or his belly. He gets a fright, he jumps up. Like this the *kapuwa* (immature small coconuts) jump up and fall down when you strike wood, when you *mwamwakuwa* (make percussion noises)."

During the whole taboo period no firelight must be seen in the village. If a firebrand has to be carried from one place to another, it must be wrapped up in leaves. If cooking has to be done at night, it must be done within the shelter of the hut, so that no fire lights up the village. If any light falls upon the *kapuwa*, the young coconuts, it illuminates, makes clear their faces, *imilakatile migisi*, and the fruit would fall down unripe. They must not be shocked with light any more than with sound.

Of course no one must climb up a palm, but neither may any other tree in the village or grove be climbed. This is a sign of respect for the *kaytubutabu* pole. The natives explained: "If a chief is present no commoner dares to climb up a tree. He would get above the chief (*ipayli guya'u*). The chief would get angry. So the *kaytubutabu* stick would get angry and our coconuts would fall down."

Thus life has to go on in a silent, quiet and subdued fashion during the period.

Every eight or ten days while the taboo lasts, the magician has to perform a simple rite: walking through the village among the palms he recites in a loud voice a spell inviting the palms to fruit and to mature:—

FORMULA 42

"Who is it walking about, turning about, in the main place of my village? (repeated).

I myself, Tokulubakiki, with my grandfather Yaurana, we two walk about, turn about in the central place of my village.

The coco palm bears fruit, the coco palm piles up one bunch of fruit above the other. The palm breaks forth with bunches of nuts. The palm bends down its feathery top under the load of fruit.

From the throat of the biggest palm in Omarakana, Dubwadebula, issues the voice, kiiiiiiii . . ."

It can be seen that the spell, though its home is Tukwa'ukwa, has been adapted to Omarakana. In each village there is a principal coco palm which has a personal name; in Omarakana it is called Dubwadebula, "grotto". Should such a tree be broken by accident or by the wind, or cut down as after a war, a new one must be planted in its place. The onomatopoetic sound, *kiiiii*, symbolises plenty, for apparently the fruit and leaves of a heavily laden palm when swayed in the wind and rubbed against each other, emit some such high-pitched sound. As a matter of fact the noise of the coconuts in a high wind is very disturbing and unpleasant.

The spell just quoted and the simple rite accompanying it is called *kayloulo* "the walking round", or *katuvisa kaykapola*, "the breaking of the coconut sprout". As we have said, it is repeated every eight or ten days, altogether some five or six times during the taboo period of two months. In a way it corresponds to the growth magic of the garden cycle. When the *kaytubutabu* period nears its end and the coconuts are plentiful the chief or the magician announces that on such-and-such a day "the coconuts will be opened". This is done by the ceremony of removing or turning away the *gam* (*kivila gam*), which is also called "the breaking of the tree-tops" (*katuvisa kaydabala*).

On the eve of the ceremony the magician walks among the palms in the village and utters the following spell over them:—

FORMULA 43

"Lory, cut off, cut off thy throat, O green coconut.
Lory, cut off, cut off thy throat, O red coconut.
Lory, cut off, cut off thy throat, O brown coconut.
Lory, cut off, cut off thy throat, O white coconut.
Lory, cut off, cut off thy throat, O pale coconut.
Lory, cut off, cut off thy throat, O orange-coloured coconut."

The charm corresponds with the readiness of the coconuts to be gathered. Perhaps the lory is invited to act as a scapegoat, as one who takes the chestnuts out of the fire, to mix the metaphor.

On the next day the men climb the palms and each brings down a few nuts. They remove the outer husk and hand over the nuts in their shells to the magician, who cuts them open with his adze and then returns them to the owners. A general eating of coconut and drinking of green coconut milk takes place, and this finishes the *kaytubutabu* period.

Although both coconut and betel-nut are said to be the subject of this magic and of this taboo, and both palms are tied round with *gam* and are treated with the same attention and respect, spell and rite alike show that the magic is obviously directed principally at the coconut.

To the native the essence of *kaytubutabu* is the magic; it is the magic which makes his nuts grow and be plentiful. More especially, he would say, it is aimed at the preservation of the *kapuwa*, the young unripe nuts, to prevent them from dropping. To him all the taboos are equally important and efficacious. The prohibition against climbing the palms for nuts is to him just one of a series of taboos. That it incidentally protects the nuts from being eaten and thus, in a natural manner, multiplies their number, they understand. That in this way many more nuts are made available for big ceremonial performances they readily admit. But the suggestion that perhaps it would have been just as efficient to keep the taboo and not to have made the magic at all was immediately rejected by my informants.

Looked at from the outside point of view, it is clear that magic is an essential part of this institution. It is in fact the real incentive to keeping the taboo, to making the natives alive to it. It is the magic which enables the chief to impose a period of irksome restraint when arbitrary authority might be insufficient. It is the belief in magic combined with the desire for a sufficient supply of nuts for ceremonial purposes which gives the motive to individuals. Magic is, therefore, here as in many other activities the organising force, both psychologically within each individual, and socially. It provides the motive, the justification and the outward actions which give validity to the prohibition.

Without magic and all its paraphernalia—the signs, the ritual, the taboos—it would be impossible to discipline a couple of hundred men, women and children into keeping off what to them has an almost irresistible charm and is within their easy reach. The very hardships which they undergo in observing the taboo supply also the moral force for keeping it completely.

Unfortunately I never saw the proceedings of the *kaytubutabu*.[1] At present it has fallen completely or almost completely into disuse. The native customs with regard to coconuts have been seriously undermined by an extremely well-intentioned and probably beneficial ordinance of the local Assistant Resident Magistrate, who, about ten years before my arrival, ordered that each village should plant a few hundred palms every year. As I was told by the natives: "We have now *kaytubutabu* all the time, we cannot eat our nuts, we have always to store them and store them again (so as to have them ready for planting). We do not want that, we have enough coconuts in our villages."

Anyhow there is not the slightest doubt that the taboo of the white resident is not as strong as the taboos of magic; for that I can vouch from my own observation.[2]

4. THE FRUITS OF THE WOOD AND OF THE WILD

Besides the cultivated crops reared in the gardens by dint of strenuous and constant labour; besides the semi-cultivated trees of the village, the grove and the reserved portions of the forest, there are a number of trees, shrubs and weeds which stand out from the confused background of the "mere jungle" (*odila wala*). These are more or less useful to the natives at all times, and become indispensable in years of scarcity.

Between the definitely planted and cultivated trees, such as the coconut and betel-nut, and the completely wild but useful growths of the *odila* there is an extended range of plants less and less appropriated individually, less and less cultivated and, as a rule, less and

[1] See also Note 42, App. II, Sec. 4. [2] See also Note 43, App. II, Sec. 4.

less economically important. A number of fruit trees are definitely common property; they sprout from some accidental shoot and are not tended. The only rights of ownership are connected with the fact that they grow within the territory of a village and on a part of the grove which belongs to one or another section of the community. Such fruit trees would be raided from time to time by a group of children or young men, and some of the native pastimes are associated with expeditions to gather the aromatic fruit of the *menoni*, or the pink and succulent *mokolu*, or the juicy *mokakana*.

Boys, and sometimes even girls, would prepare a *kaykosa*, a hooked stick or two sticks bound together at a sharp angle, and would attempt to pull down the fruit either by climbing on to one of the branches or else from the ground. Or the young people would practice throwing short stout sticks, *lewo*, to bring down the fruit.

Some of the fruit have a *kweluva* (season). The *menoni* ripens in the middle of the summer, in *Yavatam* and *Gelivilavi* (cf. Chart of Time-reckoning, Fig. 3); the *mokakana* in *Yakosi* and *Yavatakulu*; the *natu*, a small succulent fruit, appears during the harvest of taytu. The fruit of the *lawa* tree, pleasantly scented but tasteless, is ready from *Kuluwotu* to *Ilaybisila*, the *gwadila* in *Yavatam* and *Gelivilavi*. Other trees again bear fruit all the year round, as the *youmwegina* and certain of the nuts.

Nuts are an important and favourite class of food. Most of them have to be first deprived of poisonous qualities by long soaking in fresh or salt water. The kernel of the *lawa*, which is called *sasana*, the seeds of the *youmwegina* tree (*utukwaki*), the kernels of the *gwadila* (*kanibogina*), and of the *vivi* (*kwa'iga*) are very popular. The most important nut which can be eaten without any preparation is the *sayda*, a spirally coiled up nut, longish and pointed in shape, with somewhat the taste of the hazel-nut.

The edible leaves of the jungle are one of those articles of food which are common property and, except in the worst seasons of famine, plentiful. Of some of them the natives are really fond. They boil them in pots or potsherds with salt water, add the large yellow ants to give an acid flavour to the mixture, and usually eat them cold. Such leaves may also be baked in the earth-oven. As a rule a small branch will be cut and the young leaves plucked off.

Lokwa'i is perhaps the most important tree with edible leaves, but it is tabooed to the chiefs. Some of the others are allowed to every-one.

Most of the better fruit, leaves and nuts do not grow in the low bush which sprouts on cultivable soil while this lies fallow between one garden cycle and another. On such soil few of the larger trees are left, most of them being destroyed in the burning (*gabu*) each time a garden is made. The pandanus forms an exception; as we know, it is useful in many respects, and its fruit, the *vadila*, is sucked by the natives in hot weather. The fruit of the *noriu* and the *noku* (two of the hardiest shrubs) is eaten only in times of great hunger. Neither is taboo to chiefs, in spite of the original legend in which the Lukuba clan is said to have lost its rank because the dog ate *noku* (*Myth in Primitive Psychology*, Ch. II). A complete list of native fruits will be found in Part V (Div. IV, § 17).

It must be mentioned that certain kinds of yams migrate from the garden into the *odila*, though most varieties are not hardy enough to survive the struggle for existence with the wild weeds. The main species of *kuvi* which grow outside the garden are the *kwanada*, a yam which, as we have seen, plays a part in Omarakana garden magic, the *bubwaketa* and the *mumwalu*. Since, however, it is very difficult to find and to dig them up in the bush, these wild-growing varieties are of no economic importance.

The swamps (*dumya*) furnish decidedly less than the *odila*. The only popular fruit which grows here and is eaten in times of *malia* as well as *molu* is the *pipi*, which is small, yellowish, and extremely acid, but with an attractive aroma.

A very brief survey of the plants and trees which are used not as food but for industrial purposes will suffice here. The pandanus, *kaybwibwi*, is perhaps one of the most important; fibre, *im*, is prepared from its aerial roots and from this fibre the best string and rope is made. Its leaf-petals, *gayewo*, are very popular as ornaments and for their scent. Its fruit, *vadila*, is a favourite refreshment; it is sucked in hot weather. The light wood of the acacia, *vayaulo*, is used for shields. Mangroves provide material for houses, especially for the various poles used in the construction of the roof. The flowers of the *butia* tree are used at the season of *Milamala* and after for ornament and for their scent, as are the flowers of the hibiscus, *kaluwayala*. *Leya*, wild ginger, and *sulumwoya*, an aromatic mint-like weed, play an important part in magical ceremonies and customs.

5. MINOR TYPES OF CULTIVATION AND SECONDARY CROPS

It will be necessary to say a few words about certain less important crops and minor forms of gardening. The banana is a very important food to the natives, though it quickly fails them in times of drought, as it cannot survive the absence of rain. Apart from its fruit, the large variety, *wakaya*, supplies women with material for their grass petticoats. Bananas are grown round the villages and to a small extent in the gardens (cf. Pls. 30, 32 and 112). A favourite place for planting bananas is the moist fertile belt of soil to the east of the *rayboag*, the *momola*. Here clumps of bananas are planted in clearings of the jungle where, protected from too much sunshine and wind, they are said by the natives to thrive especially well (cf. Pl. 116). Some parts of the south, round Vilaylima, Okopukopu, and Obowada, are reputed to have specially good crops, and this is probably connected with the existence of a good supply of moisture there.

In these villages I am told there is a special banana magic within the official system of *towosi*, called *bisikola*. It takes its place as one of the minor rites performed during *gabu*. The magician goes into the garden and chants over a banana sapling, putting it into the ground in the magical corner, and afterwards each man plants a banana in his *baleko*.

In Kiriwina there is a private banana magic of which I obtained one formula, in the possession of my friend Tokulubakiki of Omarakana. The rite is very simple. It is performed over a banana tree at the time when the fruit is just beginning to develop from the flower. The owner chants over the palms of his hands and then touches the incipient banana bunch. This makes the bunch become large and the fruit stout. This is the formula:—

FORMULA 44

BANANA SPELL

"Dead man!
At night make turn and swell.
In daytime make large and bulky.

"Bananas, large and yellow, as the *kukuva* fruit—such are my bananas;
Bananas, big as the *bowada* fruit—such are my bananas;
Bananas, at which people exclaim with wonder—such are my bananas;
Bananas, that swell all round—such are my bananas;
Bananas, bulky and big—such are my bananas.

"They take a cutting shell to cut up my bananas."

There are plenty of such formulae. Some people chant them over their hands, as does Tokulubakiki; others will chant them into some *siginubu*, 'dried banana leaf', which is then thrust into the banana bunch. There are no taboos connected with this magic. I obtained the word *ki'ula'ola* as the name for this type of magic, which is an individual magic made by each owner over his own bananas. The main effect of it is to make the bananas swell.

The unsavoury invocation of the dead man or corpse refers to the latter's peculiarity of swelling. Perhaps the analogy between fingers and bananas may have something to do with it. I was not able to eat bananas for some time after I received this magic.

There are two more fruits which are really important as a means of subsistence and not merely as an added delicacy to the larder. These are the native mango, *waywo*, and the bread-fruit, *kum*. The mango, which ripens in the moons of *Yakosi*, *Yavatakulu* and *Yavatam*, can be very plentiful in good years. It is a favourite dish with the natives and, while it lasts, it is almost a rival of taytu and yams as a staple food. When a good season succeeds a bad garden harvest a rich supply of mango may mean prosperity, or at least ease, instead of famine. From this point of view its importance is considerable. There are several varieties of native mango; each of them has a special name; but I did not record these. It is eaten raw or cooked by any of the three methods of boiling, baking or broiling. The last is the usual way of preparing it. The native mango when raw is somewhat stringy and has a flavour of turpentine more pronounced than in the imported mango, so that it is not altogether palatable; but it is extremely good when broiled. There are no taboos connected with the eating or preparing of mango. It is collected by means of the *kaykosa*.

The bread-fruit is also a substantial item of nourishment. It ripens much earlier, in the moons of *Ilaybisila* and *Yakoki*, so that its season coincides with the native harvest. On that account it is not as important as the mango, but if for any reason not connected with drought or general fertility, such as a special blight or disease of the taytu and yam, the cultivated crops fail, the bread-fruit in conjunction with mango would yield a stand-by lasting over half the year. It is brought down by means of the *kaykosa* stick. When gathered unripe it may be broiled in hot embers or boiled. Ripe specimens are baked in hot stones. The seed, which has a very pleasant flavour rather like roast chestnut, is called *kweta* and is very popular with the natives.

Recently a number of introduced European fruits have been adopted by the natives, among them the papaia or mummy-apple, a word transformed in Trobriand to *momyaypu*. The tree grows wild and spreads with an extraordinary rapidity, and has become very plentiful on the outskirts of all villages. Remarkably enough other imported fruits are not taken up with any eagerness by the natives. Pineapple is cultivated almost exclusively by the few christianised natives, more especially the native teachers. Only one or two specimens of the cultivated mango exist, while such other tropical fruit-trees as the custard-apple or the *aguacate* (avocado pear) have, not yet made their appearance. The sweet potato (*simsimwaya*), which, I believe, was introduced by Europeans, has already been mentioned in Chapter V (Sec. 5).

I have also mentioned that sugar-cane is planted in the gardens, and a variety of pea and gourd. All these three crops have a subordinate economic importance. The sugar-cane is sucked, especially when natives go on a march in hot weather, and constitutes a favourite refreshment. Peas are boiled with other vegetables in stews; the flesh of the gourd is boiled, baked or broiled, but is not comparable in importance even with the bread-fruit. The shell of the ripe gourd, on the other hand, forms an important industrial article, as it is used for lime-pots in connexion with betel chewing. As we have already said, the native betel, the areca palm, is planted in villages side by side with the coconut. The betel plant itself, which provides the pods and leaves chewed with areca-nut and with lime, grows wild in the bush, though I think one or two varieties are semi-cultivated in the village grove.

This brings us back to the garden proper, one or two forms of which must still be mentioned, though unfortunately my information about them is anything but complete: I refer to the pieces of cultivable soil in the *rayboag* and to the occasional gardens made on the sea-board. In times of plenty and in ordinary seasons these are not very important and remain, so to speak, on the outskirts of economic activities. The pockets or holes in the coral ridge, which are filled with fertile humus, are owned individually. Only the large species of yam, the *kuvi*, are planted there. These thrive exceedingly well in the large humus holes, especially the long varieties, which attain there the extraordinary length of well over 2 m. (6·6 ft.). In order to get down the, at times, high and steep walls the natives have to lower a long curved pole, called *daga*, a word also used for ladder. As far as I know there is not very much work to be done on such small garden patches apart from planting. Occasional weeding is necessary and a support, *kavatam*, for the vine; the latter is usually planted at a slant so that it leans against the wall of the pocket. The magic performed over the gardens of the *rayboag* consists of two ceremonies: the soil is struck by the magician with the appropriate spell and at the same time the *yowota* rite is carried out, i.e. the ground is rubbed with magical herbs. The magic is not performed in all the holes, but in one especially handy and usually shallow. Again, when the main gardens are burned, a torch is set alight and thrown into one of the holes by the magician's acolyte.

The gardens on the sea-board, called *kasisuwa*, are more important, especially in times of slight drought and scarcity. I was told that the soil near the sea-level remains moist for a longer time than inland. Ordinarily these gardens are mainly devoted to the cultivation of banana, taro, sugar-cane and large yams. Villages like Wawela, Bwaga, Kumilabwaga, Okayaulo and Giribwa, villages, that is, which lie on or near the eastern shore, make some of their large gardens near the sea. These villages also perform the ordinary system of *towosi* magic over their seashore gardens; in Kiriwina only private magic is carried out on the *momola.*[1]

With this I have finished the account of Trobriand gardening within the limits of the material which I have collected.

[1] See also Note 44 in App. II, Sec. 4.

CHAPTER XI

THE METHOD OF FIELD-WORK AND THE INVISIBLE FACTS OF NATIVE LAW AND ECONOMICS

THE study of land tenure is a somewhat difficult and very interesting problem of sociological synthesis. Nothing reveals better the constructive or creative aspect of sociological observations among a native race than an analysis of how land tenure should be studied, recorded and presented.

The main achievement in field-work consists, not in a passive registering of facts, but in the constructive drafting of what might be called the charters of native institutions. The observer should not function as a mere automaton; a sort of combined camera and phonographic or shorthand recorder of native statements. While making his observations the field-worker must constantly construct: he must place isolated data in relation to one another and study the manner in which they integrate. To put it paradoxically one could say that 'facts' do not exist in sociological any more than in physical reality; that is, they do not dwell in the spatial and temporal continuum open to the untutored eye. The principles of social organisation, of legal constitution, of economics and religion have to be constructed by the observer out of a multitude of manifestations of varying significance and relevance. It is these invisible realities, only to be discovered by inductive computation, by selection and construction, which are scientifically important in the study of culture. Land tenure is typical of such 'invisible facts'.

The sanctions of law, the economic principles of production, the political institutions of a tribe, are also 'invisible facts'. A reader versed in sociological theory is well aware that a great deal of what has been discussed in the previous chapters is the result of constructive generalisations of this type. Our account of the distribution of garden produce in Chapter VI; the analysis of the influence of magic on agricultural work; the importance of mythology as an incentive and a basic charter; the aesthetics of gardening, belong one and all to this category. But it is in land tenure better than anywhere else that the method of discovering invisible facts by constructive inference can be brought out.

Land tenure also deserves some special attention because of its supreme importance in the practical applications of Anthropology. It would not be an exaggeration to say that mistakes in land policy have caused the greater part of colonial and imperial difficulties. Whether we take Ireland in the past, or present-day India, certain East African dependencies or the Union of South Africa—questions of land, of arbitrary expropriation or unwise apportionment, of sheer unnecessary chicanery or even of well meant but revolutionary reform, take a prominent place in racial and national conflicts. And yet hardly any record of anthropological field-work as much as mentions land tenure, and the books where the subject is discussed could be enumerated on the fingers of one hand. All this justifies us in making a digression on the method and theory of field-work, and in attempting to lay the foundations of a profitable treatment of land tenure.[1]

1. A PRELIMINARY DEFINITION OF LAND TENURE

The study of land tenure inevitably resolves itself into a number of questions as to how land is used by the community and its members. Hence this apparently simple problem confronts the enquirer with complications and difficulties. The straightforward approach by a few questions as to who is the owner of this plot or that, who exercises control, and who has legal claims to certain portions of territory, soon leads us into an *impasse*. At best, it can reveal to us the customary or legal system of titles—that is of rights, privileges and responsibilities attached to the soil. But this system grows out of the uses to which the soil is put, out of the economic values which surround it. Therefore land tenure is an economic fact as well as a legal system.

We could lay down at once the rule that any attempt to study land tenure merely from the legal point of view must lead to unsatisfactory results. Land tenure cannot be defined or described without an exhaustive knowledge of the economic life of the natives. This is by no means a truism, for most enquiries, especially the official ones, have been based on the fallacy that land tenure can be ascertained by a rapid fusillade of questions concentrated upon the legal aspect alone. When a commission is instructed to ascertain the system of land tenure in an East African tribe within two or three weeks, and does this by convening a few *barazas*; when in West Africa a body of eminent English lawyers, not versed in the know-

[1] The reader who is not primarily interested in method and in a thoroughgoing theoretical analysis of the principles of land tenure might find the following discussion a little involved and abstruse. In the next chapter all the evidence concerning land tenure is stated in a consecutive account. He might find it easier to read Chapter XII first and return to this methodological digression when he has the solution of the difficulties in hand.

ledge of African cultures nor acquainted with their language, investigate land tenure; when in one Oceanic colony after another the officials are advised to report on land tenure and do this by stating that land is owned communally here and individually there: that it is "clan property" in one tribe, the "chief's domain" in another, and "apportioned among the family" in a third—the results at best can give only a very rough approximation to the reality. But actually a typical enquiry does worse than this: it proceeds on the basis of a questionnaire inspired either by set and specialised European notions or else by some distinctions conceived *a priori*, such as e.g. the undying fallacy of anthropological work—the opposition between communism and individualism. Any observations thus obtained are then immediately mutilated or placed in a wrong perspective, and the result thus obtained, by giving us a fictitious solution, veils from us the real problem.

I would like to add at once that in criticising the "practical man's approach" to the problem of land tenure I, as an Anthropologist, cannot assume any attitude of self-righteousness. Not only has Anthropology so far conspicuously failed to supply any useful information on this capital problem of its field; but I personally can exemplify some of the most blatant mistakes from my own material, both published and committed to manuscript. It is always best to chastise the faults of one's profession in one's own person.

The complications of land tenure go further than this, however. As we know, the purely economic uses of land cannot be separated from rights of settlement, political claims, freedom of communication and transport; from territorial privilege connected with ceremonial, magical and religious life. No doubt the economic utilisation of land forms the solid core of all these privileges and claims. But land tenure must be conceived in a more comprehensive manner: it is the relationship of man to soil in the widest sense; that is, in so far as it is laid down in native law and custom and in the measure in which it controls political life, affects the performance of public ceremonies and gives access to opportunities for recreation and sports. Man's appointed and culturally defined place on his soil, his territorial citizenship, his type of residence, and those rights which underlie the various uses of his soil form an organic whole of which the economic exploitation is but a part, albeit the most important part.

This inevitable realisation of complexity and manifoldness in land tenure seems only to make our task vague and unmanageable rather than precise and straightforward. Land tenure enters very deeply into every aspect of human life, and it is the integral expression of all the ways in which man uses his land and surrounds it with the values of avarice, sentiment, mysticism and tradition. Have we then to be enticed into the study of everything, into endless intricacies, restatements and reinterpretations? The difficulty is real and it explains why both Anthropologist and administrator have so far failed to deal adequately with land tenure. The subject is complex and elusive, in spite of its fundamental relevance to the theorist and practitioner alike. And this makes it a suitable example on which to establish some points of principle in the functional method.

The maxim that you cannot understand the rules of the game without a knowledge of the game itself describes the essence of this method. You must know first how man uses his soil, how he weaves round it his traditional legends, his beliefs and mystical values, how he fights for it and defends it; then and then only will you be able to grasp the system of legal and customary rights which define the relationship between man and soil. Now that we have become acquainted with what the Trobrianders do with their soil, how they perform their magic over it, how their pride of lineage and citizenship, their kinship sentiments and family feelings are bound up with their gardens and garden produce—now land tenure has become both alive and real to us; its intricacies can now be mastered.

For—and this is the second principle of the functional method—we shall see that the organic treatment of a subject does not lead to chaos nor to endless repetitions. After an undoubtedly laborious process of unravelling contradictions and placing facts in their living relationship, we arrive at simplicity and unity within a manifold picture.

2. AN ANTHROPOLOGICAL EXPERIMENT IN DETECTION

The previous section leads us to the conclusion that our present task is one of scientific organisation of the evidence referring to Trobriand land tenure. The 'game' of Trobriand agriculture we know already. Also, between this book and previous publications on Trobriand life and culture, we know a great deal concerning their other 'territorial games'; their mode of settlement, their way of using roads and paths, water-holes and recreation-grounds; the significance of mythological spots and magical centres. There remains very little for me to add in the way of actual information. With all this, the task of defining what Trobriand land tenure really is still lies before us.

I have on purpose led up to the somewhat paradoxical situation in which we have all the elements of the puzzle in our hands and yet the puzzle has still to be solved. I have not done this merely to give a touch of detective interest to these two last chapters, but rather to exemplify the process of what has been just called the organisation of evidence: to give an insight into the constructive aspect of field-work. As we shall see, the field-worker in collecting his material has constantly to strive after a clear idea of what he wants to know; in this case, a clear idea of what land tenure really is. And since this idea has gradually to emerge from the evidence before him, he must constantly switch over from observation and accumulated evidence to theoretical moulding, and then back to collecting data again.

The reader who wishes to convince himself by his own efforts of the necessity of this procedure, might here attempt a little experiment. Close the book for a moment and go back in memory to the previous chapters. Make an attempt to select all the information germane to our subject: jot down all those facts which bear on land tenure, and try to sift and group them so that they make a coherent story about the relations of the Trobriander to his soil. In such an attempt you must start with some idea of what land tenure is. This idea you will constantly have to readjust to the occurrences which you witness and the organised behaviour which you observe. Your ideas, therefore, will have to be extremely plastic and adjustable, for your concrete data, of course, cannot be 'adjusted'.

Nothing will show you so well as this experiment what real field-work consists in. Nothing else will so well convince you that observations are impossible without theory; that theories must be formed before you start to observe, but readily dropped or at least remoulded in the course of observation and construction. You will realise, then, that field-work is a hard, persistent struggle for the vision of what a legal or economic institution is; how mythology integrates modes of behaviour and how this blends with magic and with practical work. It is the vision of the clear, firm outline of native institutions which brings order into the chaos of trifling happenings and details of varying relevance.

I will even give you a few hints. You have the verbal data and terminology of land divisions.[1] If you turn to my description of the garden council and the counting of plots in the village and in the garden you will find all that is necessary, not only about how land is divided, but how these divisions and subdivisions are appor-

[1] See Ch. II, Sec. 3 and Docs. VII and VIII; Part V, Div. I (esp. §§ 13–27), Div. V (esp. §§ 7–24), and Div. XII (esp. §§ 8–13, and 36–37).

tioned and used. Another set of data referring to land tenure will be found in the analysis of the harvest gift (cf. Ch. VI, Secs. 1 and 2), while in the account of harvest customs and throughout the gardening activities and magical ceremonies you have so often crossed and recrossed the native gardens and their environment that you ought to be throughly familiar with the topographical side. Again, in defining harvest duties and claims in Chapter VI, we stated the matrilineal laws of citizenship and property, the law of marriage, of parental duties and filial reciprocity; the relations between brother and sister and between a maternal uncle and his nephews and nieces. Finally, in our study of the office of the magician, its relation to chieftainship and to the community, we have come across yet another set of claims and duties referring to land.[1]

These data supply all the essentials of how land serves man, how man extracts his substance from land, what legal claims obtain and how duties are apportioned. But with all this I am certain that you do not know much more about land tenure in the Trobriands than I knew even when I was fairly familiar with their garden work, with most of their mythological ideas, and with their social organisation. The real mental effort, the really painful and uphill work is not so much to 'get facts' as to elicit the relevance of these facts and systematise them into an organic whole.[2]

[1] Some of the information relevant to land tenure, however, will have to be found by the reader in my little book on Myth, where the myths of origin are more fully described. Here a brief mention in Ch. I, Sec. 7, and Ch. XII, Sec. 1, was all that was possible. In Ch. III of the *Argonauts of the Western Pacific* a general outline of the social organisation of the Trobrianders is given; while *Crime and Custom*, Ch. III, of Part II, and Ch. VII of the *Sexual Life of Savages* deal more fully with the relations between Trobriand matriliny and paternal influence. Needless to say, I shall have, however briefly, to restate all the relevant data which are to be found in my other writings in the systematic account of land tenure given in the next chapter.

[2] There is no more difficult, elusive and, for the epistomologically sophistical reader, arguable distinction than that between "fact" or "datum" and "generalisation or construction". Anyone who wishes to apply the scholastic attitude to my argument will be able to retort: surely all the data contained in the previous chapters are also not "crude facts" but are already "constructions". The answer to this is that since crude facts do not exist, any relevant ethnographic observation is invariably a matter of elaboration, of bringing types of human behaviour, technical contrivances, elements of the environment, into relation to one another. But the data which I have discussed in the previous chapters was elaborated within different contexts. We were concentrating there on the typical round of Trobriand agriculture, on the apportionment of fields as a ceremonial act, on mythological ideas in so far as they illuminate native interests in agriculture. In discussing the relevancy of these data for land tenure, we have to reconstruct the material already obtained within a definite context: the relation of man to

Since this chapter is to a certain extent an autobiography of mistakes and failures in field-work, I might add here that I had not come to the Trobriands unprepared for the study of land tenure. My work on the southern coast of New Guinea had given me some experience.[1] There I had come to realise the enormous importance of the native's relationship to his territory, both in tribal life and in the dealings between settler, missionary, administrator and native. I had learned that more than half of the troubles in racial contact come from the white man's ignorance of land tenure.[2] Grave injuries had been done to whole tribes or villages in depriving them of land which was indispensable from one point of view or another, and doing it in a specially galling manner. I came to recognise in my early field-work that ownership was by no means a "plain fact", and that to understand how land is owned, it is above all necessary to know how it is used and why it is valued. But I was still very much enmeshed in the crude opposition between "individual" and "communal"; I still believed in the clan dogma, and glibly spoke about the "clan being the real owner of land", not properly understanding what I meant by 'clan', by 'owner', or by 'land'.

"The only correct course is to investigate all the rights enjoyed exclusively by an *individual* or by a *social group* with regard to a particular portion of land" (*op. cit.*, p. 592). "The right to make gardens is *vested in the clan. Each clan makes its gardens collectively* (*sic!*) within one enclosure, and each clan *has* its own territory where it makes its garden to the exclusion of the other clan. This seems to be the general form of garden land tenure." (The italics and exclamation are the fruits of wisdom after the event.)

soil with all its implications. That this construction is more complex, more deeply ramified and requires an even greater mental effort than the mere account of successive agricultural activities, will be clear to anyone after perusing this and the following chapter.

[1] "The Natives of Mailu" (*Transactions of the Royal Society of South Australia*, pp. 592–594, Vol. XXXIX, 1915).

[2] That this is just as often due to organised ill-will, greed and dishonesty may be quite rightly urged by many of my readers. Cases such as the recent breaking by the Kenya Government of the solemn pledges given to the natives, which could be paralleled in the history of the United States' Indian policy, or the policy of the Union of South Africa—might readily lend itself to violent attacks on the part of the "pro-natives". I am still convinced that, if those responsible or in authority had a clear knowledge of the facts, they could not so easily lull their conscience into the belief that land can be promised and taken away, guaranteed and alienated from its rightful possessors, as they have done persistently, systematically and with a considerable danger to the good name of the white race and to its less spiritual interests.

After a brief sojourn in the village of Dikoyas on Muru'a (Woodlark Island) I even ventured to make the following statement: "Among the Northern Massim, on Woodlark Island, I found a system of land tenure almost identical with that of the Koita and mainland Mailu" (*op. cit.*, p. 593 n.).

The reader of the present volume will see how empty all these three statements are, and above all how rash was my inference about the Northern Massim, of whom the Trobrianders are the most substantial portion.

The statements about the Mailu sound sufficiently plausible and have all the appearance of being precise though succinct. But after I had affirmed that the "clan owns the land and makes the garden" there would still have been a very long story to tell of how the "clan" "utilises" the land. I personified the clan which in itself is an unpardonable error, not only of expression but of thought. The clan has neither legs nor arms nor stomach. The clan is a group of people. My real task should have been to show how the individuals within the clan co-operate in production and distribution, how they are organised in work and in the consumption of their produce. And as regards land tenure, it would have been necessary to show how the legal claims to land enter as active forces into the system of production and consumption. In the study of Mailu land tenure I substituted a fiction and a metaphor for an honest piece of sociological analysis. The "clan" as a collective entity exists neither in social reality nor in native ideas.

In the second place I did not investigate how legal claims are used, waived or exchanged. I did not ascertain how, in virtue of certain claims, the titular owners of the soil are recompensed either economically or in honorific titles. Nor did I discover the fundamental ideas of the natives concerning land. At present I have no doubt whatever that the natives of Mailu have some sort of semi-mythological, semi-legal theory such as we find in all the Melanesian tribes of New Guinea.

3. AN ODYSSEY OF BLUNDERS IN FIELD-WORK

Assuming that the reader has now made the suggested attempt at constructing Trobriand land tenure for himself, I shall invite him to follow my own vicissitudes in the field. I started as all field-workers have to start, with the most superficial method, that of question and answer. I also had naturally to work first through pidgin-English, since Trobriand cannot be learned except on the spot. Thus I put some such question as, "What man belong him this fellow garden?", the pidgin equivalent of, "Who is the owner of this plot?" My enquiries, moreover, were limited to the very few natives who had even a smattering of pidgin. I was keenly aware of the lack of precision and insight resulting from this approach, and was not surprised that the results were correspondingly contradictory and vague. They varied according to whether the chief was present or not: in the former case he was ostentatiously declared to be the owner of the lands. At other times Bagido'u, whom we know already, would be pointed out. If in the absence of the chief or garden magician the heads of other local sub-clans were present, they would be styled the real owners of the soil. At times my interpreter —I was mainly working through a rascally fellow called Tom, *recte* Gumigawaya—would claim a piece of land as his own and tell me that he was just cultivating that plot. Or again I would walk with him through the gardens and map out the plots in a field and obtain a whole string of names for them. I remember writing out early in my field-work a preliminary account of land tenure which unfortunately I have never published—unfortunately, because it would have been an interesting document of errors in method. There I stated my opinion that in the Trobriands the natives do not really know who owns the land, that the chief has an over-right to the whole territory, vaguely acknowledged when he is not present, but definitely claimed by himself and admitted by the natives who are afraid of him; that the natives have a haphazard way of tilling the soil, there being no definite rules as to who is going to take over a plot. The account contained some elements of truth. What was wrong with it was the perspective in which these elements were placed.

As a matter of fact, I only formulated this statement after I had already learned to speak Kiriwinian. I knew that possession was expressed by the nominal prefix *toli-*, or by the use of the possessive pronouns 'mine', 'thine', 'his'. I mention this to stress the fact that though an approach through the native language is infinitely better than through English, this is not because the learning of native terms is in itself any short cut to "native categories of thought" or "native classifications". This important fact is one of the leitmotiv running through Part IV. There are hardly two words in pidgin and native which correspond more exactly to one another than the Kiriwinian *toli-* and the pidgin 'belong him'. Both are comprehensive terms, both stretch over an extraordinary range of homonymous usages, both can only be understood through the greater precision given by the context of speech and the context of situation. What this really means the reader will learn in perusing the Linguistic

Supplement (Part V) to this book, where the theory of contextual differentiation of meaning is given and exemplified, and the linguistic approach to land tenure developed (Div. XII, esp. §§ 8–13). For the moment it is enough to realise that the real trouble of verbal approach does not lie in the use of pidgin and that it is not overcome by the precision gained by substituting native words for bastard English. The value of using native is that it enables us better to understand, firstly what the natives talk about among themselves in the natural flow of conversation or argument; and secondly, their integral behaviour, which is a compound of verbal and manual actions. But however much you might be at home with the native, in his society and in his language, you will not understand an abstract and theoretical problem of native culture if you cannot grasp its essentials. Some of my trader friends spoke Trobriand better than any European language. They were entirely useless as informants and could not draft the constitution of any part of tribal law, even in the most sketchy manner, although they could behave as the natives did at a mortuary ceremony, at a *kayaku* or at a village brawl. It was only through the understanding of what might be defined as the economic, legal and mythological situations that I was led to grasp the reality of land tenure.

But before we come to that I want the reader to explore another *impasse* from which I had to retreat, or at least give up as a short-cut to knowledge. I suffered at that time from a belief in infallible methods in field-work. I still believed that by the "genealogical method" you could obtain a fool-proof knowledge of kinship systems in a couple of days or hours. And it was my ambition to develop the principle of the "genealogical method" into a wider and more ambitious scheme to be entitled the "method of objective documentation". After my Mailu failure—for I was aware that I had failed there to find out all that really matters about land tenure—I had developed a strategy of frontal attack on the subject. Early in my work at Omarakana I mapped out the territory (cf. Fig. 13). I plotted out the fields, made rough measurements of the individual plots and, in several cases, made a record of who was cultivating each plot and who owned it. The documents which I thus obtained were very valuable. They are reproduced in this book: the map, the terminology of fields and field boundaries, of plots and plot divisions; also the principles of inheritance, the *pokala* system and the manifold legal claims. This frontal attack, however, resulted in a multiplicity of unconnected and really unfounded claims. My documentation enabled me to draw up the list of legal titles which will be reproduced presently, and I found that this list agreed with that obtained by the question and answer method; and yet, as we shall see, neither of these approaches solves our problem.

Actually, side by side with these direct attacks, I was all this time accumulating that most valuable knowledge which comes piecemeal from the observation of facts. In looking through my field-notes I find quite early in my work, some time in July, 1915, a number of disconnected observations about the way in which native gardens are harvested—entries which bear on land tenure indirectly through the apportionment of the produce. That is, I find notes about the "own" produce (claim 8, see below), used by the gardener because he "owns" the plot during his gardening tenancy; about the tribute to the chief as owner of the soil (see claim 1); about the contribution to the sister's household (claim 9), gifts to the head of the local sub-clan (claim 4), to the kindred of the same sub-clan (claim 5), and to the relatives in the same village. But none of these data were correlated with land tenure.

On the other hand I find that, soon after, I was present at a *kayaku*, and that, though unable to follow the proceedings in native, I discussed their substance with some of my informants. I was aware that the chief acted as master of ceremonies; that, in some sense, he had the final voice in the disposal of village lands, and that his consent was necessary before these were put to any use (claim 1). I also found that headmen of certain sub-clans have to "give consent" (*tagwala*, claim 4) and that there were individual owners who were active at the council (claims 5, 6 and 7 combined); and, of course, that one person came very much into prominence in all discussions and in all references to land ownership—the magician, who, as it happened in Omarakana, was not identical with the chief (claim 3). We need not go further into detail here. Those who have been reading the previous chapters carefully and glance over them again with their eye on the problem under discussion, will realise that my subsequent experiences of Trobriand gardening led me to discover a slight shifting in the perspective. The magician came much more to the fore (claim 3) and so did the gardening team (claim 8). The chief (claim 1) who, in Omarakana, has divested himself of the functions of magician and organiser, receded for a time. The non-resident headmen of sub-clans became completely obliterated (claim 4) as well as the individual owners (claim 7) and the local sub-clan as independent units (claim 5).

From all my data, whether collected by the "question and answer method" in pidgin or in native, or obtained from "objective documentation", or by the direct observations of what was happening on the land, I compiled the Table of Claims which I here reproduce. It represents almost exactly my knowledge at a certain stage of my field-work—it was, in fact, drafted between my second and third expeditions. The gaps in it will be obvious. The references to chapters in this book have been introduced for convenience in collation.

1. *Chief of the district.*—Into this category enter political chiefs such as the paramount chief in Omarakana, his military rival, the Toliwaga of Kabwaku, who rules over the district of Tilataula, the chief of Tabalu rank in Gumilababa, who rules over Kuboma, the Tabalu ruler of Kavataria, and so forth (cf. Part I, Secs. 4, 5 and 9). He is styled *tolipwaypwaya* (master of the soil) over his whole district, and claims certain tributes from it at harvest. These tributes are most substantial in the form of *urigubu* (marriage gift).[1]

He also has a certain amount of political control over the land in the sense that his summons are obeyed in cases of war, public ceremonies and tribal gatherings.

2. *The headman of the village community.*—In a capital village of a district the chief is also the headman of the village community (cf. Part I, Sec. 9). There, in virtue of his headmanship, he acts as master of ceremonies at the village council, has the right of wielding garden magic or of disposing of it to one of his successors, or, in exceptional cases, to a son of his. In other villages the head of the dominant sub-clan (cf. Ch. XII, Sec. 3) is the political headman. He is styled *tolipwaypwaya* (master of the soil) of all village lands. He also acts as master of ceremonies in the garden council of his village, usually carries out the garden magic, distributes lands and obtains a tribute in the form of small gifts and *urigubu*. His tribute is quantitatively very much smaller than that of the chief (cf. Ch. VI, Sec. 1, and esp. Docs. II–IV).

3. *The garden magician.*—He may coincide with 2. Or he may be a different person, either because the chief or headman has transferred his prerogatives, or else because, as happens in some villages (Ch. XII, Sec. 3), the offices of political leader and garden magician still remain dissociated. He is styled *tolipwaypwaya*, and his activities, his influence over gardening, as well as the payment which he receives have been minutely described in previous chapters. It may be summarily stated that, as 'master of the soil', he obtains very

[1] Compare references to the *kayaku* in Ch. II (Sec. 3), Doc. VII and Ch. XII (Sec. 4). For the relationship between chief and magician cf. Ch. I (Sec. 6) and Ch. II (Sec. 1). See also Ch. VI (Sec. 1) for the political aspect of *urigubu* and Part I (Secs. 9 and 10).

little direct material benefit in exchange for his services, but a great deal of influence and *butura* (renown).

4. *Head of a sub-clan.*—Most villages contain one or two other sub-clans besides the dominant one. The head of such a minor sub-clan would be styled *tolipwaypwaya* (master of the soil) with reference to his special fields. At the garden council his consent is formally requested when one of the fields which he "owns" is to be put under cultivation (cf. Ch. XII, Sec. 4).

5. *Minor sub-clan as a whole.*—All the members of such a group and each one of them would be styled masters of the soil in respect of the fields which are regarded as owned by the sub-clan. It may be noted that in the case of both 4 and 5 the whole sub-clan as well as their headman may be the absentee owners. In this case they still retain some vague right of citizenship in the village from which they have emigrated, but such a right is very seldom exercised.

6. *The village community as a whole.*—All the members of such a community, irrespective of sub-clan, would generally claim the right to the whole territory of the village. Such claims are by no means empty. All the members of a village community have the right to use its public approaches, its water-holes and most of its territory in the search for wild fruits, in hunting and collecting, and to cultivate the soil.

7. *Individual members of the community.*—Every plot within a field is allotted to an individual. At times an influential headman or a chief may own all the plots in all of his fields, or again the titles of ownership to the various plots are distributed among the members of a sub-clan. The individual owner has to give his consent when his plot is gardened by someone else, and he has an unquestioned right to cultivate it himself.

8. *The actual gardener.*—Every adult male in the village community, whether a citizen or not, has the right to ask for a plot or plots on one of the fields designated as the garden site for the coming cycle. Once he has obtained the owner's permission and this has been endorsed by the magician and the chief, he is completely master of the soil on the plots which he tills during one cycle of cultivation.

9. *The gardener's sister or other female relative.*—A woman will very often speak about a plot as "my plot" when her *urigubu* portion is cultivated on it—and her claim is based indirectly on the rather complicated principles of Trobriand land tenure which we shall discuss.

Let us remember once more that what we are doing in this chapter is not to present Trobriand land tenure, but rather to retrace the steps by which, in a somewhat roundabout and blundering way, I finally arrived at an adequate theoretical grip of the problem, which in turn enabled me to collect and organise the evidence in a satisfactory manner. As I have said, the list given above represents almost exactly the state of my knowledge at a certain point. There are obvious gaps in it. Thus were I drafting it now, with my present knowledge, and wanted to cover all valid claims, I should introduce as one item, and perhaps the most important one, "the gardening team", whereas we find (under 8) only the individual gardener. Let us at once list this claim as 8*a* so as to be able to refer to it in our following analysis. One or two more subdivisions could be added under 9, so as to make the list of people who indirectly profit from gardening, under some claim or other to land, more complete.

The analysis of this table worried me in the field for a considerable time; but it did not by itself give the clue to my problem. The mere contemplation of such a document would be useless. It was its correlation with economic activities, legal claims, mythological ideas—in short, the reorganisation of evidence within the context of native agriculture, law of residence, agricultural tributes and duties towards kindred and relatives-in-law—that brought the solution of my difficulties. The reader of the previous chapters knows that all the claims contained in this list are valid. Not one of them can be dismissed as irrelevant in the light of what we know about Trobriand agriculture. The reader realises also that the list is unsatisfactory. It presents a multiplicity of claims which, without further evidence, might be either discordant and unrelated to each other or else harmonised and related.

That the latter alternative holds the reader may guess, since he knows that serious quarrels about land, dramatic upheavals connected with tenure or violent territorial changes do not exist. The natives work their land, they work it in groups which are organised, and this organisation is to a certain extent based on territorial rights. We know that the unit of gardening, territorially speaking, is the garden enclosure. The social unit of labour is, on the one hand, the body of people who work the enclosure and, on the other, the gardener's family, the members of which work a plot or several plots in conjunction. The far-reaching order and harmony which underlies the multiplicity of claims can be definitely assumed in the light of our previous knowledge. But how it is established both in native ideas and in native practice will still have to be found out by a more detailed analysis of our data.

4. FUNCTIONAL ANALYSIS OF THE TITLES TO LAND

Our task then is to organise and group the titles to land on some principles of relevancy and relationship. In an agricultural community such as that of the Trobrianders, man draws his sustenance from the soil by working it. The use of land in gardening is, therefore, obviously the most important among all the relations between man and soil. To a Trobriander, as to any other husbandman or gardener, land is as land does. In this case, more especially, it gives him his yams, his taytu, his taro, his coconuts and bananas. In these garden crops the Trobriander is interested as producer and as consumer. As producer the Trobriander values his land (1) because it supplies him with a part of the crops which he needs for his own household, and seed for the next year's gardens; (2) because from it in conjunction with his fellow-villagers he draws prosperity (*malia*) for the whole village, thus warding off hunger (*molu*); (3) because his gardens yield the *urigubu*, the portion due to the household of his sister or some other maternal relative; (4) because they enable him to pay his tribute to the chief; (5) because they give him the wherewithal to trade and contribute to tribal ceremonies; and (6) because his gardening, if good, enhances his renown (*butura*), an incentive which makes him take pride and pleasure not only in the quantity of his produce but in the lay-out and the beauty of his gardens.

Now from this point of view one entry in the list of claims among all others is obviously of supreme relevance: and this is entry 8, defining the individual gardener's claim to the land which he uses. He is entitled to claim as many plots as he and his family can work. The title hands over the soil to him and to the members of his household to clear and break, to plant, to harvest, to display the produce and then to distribute it.

But as we know, number 8 has to be supplemented by an additional entry, that is 8*a*, the claim of the gardening team as a whole. For agricultural use, in its full sense, implies not individual exploitation, nor even exploitation by families, but the co-operative work of the garden team. As we know already, and as will be specifically restated, one of the fundamental principles of Trobriand gardening is that a number of people are organised into the gardening team and thus exploit the soil conjointly.

In order to understand the relation of this social group to the soil, it will be necessary to study how the soil is permanently subdivided and how it is used. The divisions of the village territory into fields (*kwabila*), plots (*baleko*): the classification of the gardens into *kaymata* and *kaymugwa* and *tapopu* (cf. Ch. I, Sec. 3, Ch. II, Sec. 3, and Part V, Div. I, §§ 14, 19 and 27)—all these are elements which must be correlated with our knowledge of social organisation, in order to give us an idea of how titles to land function in actual exploitation. The study of how land is used leads us, therefore, to the analysis, on the one hand, of the divisions and classifications of land, and, on the other, to the study of the social organisation connected with garden work. In short, claims 8 and 8*a* stand out from all the others as the most relevant from the economic and utilitarian point of view. If we look at the tilling of the soil principally as an economic pursuit, what matters most is the fact that all residents within a village or a group of villages or a part of a village have a right to use a portion of the soil conjointly. There is a certain variety and elasticity in the way in which a village community is either subdivided into one or two garden teams, or else several smaller villages join together to co-operate in the making of one garden; so that the consideration of the garden team (8*a*) is distinct from the consideration of the village community (6). But the two are definitely related; and here we have to consider a rather complicated set of conditions which define the relationship between sub-clan, village community and gardening team. The data for the solution of the problem we have already largely in hand. For we know, in the first place, that a village community in the Trobriands is not a simple unit but is compounded of several sub-clans (cf. Part I, Sec. 9). We know also that the sub-clans vary in rank, hence in prestige and power, and that one of them takes the lead, notably the lead in gardening. But here once more the data will have to be systematically examined, integrated and considered in their bearing upon land tenure (Ch. XII, Secs. 2, 3 and 4).

Glancing at our list of claims and scrutinising the other relationships between the various entries, we see that claim 7, that of the individual owner, is also related to claim 8. At the garden council, the individual owner has to exercise his title in claiming his plots or else to surrender his title (cf. loc. cit., Sec. 4).

Three more claims are correlated to the economically relevant couple of entries, that is to 8 and 8*a*: that of the chief (1) who acts as political leader in gardening, or the headman (2) who in some communities replaces the chief, and of the magician (3) who may be identical with 1 or 2 but may also be distinct. The magician, as we know, is the real organising agent of the team, and the leader in gardening. And in those cases where the headman and the chief are not identical with him in person, their political power and leadership also enter into the activities of the gardening group.

We see, therefore, that the economic exploitation of land allows us to place an emphasis on one principle, that of the organised gardening team, as effective user of the land, and to relate to this claim a number of others.

So far we have spoken only about production. As consumer the Trobriander values the crops of the soil in so far as he uses them within his own household and in so far as he consumes the *urigubu* portion of his wife. He also uses taytu, *kuvi*, taro, bananas and coconuts, for display on various economic and social occasions and, when he is a chief or headman, he uses them for distribution and for the financing of enterprises, whereby his power is exercised and his prestige enhanced.

This distinction between the native as consumer and as producer is not mere pedantry. The interests of a Trobriander as producer are expressed concretely in all the acts of glorification associated with harvest. Consider Chapters V and VII: it is the gift or tribute taytu that occupies the central place in the garden arbour; the counting and carrying, the admiration and the bragging, refer in a large measure if not entirely to taytu which is produced by the cultivator but will not be consumed by him. This taytu, when it is finally placed in the storehouse, becomes the property of another man who approaches it with the interests and valuations of the consumer. And here hark back to Chapters VI and VIII. The *bwayma* as opposed to the *kalimomyo* is the embodiment of the consumer's attitude. The opposition between these two stands for the distinction here made between the producer's and consumer's interests.

As regards the consumer, the principle of native economics which also enters into land tenure and which plays here a paramount part, is that of the *urigubu* (number 9 of our list). Since we have discussed it fully in Chapter VI and have to return to it in Section 2 of the next chapter, we need only mention here that the *urigubu* is really a claim to garden produce, based on the woman's right to her share of the land owned by her sub-clan. But she cannot exercise this right by working the land herself because, by the patrilocal law of marriage, she has changed her residence to that of her husband. This claim then is entirely independent of the claims grouped round 8 and 8*a*. It is distinctly the claim of a consumer to something she has not produced. It is associated with the matrilineal structure of the society in combination with patrilocal marriage. But it is important to note that this principle, which gives indirect control and claims over land to certain men through marriage, also enters into claims 1 and 2 of our list. The chief's control over certain uses of land and produce in his territory is largely exercised through his marriage with women of different communities within that territory. The same applies to a much smaller extent, to a headman, who usually has two or three wives.

It is when we look at a chief, whether paramount or inferior, that the distinction between producer and consumer becomes very clear. For in the case of a chief, his great, or even—in the case of the paramount chief—immense wealth in crops is not due to his own efforts or interests as producer. And his claims on the land from which this wealth comes are not exercised directly. As overlord of his district the paramount chief has the right to marry a woman from every community, while some minor chiefs have the right to marry two or three women from adjoining villages. The kindred of these women exploit the soil for their political liege, and give him the produce in the form of the *urigubu*. Claim 1, therefore, that of a political ruler to be master of the soil of his whole district, is exercised indirectly but effectively through the mechanism of permanent marriage endowment, coupled with polygamy.

Two claims have so far been neglected: I mean claims 4 and 5. Concentrating on economic interests, we have left them on one side justly enough. Where a sub-clan has left its territory of emergence —and as we shall see (Ch. XII, Sec. 3) this is by no means an exception—claims 4 and 5 are sometimes not exercised at all economically. In other cases their economic importance is made effective through claims 2, 3 and 7: that is, in so far as the claims of a sub-clan are exercised through a leading headman or a magician, who then has a dominant place in the village community.

Let us look more closely into the matter. In discussing claims 8 and 8*a* we had to make a distinction between citizen owner and alien resident. Both of them utilise the soil as long as the alien is actually resident. From the point of view of economic use there is no real distinction, therefore, between these two categories on the one hand, and, on the other, between people who have legal claims under 4 and 5, but do not exercise them.

How does the matter look from the legal point of view? The Trobrianders have a deep conviction about the territorial rights of certain sub-clans to a precisely defined portion of territory. This conviction is expressed in a doctrine about the origins of mankind in the act of first emergence (cf. Ch. XII, Sec. 1). Legally this doctrine makes one or more sub-clans the rightful owners of a certain territory and citizens of a community. Again, custom and tradition in the Trobriands allow such primeval rights of citizenship to be handed over by one sub-clan to another under certain conditions which will presently be stated (Ch. XII, Sec. 3). The whole force of ownership—that is, the conviction that certain claims are true and valid and others are spurious and can be annulled—rests on these mythological, historical and customary ideas. If we were to use our criterion of economic relevance to the exclusion of other considerations we should have to strike entries 4 and 5 from our list as irrelevant, and we might as well obliterate the difference between the owner-citizen (claim 7) and the non-citizen claimant who constitutes a portion of 8. In doing this, however, we would do violence to the native point of view, which no anthropologist can afford to do. It is part of his duty to reproduce the attitude of the natives in matters legal and economic; and in this instance we should leave unexplained certain elements of procedure at the garden council (*kayaku*) which to the natives are of the greatest importance. Worse than this, we should neglect a factor relevant to our problem since native ideas and convictions very often supply the binding force of obligations and duties. In fact, if we were to enquire more fully why a non-citizen (8) is allowed to garden, the first answer received would be that this is because of his quality as 6, that is, as member of a village community. Now this quality, that is the right of residence in an alien village, is always based on the relationship of a man or woman to some member of the local sub-clan. In other words, 6 is derived from 5. This relationship may be two-fold: historically, as we shall show by examples in the next chapter, an immigrant sub-clan may, so to speak, grow gradually into an alien territory and be incorporated into it on the basis of adoption through marriage: a woman of high rank having married into that village and her offspring being allowed to settle there permanently. But such an historical event is only an extension and stabilisation of what happens within every household and to every man and woman: that is, lifelong residence of a woman in her husband's community by the rule of patrilocal marriage, and the residence there until maturity of her children. Ultimately, therefore, every right of temporary residence or permanent citizenship springs from the prerogative, expressed in claims 4 and 5, of the indigenous local sub-clan to control its territory or to delegate this control to others.

It has become clear that there are a few fundamental principles which underlie all claims and control all practices referring to land. These principles, or doctrines, as I prefer to call them, can now be briefly enumerated.

A. There is one main conviction, an *idée maîtresse*, which dominates the whole attitude of man to soil in the Trobriands. A man or woman have rights over a given territory in virtue of the fact that their ancestress in matrilineal filiation emerged from the soil at a definite sacred spot situated in that territory. The system of matrilineal descent, therefore, combined with what we might call the doctrine of first emergence, constitute the legal and mythological foundation of Trobriand land tenure.

B. The law of exogamous and patrilocal marriage compels a man to take in wedlock a woman who is not of his sub-clan nor yet of his clan. She joins him in his community and remains there till the end of her life or the end of her marriage. Her children, though belonging to her sub-clan and community, have the right of residence until they reach maturity. This law of exogamous and patrilocal marriage is independent of the doctrine of first emergence. It affects land tenure in a two-fold manner: (i) It separates the woman from her own land, while her rights in the ancestral soil are recognised in the institution of *urigubu*. (ii) It constitutes her and her offspring into non-citizen residents of her husband's community, and through that it entitles them to join the gardening team, i.e. the group of agricultural producers in the husband's community.

C. The doctrine of magical leadership establishes a united team who cultivate one garden within one enclosure and in one organised activity. This doctrine is perhaps not quite independent of the previous ones. In so far as magic is also a product of local emergence, the doctrine is derivative of A. In so far as the membership in the gardening team is the result of the right of citizenship and right of residence, this doctrine is derived from A and B at the same time. The organising influence of magic is moreover associated with the organising political power of the headman and of the chief. With all that, this doctrine plays an important part in native belief in the spiritual claims of the magician to be the real master of the soil in virtue of his magic, and this even when he is not the same person as the headman. This doctrine also affects land tenure in so far as it is the expression of the unity and solidarity of the gardening team.

D. The affirmation that rank outweighs the claim of local emergence. This doctrine might also be formulated—as it often is by the natives themselves—in the principle that the highest sub-clan, the Tabalu, are masters of all soil and can claim and use it wherever they like; while other sub-clans of high rank share in this prerogative according to their degree. This doctrine is expressed historically by the gradual shifting of sub-clans of higher rank from their original territory to more fertile districts, while the

Tabalu, the highest sub-clan, gradually came to occupy all the centres of agriculture, political influence and economic exploitation.

These four doctrines are not, of course, codified in any explicit native tradition. We find them here and there formulated in myths, in historical traditions, in the legal principles of marriage, with all that these mean to the Trobriander. It would be possible to show, point by point, both that the doctrines work within the institutions and also to document them by means of genuine native statements. What the sociologist has to do is to place the facts and statements in relation to each other and to extract from them the really relevant principles. I think that the four doctrines contain the full legal theory of, and sum up the most important economic usages connected with, land tenure. Doctrine A is fundamental from the legal point of view. Doctrine B, in its two-fold influence, affects profoundly the actual organisation of gardening. Doctrine C represents rather the native emphasis on the importance of magic, and it does not add very much to what has already been contained in doctrines A and B. Doctrine D affects the past and therefore the present practice. It is perhaps the least familiar to us from previous descriptions and will have to be discussed fully in the subsequent chapter.

In the following chapter, where the account of land tenure is given through the exposition of these four doctrines, we shall see how they combine and the manner in which they work. Even now, however, glancing at our Table of Claims, it is obvious that Doctrine A is expressed above all in claims 4 and 5, and in so far as Doctrine C derives from A, also in claim 3. When 1 and 2 are not derived from acquired rights, they are also based on doctrine A. *De facto*, however, claim 1 is not based on Doctrine A in the case of the paramount chief. Doctrine B finds its clearest expression in claim 6, for it underlies the constitution of the village community. The combined doctrines A and B (i) give us the *urigubu* principle which controls 9 and also, to a certain extent, enters into 1 and 2. Doctrine C is embodied primarily in claim 3. Doctrine D is not embodied directly in our table. Since it is expressed in an historical process, and our table refers only to the results of this process, it obviously cannot figure in it. It could be said that it determines the historical antecedents of claim 1, in that the chief's over-right in the district is based on the influence of rank; or else, if we formulate it as the general claim of high rank to any territory it might choose, it could also be associated with claim 1, 2 and 3, as well as 4 and 5. As such, however, it certainly is not embodied in our list of claims. We might perhaps express it by making a distinction between the following entries: 4*a* and 5*a*, "sub-clan and its head who own the land by the claim of autochthony"; 4*b* and 5*b*, "sub-clan and its head who own the land by rights of acquisition," and 4*c* and 5*c*, "absentee sub-clan and its head". This, however, would be perhaps pushing our distinctions beyond the limits of profitable precision and in the following discussions we shall drop the discriminating indices *a*, *b* and *c*.

Thus some order has been introduced into an apparent chaos of claims by the discovery of more fundamental principles underlying them, which correspond both to native ideas and sociological realities.

At the same time we have seen that another analysis, not unrelated to that of the four constituent doctrines, introduces some order into our inventory of claims. That is the consideration of the several aspects of land tenure, legal, economic, mythological and political. From the economic point of view, claim 8 and 8*a*, which are mainly based on Doctrines B (ii) and C, come into the foreground, and we find claims 1, 2 and 3 related to them. From the legal point of view, groups 4 and 5 are predominant, and Doctrines A and B are most important. At the same time, the legal concepts cannot be understood apart from the study of mythology, which forms the tribal charter of fundamental rights and principles (cf. Ch. XII, Sec. 1). To the study of mythology we have to add a consideration of past history as embodied in legends and in remembered events. This aspect is obviously connected with Doctrine D.

Incidentally a study of the legal aspect of land tenure brings us to a point which we might have completely missed if we had concentrated on its economic treatment, exclusively. For legal analysis demands that we should study inheritance, alienation and leasing, that is, transference of titles 7 and 8. We shall see that alienation is never an economic transaction, nor yet does it ever happen through conquest; but that it is dependent on the combination of Doctrine A, that is, right through first emergence, which even when surrendered acts through the act of surrender; of Doctrine D, that is, power of rank, and of Doctrine B which supplies the mechanism of surrender. Within natural inheritance, again, the matrilineal principle does not act simply but is supplemented by the usage of livehand surrender under the system called *pokala*. A study of the legal aspect, therefore, consists not in an enumeration of rules, not in an inventory of claims, but in the study of sanctions and in the discovery of the way in which the rules and the sanctions operate.

If we want briefly to summarise our results in a synoptic table, we could list in the first column the four doctrines, subdividing B according to its two sociological effects. We shall see then that we can roughly attribute a type of social grouping to each doctrine, the second one obviously having a two-fold social influence. In the third column we can list the main influences of each principle. We would then obtain the following table:—

<table>
<tr><th colspan="2">DOCTRINE</th><th>SOCIAL GROUPING</th><th>ASPECT</th></tr>
<tr><td colspan="2">A. First Emergence</td><td>Sub-clan</td><td>Legal, mythological</td></tr>
<tr><td rowspan="2">B. Law of Marriage, Matriliny and Exogamy</td><td>urigubu</td><td>Matrilineal kindred: Mother, brother, children</td><td>Legal and economic (consumption)</td></tr>
<tr><td>Right of residence of wife and children</td><td>Family and village community as agglomeration of families</td><td rowspan="2">Legal and economic (production)</td></tr>
<tr><td colspan="2">C. Magical organisation</td><td>Garden team</td></tr>
<tr><td colspan="2">D. Rank</td><td>Political units: Paramount chief's domain; Districts; Compound village communities</td><td>(Historical) Political</td></tr>
</table>

As with all such representations, it is essential not to make a fetish of this diagram. It is merely a convenient summary of some of our results. The first column shows the four fundamental doctrines, legal and mythological. It shows also that in the systematic statement of Trobriand land tenure it will be best to proceed by the analysis of these doctrines. And since these doctrines are at the same time independent of and adjusted to each other, it will be necessary to examine how they combine.

The second column shows the social consequences of each doctrine, and the integration through that doctrine of people into a specific mode of grouping. The third column shows how a doctrine of land tenure defines the mode of settlement and the legal, economic and political relations of a community to the land.

We have now laid down the complete scheme of Trobriand land tenure, albeit in a bald and abstract manner. We have indicated the four principal native doctrines which control it. We have shown how the facts of ownership, utilisation, settlement and political control combine into a system of usages which allows every Trobriander to obtain his share in the territory, to exploit this and divide the produce among a number of people who receive this share because of their claims to the land. But in order to bring out the essentials of this system, as well as the method by which my conclusions were reached in the field, I have not given the full concrete reality of mythological belief, nor details of legal claims and prerogatives, or the vicissitudes of historical development—in so far as we can reconstruct these with safety; nor yet the living reality of actual usage. To clothe this skeleton account with flesh we have once more to retell the tale of native land tenure: this time consecutively, systematically, and with full consideration of the native point of view and the details of native usage.

CHAPTER XII

LAND TENURE

In the last chapter we formulated a brief and abstract definition of land tenure in the Trobriands. We know now that into it enters a system of native ideas, more precisely four native doctrines, which control the law of tenure and the rights of citizenship. They profoundly influence the natives' idea of the relationship between man and soil, thereby integrating human beings into a number of social units and transforming the soil from a merely physical into a culturally determined object.

Yet with all this, the working of these mythological and legal doctrines refers primarily to the practical uses which the Trobriander makes of his land, to the requirements of agriculture, the rotation of crops, the manner in which land is exploited by individuals and in which the produce is apportioned to other people. All this enters into the constitution of land tenure. In this chapter, therefore, we shall have to restate our account of land tenure systematically, and put the various elements which influence it in their proper place, and in their proper perspective. It will be best to begin with a statement of what might well be said to be the fundamental doctrine defining man's relation to the soil: the myth of first emergence.[1]

1. THE DOCTRINE OF FIRST EMERGENCE AS THE MAIN CHARTER OF LAND TENURE

Every Trobriander, man or woman, believes that by birth and descent he or she is connected with a definite spot, and through this with a village community and with a territory. For everyone believes that his lineage, in the person of his first female ancestress in direct line, issued from a definite spot in the Trobriand territory.[2] The myth of first emergence is definitely a matrilineal one. It always refers to a woman, at times accompanied by a man who is her brother not her husband. This belief, combined with the principles of matrilineal descent, furnishes the charter of citizenship and land tenure to every Trobriander. For by the act of first emergence all the descendants in direct female line of the original woman have acquired the right of citizenship in the territory surrounding the

[1] See also Note 45 in App. II, Sec. 4.

[2] Cf. Ch. I, Sec. 7. The reader will have noted that the Tudava myth is in a way inconsistent with the local myths of origin.

spot of her emergence. This belief is the foundation of almost all territorial rights and claims; almost, but not all, because life is always more complicated than the strongest belief and neatest legal rule and, as we shall see, the doctrine of first emergence is combined with, and often even overshadowed by, other ideas which at times refashion it out of all recognition. But it will be clearer to state each doctrine in isolation.

The spot or hole of emergence is usually called *bwala* (house); at times more specifically *dubwadebula* (grotto), *pwana* (hole) or *kala isunapulo* ('his'—or rather 'her'—'spot of emergence'). In every case the natives can point out the precise and actual landmarks: in Omarakana, for instance, there is a water-hole called Bulimaulo in that part of the village grove which is called Obukula, from which the sub-clan of Kaluva'u came out (cf. Doc. IV, Pedigree 2). Also near Omarakana there is a little grove called Sakapu and in this grove there is a rocky scar in the earth. This hole, also called Sakapu, is the place of emergence of the Burayama sub-clan of Kwaybwaga, who still retain some claims over the lands of Omarakana. In other villages a heap of stones or a coral boulder right on the *baku* (village place) marks a spot of emergence; or else a hole or outcrop in the surrounding grove or in one of the fields. Thus the pariah village of Bwoytalu, which consists of several component parts, has emergence holes for every one of the sub-clans inhabiting the village, and each hole of emergence is to be found in the village or in the immediately surrounding territory. The same holds good with regard to the villages of Ba'u and Lawaywo, which also are of very low rank. I stress this point because, as we shall presently see, some of the high-rank communities do not trace their emergence to the territory which they inhabit.

The heads of *waya* (inlets) furnish a few sub-clans with their original hole: the Kwaku, a local sub-clan in the village of Oburaku, the sub-clan of Okopukopu, one of Tukwa'ukwa and the local sub-clan of Kwabulo.[1] Some sub-clans of the eastern villages, Kapwani, Idaleyaka, Liluta and Moligilagi, trace their descent, or rather ascent, each to a grotto in the *rayboag*. The most famous of such holes of emergence, called Obukula, differs from all the others in that from it are supposed to have emerged the original animal ancestors of the four clans which were the first to come out upon the surface of the earth. From that hole also a great many of the aristocratic or ancient sub-clans trace their original emergence—and in this the hole of Obukula is also anomalous. The highest of the high, the Tabalu sub-clan of the Malasi clan, the Mwauri, Tudava and

[1] For the associated myth of Inuvayla'u, cf. *Sexual Life of Savages*, pp. 347–55.

Mulobwayma sub-clans of the Lukuba clan, and the Kaylavasi sub-clan of the Lukulabuta clan—all these came out of the Obukula hole. The bearing of this myth on land tenure we shall have to consider later.

While speaking about myths of first emergence, I must make clear their dogmatic character. It is not, however, easy to draw a clear line between a simple assertion—the bald statement of an original occurrence—and such a statement developed and embellished by incidents. In the Trobriands only one myth of first emergence is expanded into a long and dramatised story, and that is the myth of the first emergence of the four ancestors of the four main clans. The other statements are usually very simple. They reduce themselves to a mere affirmation: "From the hole of Bulimaulo there emerged first Kaluva'u and his sister, Bokaluva'u. They brought with them the magic of our soil and many other herbs." But this assertion enters into any discussion about the pedigrees of the sub-clan, about the division of land, about the respective claims to territory of the various contiguous sub-clans. At times, when some special magic is discussed, such a myth becomes fuller. Thus in connexion with the magic of rain and drought, we have a story about the emergence from Bulimaulo of an ancestral woman, Bopadagu. We are told subsequently how she gave birth to various animals associated with rain magic, to rain itself and to one or two children, and we are told in a developed and dramatised story how she handled that important but dangerous fruit of her womb—rain. But where there is no question of special magic the myth of first emergence takes the extremely simple form of the mere mention of the spot, of the names of first ancestors, of the insignia of rank and dignity which they brought with them, and of the types of magic, sometimes reduced to gardening magic only, which they carried from underground.

It was necessary to give these concrete illustrations because the first emergence is so important in native ideas on land tenure.[1] Of particular interest for us is the fact that with such a hole of emergence there is always connected a village, or part of a village, and a territory, or what we might call an assortment of lands, both of which belong to the people who came out of the hole. As a rule this comprises some waste land, a tabooed grove or two, a portion of the *rayboag* and perhaps one or two fields in the *dumya* (swamps); in every case it includes a large portion of cultivable bush (*odila*), divided into a number of fields which are subdivided into plots. Those villages which are near the open sea own a part of the eastern seashore

[1] See also Note 46 in App. II, Sec. 4.

(*momola*) with a fishing and bathing beach and a few sheds for their canoes. On the lagoon the beach is called *kovalawa* and here canoes are kept. Thus a hole of emergence is always the centre of a contingent territory which encloses a village, or part of it, and affords the following economic opportunities to its members: access to fertile, cultivable soil, invariably; at times access to navigation and fishing areas; a certain district for recreation and, of course, a system of roads communicating with other villages.[1]

Thus we see that the combined doctrines of first emergence and matrilineal descent sanction the direct and full economic and general use and enjoyment of a given territory, ownership of the cultivable lands included in it; and invest the sub-clan with a number of traditional, magical and religious claims.[2] All members can claim this territory as their own, and every member has his share in this joint claim (entry 5 in our list in Ch. XI). These claims, moreover, have ultimate legal validity in the Trobriands; that is, wherever they have not been overruled by, or ceded to, the one influence which overrides even the doctrine of first emergence—the influence of rank. It is only the members of the sub-clan who can use the title *toli* (owner of) with regard to village garden lands and mythological spots: *tolivalu* (owners of village);[3] *tolipwaypwaya* (owners of soil); *toliboma* (owners of tabooed grove). The members of the sub-clan have the right of citizenship, that is, the absolute and unquestionable right of residence. They can also deny residence to any of the people who live in the village in virtue of one or the other of the derived rights of residence or the customary indulgences. The citizens also have the ultimate right to cultivate as many plots of the common soil as is necessary for any one of them. The right to cultivate the soil is inextricably bound up with the right to claim a village as theirs, to inhabit it and to perform magic.

[1] For the classification of lands, cf. Part V, Div. I, §§ 6–8, and Ch. I, Sec. 8; and for the parcelling out of the cultivable soil, Part V, Div. I, §§ 13–20, and Ch. II, Sec. 3.

[2] In another of my writings I have formulated this as follows: "The sociological relevance of these accounts of origins would become clear only to a European enquirer who had grasped the native legal ideas about local citizenship and the hereditary rights to territory, fishing-grounds and local pursuits." (*Myth in Primitive Psychology*, p. 54). There also additional data concerning myths of origin will be found.

[3] The word *tolibaleko* would as a matter of courtesy be used also with regard to the man who cultivates a plot for the season. If, however, you questioned in any such searching phrase as: 'Who is the true master of the plot?' (*avayta'u tolibaleko mokita?*), or 'Whose own garden plot is that?' (*avayla la baleko toulela?*), the name of the legal permanent owner would be forthcoming.

Doctrine A thus culturally shapes or frames the territory. It has its sociological consequences which are quite as important to land tenure as the apportionment and assortment of the territory. These sociological consequences culminate in the formation of the sub-clan. The structure of the Trobriand sub-clan is determined first and foremost by the principle of matrilineal filiation. To restate this briefly: descent is traced in the female line exclusively, that is, a child belongs to the bodily substance of its mother and inherits her social characteristics and claims. Membership of a sub-clan is absolutely inalienable—you cannot change it or affect it. In the case of females, the line runs straight from mother to daughter and so on. In the case of males, the succession and inheritance follow the adelphic line, that is, a man is succeeded not by his next of kin in the immediately younger generation but by his younger brother, and only after the series of brothers has been exhausted do the uterine nephews come in.

What is interesting in Trobriand inheritance, especially of land and of magic, is that although it is the legal due of the younger man it has yet to be purchased by a special system of payments called *pokala*.[1] If, for instance, a younger brother or maternal nephew of the titular owner, i.e. his direct and immediate heir, or maybe heir once removed, wishes to acquire the title of *tolikwabila* or *tolibaleko*, he would offer his senior several substantial payments and the title would be, so to speak, gradually relinquished. Such transactions usually rest on previous mutual agreement and, as already mentioned, the titles concerned are almost completely honorific and carry with them no economic benefits.[2]

The structure of the sub-clan is also modified by the principle of seniority, that is, age and superiority of generation give a man greater importance and a higher status within his sub-clan. This refers also to what might be called kindred groups or lineages within the sub-clan. All the members of a sub-clan count their descent theoretically to the common ancestress. Practically, however, not all of them can genealogically establish their relationship, for the counting of kindred in the Trobriands does not go beyond the grandparent or great-grandparent. Those who can genealogically establish their relation might be called genealogical kindred, or

[1] Cf. *Argonauts of the Western Pacific*, pp. 185 and 186, and *Sexual Life of Savages*, p. 178.

[2] The *pokala* system, which conveys hardly any benefits at all to the purchaser while it provides the elder man with a substantial present, seems to me to express much more the desire of the heir to pay tribute to his predecessor than to be an act of purchase. This would make its name, *pokala*, which is identical with "tribute", easily comprehensible.

kindred group. The various groups recognise with regard to each other a relative seniority. Thus one of them is regarded as the eldest, that is, the most important. The eldest male of the eldest lineage is the head of the whole sub-clan. For—and here we come to another principle affecting the structure of the sub-clan—in the hierarchy of the sub-clan the male element preponderates, while in filiation the female element is the determining factor. The formal title of ownership in the communal territory of the sub-clan is nominally vested in the head of the sub-clan (entry 4 in our list). He is styled *tolipwaypwaya* (master of the soil) or *tolikwabila* (master of the fields) in a more specific or personal sense than any other member of the sub-clan.

One complication enters into the picture through the fact that hardly any Trobriand community is entirely homogeneous. That is, within the same territory and in the same village community we find usually two or more autochthonous sub-clans whose territories and whose claims are not precisely marked off from one another but remain compounded. To take Omarakana first, we have there two local holes of emergence: Bulimaulo and Sakapu, and two corresponding sub-clans: Kaluva'u and Burayama (cf. Doc. VIII). Both sub-clans have the right to reside in Omarakana, and both sub-clans own land within the joint territory. Their garden plots moreover are not grouped in two separate parts of territory but scattered all over the various fields. In Omarakana, however, conditions have been complicated through the advent of the Tabalu—an historical fact which will be more fully analysed presently. In villages of low rank, where conditions are simpler, each component sub-clan of a community has either a clearly defined portion of territory or else owns two or three specific fields. Also within each village such a component sub-clan occupies either the whole of a small ring of houses or a locally contiguous segment of the large ring.

As regards the distribution of sub-clans, then, the village community as a rule consists not of one sub-clan but of several. Only a few communities,e.g. Suviyagila, Luya and Lobu'a, are self-contained villages consisting of one sub-clan only. The villages of Yourawotu, Tilakaywa and Kupwakopula consist each of one sub-clan grouped round one central place. But these villages are really so near to each other that they form one composite cluster. Large village clusters such as Sinaketa, Kavataria, Kwaybwaga or Yalumugwa consist of several component villages which may be either homogeneous or else compounded of several sections of one sub-clan each. Finally, we have big villages, of which Omarakana is one example, and Obweria, Gumilababa, Okaykoda and Kabwaku are others, where, round a common *baku* (central place) in continuous contingent circles, two or more sub-clans live side by side in more or less homogeneous sectors. In Omarakana, besides the dominant sub-clan of Tabalu, there are the Kaluva'u and the Burayama. While in Gumilababa there are the Tabalu and two or three sub-clans of the Lukwasisiga clan.

Now this complication is again simplified by the fact that among the several component sub-clans one essentially is recognised as either the elder or else as of higher rank. The leader of such a sub-clan becomes the headman of the village community, and in the Trobriands every village community has one headman and one headman only.

On this headman devolve all the honorific titles, as well as all the ceremonial functions, offices, activities and powers vested in the village community as a whole; more specifically, he would wield any magic of which the sub-clan is possessed on behalf of the whole community, but at times he would also take over the magic of some of the inferior sub-clans. A notable instance of this is the rain and sunshine magic which genealogically belongs to the Kaluva'u but is wielded by the Tabalu of Omarakana.

With all this, the sub-clans of inferior rank or more recent settlement retain their claim to land in the joint territory. Very often the head of such an inferior sub-clan has to be specifically consulted at the council meeting. His position as owner of the soil and the ceremonial rights attendant on this were well put to me by an informant of Oburaku, when speaking of Mosagula Doga, who was headman of the indigenous sub-clan called Kwaku (Text 94. Part V, Div. XII, § 36).

"(i) The master of the village (i.e. the component part of a village, in this case *de facto* the sub-clan) is also the master of the fields. (ii) The kinsmen of Mosagula Doga have lived from old in Oburaku. (iii) Many people have their plots in the field called Wagwam. This man (Mosagula Doga) is master of the field. (iv) At the garden council held in front of Navavile's (the garden magician's) house, Navavile would ask: 'What do you think? Shall we cut your fields?' and Mosagula Doga would answer: (v) 'Good, strike my garden site.' "

This sub-clan had, as a matter of fact, not only lived longest in Oburaku, but had been there from the beginning, as they had emerged from the head-pool of a local creek. But though the oldest, it had not retained either the headmanship of the whole community or the conduct of the communal garden magic. Their seniority, however, gave them the ceremonial privilege of being specially consulted at the garden council. The fact that the garden magician himself had to ask permission before proceeding to the choice of a garden site has a great importance to the natives.

Although in other communities the less important sub-clans may not possess such special claims, yet each of them owns a portion of the territory. Any adult male of such a minor sub-clan may also have specific personal claims to one or more of the plots (*baleko*) which lie within the sub-clan's field (*kwabila*) (number 7 of our list). What exactly these personal titles to a plot mean we shall see when we discuss the allotment and leasing of garden land (cf. Sec. 4 of this chapter).

The permanence of the sub-clan and the immutability in every individual of his sociological nature are expressed by what might be called a spiritual continuity which also has a territorial or local character. After death man migrates into the other world, which is situated somewhere under or on a real island, Tuma, to the north-west. There the spirits still keep together and to one locality; they till common gardens on communally owned land, much as in this world. When during the festival of *milamala* the spirits return to their own locality they are tended by their own kindred and given offerings of food grown on their own soil. The material remains of man also continue to be bound up with the soil which gave his lineage birth. A man or woman must be buried in his own village, and after the bones have been interred, dug up and have passed through several vicissitudes, they will finally come to rest in a communal cave belonging to his sub-clan and situated on its aboriginal primeval territory.

Thus the belief in after-life as well as the treatment of bodily remains—and there is no close correlation between these—establish an additional unity of a spiritual nature among the members of the sub-clan.[1] Another factor of a spiritual nature which also enhances the unity of the clan is magic. For we know already that the original ancestors brought one more gift of the highest importance from underground to bestow upon their descendants: the various magical systems. Some of these, though they came from underground with the first ancestress, have not been retained within the sub-clan. Others, such as war magic, the magic of *kaytubutabu* (see Ch. X, Sec. 3), various types of fishing magic, some forms of canoe-building magic, and one system of love magic, are still monopolised within the original sub-clan in which they were vested by ancestral emergence. The form of magic, however, which bears directly on land tenure and interests us here primarily is the magic of *towosi*. The ancestral couple, brother and sister, brought the herbs and other substances for

[1] Cf. also my article "Baloma" in *J.R.A.I.*, 1916.

the magical mixture, or at least the recipe for these; they brought the prescriptions for all magical ritual and procedure, for the taboos to be kept; above all, they brought with them the full text of magical spells (cf. Ch. I, Sec. 7). This magic is normally transmitted in direct maternal line and according to the same principles which govern the transmission of property, privilege and status. Incidentally, magic, like land titles, has to be purchased by the *pokala* system, and here obviously the purchase cannot be omitted: since the gift of magic consists in the teaching of formulae and in the instruction in ritual, it cannot be handed on by a dead man to his successors.

The doctrine that only the magic which originally sprang from a land can give it fertility is of considerable importance. But though the doctrine represents a genuine native attitude, it has obviously not been maintained as a matter of historical fact. It is enough to glance at Document V, where we can see that several communities practice the same system of magic, to realise that there must have been a great deal of shifting and borrowing. More specifically, the magic practised in Omarakana once belonged in all probability to the village of Lu'ebila after which it is named (Kaylu'ebila). Also indigenous magic is often performed by an immigrant sub-clan. Yet in spite of these exceptions there is a deep conviction that in one way or another the magic practised over a territory is profoundly associated with it and either has emerged from it with the original ancestors or has passed through a process of what might be called mystical naturalisation.

Magic, strictly speaking, should always be handled by the eldest member of the senior lineage in every sub-clan. At times he delegates this prerogative and duty to his immediate successor, or to some other younger member of his lineage on whom the dignity of headmanship will sooner or later descend. Sometimes, however, it is not given to younger brother or maternal nephew, but to a person who has no legal place in the lineage, that is, to the headman's own son. It is usual in such cases for the son to become naturalised by the act of cross-cousin marriage, which gives him almost full legal right to reside in his father's community. But to this we will have to return in the next section.

We have now stated fully the principles involved in Doctrine A; the territorial and sociological consequences of this doctrine; some of its technicalities and implications. The doctrine is not quite as simple, perhaps, as it appeared to us in the discussions of the previous chapter. But the adventitious complication which results from several sub-clans compounding to form one community is effectively dealt with by Trobriand tribal law, in that there is always a subordination to one dominant sub-clan. We have seen also that the sub-clan is not a simple unified group but that it has a distinct structure of seniority and lineage. With all this, both the tenor and the workings of this doctrine as it stands are simple enough. It creates the territory and it gives rise to the sub-clan. It unites an organised group of men—the sub-clan—with the territory which they own. It creates, in short, the Trobriand idea of citizenship—an idea which corresponds to the native terms *tolivalu* (masters of the village) and *tolipwaypwaya* (masters of the soil).

United by common mythological sentiment, united by a great many economic interests, united also in a political community of joint armed forces, not for the defence of the territory but for the enhancement of their renown, the citizens identify themselves with their territory and own it in virtue of this identification. The principle: one hole of emergence, one sub-clan, one territory—or a definite part of it—and one headman, runs through the whole social organisation of the Trobriands. To the Trobriander land, the territory, the soil he treads and the soil he works, the rocks, groves and fields where he plays and lives, are actually and not merely legally bound up with him. Land for him is the real mother earth who brought forth his lineage in the person of the first ancestress, who nourishes him and will receive him again into her womb. The first principle gives every man a right of residence, a right to a portion of land in it, a real asylum and a place from which he can never forcibly be moved.

We see, therefore, that the short mythological story of first emergence, which every person, male or female, can tell in connexion with his lineage and his land, is not an idle fairy-tale. It is a live force, active, effective, co-ordinating human work, integrating human grouping, and conferring very definite economic benefits on people. To understand the function of this belief is to gain an insight into the correlation between myth, moral conviction, legal usage and economic organisation.

With all this the very constitution of the sub-clan contains certain principles which to a very large extent break up the simple working of Doctrine A and make it by itself quite inadequate in determining land tenure. As it stands, this doctrine would afford a very clear, consistent and precise theory of legal and economic rights to the soil. Taken at its face value, it would simply mean that a group of citizens, male and female, have the right to use their territory together, to co-operate, to produce, to gather the fruits of their labours, and consume them together. And yet a very essential principle of the matrilineal doctrine of kinship makes the sub-clan into what might be styled a non-cohabitable, non-co-operative group. For to the Trobriander the relation between brother and sister—and let us remember that this relation is the very core of their kinship system—is defined, not by an injunction to co-operate, but by a rigid taboo which separates them. Brother and sister are excluded from all intimate contact; from all familiar, free and untrammelled personal intercourse. They are not allowed to share the same dwelling; they may not ever work together, for work implies free conversation, occasional bodily contact, and a certain lack of restraint—all of which is not permitted to these two persons. The rigid taboo referring to uterine brothers and sisters is also extended to further groups of kindred, weakening as it widens. Clansmen and clanswomen cannot be united in marriage, cannot therefore share households, and would not be suitable as members of co-operative garden teams.

Doctrine A, in short, in so far as it contains the charter of the constitution of the sub-clan, makes this the legal unit of ownership in land. Through the brother and sister taboo and the correlated sub-clan exogamy, it makes the sub-clan into a group which though it owns land is prevented from using it either in production or in consumption. As it stands, Doctrine A contains therefore an inner contradiction and its results would be completely nihilistic for land tenure as an effective force in the economic exploitation of the soil.

Doctrine A, however, does not stand alone. It is supplemented by the law of marriage which we have listed as Doctrine B in the previous chapter. This doctrine completely remoulds the constitution of the local group. The sub-clan remains as the core of the local group, but to this are added new members, that is, the wives and children of the citizens. These wives bring with them, in addition to their membership, an economic endowment in the form of the *urigubu*. Marriage, therefore, leads to the formation of a producing and consuming group—the household; while a number of households compound into the village community. This again, reorganised into the gardening team, becomes the effective unit of agricultural production. Also marriage, in conjunction with the doctrine of rank, introduces another disturbing influence into the workings of Doctrine A. We can see at first glance, therefore, that the final result of the working of these doctrines will give us a very different picture from the simple, legally clear, mythologically founded charter of land tenure.

2. THE LAW OF MARRIAGE AND ITS TWO-FOLD EFFECT

Man and woman must marry. Without a wife the Trobriand adult male is not complete; he has no status. A woman, in order to bear children, which is her main business in life, must have a husband. Marriage is patrilocal, in other words the husband establishes the new household in his own, that is his mother's and his matrilineal uncle's community, and his wife joins him there. Marriage is also exogamous, and therefore extra-territorial for the woman. She lives in her husband's community and this cannot be her own. Marriage at the same time is essentially matrilineal; that is, the woman retains her membership in the sub-clan and bestows this on her children.

This short statement of the main principles of the law of marriage must suffice. Marriage, as such, belongs to pure sociology. We have dealt with it, moreover, in Chapter VI (Sec. 2). There we have entered minutely into the motives for marriage on the man's and woman's side, we have shown the character of maternal and paternal relations to the children, and we have also dwelt on the most characteristically Trobriand feature of marriage—the *urigubu* endowment. At this juncture marriage interests us in that it affects the residence of the wife and of her children. Since residence is the basis for effective economic co-operation, above all in agriculture, residence is deeply connected with land tenure.

It would be perhaps more correct to say that residence, as related to titles in land, is one of the main elements in Trobriand land tenure. From this point of view, the law of marriage has a two-fold consequence with regard to land tenure. In the first place, it establishes a new economically co-operative unit—the family; and since the village community consists of a number of families, the law of marriage also is at the basis of the constitution of the village community. In the second place, while the woman becomes a resident and effective co-operative member of the village community, she legally remains a member of her own sub-clan. This membership entitles her to a share in the produce of another community—her sub-clan's—that is to *urigubu*.

How far is the *urigubu* connected with land tenure? We have analysed this gift very fully above (Ch. VI, Sec. 2), but there is one point which it was not possible to bring out in that context. We have seen that the *urigubu* contributes greatly to the position of the woman in marriage. We have seen also that it is the endowment by the wife's brother of her household in virtue of the native principle of kinship which makes a man, his sister and her offspring into the real unit of filiation.

Let us look more closely at the woman's economic position throughout her life. As a girl in her parents' household she works on her father's soil and contributes towards the raising of the crops which will be in part consumed within the parental household. When she grows up she will share in her husband's gardens and benefit by part of the produce from them. But where is her own soil? Her mother's brother holds it, her own brother will later inherit it. The former is now annually furnishing the *urigubu* gift to her parental household. She, together with her mother and father and the other children, subsists partly on the produce of the soil which is her own. When in her turn she receives the *urigubu* for her own household, this also will be raised on her own soil. We can, therefore, from the point of view of land tenure, regard the *urigubu* as the annual return from the joint patrimony, the portion which is due to the woman from her brother; because the land which he husbands is partly his own, partly held in trust for the females of each generation. The *urigubu* gift, which must be raised on this land, is given by one who uses the land to a group of people who have a right to it, but who are, under the law of patrilocal marriage, "absentee owners".

Let us also remember that the *urigubu*, although formally offered by the wife's brother to her husband, is really a gift *ad personam*, the person being the recipient's wife. At her death it automatically stops; the children should return to their own community, but if they remain in their father's community, the maternal uncle need not, though I think in practice he very often does, continue to supply them with vegetable food.

It is necessary to emphasise that *urigubu* and land tenure follow from the same set of ideas concerning kinship and the territorial apportionment of land. The natives, in describing *urigubu*, would not naturally define it as the result of a woman's title to her land. But they are quite clear that it is due to her household, because she and her children are really part and parcel of her brother's kinship group. They are also fully aware that she and her children are real owners of the soil on which the *urigubu* is raised. Thus while the juridical generalisation has to be made by the observer, all its concrete presuppositions exist in native legal theory. In reality the *urigubu* makes a great difference in the married status of the woman, in her influence at home and in her position in the village community. The *urigubu* also allows women, especially if they have many brothers, to exercise a wide choice in husbands.

We could say that the claims to land are vested in men and women; but that the men can exploit their land directly, while the women, and their children before maturity, profit by it indirectly.

The bearing of the *urigubu* upon land tenure is now quite clear. If land tenure in the widest sense be defined as the effective use of titles in the utilisation of land, the *urigubu* is a principle which annually diverts part of the crops into the hands of those members of a sub-clan who, because of their female sex, cannot reside in the sub-clan's territory and yet have a claim to part of its produce. The law of marriage establishes households with divided kinship. These households, as we know, are financed from two sources, corresponding to this two-fold kinship. The family subsists partly on the soil on which they reside, but partly also on the soil to which the alien residents, the woman and her offspring, have the traditionally founded legal claims.

The family, however, is not merely a group where agricultural produce is consumed. It is also a productive unit. And here again the law of marriage affects land tenure profoundly. Land tenure vests the title to land in the father of the family. He is the citizen of the local group, the member of the sub-clan, and as such has the fundamental right to cultivate soil. The law of marriage joins his wife to him. By marriage she acquires the right to assist him in garden work—a right which is also a duty; and she acquires a claim to part of the joint produce. In a way the claims to land of a woman resident by marriage are as firm and unquestionable as those of her husband. They are less fundamental because they depend on the contract of marriage which, in the Trobriands, can be dissolved by divorce.

It might be incidentally remarked that normally a Trobriand woman never lives in that section of the village of which she is the real citizen. As a child she lives in her father's village; he is not of her mother's clan and his community cannot be hers. The only exception—for as we found there are exceptions to every sociological rule—is when under cross-cousin marriage the arrangement is matrilocal and it is the girl who moves to her own community, which is also the community of her maternal uncle, that is, of her husband's father.

It is not necessary to insist upon the character of the family as a productive unit. Throughout the previous chapters we have been able to follow the importance of the group consisting of man, wife and children in agricultural activities. Within each family the man has to cut the scrub, the woman to do the weeding, while both men and women plant and harvest. When communal labour is not employed the individuals in a family divide the tasks according to sex, within the precincts of their own gardening plot or plots (cf. Ch. I, Sec. 8, and Ch. IV, Sec. 5).

Throughout gardening, therefore, the family is in a way the most important joint exploiter of the soil. Once a portion of land is assigned to a man at the garden council, he, his wife and his children have an unquestionable right to use it productively during the whole gardening cycle. The right of the man is primary, that of his wife and children is derived. But in practice all these claims form one unified system, indeed one claim. This allows them to cultivate the land and then to appropriate as much of the produce as is necessary for the household, the amount being limited only by the husband's pride and kinship sentiments which compel him to devote a substantial portion to the endowment of his own sister, sisters, or kinswomen.

Important as the family is in the process of production, it never remains an independent productive unit. The real unit of production is the gardening team. The gardening team is, in a way, nothing else but the village community reorganised for gardening. Since again the village community is but the sum of families resident within it, we might say that the gardening team consists of all the heads of all families, as well as of their dependants, in so far as they are constituted into a group for purposes of gardening.

Let me deal first with an apparent complication. We have seen already in the previous chapter that a village community consists often of several sub-clans. It was mentioned there also that invariably one of these sub-clans plays the leading part politically. As a rule also one of the sub-clans plays a leading part in gardening. The head of the sub-clan or his delegate would then be the communal garden magician and the whole village would, under his leadership, be recrystallised into a gardening team for the purposes of agriculture. At times, however, instead of one gardening team we have two or even three within the same local group. The joint village of Omarakana and Kasana'i, which comprises also the older settlements of Yogwabu and Katakubile, is split into two garden teams, one led by the garden magician of Kasana'i—in my time Tokolibeba—the other under the leadership of Bagido'u. In the village of Yalaka, which consists of four independent sub-clans each inhabiting a component village ring, there is one garden enclosure and one magician. In the three villages called jointly Kurokayva there are, as we know (Ch. IX, Sec. 2), two garden teams which are, however, led by the same magician. In Gumilababa and in Obweria magic is made jointly for the whole village; there is one gardening team and one *towosi*. In Sinaketa we have as many as four or five gardening teams. Oburaku, on the other hand, which consists of seven component villages, has only one team and one *towosi*. This splitting of communities into two or more gardening teams, each of which again contains two or more subdivisions of the component village, seems to complicate matters greatly. In reality, however, this complication exercises no influence whatever on production, legal conditions, or the sociological organisation of agriculture as a whole. To the villagers themselves it gives interest, variety and the additional zest of competitiveness at close range. But we are not going to introduce a new doctrine or new principle to account for it. The re-grouping of villages into gardening teams I have found as adventitious as it is little relevant. It is one of the facts which can only be explained by some accident or vagary of the past historical process, which means that it cannot be explained at all since this process is untraceable.

Let us now pass to the organisation of the gardening team and to its rôle in land tenure. In the first place then, the garden team is either a village community as a whole or else one or more self-contained parts of a village. For—and this must be made quite clear—when a compound village such as Omarakana and Kasana'i split into two gardening teams, this split follows exactly the local subdivision of the village. The whole of Omarakana and Katakubile, which is part of it, are one garden team; the whole of Kasana'i and its suburb Yogwabu are the other part. The gardening team, therefore, is a locally contiguous section of a larger community when it is not that community as a whole.

As to the inner organisation of the garden team we will only rapidly summarise the relevant points in our knowledge. The gardening team consists first of the headman and garden magician, who may be the same person. The magician is the actual leader in work as well as in magic. The headman, when he is distinct from the magician, may perform one or two ceremonial and legal acts. He opens the proceedings at the *kayaku* and directs the allotment of the plots. But after that he hands over the whole conduct of affairs to the garden magician. The rank and file of the gardening team are differentiated on several counts. First of all we have the adult males who appear at the *kayaku*, to whom plots are personally allotted, and who do the gardening in their own name. In the actual economic work this group is supplemented by their women folk, and to a smaller extent by the children. Within the first group, however, we must make a further distinction. The gardening team consists of citizen residents who own land in their own right and alien residents who are allotted one or more plots by the secondary claim of residence. To the technicalities of this allotment we shall return presently. The alien residents mainly consist of adult sons of citizens, and as we shall see, in villages of rank, the chief or headman usually keeps dependants who are not necessarily his sons.

If we observed the garden team throughout a series of years, we would find, therefore, that it is, from the sociological point of view, a changing unit. There is a permanent core, which is formed by the sub-clan. The headman, the magician, the citizen members always belong to the local group, and though they may die and be replaced by their successors, sociologically this core remains stable.[1] On the other hand, the non-citizen residents change, and change in a somewhat adventitious manner. The citizens may marry women from a great many other sub-clans. The offspring of such unions belong to an ever-shifting variety of alien kinship groups. Young boys grow up and for a time they cultivate some plots within the gardening team. At maturity or marriage they normally return to their maternal community. They are replaced by young men belonging by descent to the local group, who return from their fathers' villages to join their maternal uncles and make their permanent home with their sub-clan. These men in due course marry girls from alien sub-clans, and thus the cycle continues.

The garden team then is a special unit of social organisation which comes into being by the integrating influence of agricultural work on land tenure. This latter, as we know, resolves itself largely into the workings of the two doctrines of first emergence and of marriage. In the organisation of the garden team one more effective force enters—that of magic, or perhaps we might better say: the economic leadership of the magician. In our analysis of the previous chapter we spoke of it as an independent doctrine of land tenure (Doctrine C). We can see now clearly, however, that it does not affect land tenure in any way comparable to the influence of the two previous doctrines.

Nevertheless, it is by no means negligible. The effective use of land, the exploitation of the sources of fertility in the soil, is definitely vested in the gardening team. This is the case both in economic reality and in native ideas. For over and above the efforts of each individual and the co-operation within the family, a considerable amount of concerted work is necessary in order to make Trobriand agriculture as productive as it is. Thus there are a number of tasks which the members of the gardening team must execute simultaneously and to a certain standard in order to protect the garden as a whole. The building of the joint fence is the most important of

[1] We know already that the magician may be the son of the headman. This of course is an exception and it does not affect the general rule, for the office always reverts to the sub-clan.

such activities, but the same applies to a smaller extent to most garden work. Although it is not necessary that every plot should be completely cut down, yet when too much bush is left standing, the burning is less effective on the plots already cut, since the flames do not sweep evenly and the fire cannot assume proper dimensions. In consequence of this there is more work on the *koumwala*. One of the effects of team organisation and of the influence of the *towosi* is that people are not allowed to lag too far behind or in too great numbers. In weeding, also, when a woman is very slack and allows her plot to be overgrown with noxious herbs, these will naturally spread into other plots. Here also the natives both expect and practise a certain solidarity in garden work. At thinning, especially if there is any disease among the roots and the bad ones have to be taken out, neglect on the part of one gardener harms his neighbours. Thus the organisation of the gardening team affects the economic use of the territory—in other words, it affects the exploitation of titles to land, that is land tenure.

It is clear, therefore, that the magician's claim (entry 3 of our list) to be the master of the soil is not an idle one. This we have seen throughout the previous chapters, but it was necessary to emphasise here its effective value. We see also that the solidarity established by common work in the field—a solidarity which unites citizen members, their wives, and families, as well as non-citizen residents—is an additional force in the effective use of land. The gardening team is a highly derived and complicated product of the working of Doctrines A, B, and C. Our analysis has shown us precisely how these doctrines work together in producing the most important group in Trobriand agricultural exploitation: the gardening team. The constitution of this group, however, as well as the working of Doctrine A, is affected by one more principle, to which we now turn.

3. RANK AS A PRINCIPLE OF TERRITORIAL OCCUPATION

We have seen how profoundly modified Doctrine A becomes by the workings of Doctrine B. Yet with all this there is one point on which the law of marriage cannot overrule the working of the belief in first emergence. And that is the ultimate continuity of territorial association. Because, as we know already, there is one mechanism of readjustment which brings the males of each generation back to their ancestral home: the principle of split residence, that is, the law and usage which forces every man at maturity or at marriage to return to his mother's community and settle there for the rest of his life. Thus the male members of the sub-clan are reunited in each generation and the sub-clan becomes not merely a kinship unit but also a local one. If there were no other influence at work, the law of marriage could not disrupt the territorial unity of the sub-clan and this latter would keep its lands in perpetuity.

There is, however, one more doctrine which enters into the whole question of settlement and by influencing the balance between the paternal and the maternal principle in the family upsets the readjustment of Doctrines A and B. Doctrine A is essentially matrilineal; though overridden by marriage it reasserts itself in the *urigubu* and in the return of the male children to their maternal uncles' community. Doctrine B is in its essence patriarchal and, as we might call it, patrophilic. It establishes the patriarchal household, the patriarchal village community, of which the eldest male of the most important sub-clan is headman. It produces also the gardening team.[1]

In Trobriand society there are then obviously two forces or influences at work: the one personified by the father of the family, the other by the wife's brother. Since these two in the Trobriands as elsewhere are human beings, and not mere cyphers, they behave accordingly: each one tries to get as much as he can out of his legal claims and at the same time he tries to weaken the legal claims which work against his influence and power. The father of the family, on the one hand, is fully satisfied with the privileges which matriliny bestows on him: he gladly receives the full instalment of *urigubu* or even a surplus; he accepts the services of his wife's brother and other kinsmen; and he claims the services of his children, notably his sons, in the name of matriarchal law, by which his sons are classed with his relatives-in-law and thus are beholden to him. On the other hand, he asserts the principle of patriarchy and, more than that, he behaves as a loving father. Thus, for instance, he is not satisfied with legal prestations from his sons at a distance. He tries to keep them at home even after maturity, very largely because he is attached to them and craves their presence and company; thereby attempting to override

[1] I am using here the terms "patriarchal", "patrophilic", "matriarchal" and "matrilineal" in a sense clearly defined by the context. The intelligent reader will realise that I intend merely to define the character of certain arrangements, laws and customs. Those which enhance the power of the father I call "patriarchal", those which enhance the power of the mother's brother I call "matriarchal". He will, moreover, if he is acquainted with some of my other writings, know that in my opinion no human society is ever constituted exclusively on the principle of matriarchy or patriarchy; that even filiation is never exclusively matrilineal or patrilineal; but that there is always a compounding of the two principles, though the balance may be weighted decidedly on the one or other side.

the matrilineal principle of readjustment—the custom of split residence and of the return at maturity to the ancestral village. He attacks thus the authority of his sons' matrilineal uncle. Not only that: by keeping his sons at home he attacks the matrilineal principle in his own person, in so far as he is the head and representative of his own sub-clan. Because when the sons remain with him he always bestows on them various privileges and offices which belong by strict law to his sister's male offspring. This he can do; he is in fact aided and abetted by such customary arrangements as cross-cousin marriage, and usages of tolerance which allow a man to hand over to his sons certain personal goods, traditional privileges, the exercise of magic, which by strict matrilineal law they should never possess or exercise.

We might describe the conditions in the Trobriands as a dynamic adjustment between the patriarchal and matriarchal principle. It is not a static balance but rather a perpetual conflict—a conflict which enters into personal relations within the family, within the village community, within the constitution of the tribe as a whole. Yet it is obvious that there would be no room for conflict, for readjustments or encroachments, if naturally and normally the personal influence of the two men—the father and the mother's brother—were even. There might be the inevitable variations according to personality, but these could not produce any lasting or profound effect. In the Trobriands, however, there is one element of a sociological nature which enters here and tips the scale on one side or the other. For, as we shall presently see, this principle does not act invariably or even predominantly on the side of the patriarchal influence. The principle obviously is that of rank, and it acts in a two-fold manner.

Let us first briefly recapitulate—for with most of the facts we are already acquainted—how rank enhances the power of a man within his household and community, that is, the power of a man *qua* patriarch. A man of rank has the privilege of polygamy, to which is superadded the advantage of a more substantial *urigubu* for each wife. Thus, through the working of the matrilineal principle associated with the essentially patriarchal personal status, such a man diverts to himself a considerable amount of wealth. Speaking in terms of production and land tenure, this means that he obtains through rank, matriliny and patriarchy a system of effective claims over a more or less extended territory. These claims figured on our first list of titles as the chief's over-right in the lands of his territory. We know that in the case of the paramount chief of Kiriwina the title and the real emoluments are considerable. They are by no means negligible in the case of some of the other headmen of high rank. In this context it is necessary to emphasise that here rank, together with the other principles, does affect the ultimate utilisation of produce and that parallel to this there runs a system of official claims to land. Rank, therefore, undoubtedly enters into the constitution of Trobriand land tenure.

But this action of rank does not bear directly on the readjustment between the matrilineal and patriarchal principle. This readjustment is achieved in so far as a man of rank can easily overrule the matrilineal influence with regard to his own sons. No matrilineal uncle will contest any claims to his nephews, if the father of these nephews is a man of rank higher than his own. It is a privilege for the sons to remain in a village of high rank. Their matrilineal uncle gains personal influence through it and very often does not want to regain them; especially since, by being a chief's sons, they develop a certain arrogance and a number of extravagant claims which would make them undesirable as subordinate kinsmen. But even if a man wanted to exercise his rights as matrilineal uncle he could not do so against the wishes of his superior brother-in-law.

The position of a chief, whether paramount or not, within his own community is less untrammelled in the matter of favouring sons. To a certain extent a man of rank can overrule the claims of his own kinsmen in favour of his sons more readily than a man of low rank; for rank in the person of the eldest head of a sub-clan becomes effective political power. But the rank of the chief, say the paramount chief, is not higher than that of his brothers and matrilineal nephews.

But here the balance is more evenly distributed. When the position of the chief's sons becomes anomalously high through undue favouritism, the chief's own kinsmen not merely resent it, but they can act. I have described once or twice cases of political tension and disruption arising from this cause, notably that which culminated in the expulsion of Namwana Guya'u from the capital against the wishes of the paramount chief. In several other communities of the Trobriands similar conditions obtained. Thus in Kavataria the last effective chief of high rank, Pulitala, had elevated his son Dayboya almost to the position of his successor, against the strong opposition of all the other Tabalu in the village. I believe that, had it not been for European influence which not only generally weakens tradition but specifically favours patriliny, Dayboya would have had to leave Kavataria after his father's death. As it was, a state of tension, intrigue and anarchy came into being and Kavataria sank from a village of real importance into a chaotic agglomeration of factions. A son of Mitakata, the chief of Gumilababa, died by magic some time before I arrived in the Trobriands, owing, as I was told, to the resentment of the chief's kinsmen at undue favouritism. On the other hand, in one or two communities, the son of the chief, although raised to a very high position, was able to retain the consent and personal goodwill of his father's kinsmen. A notable case of this was Kayla'i, son of M'tabalu, the aged chief of Kasana'i. During my second expedition (in 1915–16) M'tabalu was alive and ruled over Kasana'i. But his son Kayla'i wielded the magic of the gardens and was even the temporary officiating wizard in the great magic of rain and sunshine. That is, he had in his hands the powers which by rights belong to the paramount chief. During my last visit (in 1917–18) I found that M'tabalu was dead, but Kayla'i's position in Kasana'i had remained unimpaired. Nor is such a case an exception due to the decay of ancient law. The predecessor of Bagido'u in the office of *towosi* was, as we know, his father Yowana. Yowana was the son of a reigning paramount chief of Omarakana, had received from his father the magic of gardens, the title to a few *baleko* and, as I was told, even to a *kwabila*, and became altogether a leading personality in Omarakana. He belonged to the Kwoynama sub-clan of the Lukwasisiga clan, which in a way was the foremost among the sub-clans with which the Tabalu intermarried. His position in the capital was made more secure by cross-cousin marriage to a Tabalu girl—the mother of Bagido'u, and there were apparently no intrigues against him nor any resentment at his preferment.

In all this, we see that rank overrules the workings of Doctrines A and B as regards residence, and hence in the use of the territory; even more, it may override the right of the sub-clan to exercise garden magic and to control its lands. The fact that, in the very capital of the Trobriands, the most important magical instrument of power—the magic of rain and drought—is performed by an alien, that the economically most relevant magic, that of the *towosi*, has been repeatedly in the hands of aliens, shows how powerfully the influence of rank can overrule the workings of Doctrine A through the instrumentality of Doctrine B. For, let us remember, it is always rank acting through the relationship of father to son—a relationship based in the Trobriands essentially on their law of marriage.

Rank, however, as we have said already, acts both ways. It not only overrules matriliny by allowing a father of higher rank to set aside the fundamental principles of filiation and to introduce his own sons into the community. It may also tip the balance in favour of the matrilineal principle. This occurs when a wife of higher rank marries a man of lower rank and settles in his community. Her children, notably her sons, will be personally dear to their father. As father he wishes to retain them in his community quite as much as a father of high rank wants to retain lower-rank sons in his own village. And here rank helps him, though in a slightly different manner. He is backed up, not by his own rank, but by that of his wife and her offspring. Since such a woman would normally marry a headman of the community he would have some power of his own over his own kinsmen. This power, however, would not be sufficient to allow him to keep his sons with him permanently. But since they are of a higher rank, and backed up by their matrilineal uncle, they can have their way. If they consent, they can remain in their father's village, and no one would dare to oppose them. The greater the difference in rank, the more nobly born the wife, the less the force of any objections to their permanent residence in the village. Hence, as we shall presently see, the noblest lineage, that of the Tabalu, was able to take root all over the district. The feeling of the members of the autochthonous sub-clan is complex. On the one hand they are proud of retaining a lineage of high rank in their village. It adds prestige to the whole community and it enhances the power and standing of every individual. At the same time, the autochthonous elders resent being put into a subordinate place. For whenever men of a higher sub-clan settle in the village they gradually acquire an increasing power. For when a headman retains sons of a rank higher than his own, not merely a single individual becomes definitely associated with the village community, but a whole lineage. The high rank son of a headman or notable of lower rank obtains rights of lifelong residence, not in virtue of his father's rank, power and influence, but because of his own personal status. This status allows him to remain where he wishes and makes him automatically the most important person in the community. Since the new community becomes subject to him, it also becomes the home of his descendants in mother line. From his new home he will prepare the *urigubu* for his sisters, and the sons of these sisters will naturally come back to him and reside with him, and also succeed him in his offices. The concrete examples which will be given presently will show how gradual is this process, how a sub-clan of high rank takes roots in its new community by degrees, acquiring the various offices and privileges one by one, the exercise of garden magic being usually the last to come.

It is clear now that although the adoption of a son by a chief in virtue of the latter's superior rank is on the whole more frequent and at times more dramatic, it is less important than the settlement of high-rank people in villages of low rank through the mechanism of a woman's marriage into an inferior sub-clan. The residence of a lower-rank son in a high-rank father's village is a phenomenon which never can outlast a generation. If the son has married his father's matrilineal niece, the offspring will automatically be full-titled citizens and the introduced lineage becomes eliminated. If he has remained in his father's village while marrying a woman of a sub-clan who has no right of citizenship, the offspring of that woman will belong to the mother's sub-clan and the influence of the father's father—who is the chief or headman of high rank—is never sufficient to keep his son's sons in the community.

To sum up then: rank operating through the father brings about the phenomenon of a constantly recurring but sporadic influx of alien residents. It is an important phenomenon but produces no lasting imprint on the constitution of the local community and the fundamental rights of land tenure. Operating from father to son, rank cannot override Doctrine A. Operating through the wife and allowing sons of high rank to settle in an inferior village, Doctrine D, as we have called it, that is the doctrine of rank, brings about the permanent shifting of authority and sovereignty to new territories. It also overrides and modifies powerfully Doctrine A.

It will be best now to illustrate these general considerations by concrete cases. The patriarchal operation of rank from father to son has been fully documented already. I should like only to add that, in going over the ancestral list of names in Formula 2, Bagido'u told me that besides his father, Yowana, several other names belong to members of the Kwoynama sub-clan; that, in other words, the wielding of garden magic by a member of an alien sub-clan was not an exception. Apparently only members of the Kwoynama sub-clan were ever allowed this privilege.[1]

Turning now to the other phenomenon, we know already that the spread of sub-clans of high rank, their acquisition of real and full rights by settlement, of the *toli* title over the land, of its magic and of other offices is not by any means a hypothetical reconstruction. We know already that the sub-clan of highest rank, the Tabalu, have gained footholds in all the most advantageous places in the district. Other sub-clans of rank—the Kwoynama, the Burayama, the Tudava and the Mwauri—have all moved from the places where they originally emerged to other communities. On the whole the direction of the movement is from less fertile, less advantageous positions to centres of rich districts, grouped round Omarakana. On the other hand, the sub-clans of low rank, such as the four lineages in Bwoytalu, the people of Ba'u, the original sub-clan in Omarakana, and a whole series of others who need not be enumerated here, have

[1] Cf. comm.: 2 to M.F. 2 in Part VII.

always been and remain autochthonous. The correlation between high rank and mobility, on the one hand, and, on the other, low rank and territorial permanence, is clear. The only people of high rank and importance who still dwell in their original place of emergence and residence are the Toliwaga of Kabwaku.

Since the process of territorial diffusion is best seen in the highest sub-clan, the Tabalu, and since in their case we can still study the various stages and hence the mechanism of expansion, it will be well to give a brief survey of their history. The first ancestor of this sub-clan came out of the ground in the grove of Obukula near Laba'i. This village lies characteristically on the north-western shore of the island; for all beliefs in ancestral spirits as well as most of the legends about first things point towards the north-west, while cultural and sociological spread has a tendency to take place from north-west to south-east. There are no legendary data as to when the Tabalu left Laba'i. But the present rulers of the place, although belonging to the Malasi clan, are not Tabalu but a sub-clan of low rank. The Tabalu must have once owned the village of Lu'ebila. Following the further history of this sub-clan, their main centre was for a long time the village of Dayagila on the direct line between Laba'i and Omarakana. But that village also they abandoned. Later on the capital was, according to native tradition, Omlamwaluwa—a village now extinct, which flourished on the present site of the Omarakana grove, or perhaps immediately adjoining to the north. From there they just moved a few hundred yards south to Omarakana and Kasana'i, the older lineage settling in the present capital, the junior in Kasana'i. How many generations ago this happened it was impossible for me to reconstruct. The traditions of Omlamwaluwa, and even of Dayagila, are clear but by no means concrete or detailed. The tradition about Lu'ebila is vague, while that about the origins in Laba'i is firm but distinctly mythological. Most of the further data have a much more precise historical character. While the senior line of the Tabalu remained in Omarakana, women of Tabalu rank married into Gumilababa, the capital of Kuboma; Kavataria, the most important village on the lagoon; Kaduwaga, on the island of Kayle'ula; Sinaketa and Vakuta. One centre of the Tabalu—Olivilevi—came into being, not by marriage, but by independent foundation. A number of people from Omarakana moved there, constructed a new village, and admitted other sub-clans to act as their henchmen and vassals two or three generations ago. Several settlements, notably Oyweyowa and Osaysuya, became subject to the Tabalu, who settled there through the mechanism of marrying in; but subsequently the lineage died out. In Vilaylima two or three generations ago, Kabwaynaya, a Tabalu woman, married the headman. During her lifetime she was the *de facto* ruler of the community, and after her death her sons continued to be chieftains of the community and the district. She had only one daughter, who died childless. The village was thus during two generations a Tabalu centre, and then returned to the original owners, but it was distinctly raised in rank by the process. The most recent case is the settlement in Tukwa'ukwa of a Tabalu chief, by name of Mosiribu; his mother had married a local headman. Since Mosiribu has one or two sisters who again have children, Tukwa'ukwa bids fair to become another Tabalu centre.

What interests us most in the study of these concrete data is the degree to which local claims and privileges have been acquired by the inmarried sub-clan of higher rank. Let us take the capital first. The Tabalu have been in Omarakana for generations. They have taken root firmly with a two-fold effect: on the one hand, they are now in possession of everything which the community can offer: land, all forms of magic, the paramount chieftainship of the district. Kiriwina is their land, and the central place of Kiriwina, Omarakana, is their residence, their territory, their burial-ground. The soil, in short, with which they are completely identified. On the other hand, the whole mythological glory and power of the Tabalu has been bestowed upon the capital village in the district. The importance of Kiriwina in the eyes of the whole area consists in the fact that it is the district of the Tabalu, that Omarakana is its capital. And here the sociologist has to enter more minutely into some aspects of native psychology, referring to law, tradition mythology and rank. Let us hark back to some of the data discussed in the earlier parts of this chapter. The capital concept of land tenure is that all rights and claims to a portion of territory are due to first emergence. Yet at this juncture we find that in a great number of centres, as many as fourteen or fifteen, one sub-clan claims territorial sovereignty, apparently not in virtue of first emergence, while, paradoxically, the community where this sub-clan emerged does not belong to it any more. There seems to be a contradiction, and the whole importance of Doctrine A appears grossly overstated.

Yet this is only part of the truth. In reality Doctrine A, though overruled and overshadowed by the principle of rank, is by no means annihilated by it. For—and this is a very important point in our understanding of the combination of the legal and mythological ideas of the Trobrianders—the Tabalu acquired the rights to land in every case through the surrender to them of the rights of first emergence by the original sub-clan. I was best able to study this in

Omarakana. But this is really a test case, since here the supreme clan in its most important lineage has superseded one of the humblest local groups, and in Omarakana we find a live tradition that all the privileges of the Tabalu had been acquired by voluntary and explicit surrender on the part of the Kaluva'u. In the myth of the origins of rain and sunshine magic, notably, I found that, almost every time it was told to me, the fact was mentioned that either the original woman who gave birth to rain and to its magic handed over her powers to a Tabalu or that this was done on some subsequent occasion. As to garden lands, the ownership is definitely vested in the chief. But the minor sub-clan still retains titles to a few plots. And the head of the minor sub-clan, Mema'okuwa, is invariably bracketed with To'uluwa as well as with the headman of the Burayama sub-clan as joint owner of the territory. If I had to summarise the rulings of tradition, I would put it thus: the original owners of Omarakana were highly honoured by the advent of the Tabalu; they surrendered their rights naturally and willingly. But this surrender had to take place. The present rights of the Tabalu are the result of the following legal facts: first of all the first emergence of the original sub-clan, secondly the advent and marriage of Tabalu women in the community of Omarakana, and thirdly the surrender by the original sub-clan of all the rights to citizenship, land and magic.

A complication obviously enters in the fact that the magic which is now practised in Omarakana is neither the local magic of Laba'i nor the local magic of the Kaluva'u sub-clan. It belongs, as we have seen, to Lu'ebila. When I enquired why this magic was practised there I was invariably told that it was the best of all in the Trobriands and that therefore it had been adopted by the highest sub-clan. Through a subtle mental process the conviction has developed that this magic really belongs to the soil of Omarakana. The natives were always slightly startled when I drew their attention to the fact that it was not the magic of the soil. But such contradictions exist everywhere in native belief and legal institutions. Thus in the whole process of the spread of the Tabalu we have another interesting phenomenon: invariably they bestow some of their intrinsic glory on the village where they settle; invariably, also, they take a great deal of local colour from the autochthonous inhabitants. I have already mentioned in Part I that on the lagoon they have abandoned the rigid taboos which, according to the strict ritual and prescriptions of Kiriwina, constitute the very essence of rank. There is no doubt also that by this defection the Tabalu lineages of the lagoon have surrendered some of their rank. They are in reality no longer the full equals of their kinsmen in Omarakana or Olivilevi. The spreading process is thus a process of mutual adaptation.

Comparing communities where Tabalu ascendancy is well established with more recent incursions, we can see that the process of surrender is very gradual. In Omarakana, as well as in Kavataria, Gumilababa and Sinaketa, the ruling sub-clan has taken over completely all offices and privileges. In Vakuta the Tabalu own most of the garden fields, but garden magic is still in the hands of the original sub-clan belonging to the Lukuba clan. The same was the case during the short reign of the Tabalu in Vilaylima. In Tukwa'ukwa the Tabalu chief has at present only one field. He does not practise garden magic and he does not usually act as village headman. We can safely assume that the same type of gradual acquisition took place everywhere where now we find the Tabalu firmly established. The titles to land and to other privileges are handed over one by one to the inmarried sub-clan of higher rank. And after the process has been completed the new-comers obtain absolute sovereignty, citizenship, and rights to land. The titles of the original sub-clan remain dormant and subordinate, but never completely extinct. Their existence is perpetuated in mythology and in certain secondary titles to one or two fields. It would appear that the last thing to be relinquished is the practice of magic.

One more point might here be added. We know that besides the myths of local emergence there is one important and explicit story of how the four clans came out of the soil in the sacred grove of Obukula near Laba'i. This story tells us how the rank of the four clans became graded (cf. *Myth in Primitive Psychology*, Ch. II). The Malasi finally became the most important—the ruling clan. But there is not the slightest doubt in the minds of the natives that this distinction belongs only to one sub-clan of the Malasi, that is, the Tabalu. There is no doubt that, in the minds of the natives, this general myth establishes the Tabalu as overlords of the whole district. Their right to spread, to settle wherever they like, to assume the lordship over any community, is to a large extent based on this myth. Whenever there is a conflict between two influences—and such conflicts I have found in a marked degree in Vakuta and Tukwa'ukwa and in a slightly different form revived in Kavataria—the Tabalu would refer to the fact that all the soil of that district belongs to them in virtue of their first emergence.

We can thus say that the doctrine of first emergence has two aspects: one is represented by the numerous local myths of first emergence, the other by the general or national myth of the first emergence of the four clans. Thus completed and complemented, Doctrine A in a way embraces Doctrine D. And its validity is not so much overruled as complemented by the principle of rank. Whichever way we take it, however, and whether we prefer to think of Doctrines A and D as independent or as interrelated, we can see now how Doctrines A, B and D work in together.

We can also briefly summarise the specific contribution of the doctrine of rank as regards land tenure. (1) It organises village lands into the bigger units representing a district. Within such a district the exploitation of land is associated with the substantial tribute given to the chief in the form of the *urigubu*, of *pokala* and of other gifts. The chief, on the other hand, accumulates and stores this quota of the district's yield and uses it later in connexion with tribal enterprise, warfare and public ceremonies. Besides Kiriwina, where this economy assumes the biggest dimensions, we have such districts as Tilataula, Kuboma, Kayle'ula, Luba. In Sinaketa this district economy is not very pronounced. (2) The influence of rank supplies the effective force in the shifting of sub-clans from one local community to another. Thereby it partly overrides the principle of first emergence. In reality, however, it combines with this and derives its ultimate force therefrom in so far as it becomes an effective charter of land tenure. (3) When a sub-clan of high rank settles, it does not oust its predecessors but acquires from them, through their consent, the effective claims over the land. (4) Rank also introduces the patriarchal anomaly within a matrilineal system, whereby adult members who have no legal claim to citizenship and to the lands are included in the garden team. The position of such individuals is the more remarkable in that they often play a leading part, not only in the community as such, but in the magic of gardening and the organisation of garden work.

4. LAND TENURE AT WORK

The last turn of our argument has brought us back to the gardening team which clearly belongs to the subject of land tenure. The previous discussion may perhaps have appeared to lead us away for a time into the realms of pure sociology, history and mythology. In reality land tenure remained throughout our main theme. It is well to remember that land tenure in its essence is the relation of man to soil, and this relation has entered into every detail of our discussion. We were not so much concerned with the technicalities of using the soil as with the ultimate foundations of man's claim to his territory. We discussed marriage and rank. But marriage in the

Trobriands is an essential element of residence, and rank is tantamount to power; and since land tenure is really a matter of residence and power combined, we were—in discussing marriage and rank—discussing also the foundations of land tenure. Magic also, as we know, enters into any discussion on land.

It is time, however, to pass to the actual technicalities involved in handling claims to territory and in utilising them. For now we are in a position to assess the integral working of the four doctrines. The most important factors are the soil, as it is culturally determined; the gardening team, that is, the group which uses the soil; while one activity—the gardening cycle—dominates all others in man's relation to land and, in the present context, one act in this cycle is of prime importance, namely, the apportionment of lands at the *kayaku*.

As regards the soil, the essential fact that a delimited territory belongs to one group of people who live on it is the outcome of Doctrines A and D. The further subdivisions of the territory into land used for human habitations, into sacred groves, and finally into cultivable land, is the outcome, of course, of a number of economic requirements, usages and beliefs (see Doc. VIII).[1] The subdivision into fields and plots, which we find wherever land is put under regular cultivation, is not the outcome of any of our four doctrines, but rather a suitable arrangement for the handling and apportionment of lands imposed on man by the requirements of gardening. At the same time, the fact that the bigger fields are more or less ceremonially and formally apportioned to a few heads of sub-clans, while the plots are individually allotted in formal ownership, is the result of the conviction that land can be owned in perpetuity only by people who have a traditional right to citizenship. We know now that the few leading personalities who usually have such claims to the main fields are the headmen of those local sub-clans which have emerged on the territory and the headman of a sub-clan of higher rank which has settled there. After the discussion of the previous section we can now add one more principle to the rules governing alienation, that is, the transfer of land: one or several fields and plots would be handed over by the autochthonous sub-clan to an inmarried clan of higher rank, who would keep such titles in perpetuity. Also a chief of high rank sometimes permanently vests the title to one or more plots or even to a field in the person of his son of lower rank. Such a purely formal title would expire at the death of the holder or even at the death of his father. These titles are not listed in our schedule of claims in Chapter XI, because they

[1] See also Note 47 in App. II, Sec. 4.

are very exceptional and, even when they occur, of very little relevancy.

So much for the legal aspects of the soil. Passing now to the garden team, there is little to add to the account given towards the end of Section 2. But we would not be astonished now to find within the gardening team adult members who have no right to figure there, either by the principle of first emergence or by the law of marriage—I mean the grown-up sons of a chief who have remained in their father's community by the tolerance of customary usage. We would not even be astonished to see one of them taking the lead in the proceedings and acting as garden magician. We have met this phenomenon in the persons of a Yowana, of a Motago'i, of a Dayboya and of a Kayla'i, and these, as we know, can be treated as representative of a type. We would also understand that in certain cases the gardening team would include aliens of a rank higher than that of the local citizens. These would be the sons of women of rank who had married into a commoners' village. About these we could foretell that they are the forerunners of a line of new settlers of high rank. These latter would have intrinsically sounder titles to land by their very presence in the gardening team than the sons of a chief.

Let us now turn our attention to the main act whereby lands are allotted. I mean the *kayaku*. The procedure at this gathering, described in Chapter II (Sec. 3), can be now more fully stated in terms of land tenure. We need not comment here on the leading part played by the headman, who represents either the oldest autochthonous sub-clan or the sub-clan of highest rank. Nor is it necessary to enter into the rôle of the *towosi*, who may be identical in person with the headman or the latter's next of kin or else his privileged son. We also understand now why the heads of indigenous sub-clans, no longer acting as headman or as magicians, still have to give consent before gardens can be made on fields to which they have retained the title.

Passing to the allotment of individual plots within each field, our information has to be more substantially supplemented. The real apportionment of plots takes place, not publicly at the *kayaku*, but previously by private arrangement. Each plot on the fields chosen for cultivation has an individual owner, and this owner has a preferential claim to his plot. If for general reasons it is convenient for him to cultivate it he has the right to do so without further formalities. But the owner may be an absentee owner. Or one man will have several plots in a field too far for him to cultivate himself. On the other hand, there are always a number of people in the community—alien residents, junior members of the local sub-clans, or people who, though entitled to own plots, do not own them on the field chosen for cultivation—who would have to approach the owner for every plot which they want to cultivate. The usual term for begging or asking, *nigada*, is used to describe such an act of soliciting permission to cultivate. By the time of the *kayaku* the consent of the owners has been obtained and everybody knows quite well which plots he will cultivate. But the official allotment of such plots at the *kayaku* has the legal function of clinching the matter and settling the effective use of the plots for the duration of the gardening cycle. In communities of high rank the ownership of most plots is concentrated in one hand (cf. Doc. VIII). In communities of low rank there is, on the whole, a much greater variety in ownership.

Such a lease of a plot for the gardening cycle is usually followed by an economic transaction. After harvest about two baskets of taytu, and perhaps one basket of *kuvi* in addition, are brought by the cultivator to the owner. This gift is called *kaykeda* and represents about one-twentieth of the total yield. Such a gift cannot be very well regarded as economic rent, neither in its substance which is quantitatively too small, nor yet as regards the psychology of the natives. It is rather a token of appreciation, and belongs to the type of gift which we find as a concomitant of all legal transactions. In fact the *kaykeda* gift or fee is very often repaid by the owner to the lessee with a gift called *takola* or *takwalela*, a generic term for repayment of food by a valuable (*vaygu'a*), or object of use (*gugu'a*), such as a spear, or a shield, or a small cooking-pot, or a small stone blade. Such a *takola*, which as a rule would exceed in value the original gift, would again be repaid in garden produce, and this repayment would be called *karibudaboda* or *vewoulo*: it varies in quantity, and should correspond to the conventional value of the *takola* but should not exceed it. In the North the *kaykeda* is a little larger than in the South, and at the same time is more scrupulously repaid. In Vakuta even the first payment (*kaykeda*) is sometimes omitted. Very roughly speaking then, the owner is not substantially rewarded by the user for his permission to cultivate the plot. Under modern conditions, where the value of produce has sunk in comparison to the value of manufactured commodities, the owner may even be the loser by the transaction. This is the reason why at present the *takola* are very seldom given, probably only when there is some extraneous reason why the owner of the *baleko* should make a gift to the lessee.[1]

Sometimes a whole field (*kwabila*) may be leased. Such a trans-

[1] See also Note 48 in App. II, Sec. 4.

action would take place between the headmen of two communities with the consent of all the members. Some villages have less territory than they need. Others are inhabited entirely by aliens. This was the case with Kuluwa, a colony of Okaykoda, founded some thirty or forty years after the war. Again, some communities "own" a field or two too far away from their settlement and such fields would at times be leased to neighbouring villages (cf. Doc. VIII). The leasing of a *kwabila* would entail a much more substantial gift by the lessee community to the headman or "owner" of the field. This might be repaid by a valuable, *vaygu'a*. Here again the transaction has rather the character of an exchange of gifts than of economic rent.

The arrangements preliminary to the garden council and the proceedings at this are perhaps the most important part of the legal procedure in the actual apportionment of the plots. We have discussed the legal force derived from the publicity of the proceedings, from the solemnity of the harangue and from the public allotment of lands. All the decisions and all the assignments made at the *kayaku* have the character of a public contract. The transactions at the garden council have, however, no importance beyond that of publicly declaring who will cultivate this plot or the other. Everybody has to be provided with sufficient land for his needs. The fundamental right to land is derived, not from the transactions at the *kayaku*, but from the fact of residence. Here again legal and economic considerations must be scrutinised side by side, if we want to understand the full range of forces which govern the apportionment of lands. Some of the people resident in the village have a direct legal claim. They are the citizens. Such privileged residents as chief's sons, or sons of a headman of a rank higher than his own, claim the land for the same reasons which allow them to reside in the village. Finally, the chief's henchmen or a headman's dependants are resident because they are useful in producing crops. Their claims are based partly on the economic utility of what they produce to someone in power, partly on the legal position of their patron. The smaller the positive rights of an individual, the greater is usually the influence of the person who looks after him and uses him. Land and labour, the two main elements in agricultural production, are well balanced in the Trobriands. Those who mainly supply labour are allotted land enough to make their labour useful. It is only by looking at the facts in their totality that we can understand how this system works and, as we have seen, works quite smoothly.

We are also now in a position to summarise briefly the reasons why in the Trobriands there is very little trespass and only occasional arrangements to safeguard property in land and produce against theft. The real forces protecting property in land and produce arise from the integral working of the whole economic system of agriculture; from the inherent ideas about the rôle of man as gardener, and from the manner in which the produce is used.

Land, of course, is used primarily for the production of crops, that is, for the production of food. Now to the Trobriander food is not merely a utility to be consumed and coveted in the sense in which a person might covet something which can be appropriated and used later on. As we know from our descriptions of harvest, the quantity of food produced is something which will be displayed, which will redound to the glory of the producer, which will be in part given away for the sake of personal and family pride, and in part kept. The quantity of food, whether given or received or owned, is very much a matter of personal status. It is limited on the one hand by the man's rank. As we know, it must not exceed a certain measure; but, on the other hand, scarcity is a matter for personal and poignant shame. It is a most degrading thing to beg for food, and the worst insult to be told that one is in need of food, that one has no food.

Now this attitude goes so deeply that it penetrates to the illegitimate appropriation of food. The natives distinguish between *vayla'u*, the stealing of food, and *kwapatu*, the stealing of any other object, and the first is infinitely more reprehensible and shameful. There are no arrangements for protecting the ripe crops in an open garden; during harvest, when everybody is busy taking out tubers and arranging them in heaps in the arbour, there are no arrangements for sleeping in the gardens or watching. I was often told that no one would ever *vayla'u* (steal produce). Still the very existence of the word and the very strong contempt felt by the natives at the idea of it suggests that cases must occur; but they are rare. The dangers of exposure and the shame of it are apparently a sufficiently strong protection against possible trespass. Moreover no one is in actual need of crops as food; for everybody when in need can automatically get enough from his kinsmen or relatives. In addition, to steal enough tubers to be worth taking into one's own arbour would be too difficult technically and too easily discovered. The sense of personal honour and ambition connected with good gardening is strongly developed. There is a moral attitude which would make an average Trobriander scorn the very idea of using somebody else's taytu. And it is this moral sentiment in conjunction with the actual arrangements, especially the publicity and openness of harvesting and the immediate stacking and display of taytu, which act as effective safeguards.[1]

The attitude towards produce is extended to land. A community would feel deeply insulted by the suggestion that it had insufficient lands. Such communities as Kuluwa, mentioned above, established after defeat in war on alien territory, would suffer profoundly in their personal ambition by being dependent on others for land. In olden days such a community would have returned to its own territory after a few years. The fact that Kuluwa did not return is due to the decay of the old tribal order and also to the fact that, owing to the considerable diminution in their numbers, the natives have more land now and were able to accommodate the new village. This also made possible the foundation of Olivilevi. Formerly every community would have returned to its natural and inalienable territory. The stability of territorial boundaries was regarded almost as a natural fact. Whether there ever existed any shifting of the external boundaries of village fields I was unable to ascertain. Perhaps powerful communities such as Omarakana and other villages of the Tabalu would slowly and persistently expand because no one would dare to accuse them of doing so. But as a matter of fact in normal seasons there was enough land for everybody, while in seasons of drought the question of more or less of land was irrelevant. The boundaries of plots would certainly not have been moved.

The only objects connected with land and fertility which are exposed to theft are fruits and fruit trees, above all coconut and areca-nut palms. Here certain arrangements against theft did exist. A band called *kaytapaku*, made of palm leaf or a piece of dried leaf, was bound round the trunk of the tree. Over this a protective magic would be recited which contained a curse on anybody who should touch the fruits of the tree; such a man would fall a prey to one or the other of the diseases contained in the spell. Usually such formulae contained an invocation to wood sprites (*tokway*) who would then protect the property by inflicting illness on the transgressor. The natives believe that disease invariably follows the violation of such a protective mark, and everybody pays very great respect to such "Beware of Danger" signals.[2]

[1] As far as I know, the Trobrianders have no belief in the stealing of tubers by magic—a belief which, according to Dr. R. Fortune, is very pronounced among the natives of Dobu (see the *Sorcerers of Dobu*, especially the chapter on agriculture). I made no direct enquiry on the subject, but it could hardly have escaped my notice.

[2] See also Note 49 in App. II, Sec. 4.

5. SUMMARY, AND THEORETICAL REFLECTIONS ON LAND TENURE

The subject of land tenure has been used in this book to establish one or two points of method. While it has benefited from this preferential treatment, it has also perhaps suffered to a certain extent. It was first dismembered, even dissected, in Chapter XI, then patched together in a preliminary fashion; and then, upon this scaffolding, reconstructed with somewhat fulsome elaboration. In order not to miss any aspects, the descriptions and discussions may have gone a little too far, especially into the realms of mythology, of social organisation, of history, and of legal technicalities. With all this, I think it has become clear that the discussion of property in land is one aspect of agriculture. In a way, therefore, some of the most important elements of land tenure were only briefly summarised in these last two chapters and fully discussed in the previous ones. At the same time, it cannot be emphasised too strongly that land tenure is a relation of human beings, individuals and groups, to the soil which they cultivate and use. This relation, on the one hand, transforms the land: human beings subdivide it, classify and apportion it, surround it with legal ideas, with sentiments, with mythological beliefs. On the other hand, their very relation to the soil makes human beings live in families, work in village communities, produce in teams, become organised by a common belief and common ritual of a magical character. Thus the discussion of land tenure must deal with sociology, as much as with topographical details; above all it must constantly refer to economic activities. Since possession means also security of tenure and titles, it is necessary to delve deeply into historical tradition and mythological foundations.

I do not need, then, to apologise for having enlarged upon citizenship side by side with land tenure, and given an account of legal and mythological foundations side by side with economic activities. It might be interesting, however, to make an attempt to disentangle land tenure in a narrower sense from the complicated web of ideas and practices described in the previous sections. In the narrower sense land tenure is the body of rules which govern the practice of cultivation and apportionment of produce. With the knowledge gathered we could almost visualise the agricultural processes as well as the results. Let us take a bird's-eye view of what happens in the Trobriands over, let us say, one gardening cycle. Such an imaginary Ethnographer's Eye watching the territory and integrating the facts over a span of time (or if you are mechanically minded you may imagine a super-cinema operating from a heli-costatic aeroplane) would first of all perceive the co-ordination between territory and human groups. In one way the very similarity of proceedings, and also of the occasional economic transactions, would enable it to recognise the Trobriands as one homogeneous area in which land tenure remains the same fact. But the really relevant units would be districts. Within such districts it (the Ethnographer's Eye) would note the annual accumulation of produce in a centre—the capital. It would also register the occasional redistribution of this produce among various sections of the whole district. This observable objective reality would be a summary picture corresponding to the chief's claims to the territory of his district, to the benefits which he derives from it and which he invariably redistributes.

The Ethnographer's Eye would also register within each community, the annual process of apportioning several plots to each head of the family or adult resident, the permanent division of the territory into fields, and their subdivision into plots with clearly marked indelible boundaries. If our registering apparatus could discriminate between kinship grouping, let us say, by means of coloured labels attached to each individual, the sociology of the effective gardening team would be readily defined. By such means the principles governing the distribution of the total annual yield could also be perceived. A large portion of this remains within the village community. Less than half probably could be traced to various other places where the colour of the local sub-clan or sub-clans would correspond to that of some woman of a household and her children. The rôle of the family within the gardening team would be clearly registered, in that regularly and systematically its members would co-operate on the same few plots.

We will have, however, to imagine our apparatus stationed above the soil of the Trobriands and registering what happens there for a much longer time—generations and ages—in order to reconstruct what our plausible deductions tell us about past history. Our apparatus would then show us the facts of first emergence at the beginning. What they were in reality, what corresponds in the actual historical past to the native belief about a woman and her brother, both burdened with all sorts of paraphernalia, clambering out from underground—this we can but vaguely surmise. Probably the present population of the Trobriands is the result of the blending of races and cultures. The local myths of emergence may refer to acts of settlement of an immigrant stock; they may express the claims of an autochthonous clan translated into a mythological set of ideas by the invaders. All this may be picturesque, but it is not very profitable to speculate about it. What the apparatus, however, would invariably register would be the movement of some sub-clans and the stability of others. It would also disclose to us the mechanism by which these movements took place. Normally this mechanism, as we know, is what might be called adoption by marriage. Have there been in the course of history more violent, more catastrophic movements? There is no trace of them in tradition and, as far back as Trobriand culture was what it is now, no shifting of territory by conquest could have taken place. The only two cases on record of wholesale movements of a community, that is, Kuluwa and Olivilevi, were both due not to conquest but to defeat.

Right through the process we see how an objective picture of land tenure comes into being. On the one hand, it is a system of actual and effective uses vested in the gardening team. On the other hand, it is a long list of legal titles expressing various claims, most of which carry with them some privileges, at times some burdens. The purely formal, legally valid and mythologically founded titles which, as we know, are vested in the sub-clans of emergence or sub-clans of rank, do not always go parallel with the uses of land. We know now exactly where this parallelism breaks down, why it breaks down, how it is readjusted by some super-imposed doctrines or usages. The essence of the complication in the Trobriands lies in the circumstance that land tenure is legally vested in the sub-clan, while its effective economic use is in the hands of a group of people in which the men only are bound together by ties of kinship, but which also includes the wives of these and their children. The clue to these intricacies can only be obtained by a full and clear knowledge of the sociological working of matriliny, patrilocal marriage, the limited paternal influence of the Trobriander and the various adjustments between the two principles, the matrilineal and the paternal—adjustments in which rank plays a prominent part. There is thus a permanent split—or perhaps better, there is a double facet to ownership. In its productive aspect it is vested in the men of a local community, their spouses and their children. Such an agglomeration of families constitutes also to a large extent the group of consumers. At the same time, the men and women of the same sub-clans retain jointly the legal claims to their matrilineal patrimony—if this compound term may be coined. And this unity is embodied in the institution of the *urigubu*.

We can now look back on the ground we covered in our attempt to glean all the data necessary for the constructing of the invisible fact of land tenure. We are able now to support some of our earlier inferences and methodological points. Our analysis started from a criticism of the easy technique—one might almost call it the inevitable trap for the field-worker in land tenure and similar problems. The verbal approach, the collecting of statements about who is the owner, leads to a legalistic approach natural to a political officer or a European lawyer. A schedule of titles is obtained which—as we pointed out (Sec. 3 of Ch. XI)—is useless as it stands. We have seen how some of the explicit and well-established titles to land have merely an archaic significance; others are purely honorific; others legally important but economically irrelevant; others again merely figurative. In order to define them in their mutual relation and degree of relevance we have to analyse them from the functional point of view. In studying how such titles work in production—allowing some people to claim the usual yearly marriage revenue and others to accumulate such revenue in larger quantities—and how some titles serve to buttress the unity of the sub-clan, we learned how to connect the legal and the economic aspect of the institution.

In doing this we saw why it is indispensable to study the full legal as well as mythological foundations of native prerogatives and claims. This side supplies land tenure with its foundations in native belief. For land tenure is woven by its mythological and legal fringes into the matrilineal fabric of society.

In our economic analysis we saw how the various claims, titles and honorific references work as effective motives in production and consumption. As regards production, we found that effective use in agriculture is achieved through co-operation; co-operation in tilling means residence, so that residence has been throughout the key concept in our discussions of land tenure. Whether it be residence by the inalienable right derived from first emergence or settlement by rank, or by customary toleration, residence always gives full right to the cultivation of the soil.

It may be of some interest to point out how futile the distinction between communistic and individual ownership becomes in the light of our discussion. Throughout our survey we could have shown how every claim, every relationship between man and soil is emphatically both individual and concerted. For while the very conception of first emergence implies a large kindred group—the sub-clan, this group is primevally represented by an individual—the first ancestress with another perhaps, her brother—and in its present-day constitution is equally represented by an individual—its headman. It is differentiated according to sex, stratified by age, subdivided into minor lineages. Moreover, even within the sub-clan and as regards land tenure, individual titles exist and the land is subdivided, as it were, almost in deference to the desire for individual distinctions. For though personal ownership in plots approximates in a way most to our idea of the ultimate facts in land tenure, in the Trobriands it has only the very slightest economic relevance. But it is extremely important in so far as it shows how little there is of the so-called "primitive communism" in the economic attitude of the natives. Almost to spite the anthropological theorists, the Trobriander insists on having his own plot associated with his personal name. This old opposition is a vicious and unintelligent short-cut, because throughout this discussion the real problem before us was not the either-or of individualism and communism but the relation of collective and personal claims.

Again, in the exploitation of the soil we met new groups: the family, the village community and the gardening team. Here again man often acts as an individual. He is responsible for his own garden made on several plots. If he is married, he acts as the head of the individual family. But an unmarried man may be gardening by himself and for himself until he has to dispose of his produce, when, again, he will be acting as individual head of his sister's family. But even then he never acts in isolation. His individuality is as a rule the meeting-point of two groups: his own household and the matrilineal kinship group in which he acts as brother. Again, he is a member of a co-operative group—the garden team—where as likely as not he will be engaged at certain phases in communal work and throughout will have to bear certain communal responsibilities.

The chief and headman receive their large share of agglomerated revenue as individuals. But they definitely act for their families, while a chief of a district acts for a whole group of village communities. It might be said that a representative and outstanding individual, whether he be a magician or a political chief, invariably acts in a two-fold capacity: as an individual and as a representative acting through the group and for the group.

There is no more jejune and fruitless distinction in primitive sociology than that between individualism and communism. Communism as a cultural reality is possible only through the advent of the machine. In the measure as the human being has to serve the machine, adapt his work and his mind to mechanisms and depend on machine-made goods for his existence, a new sociological phenomenon develops. On the one hand, human beings become more useful the less differentiated they are. Man has to become an interchangeable part of the vast human mechanism which is only a counterpart of the material mechanism. On the other hand, the personal satisfaction and happiness of an individual can only be achieved if he remains completely in tune with the great average of his fellow-citizens, with whom he has to consume the same goods, read the same newspapers, thrill to the same films, march to the same tune of the same hymns, whether these be sung in praise of communism or fascism.

PART III

DOCUMENTS AND APPENDICES

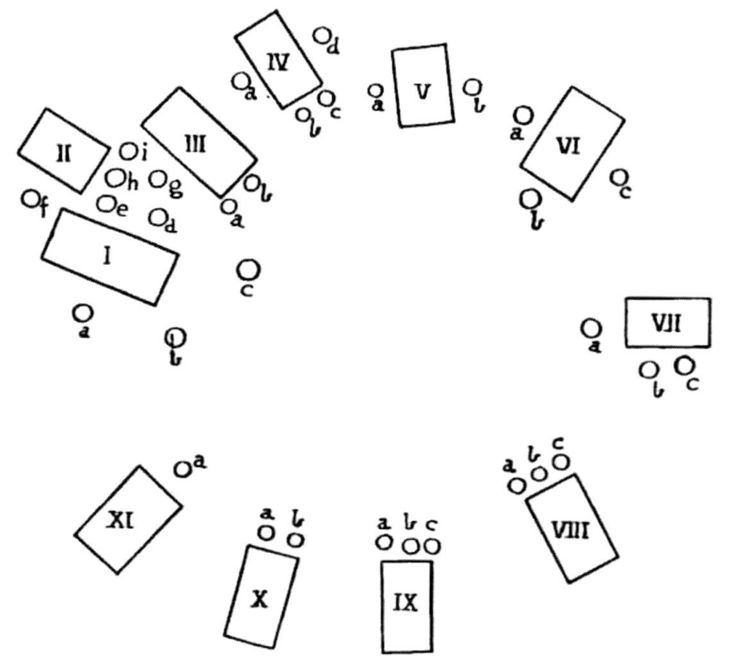

Fig. 11. Plan of the *baku* (central place) of the village of Yalumugwa, showing position of *bwayma* (storehouses) and of *gugula* (taytu heaps) at harvest, August 1915.

DOCUMENT I

THE *DODIGE BWAYMA* (FILLING OF YAM-HOUSES) IN YALUMUGWA

(Chapter VI, Section 3)

THE adjoining plan shows the central place of one of the component villages of Yalumugwa and the surrounding foodhouses with their annual harvesting heaps. This is a community of *gumguya'u* rank, that is, one belonging to an aristocratic sub-clan (*dala*), though not of the highest rank (cf. Ch. XII, Sec. 3, and Part I, Sec. 9). In principle, only members of that *dala* and the "owners" or "masters" of inferior rank, but often with more ancient rights of citizenship (cf. Ch. XII, especially Secs. 1-3), are allowed to have their *bwayma* on the *baku*. In Yalumugwa eight yam-houses are owned by members of the Lukwasisiga clan, of which seven (I, II, III, V, VI, X and XI; see Fig. 11, and table at the end of this Document) belong to the sub-clan of Kwoynama, the present owners of Yalumugwa. One belongs to a Lukwasisiga *gumguya'u* of another *dala*, married to a woman of the last-mentioned sub-clan.[1] The other three *bwayma* (IV, VIII and IX) belong to *tokay* (commoners) of the Malasi clan, who were probably the original "owners" of Yalumugwa.

On Plates 75 to 77 the reader will see the central place here described. Plate 75 was taken between *bwayma* I and XI. It shows storehouse VI to the left and VII to the right, with some smaller yam-houses of the outer ring in the background. The heaps in the foreground belong to *bwayma* I, which can just be seen on the extreme left. On the extreme right is the end of *bwayma* XI, with a man sitting on it. The heaps of *bwayma* VI had already been stored. Plate 76 was taken from the middle of the *baku*, and shows *bwayma* I, covering *bwayma* II, which is invisible on the plate. It also shows *bwayma* III, IV and V. Plate 77 is taken right in front of *bwayma* VII and shows heaps *a*, *b* and *c*.

The table given at the end of this Document, taken in conjunction with the plan and the explanatory pedigrees, illustrates the following rules: only owners[2] of the village or inmarried (usually by cross-cousin marriage, cf. Ch. XII, Sec. 2) men of rank own yam-houses in the inner ring; each *bwayma* has at harvest-time a cluster of heaps (*gugula*) around

[1] I cannot say for certain whether this marriage is one of the half a dozen unions in which the rule of exogamy is broken, or whether I made a mistake in noting that the man belongs to the Lukwasisiga clan.

[2] In Ch. XII, Secs. 1-3, you will find that there are several classes of owners, and one or two classes of inmarried men. In the component village of Yalumugwa here described, there are two sub-clans with rights of citizenship, i.e. "owners" or "masters", of which one has got the exclusive right of ownership, and there is one inmarried man.

it; each heap is on a definite spot which is assigned to one man—the donor or filler of the *bwayma*—who year after year puts the crops from his *baleko* into a heap on this spot and presently stores them into his allotted compartment.

The present headman of the village is Yovisi, who has taken over the office of headmanship from Topinata'u, his *kadala* (maternal uncle), now too old and weak to carry on. Before Topinata'u, Gumabudi, now dead, was the headman and *towosi* of the village.

The following list gives the eleven *bwayma* with the name of the contributor of each heap and the reasons why he contributes.

Bwayma I and *II* belong to the present *tokaraywaga* (headman) Yovisi (Ped. 1).[1] He has two wives and receives his nine heaps from his relatives-in-law and his son. The two pedigrees, 1 and 2, will explain the sociology of these duties. The letters against the names refer to the plan.

III belongs to Yabuna (Ped. 1), who gets two heaps each year, *a* being contributed by Bewona, brother of Tuginitu, Yabuna's wife; *b* given by Tokavataria, Yabuna's younger brother.

IV belongs to Gubayladeda (Ped. 1), a commoner of the Malasi clan, who has the privilege of being Yovisi's brother-in-law (cf. Ped. 1). He receives four heaps, *a* from Samugwa, his younger brother, *b* from Yovisi, *c* from Gapulupolu, *d* from Tobiyumi, his son. Thus two of the heaps are given by his wife's (*Aykare'i's*) brothers, one by his son (who is also counted as a relative-in-law) and one by his own brother.

V is the *bwayma* of Gapulupolu (Ped. 1), heap *a* being given him by Megalabwalita, his wife's brother; heap *b* by Mukava'u, his wife's and his own son.

VI is the *bwayma* of Topinata'u (cf. Ped. 1). Heap *a* is given him by Lubagewo, his wife's sister's son, heap *c* by Kalumwaywo, his wife's *kadala* (maternal uncle or nephew), and the main heap *b* by Yovisi, his own *kadala* (maternal nephew).[2]

VII is the *bwayma* of Mwaydola (cf. Ped. 3), the *gumguya'u* of the Lukwasisiga clan who does not belong to Yalumugwa, but settled and was accepted in the village. He receives three heaps, *a* from his son Deliviyaka, *b* and *c* from his wife's brothers Monori and Gumlu'ebila.

VIII is the *bwayma* of Inkuwa'u (cf. Ped. 3). Heap *a* is given him by Kaduguya, his wife's brother, heap *b* by his son Tokavataria, heap *c* by Moraywaya, his *kadala* (maternal nephew).

IX is the *bwayma* of Monori (cf. Ped. 3). He receives heap *a* from Tukwalapi, his son, and heaps *b* and *c* from two kinsmen (*veyola*), of whose exact relationship I am not certain. As his wife is not alive, he does not get anything from his relatives-in-law.

X is the *bwayma* of Tobuguya'u (cf. Ped. 1). He receives heap *a* from his wife's brother, the same Lubagewo who gives taytu to Topinata'u, and

[1] For pedigrees here referred to, see end of this Document.

[2] The kinship term *kadala* is reciprocal and denotes maternal uncle and nephew, who call each other by the same term.

who is a relative-in-law of both of them. Heap *b* is given by Yuvata'u, his wife's brother's son. Yuvata'u has replaced his father on the latter's death. Thus this belongs to what might be called a spurious type of *urigubu* (cf. Doc. II).

XI is the *bwayma* of Togabutuma, elder brother of Tobuguya'u (cf. Ped. 1). He is a widower not yet remarried, and apparently very unpopular, as he is the putative pretendent to headmanship in the village. He receives only one heap from his younger brother, Tobuguya'u.

Glancing at the Table of Recipients and Donors, we can see that there are thirty-two heaps to the eleven storehouses. Although these heaps are in theory provided every year in the same quantity, in certain years one or two of the regular donors may fall out through illness or absence or, in a very bad year, through inability to provide a sufficient amount to make a full *gugula* (heap). Out of these thirty-two there is only one heap, Xb (see Fig. 11), which is definitely irregular, in that it is given not by a *veyola* (blood relative) of the wife, but by her brother's son. This is one of the encroachments of the patrilineal principle upon the system of maternal descent. Out of the remaining thirty-one, six contributions are made by younger kinsmen to elder kinsmen; they belong to the *kovisi* type of gift. The remaining twenty-five, an overwhelming majority, are of the orthodox *urigubu* type of gift, given by the wife's kinsmen to the husband. It must be remembered that in the Trobriands the son is the wife's kinsman and that his contribution falls within the strict definition of *urigubu*.

TABLE OF RECIPIENTS AND DONORS

**I. Yovisi.* (†*Bwodamayla's bwayma.*)

†a Kabitala (W. Br., Ur.).
†b Samugwa (W. Br., Ur.).
†c Mokalagaluma (W. Br., Ur.).
*d Gubayladeda (W. Br., Ur.).
†e Toluguya (W. Br., Ur.).
†f Kadakayauli (W. Br., Ur.).

**II. Yovisi.* (†*Nu'erebe's bwayma.*)

†g Togaluma (W. Br., Ur.).
†h Gumilabu'a (W. Br., Ur.).
*i Nugulava'u (son, Ur.).

**III. Yabuna.*

*a Bewona (W. Br., Ur.).
*b Tokavataria (Y. Br., Ko.).

**IV. Gubayladeda* (Malasi).

†a Samugwa (Y. Br., Ko.).
*b Yovisi (W. Br., Ur.).
*c Gapulupolu (W. Br., Ur.).
*d Tobiyumi (son, Ur.).

**V. Gapulupolu.*

*a Megalabwalita (W. Br., Ur.).
*b Mukawa'u (son, Ur.).

**VI. Topinata'u.*

*a Lubagewo (W. S. son, Ur.).
*b Yovisi (own *kadala*—S. son, Ko.).
*c Kalumwaywo (W. S. son, Ur.).

‡VII. Mwaydola. (Ls. *gumguya'u* not Kwoynama.)

‡a Deliviyaka (son, Ur.).
‡b Monori (W. Br., Ur.).
†c Gumlu'ebila (W. Br., Ur.).

‡VIII. Inkuwa'u (Malasi).

‡a Kaduguya (W. Br., Ur.).
‡b Tokavataria (son, Ur.).
‡c Moraywaya (S. son, Ko.).

‡IX. Monori (Malasi).

‡a Tukwalapi (son, Ur.).
b and c (*veyola*, Ko.).

**X. Tobuguya'u.*

*a Lubagewo (W. Br., Ur.).
*b Yuvata'u (W. Br. son, spur.).

**IX. Togabutuma.*

*a Tobuguya'u (Y. Br., Ko.).

The *bwayma* I, II, III, V, VI, X and XI belong to members of the Kwoynama sub-clan of the Lukwasisiga clan. The clanship of other owners is indicated after each name.

* = on main Pedigree, 1.
† = on Pedigree 2.
‡ = on Pedigree 3.
Ur. = Urigubu.
Ko. = Kovisi.
Spur. = spurious Urigubu.
W. = wife's.
S. = sister's.
Y. Br. = younger brother.
Ls. = Lukwasisiga.

Roman numerals = *bwayma*; letters = *gugula* (heap). (See Fig. 11.)

PEDIGREE I

Italic numerals = *bwayma* owned.
Italic type = headman of village.

Roman numerals and letters = *bwayma* and heap given.
M.t. = Malasi *tokay*.

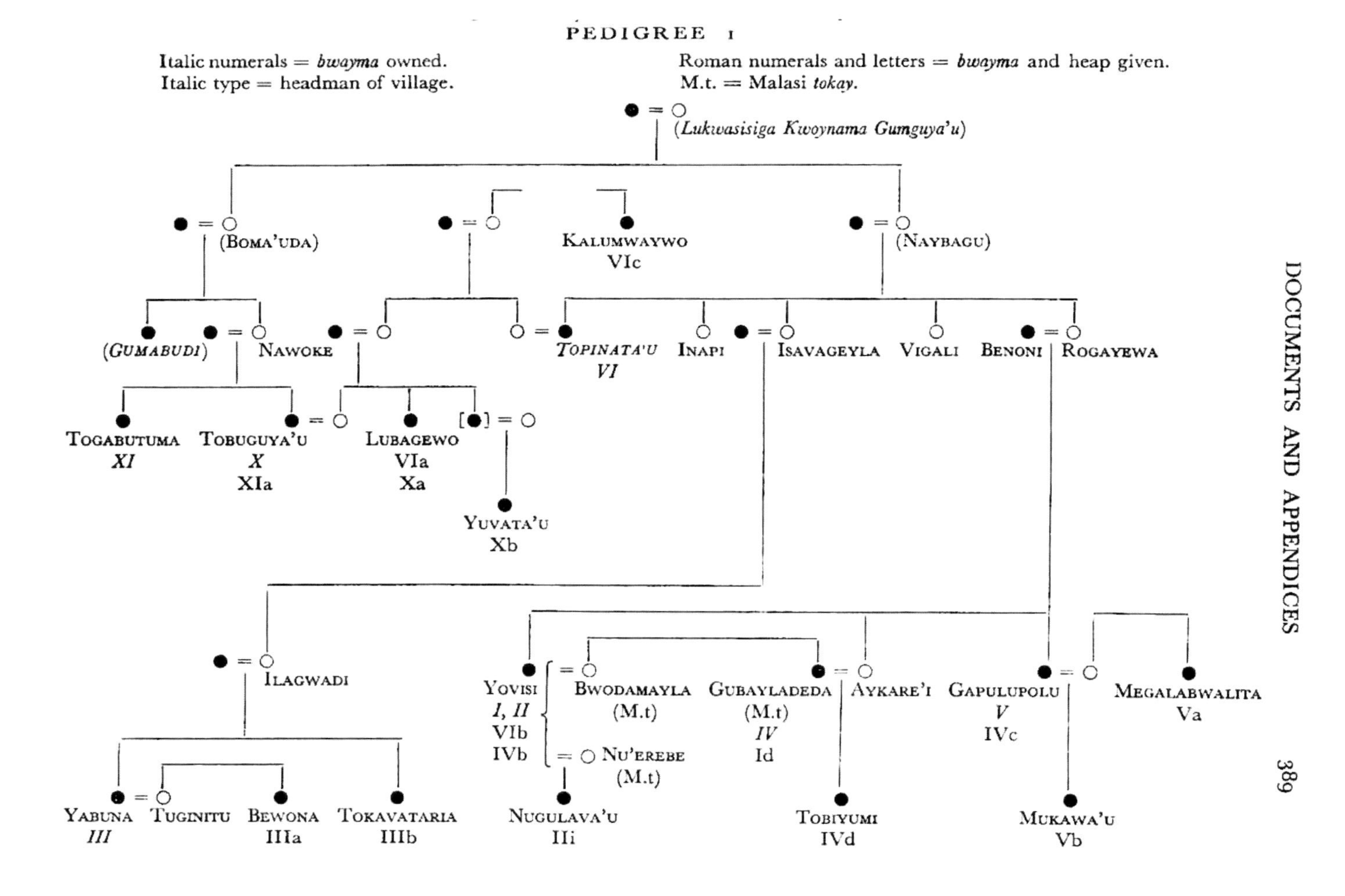

PEDIGREE 2

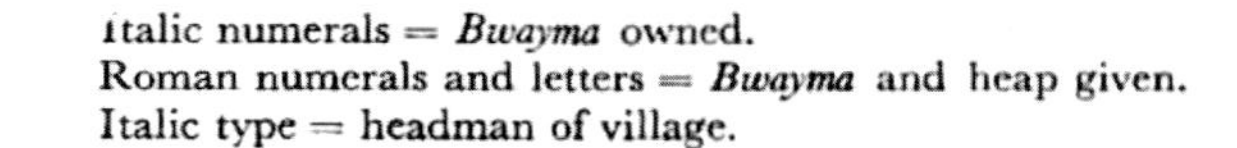
Italic numerals = *Bwayma* owned.
Roman numerals and letters = *Bwayma* and heap given.
Italic type = headman of village.

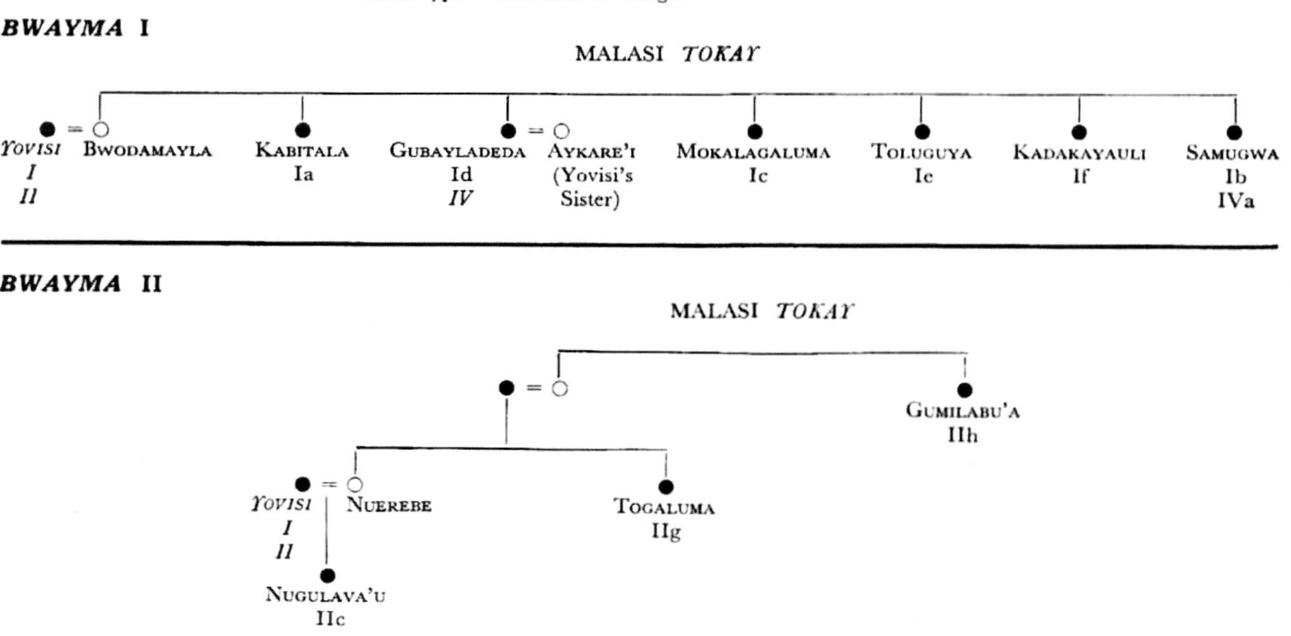

PEDIGREE 3

Italic numerals = *Bwayma* owned.
Roman numerals and letters = *Bwayma* and heap given.

BWAYMA VII, VIII & IX

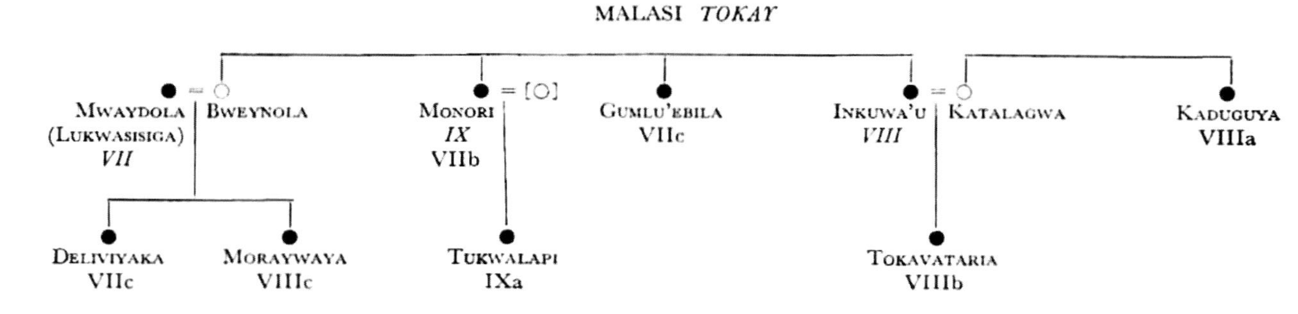

DOCUMENT II

COMPUTATION OF THE HARVEST GIFT PRESENTED AFTER THE *KAYASA* IN OMARAKANA IN 1918

(Chapter VI, Sections 1 and 3)

THIS document consists of a plan showing the distribution of harvest heaps given to To'uluwa in 1918, a table with the details concerning each heap (*gugula*), and the analysis of the table.

A big competitive harvest was organised in that year. The result was that the donors, for reasons specified in Chapter VI (Sec. 3), especially those resident in the villages of Kwaybwaga, Liluta and M'tawa, strained their resources to the utmost. The total amount of the gift that year was just over 20,000 baskets, which is about four times as much as the ordinary quota. From direct statements of the natives it seems that this yield was about the same as the average gift in olden days. This was confirmed by a number of indirect indications. For instance, several old men told me, when I was making a record and computation of an ordinary harvest gift in 1915, that only one-quarter or one-fifth was put in each yam-house of what used to be given. Discounting retrospective optimism, I think we shall not be wide of the mark if we assess 20,000 to 25,000 baskets as the average amount of the chief's *urigubu* in olden days. We have only to remember that whereas at present (1918) the chief has twelve wives, in olden days he had from sixty to fourscore.

The approximate position of each heap with reference to the yam-house can be seen on Fig. 12. In the first column of the table we have the name of the giver; in the second his village community; in the third his sociological description; in the fourth his relationship to one of the chief's wives in virtue of which the harvest gift is given; in the fifth the number of baskets offered; and in the sixth a brief indication as to the sociological character of the gift. In connexion with the last column it is necessary to introduce the following categories:—

A. is *genuine urigubu*, that is a gift offered by real, i.e. matrilineal, kinsmen of the wife.

B. is a *pretence urigubu*, that is a gift offered by one who though not a kinsman is made to figure as a kinsman.

C. is an *agnatic urigubu* offered by the wife's patrilineal (agnatic) relatives.

D. is *affinal urigubu* offered by a man who is married to the chief's wife's sister.

E. is *spurious urigubu* offered in virtue of a purely artificial relationship, not even pretending to establish either a bond of kinship or any other.

F. is *tributary urigubu* levied as a tribute on a local resident or vassal.

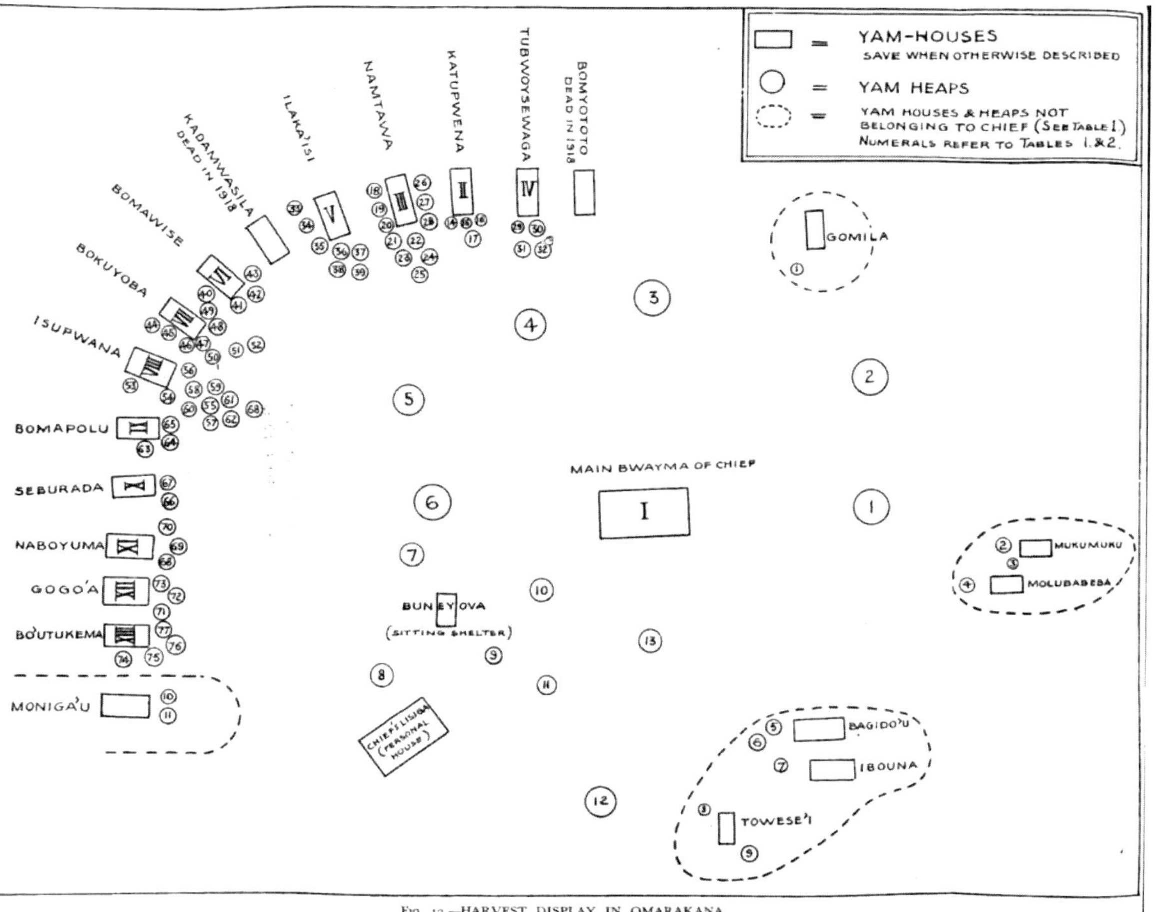

FIG. 12.—HARVEST DISPLAY IN OMARAKANA

G. is a gift from the chief's own kinsman to one of the chief's wives. Whether this would be called by the natives *kovisi* or simply *dodige bwayma*, I am not quite certain.

The first thirteen entries in our table refer to the givers who filled the main *bwayma*. Roughly speaking, the thirteen entries each correspond to one of the chief's wives. In olden days when the chief had as many as fourscore wives, there was a compartment for each of them in the central storehouse. This latter is said to have been much bigger then. The principal matrilineal relatives of each wife filled one compartment. Nowadays when the chief's matrimonial list has shrunk to only twelve wives, even this rule has become modified. Thus for one wife (Gogo'a) there is no contribution listed among the thirteen first entries on our table. One recently dead wife (Bomyototo) has still one entry, though the actual taytu is stored in another wife's compartment, that of Isupwana. And again, Bokuyoba, who might be described as the *doyen* of the chief's harem, has two contributors, who put their crops into her compartment.

These thirteen contributors are the remnant of what was the most important body of notables in the whole area of the Northern Massim. In those days the chief had one wife from each of the tributary communities, rising, as already said, perhaps to fourscore or so. But as some communities died out, others were merged, others ceased to provide wives, and finally the joint influence of Christianity and Government made the last two or three chiefs afraid to marry new wives, the number shrank gradually to forty (Bugwabwaga), thirty (at the beginning of Numakala's reign), twenty-four (the number with which To'uluwa started), and twelve in 1918.[1]

In old days, moreover, not only was each community represented at one of the chief's matrimonial establishments, but her most important cognatic kinsman provided each year a large heap for the central storehouse. Even now, among the thirteen contributors, there are representatives of most of the important villages or sections of villages surrounding Omarakana.

Among the thirteen givers, there are eight unquestionable, straightforward *urigubu* gifts of the type we defined as A. There are two cases (Nos. 4 and 13) of pretence *urigubu* in which the matrilineal kinship is assumed to exist, but is merely the relationship defined by the term, *kakaveyola*, i.e. identity of clan, not even of sub-clan. No. 7 is a spurious type of *urigubu*, since it is based on adoption not even within the same clan. No. 10 is an agnatic gift, for it is the "father" who gives. No. 2 might

[1] My reconstructive attempts at establishing the real average number of wives in the past, as well as the maximum, lead me to the following results: in pre-European days, that is, some hundred years ago, the average number was probably sixty; the maximum might have reached eighty. The immediate predecessors of To'uluwa had probably as many as thirty wives and not more than forty. The contact with Europeans soon led to a reduction of numbers, mainly through epidemics in most of the communities.

be called a straightforward *urigubu* if we admit that the gift is offered to the shades of the recently dead Bomyototo and only stored in another wife's compartment; otherwise it is again a case of pretence *urigubu*, B, since the giver stands in no relationship except sameness of clan.

Passing now to the other storehouses, each of which is allotted to one special wife. I am giving them in the—I fear—somewhat arbitrary order in which I had listed them in the field and which I found it inconvenient to change.

Katupwena. Here we have one case of genuine *urigubu* and three cases of agnatic, two of which, 14 and 15, are moreover uncertain cases, since the givers are not real fathers but merely clansmen, *kakaveyola*, of the father.

Namtawa. Here, out of the eleven donors, we have five genuine *urigubu* (18, 19, 26, 27, 28); two cases of pretence *urigubu*, given by clansmen of the wife (20 and 21); one affinal (25), one agnatic (22), and two spurious *urigubu* (23 and 24), given by sons of the dead wife as an expression of their personal devotion to Namtawa and the friendship between the two women.

Tubwoysewaga. Here we have only one real *urigubu* (29), and three agnatic; but even these are spuriously agnatic, since the donors are merely clansmen of the father. In such a gift as this we find exemplified the working of other forces besides those of the real *urigubu* duty. The village of Kaybola represented by Tubwoysewaga has to contribute to the chief's wealth and any spurious or indirect relationship to the chief's wife is taken as a pretext as long as her *bwayma* is filled.

Ilaka'isi. Out of the seven givers, four present genuine *urigubu* (33, 34, 38, 39). One gives a tribute in virtue of his residence (35). Two (36 and 37) are the chief's sons by another wife and present what can be regarded either as pretence *urigubu*—they are clansmen (*kakaveyola*) of Ilaka'isi—or, in so far as they are the sons of the late Kwadamwasila and their gifts are offered to the shades of the deceased mother and only disposed of by Ilaka'isi, as genuine *urigubu*.

Bomawise. Of her four contributors, two are genuine *urigubu* (40 and 42); one is "father", that is the father's sister's son (41); and one gives an entirely spurious *urigubu* in which the relationship is very complicated and based on adoption (43. Cf. diagrammatic pedigree.)

Bokuyoba receives nine harvest presents of which only two are genuine *urigubu* (44 and 50). Two are agnatic, neither of which, however, is a genuine gift from father: 46 is from the paternal half-brother which to the native is no relationship at all, and 47 is from a very fictitious agnatic relative. Both gifts could, therefore, also be classed as spurious *urigubu*. One gift (48) is received as tribute from a resident in Omarakana, and three are spurious *urigubu*; based in one case (45) on personal sympathy between the chief's favourite son, Namwana Guya'u, and Bokuyoba. In the second case (49) a gift offered to the shades of Ilabova is placed in Bokuyoba's *bwayma* as she is the giver's adopted mother. In the third case (52) an even more involved relationship exists, as the donor is the younger brother of an adopted son of Bokuyoba—this adopted son being the maternal nephew of another of To'uluwa's wives who died. Lastly there is an instance of *kovisi* (51), a gift by the husband's own kinsman, at times called *kovisi*, at times simply styled *dodige bwayma*.

Isupwana has ten contributors of which three (56, 60 and 61) are genuine *urigubu* givers. There are one agnatic (62) and two affinal gifts (57 and 58); the latter not even a genuine affinal relationship, as the donor is married only to the half-sister of Isupwana on the paternal side. Of the four spurious *urigubu*, two (53 and 54) are not of the same sub-clan or even clan, so the relationship is a distant one on the paternal side. The next (55) gives on the pretext that in ancient times some relationship existed between the giver and Isupwana's father, and the last (59) because the villages of the donor and of her father are contiguous and have the same local sub-clan.

Bomapolu's three gifts are offered by two genuine *urigubu* donors (63 and 65) and one agnatic (64).

Seburada has only two fillers of her *bwayma*, one genuine *urigubu*, not given this year, and one agnatic.

Naboyuma receives two real *urigubu* presents (68 and 69) and one affinal (70) which is really spurious, as the giver did not marry her real sister but only a woman of her sub-clan.

Gogo'a has three contributions, two genuine *urigubu*, and one (71) which might be called agnatic, as it is given by a man styled her father, though really the relationship is much more remote, or it might be called *kovisi*, as the donor is To'uluwa's kinsman.

Bo'utukema's four contributors consist of two genuine *urigubu* givers (74 and 77), one agnatic (76), and one spurious agnatic (75), as the giver merely married a kinswoman of Bo'utukema.

Computing the number of baskets given to the chief from each of the several sociological categories into which we divided the offerings, we arrive at the following table:—

A Genuine urigubu		35	donors	9,444	baskets
B Pretence	,,	4	,,	1,003	,,
A or B	,,	3	,,	2,000	,,
C Agnatic	,,	8	,,	1,430	,,
D Affinal	,,	2	,,	210	,,
E Spurious	,,	11	,,	4,114	,,
C or E		8	,,	1,090	,,
D or E		2	,,	251	,,
F Tributary donation		2	,,	272	,,
G Kovisi gift		1	,,	50	,,
C or G		1	,,	150	,,
		77	,,	20,014	,,

In looking at this little computation, we see at once that, with the exception of an almost negligible proportion, all the gifts belong to the *urigubu* class. For whether genuine or spurious, agnatic or affinal, real or pretended, the bulk of the offerings in categories A to E consists of *urigubu*, which means gifts to the chief from people who either are or claim to be related to his wives. In the three last entries which are not *urigubu*, and one of these is doubtful, we have only 472 baskets, barely a fortieth part of the whole yield, or 2½ per cent. This shows that whatever the real motives and social forces which compel the vassal to give an offering to the chief, the proper channel for such a gift would be the *urigubu*, a harvest gift based on relationship by marriage.

The hold which the principle of *urigubu* has over the native is proved by the fact that in old days the chief used to have, in continuous succession, a wife from every tributary community. It was not proper to give tribute except as a periodical endowment of one's kinswoman. The fact that now, when owing to extraneous and artificial conditions the chief does not dare to marry so many women, the tribute is paid to him at all is a remarkable sign of loyalty, devotion and fidelity to tradition among the Trobrianders.

Even more remarkable is it that they should feel it necessary to invent all kinds of fictitious and roundabout ways of justifying the tribute by making it appear like *urigubu*. Whether agnatic and affinal *urigubu* were practised to any large extent in olden days it was impossible for me to ascertain. Probably a father would even then give presents to the chief on account of his daughter, since it was an advantage to be related to the chief by marriage, and since the attachment of a man to his daughter in the Trobriands is considerable. But it certainly was not practised to such a great extent as now, since almost everybody had a matrilineal kinswoman in the chief's establishment.

The importance attached to the fact that all gifts should be given in the guise of *urigubu*—a fact which proves its extraordinary tenacity—is shown by the figure in E, which is the highest next to the *urigubu* itself. For spurious *urigubu* demonstrates that where there is no possibility of establishing any blood relationship this relationship is still assumed to exist.

The following data are not really relevant to the subject-matter of the present document. They have some documentary value as regards the substance of Chapter VI. They show all the heaps that were allowed to be displayed on the central place of Omarakana at the same time as the chief's *urigubu*. They also show the relative insignificance of the total gift to these eleven residents in the capital—the sum of the baskets received amounts to 1,003, that is less than 5 per cent of the chief's tribute—and that the gifts with one exception are *urigubu*.

TABLE 1

URIGUBU GIVEN TO OTHER RESIDENTS IN OMARAKANA

1. Gomila, given by his wife's brother, 158. (Urigubu.)
2. Mukumuku, given by Tokulupa'i, 100. (Urigubu.)
3. Molubabeba, given by Tokulubakiki (his son), 50. (Urigubu.)
4. Molubabeba, given by Mwataniku, 60. (Urigubu.)
5. Bagido'u, given by Mitakata, 52. (Kovisi.)
6. Bagido'u, given by Topeulo, 150 or 170. (Urigubu.)
7. Bagido'u, given by Towese'i, Ibouna's brother, 53. (Urigubu.)
8. Towese'i, given by Yatalisi, 150. (Urigubu.)
9. Towese'i, given by Monobogwo, 40. (Urigubu.)
10. Moniga'u, given by Muduwa'u, 120. (Urigubu.)
11. Moniga'u, given by Moyadeda, 70. (Urigubu.)

TABLE 2

TO'ULUWA'S HARVEST GIFT AT THE *KAYASA* HARVEST IN OMARAKANA IN 1918

List of Abbreviations and Native Terms:

g. = *guya'u*—Chief of high rank
gg. = *gumguya'u*—Chief of lower rank
Lb. = Lukuba clan
Ls. = Lukwasisiga clan
Lt. = Lukulabuta clan
M. = Malasi clan

t. = *tokay*—Commoner
Osisuna = Immediate environment of a village
Tokaraywaga = Headman
Toliwaga = Chief of inferior rank
Veyola = Kinsman

I Donor's Name	II Village Community	III Sociological Description of Donor	IV Relationship to Chief's Wife	V Number of Baskets	VI Socio-logical Definition of Gift
		I. MAIN BWAYMA			
1. Kwoyavila	Suvayalu, suburb of Liluta	Karaywaga of Suvayalu, Ls. Kwoynama	Brother of Isupwana	1,514	A
2. Simdarise Wawa	Mtawa	Tokaraywaga, Ls. t. dala: Metilawaga	Son of Bomyototo's elder sister. Contribution allotted to Isupwana	1,900	A or B*
3. Kumatala	Yaluwala, suburb of Kwaybwaga	Tokaraywaga, towosi. Ls. t. Came out osisuna Kwaybwaga	Mother's brother of Ilaka'isi	1,454	A
4. Mwagwaya	Oylova'u, suburb of Dayagila	Tokaraywaga of Oylova'u. Lb. t.	Kakaveyola of Namtawa	443	B
5. Walasi	Kaybola	Tokaraywaga of his suburb. Ls. t. Kwoynama	Brother of Tubwoysewaga	500	A
6. Tokwamnapolu	Kuluvitu (lives in Obulabola, suburb of Kwaybwaga)	Becomes karaywaga of Obulabola on his father's death. Ls. t.	Brother of Bomawise	1,120	A

7. Tokunasa'i Guya'u	Kaytagava, main suburb of Kwaybwaga	Tokaraywaga or Kwaybwaga. Ls. gg. Burayama	Bokuyoba's "adopted" son. It is not a case of real adoption: Bokuyoba looked after him on the death of his real mother, Ilabova	802	E
8. Tourapata	Wagaluma	Tokaraywaga (subsidiary one). Ls. t. (half toliwaga). Dala: Udokakapatu, came out in Sakapu with Burayama	Son of Bomapolu's elder sister	200	A
9. To'uma	Okayboma, real village Yalumugwa	Karaywaga part of Okayboma. Ls. t. came out near Yalumugwa	Brother of Naboyuma	100	A
10. Mkwaysipu	Suburb of Mtawa	Karaywaga of his suburb. Lt. t. Came out Laba'i	"Father" of Seburada. Actually her real father's maternal nephew	100	C
11. Monumadoga	Tilakayva	Karaywaga of Tilakayva, Lb. t. Came out osisuna Tilakayva	Brother of Bo'utukema	120	A
12. Kani'u	Liluta	Karaywaga, Lb. Mwauri	Brother of Katupwena	1,054	A
13. Pilu'u'ula	Dayagila	Karaywaga, Lb. t.	Clansman of Bokuyoba	350	B
		II. KATUPWENA'S BWAYMA			
14. Tagona	Kudokabilia	Tokaraywaga of a suburb. M. t. Came out locally	Clansman of Katupwena's father	350	C or E
15. Meligata	Kudokabilia	Karaywaga of suburb of Kudokabilia. M. t. Veyola of No. 14	Clansman of Katupwena's father	100	C or E
16. Menanagwa	Liluta	Mwauri, Lb. gg.	"Brother" of Katupwena. Mother's sister's son	100	A
17. Gumakawa'i	Lives in Liluta	M. t.	Own father	120	C

* No. 2 is A if counted to Bomyototo; B if counted to Isupwana.

TABLE 2—*Continued*

I Donor's Name	II Village Community	III Sociological Description of Donor	IV Relationship to Chief's Wife	V Number of Baskets	VI Socio-logical Definition of Gift
		III. NAMTAWA'S BWAYMA			
18. Tobwa'uli	Kapwani	Karaywaga, suburb of Kapwani. Ls. t. Kapwani	Mother's brother	130	A
19. Mwanabugwa	Kapwani	Karaywaga, suburb of Kapwani. Ls. t. Veyola of No. 18	Son of her elder sister	140	A
20. Togo'ula Liku	Idaleyaka	Tokaraywaga of village and towosi. Ls. t.	Clansman	100	B
21. Mokalayma	Idaleyaka	Ls. t. Veyola of No. 20	Clansman	110	B
22. Wagabwalita	Yuwada (lives near Idaliyaka)	M. t.	"Father", real father's kinsman	100	C
23. Tomwako'u	Mtawa	Tokaraywaga of suburb. Ls. t. Metilawaga	Son of Bomyototo and gives vicariously to Namtawa	250	E
24. Setukwa	Mtawa	Ls. t. Metilawaga	Brother of No. 23	80	E
25. Toriyova	Lives in Omarakana, real village Kaybola	Lives in Omarakana because married to Bodoyuwa, younger sister of Namtawa	He fills Namtawa's bwayma on that account	60	D
26. Mwaydayli	Omarakana	Ls. t. Mwane'i	Son	100	A
27. Moludobu	Okayboma	Ls. t. Mwane'i	"Brother", member of same sub-clan and generation	200	A
28. Nabwasuwa	Omarakana	Ls. t. Mwane'i	Son	50	A

IV. TUBWOYSEWAGA'S BWAYMA					
29. Giyo'um	Kaybola	Ls. t. Kwoynama	Sister's son	100	A
30. Woynama	Kaybola	M. t. Came out somewhere in Kaybola	"Father", father's clansman	200	C or E
31. Ginukuwa'u	Kaybola	M. t. somewhere in Kaybola, but not same dala as No. 30	"Father", father's clansman	100	C or E
32. Mokilavala	Lu'ebila	M. t. Came out Laba'i	"Father", father's clansman	100	C or E
V. ILAKA'ISI'S BWAYMA					
33. Giyotala	Lives in Kaybola, real village Yaluwala	Ls. t. Came out osisuna Kwaybwaga. (Same as No. 3.)	Her maternal uncle	100	A
34. Towosawa	Yaluwala	Elder brother of No. 33	Her maternal uncle	50	A
35. Moniga'u	Omarakana	Fills her bwayma because he lives in Omarakana. Ls.	Clansman	100	F
36. Yobukwa'u	Omarakana, real village Liluta	Ls. gg. Liluta	Sons of Kwadamwasila, who is dead	50	A or B
37. Kalogusa	—	Brother of No. 36		50	A or B
38. Monitaba'i	Lives in Kammamwala, real village Yaluwala	Ls. t.	Brother	100	A
39. Bwaylusa	Lives in Omarakana, real village Yaluwala	Ls. t.	"Grandfather", brother of her maternal grandmother	50	A
VI. BOMAWISE'S BWAYMA					
40. Bayoli	Kuluvitu	Ls. t. Younger brother of No. 6	Own brother	490	A
41. Manimuwa	Obulabola	M. t. Came out in Obulabola	"Father", real father's maternal nephew	350	C
42. Bulubwaysiga	Lives in Obulabola, real village Kuluvitu	Ls. t.	Mother's brother	100	A

TABLE 2—*Continued*

I Donor's Name	II Village Community	III Sociological Description of Donor	IV Relationship to Chief's Wife	V Number of Baskets	VI Sociological Definition of Gift
		VI. BOMAWISE'S BWAYMA—*Continued*			
43. Tawa'i	Lives Kaytagava, real village Madoya, part of Wakayse	Lb. t. Came out near Mwadoya	The donor adopted Bomawise's brother's son, to whom she stands in relation of *tabula*—joking relation	150	E
		VII. BOKUYOBA'S BWAYMA			
44. Mokaypwes	Kammamwala	Tokaraywaga of Kammamwala, towosi and principal sorcerer of Kiriwina. Lb. Tudava	"Brother", common maternal grandmother	380	A
45. Namwana Guya'u	Osapola	Tokaraywaga Osapola. Ls. Kwoynama. Eldest son of recently deceased wife, Kadamwasila	No relation to Bokuyoba. Only in his childhood she was specially good to him	250	E
46. Yadiwoyga	Kwaybwaga	M. t. Came out near Kwaybwaga (Mkunela)	Half-brother, same father, hence not real kinsman	100	C or E
47. Kagidakwa	Obulabola, suburb Wakayse	M. t. dala of Mweydo'u	He is "brother" of Bokuyoba, as his father and hers belonged to same dala	100	C or E

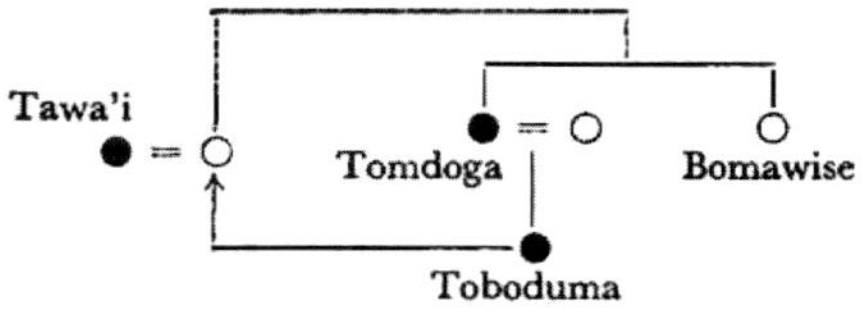

Tawa'i — Tomdoga — Bomawise — Toboduma

48. Kalumwaywo	Vilomugwa (resident vassal in Omarakana)	Ls. Moyovisi; hole of emergence Mukunela	Kalumwaywo's father used to fill the same storehouse before him. Gives urigubu as patrilineal hereditary duty	172	F
49. Tonuwabu	Kwaybwaga (same as No. 7)	—	Real mother Ilabova. Adopted son (cf. No. 7)	252	E
50. Buko'u	Lives in Yalumugwa	Lb. Tudava	Her "brother" in so far as belongs to same sub-clan	40	A
51. Molubabeba	Omarakana	Malasi, Tabalu	Kinsman of To'uluwa	50	G
52. Giyolikuliku	Okayboda	Ls. toliwaga; came out Bwaydaga	Younger brother of Yabugibogi, who was adopted by Bokuyoba. Hence styled her "son"	100	E
		VIII. ISUPWANA'S BWAYMA			
53. Mwabuwa	Yuwada	M. t. local sub-clan of Yuwada	Styled her "son" but not same sub-clan or even clan; relationship on paternal side	250	E
54. Muramata	Mtawa	Ls. t. Metilawaga	Maternal nephew of 2. Hence gift really given on account of deceased Bomyototo	1,380	E
55. Tonuviyaka	Mtawa	M. t. Mtawa	Entirely fictitious. "Father." Because in olden days ancestors had founded this relationship	250	E
56. Tokwoyoulo	Osapola	Ls. Kwoynama.	"Brother." They had common maternal grandmother	450	A
57. Buyavila Kiriwila	Yogwabu (resident in Liluta)	M. t. sub-clan Kaluva'u; hole Bulimaulo	"Elder brother" because married to Isupwana's elder sister	150	D
58. Toboyowa	Wakayluwa	M. t. Wakayluwa	"Younger brother" because married to half-sister of Isupwana	201	D or E

TABLE 2—*Continued*

I Donor's Name	II Village Community	III Sociological Description of Donor	IV Relationship to Chief's Wife	V Number of Baskets	VI Socio-logical Definition of Gift
		VIII. ISUPWANA'S BWAYMA—*Continued*			
59. Kabaykula	Kaulagu	M. t. Kaulagu	"Son" of Isupwana on entirely fictitious grounds, because his village and her father's are contiguous and have same local sub-clan	350	E
60. Gomila	Omarakana	Kwoynama. Ls.	"Brother", they had common maternal grandmother	180	A
61. Tomyova'u	Okayboda	Kwoynama. Ls.	"Brother", very distant but real kinsman	160	A
62. Giyokaytapa	Yourawotu	M. t. Yourawotu	"Father", because maternal nephew of real father	120	C
		IX. BOMAPOLU'S BWAYMA			
63. Inosi	Omarakana	Ls. t. Sakapu	Son	120	A
64. Bwaysa'i	Obowada	M. t. Obowada	"Father", maternal nephew of real father	240	C
65. Kalubaku	Wagaluma	Karaywaga of Wagaluma. Ls. t. Bwala: Sakapu	Real brother (old man)	50	A
		X. SEBURADA'S BWAYMA			
66. Mosilapela	Mtawa	Lt. Came out Mtawa	"Father", younger brother of real father	380	C
67. Kayvala	Lobua	Lb. gumguya'u	Mother's brother	He did not give this year as wife died	A

		XI. NABOYUMA'S BWAYMA			
68. Dido'i	Yalumugwa	Ls. t. Yalumugwa	"Brother", same maternal grandmother	50	
69. Kadinaka	Kaytuvi	Ls. t. same dala as No. 68	"Brother", member of same sub-clan and generation	10	A
70. Kwaluma	Kwaybwaga	M. t. Kwaybwaga	Styled "elder brother" because married "elder sister", woman of same sub-clan, ungenealogically related	50	D or E
		XII. GOGO'A'S BWAYMA			
71. Mukumuku	Omarakana	M. Tabalu	Styled her "father" because she was the daughter of Taurisi Guya'u, who had same grandmother as Mukumuku	150	C or G
72. Mokaywori	Vilaylima	Ls. gg. Kwoynama	Mother's brother	60 (or less)	A
73. Kwayva'u Guya'u	Vilaylima	Ls. gg. Kwoynama	"Son", i.e. son of elder sister	50	A
		XIII. BO'UTUKEMA'S BWAYMA			
74. Toviyama'i	Tilakayva	Lb. t.	"Brother", they had same maternal grandmother	110	A
75. Towolila	Vakayluva	M. t. local sub-clan of Vakayluva	"Father", because he married a woman standing in relation of "mother"	40	C or E
76. Monobogwo	Vakayluva	M. t. Vakayluva	"Father", kinsman of Bo'utukema's father	20	C
77. To'ulobu	Tilakayva	Lb. t.	Mother's brother	12	A

DOCUMENT III

THE DECAY OF THE CHIEF'S HARVEST GIFT

(Chapter VI, Section 1)

It is impossible, unfortunately, for me to give as full a description of the ordinary harvest gift received nowadays by the chief, since in 1915, the only occasion on which I witnessed it, I had not yet come to understand the sociological groundwork of the harvest contribution. I was able subsequently to correct my mistakes to a certain extent, and I think I have arrived at a fairly accurate reconstruction of the ordinary harvest contribution, in one of its aspects at least, as it is now received by To'uluwa. But nothing can replace a reliable document drawn up while a social act is in progress.

I shall, however, make use of my initial mistakes to illustrate some of the difficulties in field-work, to show how from a welter of dissociated facts—scientifically worthless—the Ethnographer gradually arrives at the principles which underlie these and introduces meaning and order into his observations (cf. App. II).

When I first came to Omarakana in June, 1915, I not unnaturally started with the assumption that the harvest produce given to the chief was a "tribute" from his subjects. I also received the impression that this tribute was given as from village communities and not from individuals. Having been led to believe that the chief "owns" the large yam-house in the central place, while surrounding storehouses "belong" to his wives, I tried to compute a list of those communities which were tributary to the chief in filling his central *bwayma* (yam-house). In this enquiry, however, I was puzzled by the fact that the natives in speaking of a community always gave me the name of one individual or two, and invariably stated the relationship of these to one of the chief's wives. This is a quotation from my field-notes:—

"To'uluwa has one big *bwayma* in the middle of the *baku*. The following fill it with *kaulo*:—

Village	*Men*
Kaybola	Giyotala, *kadala* of Ilaka'isi.
	Mokilavala, father of Tubwoysewaga.
Dayagila	Pilu'u'ula, brother of Bokuyoba.
	Mwagwaya, *kadala* of Namtawa.
M'tawa	Simdarise Wawa, father of Bomyototo.
Liluta	Koyavila, brother of Isupwana.
Kwaybwaga	Tawa'i, brother of Bomawise.
Omarakana	Namwana Guya'u, To'uluwa's son.
Tilakayva	Toviyama'i, brother of Bo'utukema.

Village	*Men*
Wakayluwa	Monobogwo, father of Bo'utukema.
	Toboyowa, father of Isupwana.
Wagaluma	Kalubaku, brother of Bomapolu.
	To'urapata, son of Bomapolu.
Kaytuvi	Leydoga, brother of Naboyuma.

These represent the village communities and the village headmen who supply the chief with the yearly tribute of taytu."

It is clear in the light of what we know that the above list, on the one hand, gives an entirely wrong impression of the chief's yearly harvest income; and, on the other, is incorrect in details to the extent that only two entries (Kalubaku and Giyotala) are completely accurate as regards the sociological definition of the principle involved and the general character of the gift.

In most of the other entries the kinship definition is approximate, and though at that time I was perfectly well aware of the need for enquiring into exact genealogical relationship, I did not have sufficient data to establish it in each case. Barring that, the list is almost correct. Thus Mokilavala, the "father" of Tubwoysewaga, usually does contribute at harvest, but he does it indirectly by assisting her actual brother, Walasi, who is also his son. Pilu'u'ula of Dayagila is the main contributor on account of Bokuyoba, but he is not her brother, only a member of the same clan. Mwagwaya, who is here described as maternal uncle of Namtawa, is only her distant relative (see table below). Simdarise Wawa, who supplies Bomyototo regularly with food, is not her father as I say here, but her 'son', that is, the son of her elder sister. This mistake arises out of the difficulty a European finds with the possessive pronouns in kinship terms. I was told: "Bomyototo child his," which really meant: "he is the child of her, Bomyototo," but I understood this, as in many other cases, to mean: "Bomyototo is his child."

Kwoyavila does provide Isupwana, his real sister, with *kaulo*, but he is not the headman of Liluta, only of one of its suburbs, Suvayalu. On the other hand, Tawa'i is not Bomawise's principal contributor; he is in fact a very distant relative of hers, and not a kinsman by blood. Her real brother, Tokwamnapolu, is the main donor of her *urigubu*, but no doubt some near kinsman or friend of Tawa'i was among my informants, and in a tentative collecting of information before the natives really understood that I was after all the sociological details such a slip would have occurred easily.

Namwana Guya'u regularly brings a harvest offering to his father and we find him on the list of 1918; but he does not represent Omarakana, which he left in 1916, but Osapola, a suburb of Liluta. The titular donor of this *urigubu* is Piribomatu, kinsman of Namwana Guya'u and headman of Osapola.[1] Again, instead of Toviyama'i, I ought to have listed

[1] In 1918 Kadamwasila was dead, hence Piribomatu does not appear on the list of contributors to the main *bwayma*.

Monumadoga (No. 11 of Doc. II), the elder brother, who is also the real brother of Bo'utukema. Monobogwo (No. 76) is a distant agnatic relative of Bo'utukema and does not figure in a prominent place among the givers. Toboyowa (No. 58) is not the father of Isupwana but her affinal relative through marriage with her half-sister. He also plays an entirely subordinate rôle among the chief's contributors.

Kalubaku (No. 65), as I have said, is a real brother of Bomapolu, and in 1915 was among the chief contributors. In 1918 he had become so old that he was replaced by Tourapata (No. 8) as regards the main contribution. Leydoga does not figure on our big list of Document II at all. I was told that he sometimes acts in conjunction with Kadinaka (No. 69), who is a real relative.

So much for the details. As regards the general perspective, the above short list, even if corrected in detail, still gives a wrong impression. For it implies that each gift is given on behalf of his community as a tribute from headman to chief. Practically this is almost correct, for when a headman has to give an *urigubu* present he is often helped by other members of the community. But the legal principle of the whole transaction is distorted by such a presentation. The *urigubu* is the duty of one individual to another, in virtue of their relationship by marriage. *Urigubu* is not a tribute, and it amounts to a tribute only in that it is integrated over a whole district through the chief's polygamy and multiplied by the assistance given by many men to the most important notables in each community.

It was through the very mistakes I committed in constructing this document and through further enquiries that I discovered the real nature of the *urigubu*. I was made to understand that the enormous gifts received by the chief are marriage contributions and that they are of the same nature as those given by every man to his sister's husband for the maintenance of her household. One of the most important corrections to be made in the above list was the differentiation of each cluster of villages into its component parts, instead of treating each such community as an integral whole.

The real connecting link between the fact that the chief's revenue is a personal marriage gift and at the same time is in a way a tribute consists in the fact that the chief marries a girl from the ruling family in each hamlet, so that each headman is his contributing brother-in-law; and since the headman receives substantial assistance from most of his villages, the chief draws upon the economic resources of the whole territory (cf. Doc. II). The chief's revenue, as is explained elsewhere, is not limited to harvest gifts, but consists also of tribute in coconuts, betel-nut and pigs; also in certain levies on several branches of industry and on the economic exploitation of natural resources (cf. Part I, Sec. 10).

Another mistake which I made in computing my original table was that I missed three wives completely: Gogo'a, Seburada and Katupwena. The first had no contributor filling the main storehouse, the reason being that she came from Vilaylima, a village some fifteen miles away from Omarakana by a very bad road, so that it was not easy to convey a big contribution. In olden days I was told that every year the headmen of Vilaylima would carry a few hundred baskets of taytu packed in enormous prism-shaped receptacles, *pwata'i*, all across the island and over the swamps and coral rocks to Omarakana. But nowadays her main contributions from home, as can be seen (Nos. 72 and 73), consist of two small gifts estimated at 50 and 60 baskets respectively and amounting probably to much less.

Seburada and Katupwena have their regular shares given to the chief and put in his main storehouse.

The corrections in detail and in general perspective which I have briefly summarised here gradually dawned on me during my field-work, later in 1915 and during 1916. Before I left Kiriwina in 1916 I was able to reconstruct the following table which, with one or two minor corrections in the spelling of names, I reproduce here. It corresponds fairly closely to the first thirteen entries of Document II.

LIST COMPUTED IN 1916

Wife	Donor	Relation to Wife	Community
1. Bokuyoba	Pilu'u'ula (No. 13)	"Brother" (really only a clansman)	Dayagila
2. Namtawa	Mwagwaya (No. 4)	Real brother (really only a clansman)	Oylova'u
3. Bomyototo	Simdarise Wawa (No. 2)	Her "son"	M'tawa
4. Kadamwasila (died before 1918)	Piribomatu	"Brother" (her mother's sister's son)	Osapola
5. Bo'utukema	Monumadoga (No. 11)	Brother	Tilakayva
6. Bomawise	Tokwamnapolu (No. 6)	Brother	Obulabola, suburb of Kwaybwaga
7. Bomapolu	Kalubaku (No. 65)	Brother	Wagaluma
8. Isupwana	Kwoyavila (No. 1)	Brother	Suvayalu
9. Naboyuma	To'uma (No. 9)	Brother	Okayboma
10. Tubwoysewaga	Walasi (No. 5)	Brother	Kaybola
11. Ilaka'isi	Giyotala (No. 33)	*kadala*	Yaluwala
12. Seburada	Kayvala (No. 67)	*kadala*	Lobu'a
13. Gogo'a	Mokaywori (No. 72)	*kadala*	Vilaylima
14. Katupwena	Kani'u (No. 12)	Brother	Liluta

(The figure in brackets beneath donor's name shows his number in Document II, Table 2.)

Comparing this list with that of Document II, it is remarkable that they differ in only four points, which can be easily accounted for.

Thus in 1918, owing to the death of Kadamwasila, Piribomatu was no longer among the contributors, and his disappearance reduces the number from the fourteen entries of the 1916 list to the thirteen fillers of the main *bwayma* in 1918.

The share of Ilaka'isi was given in 1918 by her mother's brother, Kommatala, because that was a very great occasion. Ordinarily her main donor would be Giyotala, who is Kommatala's younger brother. In olden days the duty would probably have fallen regularly to Kommatala. But under white man's rule a certain decentralisation has made the minor chief somewhat negligent of his duty to the paramount chief.

Seburada's share in 1918 was given by her "father" (cross-cousin, father's sister's son) because he lives in M'tawa and the keenest competitors in the *kayasa* were the people of M'tawa, Liluta and Kwaybwaga. Normally her real maternal uncle, Kayvala, resident of Lobu'a, would fill her compartment in the main *bwayma*.

Gogo'a, who is not represented in the first thirteen entries of Document II for reasons mentioned already, had yet a definite place assigned to her and her lineage in the main *bwayma*, because Vilaylima was a community of rank which from old days had regularly supplied wives for the chiefs.

My greatest mistake, however, and the one which makes both lists of 1915 and 1916 interesting only as curiosities, was that I did not realise till well into my third expedition that in Omarakana as elsewhere only part of the chief's *urigubu* is stowed away in his own foodhouse. When, during the harvest of 1918, I realised that what was being stacked into his wives' storehouses has also to be counted, it was too late to see exactly what happened in ordinary years. I reckon, however, that the amount of taytu given in 1918 was roughly four times as great as in ordinary years—in 1918 the big *bwayma* was filled to overflowing and in 1915 only about one-fourth was filled. Therefore I assess the yield in average years at about 5,000 baskets, of which 2,500 reached the main storehouse and another 2,500 the small *bwayma* of each wife.

There is one more point to be made with regard to the chief's harvest. In old days, as in 1918, a great many of his sons continued to live in Omarakana. There are at present (1918) about eighteen sons of To'uluwa living with him, and all of them work for their respective mothers; in Document II we can see that eight of the sons are represented. As a matter of fact, in normal years there would be a larger quota, because some of the sons had in 1918 joined with their matrilineal relatives and contributed to the very large heaps represented by Numbers 1, 2, 3, 6, 7, 12 and 54.

In ordinary years all of the sons work with their mothers and for their mothers, and here another complication comes in: a number of *baleko* are allotted each year at the *kayaku* to each of the chief's wives.

These *baleko* are called *gubakayeki*, and their produce, *taytumwala*, goes to wife concerned. In the production of these crops the sons of each wife play a considerable part. Much work is also done by communal labour, to which the chief summons his vassals. The produce of such gardens would not be regarded as *urigubu* and it would not figure on any list. The chief's sons, therefore, contributed to his wealth in a two-fold manner, first of all by harvesting one or two *urigubu* plots which they directly presented to their mothers' households, and in the second place indirectly, by assisting their mothers in the work on *gubakayeki* plots.

In estimating, therefore, the chief's total revenue, we must remember that the direct contributions of his sons have to be added to the *urigubu*. At present I do not think that this contribution exceeds 100 baskets per son, that is about 2,000 baskets. In olden days when the race was far more prolific, when to judge from old genealogies the average chief's wife might have as many as four or five sons and there were some sixty wives, the number of adolescent boys and young men working directly for the chief would run into a couple of hundred. That was when the village of Omarakana was twice the size it was in 1918, when side by side with it was another big settlement, Omlamwaluwa, now extinct, and when Kasana'i, to-day a mere fragment of a village, formed an enormous ring of houses and storehouses. With all this, the total income of the Tabalu of this powerful settlement might have been ten or even twenty times as big as it was in 1918, but this is no more than conjecture. And it must also be remembered that the total population of the Trobriands must have been twice or three times as numerous then as it is to-day. Hence also the demands on the chief's exchequer were very much heavier than they are now.

PEDIGREE I

(For Abbreviations, see Document II, Table 2)

PEDIGREE OF BUKUBEKU

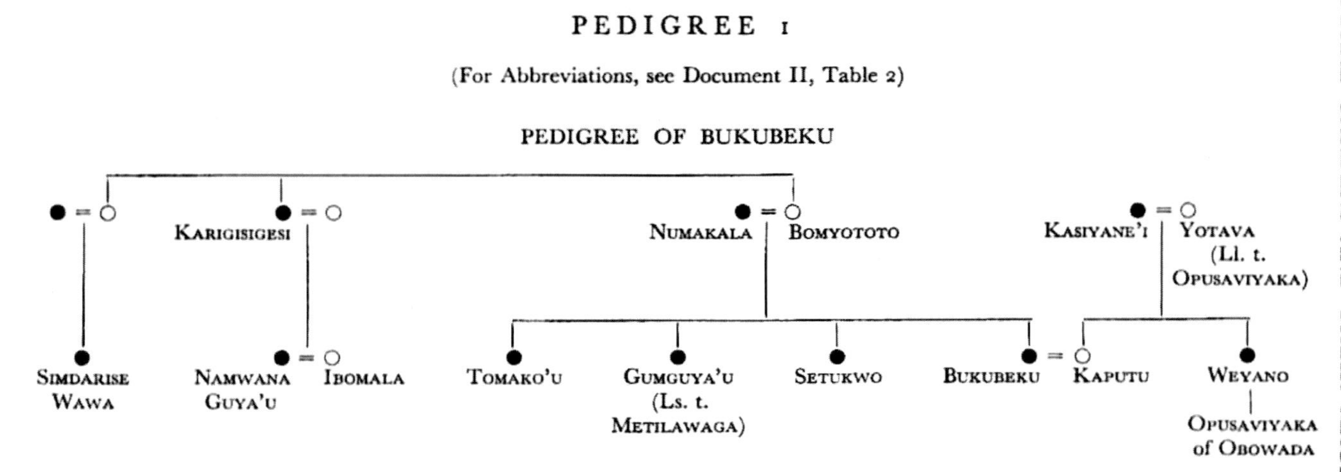

DOCUMENT IV

A FEW REPRESENTATIVE EXAMPLES OF AVERAGE *URIGUBU*

(Chapter VI, Sections 1 and 3)

I. TOKULUBAKIKI is a man of the Lukwasisiga clan, sub-clan of Tukwa'ukwa. He is the son of Molubabeba, a Tabalu of Omarakana (cf. Ped. 2). He lives in his father's village and will be familiar to some of my readers[1] as one of my best informants and as taking an active part in many Kiriwinian incidents. In 1914–15 he was cultivating three *baleko*; one in the *kaymugwa* (early gardens) and two in the *kaymata*. The former, as is customary, was regarded as *gubakayeki* and the produce called *taytumwala*; a small part of it went to Motaniku, his maternal uncle, as *taytupeta*, and the bulk to himself. The taytu grown on his two plots in the *kaymata* were given to his mother as *urigubu*, that is, stowed in his father's storehouse. Tokulabakiki keeps a common household with his parents, so that the *urigubu* which he gave was consumed not only by his father's household but his own as well.

The gift to his maternal uncle had to be made because Motaniku, an old man and a widower, had no sons.

Normally Tokulubakiki would have had to live in Tukwa'ukwa and from there raise crops to fill his parents' storehouse and to supply himself and his maternal uncle with food. Instead as a chief's son he is able to remain in Omarakana.

Tokulubakiki's *bwayma* is filled by Tobutusa'u, brother of Kuwo'igu, Tokulubakiki's wife. Tobutusa'u was only a boy of eleven in 1915 and his gardening did not amount to very much. However his *kadala* (maternal uncle), Mitayuwo, already mentioned as a great gardener,[2] comes to his rescue and gives the *urigubu* from one of his *baleko* to Tokulubakiki. This latter receives also small contributions from Tomnavadila, from Karisibeba and from Kaututa'u, who are all kinsmen of Kuwo'igu (cf. Ped. 2).

After Tokulubakiki has received his *urigubu* gift, he apportions off a number of baskets and gives them to Gubilakuna, his mother's clanswoman, whom he also calls "mother" because of the common clanship with his mother and because she is married to his father's younger brother, Mukumuku. Another gift goes to Tokolibeba, his father's brother. These gifts, however, might be regarded as given by Tokulubakiki to Mukumuku and to Tokolibeba on his father Molubabeba's account. In this case they would be orthodox *urigubu* gifts, given from son to "father". As always in native custom the orthodox pattern can be

[1] Cf. my *Sexual Life of Savages in N.W. Melanesia.*
[2] Cf. Ch. II, Sec. 3.

established by a little adjustment of facts, but the realities of life demand very often a substitute action.

Tokulubakiki gives a few baskets of yams, under the title of *taytupeta*, to the household of Gumigawaya, whose wife, Kamtula'i, he calls "mother" because his father Molubabeba and Gumigawaya are "clansmen". The entry in my original notes on this point is interesting—I have written there:—

> "Tokulubakiki gives some five to twenty baskets to Gumigawaya because Kamtula'i, Gumigawaya's wife, is his 'mother'. This relation is established because Gumigawaya and Molubabeba are of the same clan, hence regard themselves as brothers."

It is significant that writing as I did almost under the dictation of the natives, I emphasised the fact that the gift is given on account of the wife, Kamtula'i. This shows how far harvest gifts are determined by the relationship of the donor to the wife rather than to the husband. The statement then received that Gumigawaya and Molubabeba regard each other as brothers is certainly erroneous. The latter, who is a *guya'u*, would regard himself as infinitely superior to the low-caste commoner Gumigawaya, who in fact is resident in Omarakana as a vassal in special attendance on Molubabeba. The gift is really a repayment of these services by the *guya'u's* son, but a fictitious and spurious relationship has to be established to make the gift conform to one of the customary patterns.

II. Karisibeba, mentioned in Pedigree 2, is a "native" of the soil of Yogwabu in the strictest sense of the word according to native ideas and traditions, because his mythical ancestor, Kaluva'u, came out of the soil in Omarakana (at a hole called Bulimaulo, now a half-dried well), or rather in that extinct part of Omarakana called Yogwabu (cf. Ch. XII, Sec. 1). He is a sickly man and in 1915 has been cultivating only two *baleko*; from one the *urigubu* goes to Gomila, married to Malelo'i, the donor's sister; from the other to To'uluwa, the *urigubu* being put into Isupwana's *bwayma*. It will be noted that in 1918 Karisibeba, who was then seriously ill, made no contribution to To'uluwa's *kayasa* harvest. From his own sub-clan, however, Buyavila Kiriwila (No. 57, Doc. II, Table 2), who resides in Liluta (matrilocal marriage), contributed 450 baskets to Isupwana's storehouse. In the case of Karisibeba this is simply a gift to the chief from a commoner belonging to the same village. In the case of No. 57 it was an affinal *urigubu* due to the fact that Buyavila Kiriwila is married to Isupwana's sister. This is an example of how whenever possible some plausible relationship is sought for in order to make tribute look like a marriage gift. With the Kaluva'u sub-clan (Ped. 2) harvest gifts to the chief were an hereditary duty, paid by the ancestors of Karisibeba and Buyavila Kiriwila to the chief and always stacked into the storehouse of the wife of the chief who came from that sub-clan.[1] Thus Labelaba, the present donor's maternal uncle, used to fill the *bwayma* of Viyoulo, a Kwoynama woman, married to the former chief of Omarakana, Numakala. When Viyoulo died, Numakala married Isupwana. This latter was inherited by To'uluwa together with the privileges devolving on her, among which was Karisibeba's duty to supply an *urigubu* gift.

Karisibeba, being a kinsman of Kuwo'igu (cf. Ped. 2), gives a small gift of the *urigubu* type to Tokulubakiki. He regularly contributes a small gift of the *taytupeta* type to Setukwa, who married Ilakasila, Karisibeba's agnatic relative; and to Kaututa'u, his real brother, he gives a *kovisi*. There is no doubt that in these gifts considerations of friendship and of neighbourly relations are quite as important as kinship. A *taytupeta* is given by Karisibeba to Ilagaulo, his agnatic relative (a woman).

Karisibeba's *bwayma* is filled by Nabwasuwa and Mwaydayli, two sons of To'uluwa and Namtawa, and kinsmen of Karisibeba's deceased wife. As a rule after a man's wife dies, all duties of her relatives to this man become extinct. But in the case of a sickly man and one who is a close neighbour in the same village and who also gives a gift to the chief the *urigubu* obligations would be kept up.

III. As an example of the dropping of duties, immediately after the death of a wife, Bukubeku, one of To'uluwa's many "sons", in fact his elder brother's son, may be quoted. His *bwayma* had been filled up to 1914 by Weyana (Ped. 1), the only brother of Ilapotu, Bukubeku's wife. But she had been bewitched by a *bwaga'u* (sorcerer) some time before I came to Kiriwina and died during my stay in Omarakana. The next harvest Bukubeku was already deprived of his usual *urigubu*. He was my immediate neighbour in Omarakana and for some six weeks after his wife's death I could hardly sleep for the continual wailing in which he indulged. I had a great respect for his faithfulness and never interfered with the loud expression of his feelings. When he first came to my tent I gave him a stick of tobacco to express my sympathy and approval of his depth of grief. Great was my disappointment, therefore, when Bukubeku, still beaming at the sight of the tobacco, broached the object of his visit. He had heard that the *gumanuma* (foreigners, now especially white men) had powerful love charms and wished me to give him some, as he had to consider the future and would soon have to begin a new courtship. He stated emphatically that it was only his *bwayma* of which he was thinking, for what should he eat next year? He probably spoke the truth as, had his needs been other than strictly economic, he could have found plenty of unmarried girls by no means averse to an intrigue with a recent widower.

Bukubeku himself, a son of the chief, gives the harvest of two *baleko* to his mother, Bomyototo, who, however, died between 1915 and 1918.

[1] At present and for some time past there has been no chief's wife from the Kaluva'u sub-clan. For some reason, members of the sub-clan were contributing conjointly with the Kwoynama sub-clan of Liluta.

PEDIGREE 2

A. PEDIGREE OF THE KALUVA'U (YOGWABU) SUB-CLAN OF THE MALASI CLAN

Names in italic type = members of Kaluva'u *dala* of Malasi.
Names in clarendon type = men for whom Kaluva'u sub-clan have to work.
For Abbreviations, see Table 2 of Document II.

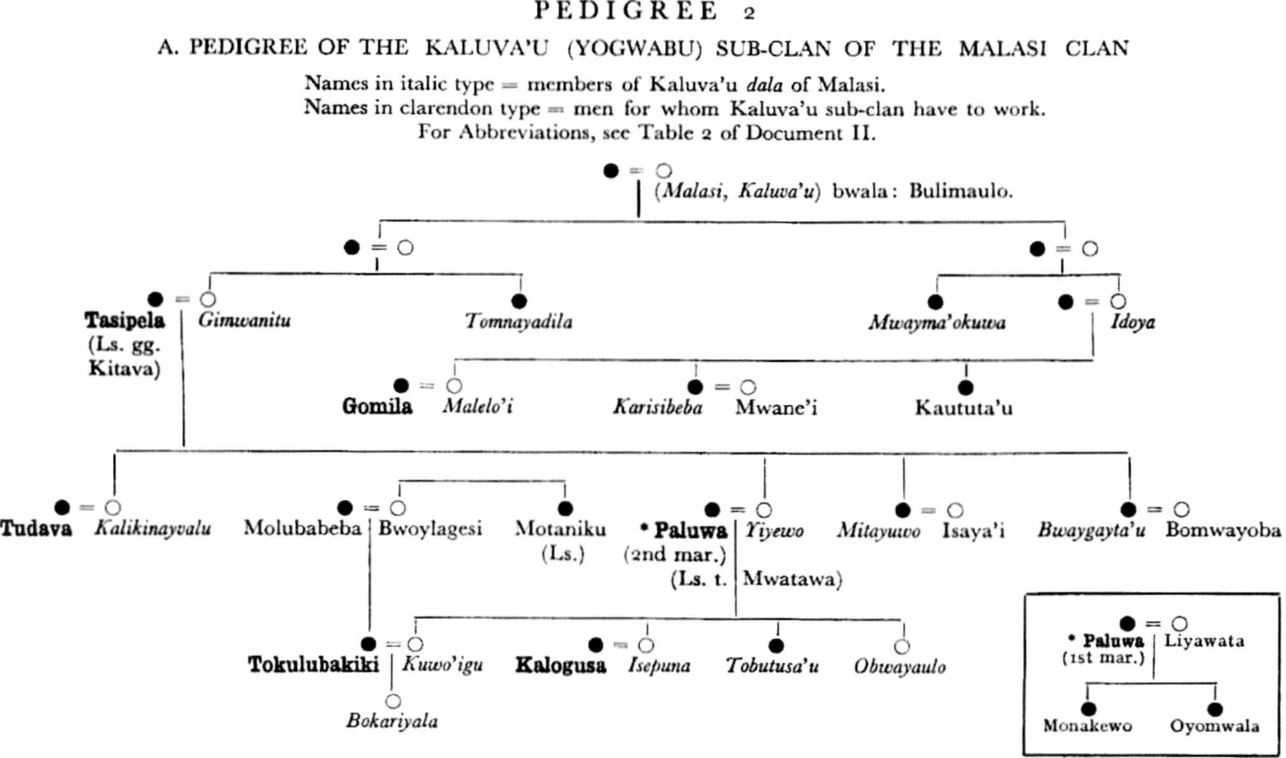

He cultivates another *baleko* for his first cousin on the maternal side, who is married to Namwana Guya'u, as shown in Pedigree 1. One more *baleko* is gardened for Setukwa, his elder brother, who lives in Omarakana.

He gives some *kovisi* to his *veyola* (maternal kinsmen) in M'tawa: one gift to an old woman, his maternal aunt (*kadala*); and another *kovisi* to his eldest brother Tomako'u.

In 1918 Bukubeku had migrated from Omarakana to his native or real village, M'tawa, and he did not give a special gift in his own name to the *kayasa urigubu*. I did not ascertain whether he was working together with his elder brother Tomako'u (No. 23 on the list of the *kayasa* contributors) or whether he joined his "brother" (mother's eldest sister's brother), Simdarise Wawa, who made the biggest contribution among all the donors on the list.

IV. Namwana Guya'u, a conspicuous personality and a typical chief's son, belongs to the Kwoynama *dala* of the Lukwasisiga clan and lived in Omarakana till January, 1916, when he was driven out by his father's kinsmen. He then moved to his own village, Osapola (a suburb of Liluta). In the year 1915 he was cultivating five *baleko*; one of them for the main *bwayma* of his father; three for his mother's *bwayma*; and one, a *gubakayeki baleko*, for himself. Every year he gave a small *urigubu* of ten baskets to Mitakata of Omarakana, who married Orayayse, his "sister" (mother's sister's daughter). After his rupture with Mitakata, this gift was stopped.[1]

Namwana Guya'u's *bwayma* is filled by Setukwa and Bukubeku, who were "brothers", mother's sister's sons, of Ibomala, Namwana Guya'u's wife. He used to receive *kovisi* from his younger brothers, Kalogusa and Dipapa.

VI. Tovakakita, younger brother of Giyokaytapa, a Malasi *Tokaraywaga*, whose sub-clan emerged (cf. Ch. XII, Sec. 1) at Gawa'i, gave me a good example of the attitude of a commoner or man of lower rank who lives in a village next door to the chief and is very much in the chief's entourage. Tovakakita is married to the younger sister of Bo'utukema, one of To'uluwa's wives, and he has therefore the right to call To'uluwa "my elder brother", though actually, of course, he would only address him as *guya'u* (chief) or else by the personal name To'uluwa. When I enquired—he was in my tent at Omarakana in the presence of a number of others—how much he received as *urigubu*, he was genuinely reluctant to speak of it. This incident is recorded in Chapter V (Sec. 5). He wanted to give me the information required, but before he gave it he found it necessary emphatically to declare that his harvest gift amounted really to nothing. At last he told me that he received for his wife about 50 baskets a year, which I think was a correct statement. Of course he probably produced as much or more for himself.[2] Tovakakita himself gave some 100 baskets a year as *urigubu* to the husband

[1] Described in *Crime and Custom* and *Sexual Life of Savages*, Ch. I, Sec. 2.

[2] For the difficulties of ascertaining the proportion of *urigubu* to *taytumwala*, especially in the case of commoners, cf. Ch. VII, Sec. 5.

of one of his maternal relatives. He does not contribute directly to the chief's harvest gift, but shares in the contribution of his elder brother, Giyokaytapa, No. 62 in Document II (Table 2).

Giyokaytapa, the headman of Yourawotu and elder brother of Tovakakita, received five heaps in 1918, in sum about 280 baskets. I made only a rough survey of the heaps as they were disposed in the village, and did not take the names of the givers, only the amount of each heap. They contained 80, 70, 60, 40 and 30 baskets.

It will be well in concluding this statement of concrete facts to give a rough estimate of the number of baskets contained in each heap. It will be seen in Document I that I failed to count the number of baskets in each heap; this document was compiled in 1915 and it was impossible for me to supplement the data in 1918. The computation of Document II in 1918 took me so much time that I was not inclined to collect more of this type of evidence. A rough survey of a number of *urigubu* displays led me to the conclusion that whereas the average number of baskets per heap in the case of a chief in olden days, as to-day at a *kayasa* harvest, was about 250, the *gugula* (heap) of a lesser chief, a *gumguya'u*, would consist of some 75 baskets and the average heap of a commoner of 50.

In Yalumugwa, therefore, counting some 10 heaps of the second category and 20 of the third, the total yield in 1915 would have been 1,750 basketsful, which might have been distributed as follows. Some 600 or 700 baskets to Yovisi, the headman, who had two wives, and received nine heaps, about 200 to the more important notables, and 100 to the commoners. The total yield of a smaller, lesser headman, as we have seen, was 280 baskets. The *urigubu* received by To'uluwa's sons varies between 100 and 500 baskets.

DOCUMENT V

LIST OF SYSTEMS OF GARDEN MAGIC

(Chapter IX, Section 1)

System	*Villages*	*Number using System*
Kaylu'ebila	Omarakana, Kuluvitu, Kapwani, Laba'i, Lu'ebila, Olivilevi, Omlamwaluva	7
Silakwa	Kasana'i, Osapola	2
Momtilakayva	Kurokayva, Tukwa'ukwa, Suviyagila	3
Gawa'i	Kammamwala, Oyliyesi	2
Giyulutu	Wakayluva, Kabwaku, Kaurikwa'u, Tubowada, Wakayse	5
Bisalokwa	Obweria, Kudukwaykela, Kavataria, Kuluwa, Kabululo	5
Gayga'i	Bwoytalu, Yalumugwa, Moligilagi	3
Kimdoga	Liluta, Kudokabilia	2
Ilakaygulugulu	Yalaka, Lobu'a	2
Molubokata	Idaleyaka, Bwoytavaya	2
Okalawaga	Dayagila, M'tawa	2
Yaluwala	Kwaybwaga, Kaulagu	2
Kayuwada	Yuwada	1
Kaykebola	Kaybola	1
Mokalawaga	Okobobo	1
Siwapula	Mwatawa (cluster of 4 villages)	1
Palela	Okaykoda	1
Siribomatu	Obowada	1
Tokavasili	Gumilababa	1
Tovanana	Luya	1
Kaykulasi	Buduvaylaka	1
Kwaypwapwatiga	Oyveyova (Osaysuya)	1

DOCUMENT VI

GARDEN MAGIC OF VAKUTA

(Chapter X, Section 1)[1]

VAKUTA, the main village on the island of that name, differs in several minor respects from the rest of the Trobriands, but belongs to the same type of culture. I have not been able to observe any gardening ceremony in Vakuta, and altogether I only spent a fortnight on the island. The information here given was obtained by the somewhat unsatisfactory question and answer method, which only permits of a rather bare and lifeless outline of customary usage and ritual. I shall emphasise the differences only between these and Kiriwinian customs, as the bulk of them are the same.

Land tenure in Vakuta does not seem to differ at all from that obtaining in the other parts of the Trobriands (cf. Ch. XII). Here is a list of fields, *kwabila*, owned by the village.

LIST OF *KWABILA* IN VAKUTA

Molayla	Karikawala
Pwaka	Kilato'u
Wabutuma	Lubwama
Pokayli	Okubusi
Kalakamtuya	Kilawota
Kaulukuta	Ilumwagala
Kabawala	Iluwaywaya
Kubwaboma	Pakulasi
Muluwata	Ore'ore
Osirisiri	Okumkumlo
Magigila	Kupiluma
Kulawotu	Dikusola
Yatawa	Duweta
Oluwa	Leleria
Wamo'i	Okayyasila
Owaywo	Ogaku
Tigitagega	Gagaywo
Ya'uyu	

Each such *kwabila* is "owned" by the headman of a sub-clan, and is subdivided into smaller plots, *baleko*, which are "owned" by individuals. Such "ownership" has small economic significance. Exactly as in other parts of the Trobriands, it is rather a ceremonial prerogative than the exclusive right of economic use.

[1] Cf. also Part V, Div. IX, §§ 16–34.

The ceremonial garden council, the *kayaku*, differs slightly from that of Omarakana. First of all there is a small gathering in front of the garden magician's house; and usually the *sousula*, the initial payment to the garden magician, is brought by the villagers at this time. The garden magician has a number of coconuts ready, and instructs his nephew or younger brother to give a coconut to each man present. At the meeting a general decision is made as to when the gardens will be started and on what field. A day or two after, the younger brothers, matrilineal kindred and other relatives of the garden magician, go fishing. The catch is brought to the garden magician's house and his wife cooks it. The members of the community forgather again and the real or big *kayaku* is held. The fish is distributed among the men, who eat it on the spot, and yams, which have been baked at the same time, are distributed throughout the village and eaten by the women and children. At this *kayaku* the garden plots are allocated.

On the afternoon of that day the garden magician ceremonially offers a small portion of fish and of vegetable food to the ancestral spirits and utters a spell in his house over the magical substances.

In Vakuta eleven substances are used—the leaves of the following: *lawa* (a tree with scented, tasteless fruit), *kaga* (a ficus), *kirima* (a ficus), *lileykoya* (an aromatic plant), *youla'ula* (a creeper), *dadam* (a tall reed growing on swampy soil), *gutaguta* (plant with edible leaves), *kaluluwa* (an unidentified plant), *seuse'u* (also unidentified) and a plant the name of which I cannot decipher, which grows on coral rock and has a red flower. He also scrapes some chalk from a large coral boulder (*kaybu'a*). With all of these save *lawa*, *gutaguta*, *kaluluwa* and *seuse'u* we are familiar in other types of magic.

A part of the herbs is left loose; a part is charmed and tied round the blade of an axe by means of a dried banana leaf (cf. Ch. II, Sec. 4); and a part is made into small bundles after having been boiled in coconut oil. I think that the same charm is repeated twice: once over the mixture when it is boiling in the oil in the *kwoylabulami*, the ceremonial cooking-pot, and again over the substances laid together between two mats.

The next morning the big procession goes into the gardens. The rank and file of gardeners rapidly cut the belt round the garden-site (*tavalise kali*—they plant the fence), and then the junior magicians or assistants of the *towosi* make holes with the magical wand in the magical corners of the two *leywota* nearest the village, and insert therein bundles of the herbs boiled in coconut oil. Then the *towosi* himself takes the medicated axe and ceremonially cuts a tree in the corner of the *reuta'ula* and utters a spell, different from the one spoken on the previous day. This is called *italala leywota* (he makes flower the *leywota*). The men then proceed to cut a road which will give access to the garden plots, an act which is described as *biyowota lopoula buyagu*, "he", i.e. the magician, "cuts open the belly of the garden" (cf. Part V, Div. VI, §§ 30 and 31; Div. VIII, §§ 9 and 10; and Part VII, M.F. 2). The *towosi* then visits each *baleko* and strikes a tree with a medicated axe and utters the same short spell as is spoken at the *talala leywota*. Immediately after he squats down, tears out a few weeds, mixes them with some of the medicated herbs and, uttering a spell, rubs the ground with them. This is the *yowota* proper.

When this is over, the magician returns to the main *reuta'ula* and, taking his axe, cuts vigorously with it into the trunk of a tree, where it is left overnight. The whole company then return to the village and are given some food cooked by the magician's wife, which they eat. Next day the *reuta'ula o kakata* (cf. Part V, Div. VI, § 29) is cut (*takaywa*) by communal labour. The owner of it prepares food which is given to the workers after the cutting is finished, while light refreshments, such as green coconut, betel-nut and sugar-cane, are given during the work.

After this the *towosi* pulls the axe out of the tree where it has remained for twenty-four hours and cuts into another tree, this time on the *reuta'ula o kikivama*. The *kema* is again left in position over-night.

Next day the second plot is cut by communal labour, and this procedure is repeated daily until all the standard plots are finished. The ordinary plots, called in Vakuta *kaymwila*, are cut by the owners working on their own plots.

The cut scrub lies for two or three weeks to dry, and then follows the *gabu*. This in Vakuta as elsewhere consists in a series of ceremonies: first a wholesale burning, *vakavayla'u*, and then the compound ceremony of *gibuviyaka*, *kalimamata*, *bisikola* and *pelaka'ukwa* (cf. Ch. III, Sec. 1). The first is different in Vakuta in that the magician begins by medicating the garden-site, which is not done in the other systems. Afterwards he returns home and utters a spell over the *kaykapola* (coconut-sprout torches). Apparently he charms these just before they are used, and not at the previous harvest, as is done elsewhere.

On the next day, about noon, the dried cut litter is set on fire by means of the coconut torches. On the day after the magician charms more *kaykapola* together with some herbs: *borogu* (croton), *peraka* and *seuse'u*. The procedure is then the same as in other communities: small heaps of refuse are made in the garden, the herbs are placed in them, and they are fired with a torch. The *gibuviyaka* is followed by the three minor ceremonies: *kalimamata*, *bisikola* and *pelaka'ukwa*. I was informed that in Vakuta the first is more particularly a magic of yams (*kuvi*), the second of bananas and the third of taro. During the third, the *pelaka'ukwa* rite, the magician erects a symbolic fence, planting a couple of sticks in the ground. This rite is called *iwaya'i gado'i* (he strikes the stake). During the four days of the ceremonies and for one day after, there is a taboo on garden work, making altogether a five days' taboo.

In Vakuta whenever a *leywota* is being gardened, other plots are under a taboo as regards work, and all are under a taboo on any day on which a ceremony is in progress.

After this rite the work of *koumwala* begins, concerning which I received the following detailed account: the word *koumwala* applies to the whole process of clearing the ground, but more specifically it denotes the cutting down of previously unnoticed saplings and the burning of litter, especially of larger sticks; and this is men's work. *Tubwalasi*, which is women's work, consists in clearing away the smaller sticks and weeding. *Yolukula*, gathering the debris into heaps, is also done by women.

After such heaps have dried, each is fired, and this is called by the generic term *supi*. Finally the ground has to be swept clean with brooms made of twigs, and this again is called by the generic term for sweeping, *nene'i*. Men then put the *tula* sticks in position, *saytula* (*say-* to place, *tula* boundary-sticks). The laying of the transversal sticks is sometimes described as *giyauri*.

The *kamkokola* are, of course, also erected in Vakuta (cf. Ch. III, Sec. 4). A taboo is laid on garden work, and stout poles are procured. Then the structures are put up with some magic of which I have, however, no details, except that the aromatic plant, *lileykoya*, so often met with in magic, is used in the rite. Then small sticks, called *kubwaya* or *kaykubwaya*, are erected which are supposed magically to induce the taytu plants to wind round the *kamkokola*. I was told by Towese'i, Bagido'u's brother, that this rite is never met with in Omarakana. He himself saw it carried out in Vakuta and remembered it as something strange.

In connexion with the *kamkokola* ceremony, we find the rite of charming over seed taytu, the *yagogu*, which is called *ulapa yagogu*. I was told in Vakuta that the *kamkokola* magic is the magic of planting, and that the charming of the *yagogu* enabled it to be used effectively as seed. There is also a rite directed towards the yam poles, called *katakwabula kavatam*. In Vakuta the *kamkokola* complex of rites is both a magic of planting and of yam poles.

The "growth magic" is apparently more complex in Vakuta than in any other community. The series starts with the rite of *vaguri*, as in Omarakana and other places (cf. Ch. IV, Sec. 1). The magician simply goes into the garden enclosure and, starting with the *leywota*, utters a spell over every *baleko*. If a *baleko* is very advanced, he does not chant over it very long: if a *baleko* is much behind, he lingers on it and charms it with greater persistence. I was told that all the rites of growth magic are carried out not once only, but day after day till the next stage is reached. After the *vaguri* comes the *vasakapu*, which makes the sprout emerge exactly as in Omarakana. It is chanted freely over the gardens. Some time after this the magician plants (*i-gilulu*) several small sticks against the *kamkokola* and, following this, the men put similar sticks against the *kavatam* so that the young taytu sprouts can climb from one to the other (cf. Ch. IV, Sec. 1). Then the magician performs the rite of *kala'i*, which consists in rubbing the ground with herbs to make the vine creep towards the supports. This rite is not known in the Kiriwinian systems.

After the *kala'i* comes the *vakwari*, a formula spoken over the gardens which makes the taytu get hold of the supports and wind round them. Then come the rites of *lasawa*, making the tubers settle down in the ground that they may multiply; *talova* or *valuvalova*, producing an abundance of foliage; *yo'uribwala*, a ceremony in which a diminutive "spirits' house" is erected, making the foliage still more luxuriant;

yobunatolu, by which the foliage is made to close up "so that the inside of the garden becomes dark".

After that a rite is performed described by the natives as *tata'i tageguda*. I am not absolutely certain about the aim of this ceremony or that of the *tata'i tamatuwo* which follows the *vapwanini* rite, but I think they refer to the cutting or trimming of the yam supports for young and for mature vines respectively. The names 'we cut the unripe thing that is cut' and 'we cut the ripe thing that is cut', fit this explanation. Then the magician "throws the final stick" (*ilova kayluvakosi*), the stick, that is, on which the taytu climbs from one support to another. Then comes a rite named in the same manner and performing the same function as a ceremony in the growth magic of Omarakana, the *vapuri* (Ch. IV, Sec. 2). This rite, as we know, aims at the multiplication of tubers. A similar rite called *vapwanini*, not known in Omarakana, makes the taro roots swell.

Tasasali is a ritual act in which a small stick is inserted into the soil near the *kamkokola* and some herbs put on to it. This is the signal for the ceremony of *isunapulo*, the harvesting of taro and large yams.

The harvest of taro is associated with the ritual eating of fish, *tavakamsi yena*. The fish is caught by the younger men of the village and some of it is offered by the magician to the ancestral spirits as *ula'ula*. Then there is a festive meal in each household, after which taro can be eaten.

After the ordinary ritual opening of the taytu harvest by the *okwala* and *tum* ceremonies (cf. Ch. V, Sec. 3), the magician ritually eats taytu (*ivakam taytu*) very much as in Omarakana, and then the new taytu may be eaten as well as harvested by all the members of the village, including the *towosi*. I am not quite certain whether, in Vakuta, there is a general taboo on eating the new taytu till the magician has partaken, but I think this is so.

My information on the garden magic of Vakuta is not comparable, of course, with my information about the Kiriwinian systems. As is always the case when one has to rely merely on verbal statements made by the natives, there may be confusion between real magic and merely accessory acts, between activities which are partially ritual and partially practical, and those which are wholly ceremonial. In order to make the above account somewhat clearer, I append a brief comparative table of the Omarakana sequence and that of Vakuta.

OMARAKANA[1]	VAKUTA
I. *Yowota.*	I. *Yowota.*
Ula'ula (oblation to spirits), charming of axes and herbs in *towosi's* house (*vatuvi* spell).	*Ula'ula* (oblation to spirits), charming of axes and herbs and mixture boiled in coconut oil in *towosi's* house; mixture and spell different from Omarakana.
Next day in garden, *towosi* cuts two saplings, reciting spell over each—rite called *talala*.	Next day in garden, holes dug and bundles of herbs inserted in *leywota*. *Towosi* cuts sapling and utters spell (*italala leywota*).

[1] Cf. also chart of Magic and Work, App. I.

OMARAKANA	VAKUTA
II. *Gabu.* 1st day, *vakavayla'u*, firing of dried scrub with *kaykapola* medicated at previous harvest. Same day collects herbs for next ceremony. 2nd day, *gibuviyaka*, burning piles of debris in which herbs inserted. 3rd day, *pelaka'ukwa*, planting and charming of taro top. 4th day, *kalimamata*; inaugural planting of *kwanada*. *Bisikola*, growth magic for taro, and erection of *sibwala baloma*.	II. *Gabu.* *Towosi* medicates garden-site first and then charms over *kaykapola*. Next day *vakavayla'u*. Sets fire to scrub with *kaykapola*. 2nd day, *gibuviyaka*, burning piles of debris in which herbs inserted. *kalimamata*, inaugural magic for yams. *Bisikola*, inaugural magic for bananas. *Pelaka'ukwa*, planting and charming of taro top. Magician erects small fence, *iwaya'i gado'i*.
III. *Kamkokola* (Taboo on gardens.) *Towosi* charms axes at home. 2nd day, erects *kamkokola* and rubs the upright with mixture. 3rd day, charms herbs at foot of *kamkokola* and utters *kaylola* spell. Magician strikes *kamkokola*. Herbs burned in magical corner with *vakalova* spell.	III. *Kamkokola* (Taboo on gardens.) *Kamkokola* erected and spell recited. Seed yams charmed over—*ulapa yagogu*. Yam pole rite, *katakwabula kavatam*.
IV. *Growth Magic.* 1. *Vaguri sobula* or *vavisi sobula*—"awakening of the sprout" or "cutting through of the sprout". Simple rite of charming over each plot.	IV. *Growth Magic.* 1. *Vaguri* (same as in Omarakana).
2. *Vasakapu*—the magical conjuring up of the plant from the soil. Simple charming of the plot.	2. *Vasakapu* (same as in Omarakana).
	3. *Gilulu* — sticks put against *kamkokola* and spell uttered.
V. *Magic of Weeding* (*Kariyayeli sapi*).	4. *Kala'i*—ground rubbed with herbs to make vine creep towards supports—spell recited.
IV. *Growth Magic* (continued). 3. *Kaydabala*—small stick put horizontally between two yam poles to delop leaves of taytu. 4. *Kaylavala dabana taytu*—spell chanted over each plot to produce rich foliage.	5. *Vakwari*—making vine wind round supports. 6. *Lasawa*—spell inducing tubers to settle down in ground. 7. *Valuvalova*—spell creating abundance of foliage. 8. *Youribwala*—minor rite consisting in putting up small spirits' house to make foliage still more luxuriant.

OMARAKANA	VAKUTA
5. *Kasayboda*—simple spell to make the vines close up and produce shadowy interior.	9. *Yobunatolu*—spell making foliage close up.
VI. *Magic of Taro and Yam Harvest* (*Isunapulo*).	10. *Tata'i tageguda*—rite probably connected with cutting of new yam supports. 11. *Ilova kaluwakosi*—spell making vine climb from one support to another.
IV. *Growth Magic* (continued). 6. *Vapuri*—simple spell producing clusters of new tubers. 7. *Kammamala*—spell making the tubers break forth again. 8. *Kasaylola*—spell of anchoring of new tubers.	12. *Vapuri*—spell producing clusters of new tubers. 13. *Vapwanini*—spell making taro roots swell. 14. *Tata'i tamatuwo*—"cutting of ripe"—probably final trimming of yam poles. 15. *Tasasali*—insertion of stick near *kamkokola*—signal for *isunapulo*.
VII. *Magic of Thinning* (*Mom'la*).	V. *Isunapulo*. Rite inaugurating taro and *kuvi* harvest. *Tavakamsi yena*—ritual eating of fish at taro harvest.
VIII. *Magic of Harvest* (*Okwala.*) First ceremony of taytu harvest. *Tum*—second ceremony. *Vakam taytu* rite.	VI. *Okwala*. First ceremony of taytu harvest. *Tum*—second ceremony. *Vakam taytu*, no spell or rite.

This table contains a summary of notes taken in Omarakana and Vakuta respectively, and the two columns have been drawn up independently. Placed thus side by side, we see the essential similarity between the two systems; the main rites (roman figures) are almost identical. I think the absence of a rite for weeding and a rite for thinning in Vakuta is not due to a faulty record, but is a fact. At the same time I am not quite sure whether the second point of difference, the much later stage at which *isunapulo* appears, is not due to a confusion in my notes. It is notoriously impossible to trust natives as to order of sequence when they are giving an account from memory. On the other hand, the greater complexity of Growth Magic in Vakuta may be regarded as a well-ascertained fact.

This table interests us not only as a document of comparison. The Omarakana column, which suffers from none of the defects of its companion, gives a synoptic summary of the facts contained in the narrative description and brings out the salient features in the Omarakana system. It is thus a useful supplement to the much more detailed table found in Appendix I.

DOCUMENT VII

GARDEN MAGIC IN TEYAVA

(Chapter X, Section 1)

TEYAVA is a village on the shores of the lagoon. I assisted there at the garden council, at the first inaugural rite and at the harvesting ceremonies.

The *kayaku* takes place during the harvesting of the old gardens or before. On the day of the garden council, the wife of Mwadoya the magician prepared the earth oven and baked some taytu and other vegetable food. In the early afternoon the magician summoned all the men to the meal, which was served on the central place in front of his yam-house. He addressed them in these words. (For this and the following quotations, see Part V, Div. III, § 19, Text 28b.)

"Boys, if it is your wish, come here, let us talk: which field shall we strike this year? Shall we strike the field which is near the creek, or the field which is towards the coral ridge?"

He then mentioned a series of names, which I did not note down, and one of the men, obviously by previous agreement with him, or he himself, would say, "No, not this. Let this field remain (fallow)"; or "We others do not want this field"; or "This field better remains on one side."

Then the magician suggested the field called Odaybayabona, to which everybody agreed in a chorus, saying: "Yes, good, that's right! It is our wish that we others should cut the scrub on this field." There was no doubt whatever that the field had been agreed upon at the beginning, but he still went on, "Well, if it is your wish to cut the field called Odabayabona, good. Your minds have settled it."

Then the magician began to allot the plots by enumerating their names. He started by saying: "So-and-so will cut the standard plot," and then passed on to the plots round the fence, naming or describing them. As each was mentioned a man would say: "This plot I shall choose," or "This plot I am taking," or "This plot myself." The garden magician then said: "We have now distributed the fence, the inside of the field still remains." Then they proceeded less carefully and in a rougher manner to distribute the plots remaining.

On the second day the garden magician went and collected the herbs. I was told in Teyava that he is not allowed to dally over the business: "If he remained and dallied, the spirits would strike him and he would fall ill" (Text 41a in Part V, Div. VIII, § 14). The herbs which he collects here are *yola'ula*, *kotila*, *lubiyayaga*, *yokunukwanada*, *yayu*, *silasila*, *guvagava*, and he also brought *ge'u*, stuff from the bush-hen's nest. These herbs are wrapped up, not in the ordinary dry banana leaf which has been bleached in the sun, but in the green leaf toughened over fire. Such leaf is called *sasova*. In Teyava the herbs are not medicated between two mats, but are spread on a large portion of toughened banana leaf, which afterwards is tied up and left on the shelf in the house, which from then becomes tabooed to child or woman till the following morning. The axe and the *kaylepa*, magical wand, are also medicated in the village, and the latter is wrapped up in another piece of *sasova*. Next morning an acolyte carrying the *kaylepa*, the *towosi* carrying an oblong basket with the herbs in it, and all the men go to the garden. This is done before sunrise, at the time called by the natives *gabogi*. First of all, the acolyte hits the ground with the wand without any further speech or ritual. Then the *towosi* takes the medicated herbs and rubs the soil with them, which is the *yowota*. After that he cuts a sapling called *kayowota* with the medicated axe and sticks it in the ground. There are certain differences therefore in the ritual of this ceremony as compared with the magic of Omarakana and Kurokayva.

Next day, which is the fourth day of the ceremony, the magician "walks with the wand", *bilolo kaylepa*. During this day no clearing may be done in the gardens which are under a taboo. This taboo lasts for three more days, and on the eighth day the *towosi* addresses the men, saying: "Bring me all your axes, O men, so that they may lie in my hut." Then, referring to the coming rite, he said: "I shall wrap up the axes, I shall charm over all the axes" (Text 41a, loc. cit.), and in the evening he put the axes between two mats and medicated them. The next day, the ninth day, the magician remains in the village while all the men go out and each makes the *talala* rite on his plot. In Teyava this rite consists in cutting a few trees on each *baleko*, after which the real work of clearing begins.

I have no details about the intermediate ceremonies. I was told that, after the *kamkokola* rite, there is a special magic of yam supports, *kavatam*, in which the magician medicates the axe and, striking the support, "anchors it firmly" in the ground. Then comes the magic of weeding. This in Teyava is done in two instalments. First the magician medicates the point of his digging-stick, and then going to the standard plots, the *leywota*, he inaugurates the removal of the big weeds, the *pwakova* proper. A few days later he chants another spell over the point of his digging-stick and gives the magical blessing to the *sapi*, the cleaning or brushing away of small weeds (cf. Ch. IV, Sec. 2).

The sequence of growth magic contains certain characteristic rites. We find the "waking of the sprout" as in Omarakana (cf. Ch. IV, Sec. 2), but after that there is a magic performed over an axe which is called "the cutting of the unripe". This magic refers to the growth of the plant before the tubers are ripe. After a ceremony corresponding to the rite of "closing up the leaves" in Omarakana, the magician "cuts the ripe", a ceremony concerned with ripening tubers (cf. also Doc. VI). The complex ritual of harvesting the taro and the large yams follows—the *isunapulo* as it is universally called in the Trobriands. In the course of it the magician "cuts the digging stick", that is, charms over its point, and later in the same day "splits the tuber", that is, takes a tuber and splits it with his adze against the pole of the *kamkokola*. He carries the digging-stick to the garden and leaves it there till the evening, when he digs up a few large yams and taro. These are brought to the village and next day laid on the graves of those who have died since the last harvest. It is called "the sacrificial offering to the graves" (*ula'ula walaka*). When there are no new graves a ceremonial exchange is made with the village of Tukwa'ukwa, and the festive consumption of fish follows.

Only after the *isunapulo* does the *basi*, the thinning of the taytu, take place. Still later (cf. Chart of Time-reckoning, Fig. 3) comes the ceremony of *okwala*, that is, the first harvest ceremony. The acolytes gather large bunches of *gipware'i* (lalang grass), some of which are made into knots and distributed along the road to the gardens near the *reuta'ula*. In the magical corner of the *reuta'ula* (main *leywota*) the magician medicates a heap of *gipware'i*, a piece of which is put by the acolytes on each *baleko*. This imposes a taboo. The bands of *gipware'i* on the way to the garden indicate that no one must go there to work.

A few days after this the ceremony of *tum* takes place. I witnessed it personally in Teyava, though I was not able to record the spell. Some leaves of the aromatic *lileykoya* are wrapped by means of dried banana leaf round several adze blades and medicated by the *towosi*. On the day I witnessed it, several adzes were thus prepared early in the morning in the magician's house. About eight o'clock we went into the gardens. Some boys went ahead to find out the exact spots on which the yam vine was ripe for harvesting. The chief magician, Mwadoya, accompanied by a younger brother and other acolytes, each carrying a medicated adze on the shoulder, repaired to the magical corner of the standard plot, dug up one of the round yams called *kasiyena* and split it with his charmed adze against a *kavatam* pole, uttering these words:—[1]

> "I shut up thy mouth, O bat, flying-fox,
> thief, bush-hen. Go down, go away."

This was, of course, not the main spell, which had been uttered over the adzes in the house and was long and full of repetitions. After the splitting of the yam, he pulled some weeds out of the soil, made them into a small heap, and "weighed it down" with the split yam. Then he chose an exceptionally good taytu plant on the *leywota*, cut the stem with his medicated adze and dug out the tubers, which were collected by the boys who attended us. The magician's assistants performed the same ceremony on the ordinary plots and collected their share of tubers, which amounted to quite a fair number. The taytu was put in baskets (*vataga*) covered with taytu leaves, which were slung on carrying poles, a basket at each end. The tubers were displayed on the central place of Teyava and then divided among the several component parts of the village. Some of them were afterwards exposed on the graves of the recently deceased.

[1] Cf. Ch. V, Sec. 3, and Pls. 51 and 52.

DOCUMENT VIII

THE GARDEN LANDS OF OMARAKANA

LOOKING at the attached map, on which the garden lands of Omarakana are given in outline, we see a self-contained territory, comprising a village, its system of communications, its water-hole, its sea-beach, its one or two tabooed groves, its portion of sea-board, coral ridge and garden lands.

It must be kept in mind, of course, that Omarakana is unquestionably *the most important settlement in the Trobriands. It is its political capital* and it is situated in a most fertile belt of agricultural territory. Seven or eight roads converge on the joint settlement of Omarakana and Kasana'i. Five villages are to be found in its immediate neighbourhood, and are indicated on the map—that is, if we count the three small component villages of Kurokaywa as one.

As regards the assortment of various lands, we see first of all that the twin settlement of Omarakana and Kasana'i is surrounded by a large grove which expands especially towards the north-east. There the spring or rather water-hole of Bulimaulo is situated. At present it cannot be used to supply the village with water as it is muddy and almost dry. But it plays an important part in the magic of rain and drought. The women of Omarakana and Kasana'i have to go every morning or afternoon to the water-hole of Ibutaku, over a mile away along the road called Lomilawayla, which passes by the three tabooed groves.

On this map only the garden lands of Omarakana have been plotted. The territory of Kasana'i lies directly to the north, and this village has its own road to the sea which runs through the lands of Omarakana. As we know, the people of Kasana'i make their gardens separately from Omarakana and have a distinct territory.

Let me recapitulate briefly (cf. Ch. XII) the general principles of land tenure, citizenship and rights of residence—these three facts being inextricably connected—on the example of Omarakana. First as regards history: there are two autochthonous sub-clans who still possess and exercise claims to the territory of Omarakana. One of them, the sub-clan named after the first ancestor, Kaluva'u, lives in the village section called Yogwabu and belongs to the Malasi clan. It originally owned the rain magic but has, in the course of past history, surrendered its claims to this, to most of the soil and to local sovereignty to the sub-clan of Tabalu. The Kaluva'u sub-clan is of low rank. Secondly there is a sub-clan of higher rank, named after its first ancestor, Burayama, who emerged in the grove of Sakapu, and who in former times occupied Katakubile—a now extinct village section of Omarakana. This sub-clan has since migrated to a section of Kwaybwaga, of which village cluster they are the actual head-men or rulers. The Tabalu, as we know, obtained possession and sovereignty of Omarakana through settlement by marriage

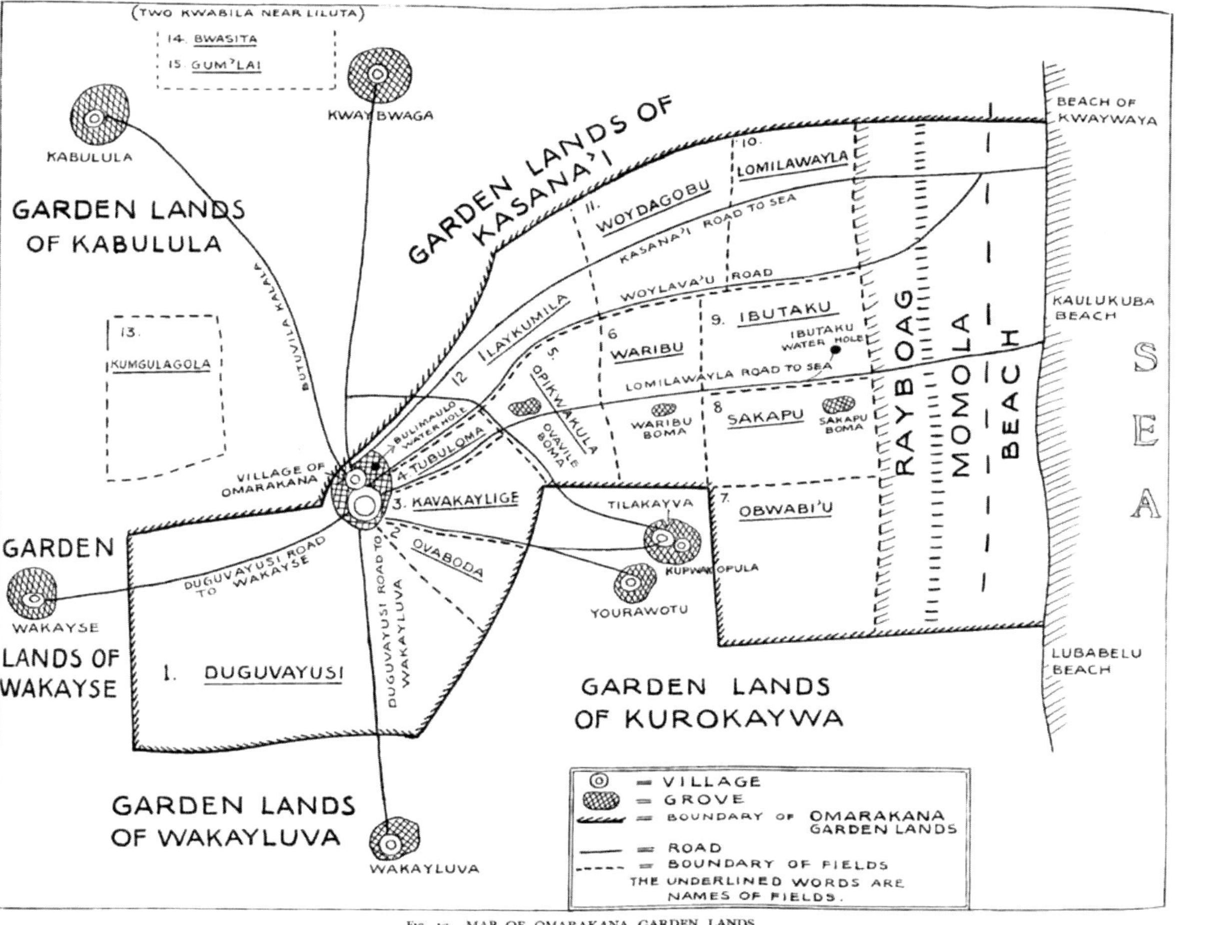

Fig. 13.—MAP OF OMARAKANA GARDEN LANDS

(Ch. XII, Sec. 3). In the course of time, either because the senior lineage of the Tabalu settled here or for some other reason, this village became the premier settlement of the whole archipelago and the acknowledged capital of the district. The real cause of this may have been not so much seniority, as the fact that here the Tabalu obtained among other privileges that of rain magic and thus became the wielders of rain and drought, hence of fertility (Ch. I, Sec. 8; Ch. V, Sec. 1).

Let us start with the village site. Omarakana itself is not altogether a representative village, because the importance of the paramount chief himself, and the wide difference in rank between him and his kinsmen on the one hand, and the commoners on the other, completely colours and even distorts the whole character of public life on the central place. The chief's personal dwelling stands in the middle of the inner ring, the only instance in the whole of the Trobriands. (As far as I remember even in Olivilevi—which is a recent offshoot of Omarakana and the capital of the junior branch of the Tabalu who rule over Luba—the chief's personal hut does not stand inside the inner ring but is in line with the yam-houses of the inner ring.) In other villages ruled by a man of high rank his personal hut would stand among the yam-houses of the inner ring.

The public life of Omarakana is also dominated by the fact that the chief holds his Court in the *baku*, that only people who are invited as his guests are admitted to the portion of the central place situated in front of his personal hut, that is, between this and the large yam-house. In pre-European days, this was also the burial-ground for the Tabalu. Here also the very high platform for the spirits would be erected, and here was found a tabooed heap of stone which no one must tread on nor even approach. The eastern part of the *baku* was the dancing-ground of the village; here also at night children and adults would play their games and in the day-time commoners would forgather. As long as a man was resident in the village he had the customary right to use this place.

The outer ring of the village can be roughly divided into three sections. On the south-west there was a row of huts and yam-houses belonging to the chief's wives. The chief's kinsmen lived to the south-east where in olden days the small village called Katakubile used to stand. The north-eastern part of the village, called also Yogwabu, was inhabited mainly by residents of lower rank and included a few huts of the Kaluva'u sub-clan. Every resident has the right to a site for his living-house and for a yam-house, and he and his family naturally used the adjoining portions of ground for such domestic purposes as cooking, preparation of food, manual work and sociable forgathering (cf. Part I, Sec. 7).

In villages of lower rank the whole of the central place would be used by residents. The difference between non-citizen residents and real citizens consists in the latter's right to be buried there and to take a leading part in ceremonies and communal meetings.

In the grove of Omarakana and Kasana'i, all villagers would have the right of gathering wood and wild fruit. Some trees, especially coco and areca palms, mango and bread-fruit were individually owned. Only citizens would have the right of planting and owning such trees, but a non-citizen resident would usually be given the use of some of them by the citizen through whom he resided; that is, a man would give the usufruct of such a tree to his son, or a chief to his dependant. All the residents in Omarakana, as in other villages, have the free use of such public places as roads, groves, the water-hole, the *rayboag*; and especially of favourite spots for games and recreations which are often held round the water-hole or round some of the fruit trees growing in the *odila*.

The uncut jungle on the *rayboag* could also be scoured and used for any economic purpose by any resident in the community. The holes in the *rayboag*, as we know, are allotted to individuals and yams would be cultivated there and owned by the proprietor. On the portion of fertile soil adjoining the beach there are some four to six fields owned nominally by the paramount chief. I did not enquire into this matter very carefully, but I think that regular gardens would be made there only in times of drought or when for other reasons the cultivable soil on the near side of the *rayboag* was exhausted or unusable.

We pass now to the soil of the *odila* (uncut bush) used periodically for gardens. All this ground, as we can see on the map (Fig. 13), is subdivided into *kwabila*. The boundaries (*karige'i*) are fixed and definite; some of them are marked by roads, others by natural features such as the outskirts of the *rayboag* or of a swamp (there are, however, no swamps on the garden lands of Omarakana—the swampy district lying to the east and north-east of this settlement). Some of the *karige'i* are artificially made by a long solid heap of stones or marked by a row of large trees. As can be seen from the map, the size of the *kwabila* varies considerably. I was unable to survey the ground and I cannot assess their exact size, but it is possible to calculate these roughly by taking as the basic unit of measure the distance between the villages of Omarakana and Wakayluva, which was almost exactly a mile. I drew the map by estimating the longer distances by the time it took me to walk them in a straight line, by measuring the shorter distances in paces and by taking the angles with a compass. The size of the plots (*baleko*) is, roughly speaking, constant. I calculated that four *baleko* would form one morgen (about two acres). Since the *kwabila* vary from between six or eight to fifty or sixty *baleko*, the difference in size will be apparent.

A brief survey of the various fields would add a concrete touch to our discussion of land tenure:—

1. Duguvayusi is the largest *kwabila*, including some fifty or sixty *baleko*. When its turn comes for tilling, it is nowadays almost sufficient by itself for the main gardens of Omarakana. In olden days, however, when the village numbered probably the double of its present complement, twice as many *baleko* would be cultivated. In 1915, for instance, the main garden was made on Duguvayusi, and to this were only added the

fields of Tubuloma and Opikwakula, on which the early gardens were made. With the exception of some ten *baleko*, which belong to the sub-clan of Burayama, the rest of this field is "owned" by the Tabalu.

2. OVABODA is a small *kwabila* measuring some ten to fifteen *baleko*, of which perhaps one-third belong to the Burayama sub-clan and the rest to the Tabalu. This field lies quite close to the village, is extremely fertile and is used as a rule for early gardens.

3. KAVAKAYLIGE is a little larger than Ovaboda and is divided from it by the road going from Omarakana to Yourawotu. It numbers fifteen to twenty plots, five of which belong to the Burayama sub-clan and the rest to the Tabalu.

4. TUBULOMA is the smallest field, numbering about six or eight *baleko* all of which are owned by the Tabalu. It is used as a rule for early gardens.

5. OPIKWAKULA numbers some twenty plots, of which five belong to the Burayama, the rest to the Tabalu. On this field is situated the tabooed grove of Ovavavile, which plays a part in the Kurokaywa garden magic (Ch. IX, Sec. 2).

6. WARIBU numbers about thirty plots, of which about half are owned by the Burayama and the rest by the Tabalu. It includes a tabooed grove of the same name and like the previous field it is cut in half by the road leading from the village to the water-hole and to the beach.

7. OBWABI'U is a field of the same size as the previous one, numbering some thirty plots, of which perhaps one-sixth are owned by the Burayama and the rest by the Tabalu.

8. SAKAPU, on which there is the tabooed grove of the same name from which the first ancestor of the Burayama clan emerged, is of the same size as the two previous fields, numbering about thirty plots, of which one-half belongs to the Tabalu and the other to the Burayama.

9. IBUTAKU is named after the water-hole of Omarakana which lies in this field. This is really a large pool used by the natives not only to draw water but also to wash, bathe and frolic. (This does not spoil their pleasure in drinking the water, but I used to draw my water from a small and despised hole which was, however, not large enough for bathing.) On this field the sub-clan of Kaluva'u as well as the Burayama and the Tabalu own some plots, of which there are about thirty.

10. LOMILAWAYLA is of the same size as the previous fields, that is, about thirty plots. The Tabalu own about half and the other clans share the remainder.

11. WOYDAGOBU consists of thirty plots, of which about fifteen belong to the Tabalu, ten to the Kaluva'u and five to the Burayama.

12. ILAYKUMILA. The same conditions as the preceding one.

The following three fields are detached from the main block of Omarakana garden lands. The soil on them belongs to the Tabalu exclusively. They are all three large fields, bigger than Duguvayusi, and even in old days, I was told, each of them would be enough to accommodate the bulk at least of the main garden enclosure of Omarakana.

They number probably sixty to eighty *baleko* each, perhaps even a little more. The first one, No. 13, KUMGULAGOLA, lies within the grounds of Kasana'i towards the village of Kabulula. The two others, 14, BWASITA, and 15, GUMLA'I, are within the garden lands of Liluta. All three are thus situated within a territory over which the rulers of Omarakana have direct sway and control; and in the immediate neighbourhood of villages which were their permanent subjects and allies. Thus gardens could be made on them even in olden days when warfare might have separated any two communities not permanently friendly.

In estimating the value of the data here given it must be remembered that I had to rely on the statements of my informants as to the number of plots contained in each field. Since in calculating the plots they usually try to remember their names and go over them concretely, the figures given must be regarded as approximate. If anything, they are an understatement, since some of the plots might have been forgotten by my informants.

We can summarise the data here given in the following table:

Name of Kwabila	*Approximate Number of Plots*
1. Duguvayusi	50–60
2. Ovaboda	10–15
3. Kavakaylige	15–20
4. Tubuloma	6–8
5. Opikwakula	20
6. Waribu	30
7. Obwabi'u	30
8. Sakapu	30
9. Ibutaku	30
10. Lomilawayla	30
11. Woydagobu	30
12. Ilaykumila	30
13. Kumgulagola	60–80
14. Bwasita	60–80
15. Gumla'i	60–80
The approximate total of Omarakana *baleko* is therefore	491–573

We see thus that roughly the total number of plots at the disposal of Omarakana is 500 to 560.

APPENDIX I

ANALYSIS AND CHART SHOWING THE ORGANISING FUNCTION OF TROBRIAND GARDEN MAGIC

MAGIC plays such an important part in Trobriand agriculture that, besides meeting it in the title of the book, we were confronted by its relationship to gardens at every step of our description. Here I want in the first place to say something about Trobriand magic in general; secondly to summarise and thus expose the relation of magic to work by giving a synoptic chart of their temporal, spatial and sociological interrelations. I have on purpose not dealt very fully with the subject of magic in Part I, partly because I thought it better for the reader to become acquainted with it in its concrete manifestations and associations with gardening, partly because I wished to treat it systematically, fully and analytically, and in conjunction with the present table; and this is better done in an appendix. Also I am going to a large extent to repeat matter that has already been published (*Argonauts of the Western Pacific*, Chs. XVII and XVIII; "Magic, Science and Religion", in *Science, Religion and Reality*, edited by J. Needham, 1925; *Sexual Life of Savages*, Ch. XI; and *Encyclopaedia of Social Sciences*, edited by Seligman and Johnson, *s.v.* "Culture").

A. SOME GENERAL CHARACTERISTICS OF TROBRIAND MAGIC

To the Trobriander magic is an intrinsic element in a number of important pursuits. In this book we have met it in connexion with all the principal stages and aspects of gardening. We have also found it playing an essential part in the stacking up and storing of produce and in the magic of *kaytubutabu*, the ceremonial "closed season" for coconuts. We also had to mention the magic of drought and of rain, and touch upon the protective magic of property in trees and fruits, the *kaytapaku* (Ch. XII, Sec. 4). If we were to turn to other food-producing pursuits, we should find that the second in importance—fishing—is also controlled by magic. Here, however, any observer would be impressed by the fact that the most lucrative, most reliable and probably quantitatively the most productive method of fishing—that by poison—has no magic whatever, whereas the fishing with seine nets on the lagoon is associated with a very definite ceremonial of a magical nature, including spells and rites; and the fitful fishing for mullet as well as the dangerous fishing for shark are surrounded with a regular public ceremonial. Such relatively unimportant activities as snaring and hunting are associated with a few minor acts of private magic, but have no developed systems of public ceremonial. Again, we shall see that neither the building of houses nor the building of stores has any magic of construction. A new yam-house will be medicated with *vilamalia* magic, while a new dwelling will have some rites

Fig. 14.—CHART OF MAGIC AND WORK

In Column 3, Italic type indicates economic activities; Clarendon,

TIME	PLACE	ACTIVITY *Technical and Social*	PERSONNEL
An afternoon in the moon of YAKOKI	In front of TOWOSI's house	**KAYAKU**	Garden team
A morning some weeks later in KULUWASASA	Bush, RAYBOAG, MOMOLA		
	Coastal fishing community	**Obtaining of fish**	Younger men
Afternoon	Front of TOWOSI's house		
		Offering of fish to Magician	Men of community
Evening	In each household	**Meal from distributed fish**	Members of each family
	In the magician's house		
Next morning (in succession)	Front of TOWOSI's house	**Starting for BUYAGU**	Garden team
	On main LEYWOTA		
	In turn on each BALEKO		
Next day	LEYWOTA and chief's BALEKO	*Cutting the scrub*	All men (communal labour)
Subsequent days	Ordinary BALEKO	*Cutting the scrub*	Owner and relatives
MILAMALA (next moon in native calendar)		Complete lapse of all work for a fortnight (scrub left to dry)	Whole community

Fig. 14.—CHART OF MAGIC AND WORK

non-economic activities; and Roman, taboos or abstentions from work.

PURPOSE	MAGIC		
		PERSONNEL	
Choice and allotment of lands			Ch. II, Sec. 3
Preparatory for ensuing rite	(Collecting of ingredients)	Magician and acolytes	Ch. II, Sec. 4
Provision of ULA'ULA			
Preparatory for second stage	(Preparing the magical mixture)	Magician and acolytes	
Payment for magician and wherewithal for oblation			
Festive meal			
Inaugurating of gardening in general, and of cutting the scrub Imparting fertility to soil	Oblation to spirits M.F. 1 Charming of axes and herbs M.F. 2	Magician	M.F. 1, 2
	KAYLALA M.F. 4 YOWOTA (KAYGAGA and KAYOWOTA) M.F. 5 KAYLEPA M.F. 6	Magician and acolytes in presence of garden team	M.F. 4, 5, 6
	Partial repetition of previous ceremony	Owner of BALEKO with magician's acolyte	
			Ch. II, Sec. 5
Visit of spirits			

Fig. 14.—CHART OF MAGIC AND WORK

In Column 3, Italic type indicates economic activities; Clarendon,

TIME	PLACE	ACTIVITY TECHNICAL AND SOCIAL	PERSONNEL
End of MILAMALA or beginning of YAKOSI	Garden site as a whole	General taboo on garden work	Observed by whole community
Same day (afternoon)	Seashore and DUMYA		
	In TOWOSI's house		
Next morning	LEYWOTA, and each garden plot in turn		
Immediately after, on three consecutive days	Magical corner of LEYWOTA and each BALEKO		
Beginning next day, lasting over the moon of YAKOSI	LEYWOTA and each BALEKO	*KOUMWALA (clearing), early planting, making TULA, fence, and small supports*	Owner and family
A day at the end of YAKOSI or beginning of YAVATUKULU	All BALEKO		
Four or five subsequent days	Garden site	*Bringing in of LAPU* and taboo on other garden work	Men
	Bush, RAYBOAG, MOMOLA		Whole community
	In TOWOSI's house		
A morning soon after	Magical corner of LEYWOTA and of each BALEKO		
A day or two later			

Fig. 14.—CHART OF MAGIC AND WORK

non-economic activities; and Roman, taboos or abstentions from work.

PURPOSE	MAGIC		
		Personnel	
Ritual execution of the practical act of burning the scrub	Vakavayla'u (First ceremonial burning) (Torches charmed at previous harvest)	Magician and assistants	Ch. III, Sec. 1
Inaugurating the Koumwala	(Collecting of ingredients)	Magician and acolytes	
	Charming of torches M.F. 2	Magician	M.F. 2
	Gibuvtyaka (second ceremonial burning)	Magician and acolytes	M.F. 7, 8, 9
To keep the bush-pig from the gardens	Pelaka'ukwa M.F. 7 (planting a taro)		
To promote growth	Kalimamata M.F. 8 (planting a Kwanada)		
To make the taro grow	Bisikola M.F. 9 (building the spirits' hut)		
This stage of gardening is inaugurated by the joint ceremony consisting of the five foregoing rites			Ch. III, Sec. 3
To signify Taboo on garden work	Kayluvalova inserted	Magician and acolytes	Ch. III, Sec. 4
Preparatory to Kamkokola ceremony	(Collecting of herbs)	Magician and acolytes	
	Charming of axes M.F. 2	Magician	M.F. 2
To strengthen the growth of Taytu vines	Erection of Kamkokola M.F. 10, 11	Magician and acolytes	M.F. 10, 11
	Vakalova M.F. 12		M.F. 12

FIG. 14.—CHART OF MAGIC AND WORK

In Column 3, Italic type indicates economic activites; Clarendon

TIME	PLACE	ACTIVITY TECHNICAL AND SOCIAL	PERSONNEL
Next day, and through the moons of YAVATUKULU and TOLIYAVATA	LEYWOTA and each BALEKO	**Erection of three remaining KAMKOKOLA and of KARIVISI (building the magical wall)** *Full SOPU (planting of taytu)*	Owner and family, or communal labour
In stretches of several days, determined by the growth of crops, during the moons of YAVATAM and GELIVILAVI	Gardens	Slack season in garden work: *Inserting supports and training the vines*	Men of community
		Taboo on weeding	
		PWAKOWA (weeding)	Women (often communally)
In the moon of GELIVILAVI	In magician's house		
Next morning	LEYWOTA and each BALEKO		
Next day			
Next day	Gardens	**Digging out a few yams and taro**	Men of community
During the moon of BULUMADUKU	Gardens		
In the moon of KULUWOTU	Bush, RAYBOAG, MOMOLA		
	Gardens		
Next day	On each BALEKO	*BASI (thinning)*	Men of community

Fig. 14.—CHART OF MAGIC AND WORK

non-economic activities; and Roman, taboos or abstentions from work.

PURPOSE	MAGIC		
		PERSONNEL	
	(Working screams)	Men at work	Ch. III, Sec. 5
Promotion of growth, referring to shoots and leaves	VAGURI SOBULA M.F. 13 VASAKAPU SOBULA M.F. 14	Magician or acolyte	Ch. IV, Sec. 2
Inaugurating weeding	KARIYAYELI SAPI M.F. 15		M.F. 13, 14, 15
Further growth magic	KAYDABALA M.F. 16 KAYLAVALA DABANA TAYTU M.F. 17 KASAYBODA M.F. 18		M.F. 16, 17, 18
Preparatory for ISUNAPULO	Charming over a pearl-shell M.F. 25	Magician	Ch. V, Sec. 2 M.F. 25
Offering to ancestral spirits	Cutting off a taro plant (ISUNAPULO)	Magician or acolyte	
To signify taboo on work	Stick and leaves placed on each KAMKOKOLA		
First fruits to be displayed on BAKU and laid on the graves			
Growth magic, second cycle, referring to tubers	VAPULI M.F. 19 KAMMAMALA M.F. 20 KASAYLOLA M.F. 21	Magician or acolytes	Ch. IV, Sec. 3 M.F. 19, 20, 21
Preparatory for ensuing rite	(Collecting of herbs)	Magician	
Inaugurating BASI	MOMLA (BASI magic) M.F. 2	Magician or acolyte	M.F. 2

Fig. 14.—CHART OF MAGIC AND WORK

In Column 3, Italic type indicates economic activities; Clarendon,

TIME	PLACE	ACTIVITY TECHNICAL AND SOCIAL	
			PERSONNEL
Kuluwotu and Utokakana	Gardens and house	Slack season in garden work	
End of Utokakana (morning)	Leywota and each Baleko	Taboo on garden work	Observed by whole community
Noon of same day			
Early in the moon of Ilaybisila	Magician's house		
Next morning	Leywota		
In the moons of Ilaybisila and Yakoki	Garden site	*Main harvest: digging and cleaning the tubers; building KALIMOMYO; display and sorting of crops*	Each family
During Kaluwalasi	Garden to village	*Bringing in of crops* and **display of heaps in village. This is the period of Buritila'ulo**	Gardeners, with kinsfolk and relatives-in-law
A day late in Kaluwalasi	Bush, Rayboag, Momola		
Next morning	Chief's Bwayma and other Bwayma		
Same morning, immediately after Tum Bubukwa	Village	*Filling of the BWAYMA*	Donors and helpers
A day or two later (morning)	Bush		
(Noon)	Magician's house		
Sunset (same day)	Village (main storehouse first)		

Fig. 14.—CHART OF MAGIC AND WORK

non-economic activities; and Roman, taboos or abstentions from work.

PURPOSE	MAGIC		
		Personnel	
	This is the season of private magic:— Spell recited over Yagogu M.F. 22 The magic of the Tula M.F. 23 The magic of the Dayma M.F. 24	Magician or gardener	Ch. IV, Sec. 4 M.F. 22, 23, 24
To signify taboo	Strewing the ground with leaves	Magician and acolytes	Ch. V, Sec. 3
To make the tubers mature	Charming each Baleko with Okwala spell M.F. 27	Magician	M.F. 27 M.F. 2
Inaugurating the Tayoyuwa (main harvest)	Tum: charming M.F. 2 of adze, and charming of torches for next Vakavayla'u M.F. 2		
	Cutting the taytu stalk and pressing down weeds		
Division of Urigubu from Taytumwala			Ch. V, Sec. 3
			Ch. V, Secs. 4, 5
Preparatory for Vilamalia	(Collecting of herbs)	Magician	Ch. VII, Sec. 1
To give stability to the accumulated crops	Tum Bubukwa M.F. 28		M.F. 28
Presenting the Urigubu			Ch. VII, Sec. 1
Preparatory for Basi Valu	(Collecting of herbs)	Magician	Ch. VII, Sec. 3
	Charming over the leaves M.F. 29		M.F. 29
To make the crops last	Basi Valu		

performed over it, directed to the prevention of sorcery. Quite different, however, is the magic performed over a new canoe. The small craft used for ordinary, safe, easily controlled sailing in the estuaries of creeks or on the lagoon command no magic whatever; the larger boats used for communal fishing are charmed over with the magic of fishing; while the biggest overseas canoes have a magic of safety, speed and efficiency, elaborately and punctiliously performed over them during the process of building (*Argonauts*, Chs. IV and V). With regard to these canoes there is also a magic performed at sailing, a protective magic performed at shipwreck, and in connexion with their building and use there is the magic of *kula*. The magic of spondylus fishing and the magic of safety in foreign parts are performed during the expeditions. There is also a magic of wind and weather.

The processes of industrial manufacture have no magic at all. Neither in the making of valuable objects nor for the highly skilled industries in which the districts of Kuboma specialise do the natives ever resort to it —with one exception. For artistic carving inspiration is sometimes sought by way of magic. Even this rough survey shows that the Trobriander appeals to magic in pursuits where—as in fishing—there is an element of chance which he cannot control; or where—as in overseas sailing—these are partly under his control but still contain elements of danger and accidental mishap.

Magic is prominent also in all matters of health, bodily welfare, dangers from accident, and in the gravest issue of all—that of life and death. Mishaps to the human organism are conceived by the Trobrianders as the result of sorcery. They believe that, by means of rite and spell, those who have the knowledge can harm other people; and they have a corresponding belief that by counter-magic the harm can be undone. Similarly there is a magic whereby love can be produced and destroyed by spell and rite. Personal attractiveness, beauty, success in dancing, form another important department (see *Argonauts of the Western Pacific*, for the magic of beauty and persuasiveness at *kula*, Ch. XIII, and *Sexual Life of Savages*, Ch. XI, for the magic of love and beauty). Warfare was accompanied by specific types of magic.

The reader of this book and of my previous accounts of several aspects of Trobriand magic could easily frame a number of generalisations concerning the technique of magic and its constitution. I shall present here a brief outline of the essentials of magic, summarising what I have already stated in the chapters and passages on Trobriand magic referred to above.

Magic is conceived by the Trobriander as an intrinsic element in everything which vitally affects man and his destinies. Magic is not a thing which could ever have been invented by man. It is believed to have emerged with the first ancestors of man from underground. There it always existed. Its origins are as little a matter for speculation as the origins of mankind or of the world. The words of the spell, the form of the ritual, the very substances used in it, are coexistent from the very beginning with the things or natural processes over which they exercise a power. Magic then is a traditionally established power of man over certain natural processes, over some human activities or over other human beings.

Every magical act consists of a spell and of manual or bodily behaviour. This latter at times is reduced merely to the magician's repairing to the appropriate place and facing the appropriate object. Such is the case in most rites of growth magic. When the rite is more complicated, it may either consist in the chanting of the spell over some object which afterwards is used in the pursuit; or, in more complicated rites, the magician may invite the presence of spirits with an accompanying offering or display of food and valuables. We have met with one or two such rites in the foregoing chapters, notably the first rite in Omarakana garden magic and the *kamkokola* ceremony in the Momtilakayva system (cf. Ch. IX, Sec. 2).

Even from this brief definition it is quite clear that magic in the Trobriands is conceived as essentially the quality or attribute of man. It is not conceived as a force which can reside in an object and transmit from it to another. The part played by sympathetic substances is always indirect and vicarious. The virtue of a spell may be reinforced by being chanted over some herbs, or earth taken from the bush-hen's nest, or chalk scraped from a coral boulder. But such substances are never active alone. The spell is an essential ingredient in Trobriand magic. The spell is the most esoteric part of magic. The effective use of spells always constitutes the exclusive prerogative of the magician, whether the words are secret or not. In the case of public magic it constitutes in every case the exclusive prerogative of the magician. That is, even where magic is public and known to everybody, it may not be performed by anyone but the accredited, traditionally determined public magician. This obtains with regard to garden magic, fishing magic, war magic, *kula* magic and similar types of public ritual. In sorcery, in the magic of love, in several forms of black magic, the spells are kept a secret; they may be handed over only against payment or to persons whom the giver wishes to endow for reasons of sentiment or duty. The magical power is acquired primarily by learning the spell. This has to be memorised with absolute accuracy; any change in the wording, unauthorised curtailment or wrong method of recital being regarded as either completely fatal to the results or else diminishing their power considerably. I have already described the myth of canoe magic, in which the full spell could move a canoe to flight through the air, while part of the spell adds but to its velocity on the sea. When the spell is acquired by learning it word for word, it sinks down into the abdomen and there takes residence. When the magician recites it the action of the throat, which is the seat of the human mind or intelligence, imparts the virtue to the breath of the reciter. This virtue is then transmitted in the act of recital directly to the objects charmed or to the substances which will be afterwards applied to the objects charmed.

With this conception of magic as something residing in man, traditionally grafted from one individual to his successors in magic, produced through the use of the human larynx and the human voice—there is naturally connected a belief that a magician must keep certain taboos, submit to certain rules of conduct and be in his bodily nature appropriate to the performance of his magic. We are already acquainted with the taboos of the garden magician (cf. Ch. II, Sec. 7), and by looking up the other books on the Trobriands similar taboos will be found in connexion with *kula* magic, the magic of canoes, and the magic of love and beauty.

We know also what is meant by the bodily appropriateness or status of the magician mentioned above. The garden magician must not only belong to a certain sub-clan, but in principle he is the head of that sub-clan, or at least his accredited deputy. We know also from this book that perhaps the most momentous magic of the whole area, that of rain and sunshine, can only be performed by the paramount chief or his deputy. The local magic of organised fishing, such as that for shark, mullet and in the lagoon, must be performed by a headman of the sub-clan who are regarded as the owners of the whole enterprise and all that it involves. There are only a few forms of magic which can be purchased by anyone from anyone and carried out by a member of any sub-clan on behalf of anybody. To a certain extent, sorcery falls into this class, but the matter is too complicated to be stated briefly. Other forms of occasional or independent magic, formulae of love magic, private garden magic, minor rites of fishing, hunting or warfare also belong here. In this connexion we have now to pass to a definition of "systematic magic" as distinct from "independent" or occasional magic.

B. GARDEN MAGIC AS A SYSTEM OF RITES RUNNING PARALLEL TO AN ORGANISED PURSUIT

The distinction between systematic and independent magic is very clear in the Trobriands. Some rites and spells can be used whenever the occasion arises, and they are not embedded in a sequence of connected activities. When during an overseas expedition the wind drops, a man from the crew would invariably get up and chant a wind spell. When a person has an attack of severe toothache he or she will go to some neighbour and a magic for toothache will be performed. There are minor rites of love magic which can be performed whenever a Trobriander feels that his passion is insufficiently regarded or rewarded. But then again there is magic which is interwoven, in the form of a connected and consecutive body of rites and incantations, with a sequence of economic activities, or with such an enterprise as war, or with a season of ceremonial dancing. The magic which punctuates an overseas expedition and which really starts with the felling of a tree for the canoe and ends with some final rites of farewell is a good example of such a system. Even more convincing is the magic here given with reference to gardening. The first point which strikes us in this magic is the fact that it is regarded as an indispensable ingredient in the activities. As I have insisted, the Trobrianders would never start a garden—any more than they would think of embarking upon a *kula* expedition or a period of shark-fishing or a large dancing ceremony—without inaugurating the work by appropriate magical ceremonial and punctuating it thus throughout the season. Some minor formulae might be omitted, though I doubt very much whether they ever omit any in gardening. But the main rites may not be passed over or skimped (see the incident of lay burning of the gardens, reported above, Ch. III, Sec. 1).

The association between the technical pursuit and its magical counterpart is, as we know, very close, and to the natives essential. The sequence of technical stages, on the one hand, and of rites and spells, on the other, run parallel. The place of a magical act is strictly determined. There are inaugurative rites, such as the *yowota* or *gabu*, there are concluding rites such as the *vilamalia* and the last act of *kaytubutabu*. There are rites in which the ceremonial and the practical are interwoven, such as the *vakavayla'u* and *gibuviyaka*. Magic is as indispensable to the success of the enterprise as manual activity. In reading the foregoing account of work in the gardens and the ceremonial of *towosi* we have throughout been impressed by the fact that both are directed towards the same end: to ensure a rich yield and thus to establish prosperity for the ensuing season or year. We have also seen throughout that the contribution of magic and the contribution of work are not equivalent. Work is done to cut down the scrub, to burn it systematically, to clear the ground, to plant the crops, to protect them from the attack of bush-pigs and wallabies, to weed and then to harvest the crops. Magic adds to this by summoning the fertility of the soil, exorcising pests and blights against which no practical devices exist; frightening away the bush-pigs and invoking protective animals, magically assisting the growth of the plants underground and stimulating the foliage; finally, at *vilamalia*, by undermining the human appetite and strengthening the spiritual foundations of the yam-house.

But though different in their scope and technique, both ceremonial and practical work form part of one enterprise, are directed towards one end, and progress in a consecutive series of performances which depend one upon the other. The proper understanding of the manner in which systematic magic and the progressive body of practical activities integrate, is very important for our anthropological theory of magic and of work. It reveals the nature of the relation between magic and economic pursuits. In a way it determines the function of magic.

Now in order to bring out this relation—however clear it may already have become from the narrative description—I have found it necessary to make a comparative chart. This I had already done in the case of garden magic in the field, and I am reproducing here such a chart completed and made as transparent as accuracy and fullness allow. The chart is frankly based on the garden system of Omarakana only. A compound synoptic chart would be impossible. This the reader will see for himself if he compares the data of Chapters II to VII with those of Chapters IX and X. We might have constructed similar charts for the garden magic of Kurokaywa and Vakuta. That would have been unnecessary even if we could have made the other charts as full and well documented as the present one. What interests us in the chart from the theoretical point of view is obviously not whether a dry stick is put into the soil three days before or three days after a certain act. The minutiae of native belief and custom interest us because they document and reveal to us the native attitude towards magical act and work respectively and because they enable us to understand the real relationship between these two orders of cultural reality. By studying the present chart, comparing it with the table of Omarakana and Vakuta magic, and collating these with the additional evidence of Chapters IX and X, the sociologist will have all the data necessary to frame his conclusions.

A brief letterpress, however, will be useful. The table has two dimensions. Reading down each vertical column, we progress in time. This time progression is definitely stated in the left-most column. From left to right, along the horizontal ordinate, we have listed side by side the simultaneous facets of an event. The place where it happens, its practical nature, its magical counterpart, its personnel and, in the column inserted between secular activities and rites, its purpose.

The two main headings are obviously those of activity and magic; activity standing here for the cycle of technical work which forms the backbone of this column. But here I have also listed the ceremonial, the social and the legal events connected with gardening. For reasons of space and clarity it was expedient to place these in the same column. I have distinguished them, however, by printing those which are not economic, i.e. ceremonial, legal or social, in clarendon; the economic or technical ones in italic, and abstentions from or taboos on work in roman type. Thus *kayaku*, "obtaining of fish", "offering of fish to the magician", "meal from distributed fish", the ceremonial "starting for *buyagu*", etc., will be found printed in clarendon. They are all activities which form an integral part of the whole gardening system. But they stand outside the consecutive series of practical work. They are partly legal, partly ceremonial acts, connected at times more directly with magic, or with the organising of the whole community for work, than with the work itself. We first meet a strictly economic activity in the tenth pigeon-hole (sixth entry) of Column 3—"cutting the scrub"; next comes the *koumwala* (clearing), "early planting", "making *tula*, fence and supports", then "bringing in of *lapu*", and so on, up to the final economic act—the "filling of the *bwayma*". All these have been printed in italics. The seventh entry (eleventh pigeon-hole), "complete lapse of all work for a fortnight (scrub left to dry)," is printed in ordinary type. It is but a negative statement of what happens to the gardens during the *milamala*. Similarly with the entry which follows, "general taboo on garden work", which occurs again later on during the three harvesting ceremonies. The "erection of three remaining *kamkokola* and *karivisi* (building the magical wall)", which we have printed in clarendon, might perhaps have been put under magic as it is a counterpart of a magical rite and has only a secondary practical purpose or aim. But since there is no rite involved in this act, and it entails a good deal of work, it was probably more in accordance with our scheme to place it in the third column. The same refers to the entry "digging out a few yams and taro". In one or two mixed entries I have distinguished the sentence referring to work from that which designates either a ceremonial activity or an abstention. Thus in "***bringing in of lapu*** and taboo on other garden work"; the first part is in italic, the second in roman type; so also the entry "***bringing in of crops*** and display of heaps in village".

The last column contains acts of magic or acts preparatory to magic. These latter, when there is no rite or spell involved, have been placed in brackets. They are almost all covered by the entry "collecting of ingredients" and "collecting of herbs". Such acts as "strewing the ground with leaves", "stick and leaves placed on each *kamkokola*", though perhaps not strictly magical, since there is no spell involved, have been listed as such because they form part of a bigger magical ceremony.

Both columns 'Activity' and 'Magic' contain a subdivision each on 'Personnel'. This gives the sociological dimension of the act, or, more simply, shows whether the ceremony is performed by the magician alone or with the assistance of his helpers, or even with the wider participation of the whole community. In work or ceremonial it gives us the play of the various groups. Here a cross-reference between this table and the sociological analysis of Chapter XII will be useful. It will be quite obvious that in garden work the clan and sub-clan play no part. The groups which figure are only 'garden team', 'village community', the men workers as opposed to the women workers, the family; and if we analyse certain entries more fully we would find that differences in relationship to land would come in, especially at *kayaku* and at harvest.

Perhaps the most important column in the whole table is the one entitled "purpose" which, however, must be understood in a fuller way. In this column we have placed the entries which define the correlation between the simultaneous activities of technique and magic, or between the entries on the same line with it and those on the horizontal line or lines below. In order to gain a clearer idea of the nature of the entries in this column, and thus of the whole table, let us give a few readings of the table from left to right and from the top downwards. In the first line we find five entries only: none under magic—which simply means that in this act magic does not figure. The successive entries juxtaposed almost make sense, though in a somewhat telegraphic style: TIME "An afternoon in the moon of *Kuluwasasa*"—PLACE "in front of *towosi's* house"—ACTIVITY *kayaku*—PERSONNEL "garden team"—PURPOSE "choice and allotment of lands". The meaning is obvious: "On an afternoon in the moon of *Kuluwasasa* (and glancing at our Chart of Time-reckoning we see that this is July) the members of the garden team (Ch. XII, Secs. 2, 3 and 4, and Ch. II, Sec. 3) assemble in front of the *towosi's* house to hold the *kayaku* (inaugural garden council) in order to choose the fields and allot the plots." This gives you in a nutshell the summary of the third section of Chapter II, as is marked on the chart.

Next entry: TIME "A morning shortly after *kayaku*"—PLACE "bush, *rayboag, momola*"—PURPOSE "preparatory for ensuing rite"—MAGIC "collecting of ingredients"—PERSONNEL "magician and acolytes". This obviously means that "on a morning shortly after the garden council, the magician and his acolytes repair to the bush, the *rayboag*, and the *momola* in order to collect the magical ingredients preparatory for the rite which will follow." The reference shows you where you will find the elaboration of this summary in the body of the book.

In the same way, the next entry tells you that "on the same morning (for here the time entry has been placed against both activities) the younger men of the village repair to a coastal fishing community and there obtain fish in order to provide for the *ula'ula* offering". And the one after that "on the afternoon of the same day the magician and his acolytes prepare the magical mixture for the second stage of this magic, while the men of the community offer the fish to the magician in payment for his services and to give the oblation to spirits".

In short, glancing rapidly over the time column we see that, from one entry to another, there are at times long intervals, at times but a few hours. The simultaneity of two activities has been marked by placing them both against the same time entry. The entry under Milamala in the 'time' column marks really the pace in gardening and is related to the scheme by the indication that the scrub is drying, between cutting and burning. In most cases, when there is an entry under 'Magic' there is none under 'Activity' and vice versa, as work and magic are seldom allowed to proceed simultaneously. They are linked up by the middle column which very often indicates that a certain magical activity inaugurates the immediately following horizontal entry, which would be found among activities. Thus, corresponding to the series of ceremonies from "oblation to spirits" to "partial repetition of previous ceremony", we have the entry in the middle column: "inaugurating of gardening in general and of cutting the scrub", which bears directly on the next horizontal entry "cutting the scrub". Finally, the entry under 'Purpose' may define a practical activity as a preparation to the magical rite (e.g. bringing in of *lapu*).

It is hardly necessary perhaps to indicate what the practical value of the present chart is. It gives in a clear synoptic manner the gist of our long and complicated descriptions in Chapters II to VII. It analyses the data given there and regroups them under their several aspects so that their interrelations can be seen at a glance. Since our column 'activity' has been subdivided by differences in type into 'economic', 'legal, social and ceremonial' and 'work abstentions and taboos', and since in our column on magic it is easy to distinguish the one or two semi-religious ceremonies in which reference is made to spirits, our chart contains a *de facto* analysis into far more aspects than the two main headings might indicate. To this we may add also the indications as to how an activity is organised contained in the sub-columns on personnel. We have therefore 'economic work', 'legal and social activity', 'taboo and abstention from work', 'magic', 'religious act' and 'social organisation'—roughly speaking six different aspects of the whole process. These aspects are related to each other temporally and spatially and linked up by the column 'Purpose' into their relationship of sequence, motive and dependence. Practically the chart will enable you to place formulae and rites within their context, whether to have a rapid and handy reference—in some cases more useful than an index—or a table of contents. Theoretically it shows the relationship of all the aspects of gardening to each other. Above all it shows clearly the relation between work and magic and substantiates a theoretical principle which the facts of this book have throughout proclaimed—namely that magic in that part of Melanesia has a very definite organising function. We can see that magic in most instances inaugurates practical activities, we can appreciate the lead given by various ceremonies, the rhythm which they introduce into garden work. This, however, is not the place to enter into the theoretical conclusions which could be drawn by a sociologist or student of comparative anthropology from the material here presented. I only wanted to show that this synoptic table may prove a valuable tool for a further theoretical working out of the facts.

APPENDIX II

CONFESSIONS OF IGNORANCE AND FAILURE

1. "NOTHING TO SAY"

PERHAPS the greatest difficulty in handling a record of field-work consists, for the theoretical student, in forming some judgment as to the nature of the lacunae in which every record naturally abounds. Are they due to negligence? Or to lack of opportunity? Or to the fact that there is really "nothing to say" about the subject? This last possibility we can in fact rule out from the outset. I well remember discussing some points about an ethnographic area which I was studying with one of the most distinguished workers in that field. That was a few years before I had been in the field myself. I drew my friend's attention to the fact that on a certain subject there was no information whatever in his works. "I had nothing to say about it", was the answer; and when I enquired why, he simply glared at me. An undiluted Slav at that time, I pressed my Anglo-Saxon friend further and tried to point out that a field-worker has no right to have "nothing to say" on any relevant subject. The subject here was the family and the area was aboriginal Australia. I insisted that there either was a household or not, that husband, wife and children either lived together, slept together and ate together or not. Finally, with his back to the wall, my friend concluded: "Well, I found out nothing about it." I failed to retort: "But it was damn well your duty to find out all about it"; none the less, this is what ought to be said, more politely, on the matter. The anthropologist must at least state whether he was looking for a certain phenomenon and failed to find it or else that he failed to look for it.

The "nothing to say about it" principle is perhaps the main reason why anthropology has not made sufficient progress on its empirical side; and it is the duty of the field-worker to render a careful and sincere account of all his failures and inadequacies. Perhaps the first record where this was done in a really scientific spirit was in Radcliffe-Brown's *Andaman Islanders*. In my *Argonauts of the Western Pacific*, which was published at the same time, I briefly outlined the methods by which I obtained my evidence. But in the course of my narrative I did not go, with anything like enough conscientious thoroughness, into the delimitation of my knowledge and the probable existence of certain lacunae. In writing this book, I determined not to neglect this matter. Finding, however, that the caveats and negative digressions became so frequent that they interrupted the course of the narrative, I determined to substitute a brief summary of "gaps, failures and mix-ups", in so far as I am aware of them. Since the matter is of interest to the specialist rather than to the general reader, I have relegated this into an appendix. But before indicating those lacunae of which I am myself clearly aware and giving a few hints of how I came to miss certain important facts or even aspects in my field of study, it will be necessary briefly to state my credentials in field-work and to summarise my methods which have been already fully discussed in Chapter I of *Argonauts*.

2. METHOD OF COLLECTING INFORMATION

My field-work in Melanesia consisted of three expeditions; the time spent actually among the natives was about two and a half years (cf. the chronological table in *Argonauts*, p. 16). Counting the time given in between the three expeditions to sorting my notes and writing them up, to formulating problems and doing the constructive work of digesting and recasting the evidence, my field-work can be said to cover over four years (early September 1914 to end of October 1918). I am laying some stress on this because I firmly believe that a few months' interval between two expeditions of one year each gives infinitely greater opportunities to the anthropologist than two consecutive years in the field. Of my sojourn in native New Guinea, I spent six months on the south coast, and the rest in the area of the Northern Massim. In the latter I made only a short visit to Woodlark Island (Murua) and two long stays in the Trobriands.

I had studied the structure of Melanesian languages theoretically, and was acquainted with one of them (Motu), when I arrived at Port Moresby in the first days of September 1914. This language, and this language exclusively, I had to use in my field-work among the Mailu. When on my second expedition I arrived in the Trobriands (June 1915) I had not prepared myself for work in that language, because I did not intend to settle in that district for any length of time. By September of that year, however, I found that I could use the language readily in conversation with my informants, though it was much longer before I could follow easily conversations among the natives themselves. In fact I do not think that I reached this stage until I had made a very thorough study of my recorded linguistic material during the subsequent interval (Melbourne, May 1916 to August 1917), and had had a month or two's practice on my third expedition. From that time I had no difficulty in rapidly taking down notes in Trobriand and in following general conversations among the natives. As every reader of Part V will see, the difficulty lies in quickly filling out the gaps in a statement from contextual data. In other words, I think that a complete knowledge of any native language is much more a matter of acquaintance with their social ways and cultural arrangements than of memorising long lists of words or grasping the principles of grammar and syntax which—in the case of Melanesian languages—are appallingly simple.

As to my mode of residence, I have several times insisted on the fact that satisfactory field-work can only be done by one who lives right among natives. Only for brief intervals, all in all not more than six weeks, did I enjoy the hospitality of my friend Billy Hancock of Gusaweta, and of M. and Mme Brudo of Sinaketa. The rest of the time was spent right among the native huts where I used to pitch my tent.

Since gardening is an activity which permeates native life and penetrates right into the village, materially as well as spiritually, I had no difficulty in seeing every phase of it as often as I cared to. By a number of coincidences, however, I saw much more of the early stages of gardening and of the facts relating to harvest than of the intermediate phases. As the reader who glances at the Chart of Time-reckoning (Fig. 3) can see, the *kula* activities to which I devoted a considerable amount of attention, fall at the intermediate stages of gardening. Again, by glancing at page 16 of *Argonauts*, the reader will see that, especially during my third expedition, I was away from the agricultural districts during March and April.

In Part I (Sec. 3) I have described my first impressions of agricultural activities. This was the stage of chaos and of piecemeal observations. But even then I was able laboriously to collect several "solid and well documented" data and to begin to penetrate into the inwardness of the native attitude towards gardening. In Chapter I of *Argonauts* (especially Secs. 2–9), I have pointed out that field-work must always consist in (i) statistical documentation by concrete evidence, (ii) the collecting and recording of the "imponderabilia of actual life", and (iii) linguistic data. These three classes of evidence are perhaps even more palpably differentiated in the present record than in my earlier books. To the linguistic evidence I have assigned a separate part: it occupies the second volume. Objective documentation has again largely been collected in a special place (Part III). The imponderables of behaviour are knit into the main narrative. As far as it was possible to do so without confusing the trend of the story, I have tried to indicate in each special case the way in which I have arrived at a certain psychological interpretation or integrated a multitude of small symptoms into a generalisation concerning native attitudes and mannerisms of behaviour.

With regard to the progress of knowledge and understanding of native life, I should distinguish what might be called the surface attack on a cultural phenomenon such as agriculture from two further stages in analytic penetration. By "surface attack" I mean the collecting of well crystallised, clearly defined facts concerning the activity in question and the recording of relevant points in native law, in economic practice and in belief, as more or less self-contained isolated items. Thus the subject of this volume would give us the following headings: land tenure, technique of gardening, the treatment of the harvested crops, the subsequent use of these crops, the mythology of gardening, the ceremonies of magic. Advancing to the second line of attack in field-work I would consider what relations the various institutionalised facts bear to each other. In land tenure, for example, the really fruitful research begins when, given the purely formal apportionment of titles (cf. Ch. XI, Sec. 3), we enquire what rôle each of these titles plays in production. This enquiry is tantamount to an analysis of the relation between legal ownership on the one hand and organised production on the other. Again, the question "How is land tenure related to native traditions concerning man's connexion with the soil?" leads towards the whole mythological and legal foundation of land tenure (Ch. XII). I need not emphasise here that the most important information contained in this monograph consists not so much in a statement of isolated facts and aspects as in the analysis of the interrelation and interdependence of these. The reader will have seen that the significance of agricultural magic in the Trobriands lies in its organising influence upon native production and in its connexion with the mythology of gardening, of land tenure and of local citizenship (cf. Doc. VIII). The place of agriculture in tribal life is not defined merely by the study of the technique of tilling. The driving forces to effective tilling cannot be understood unless we realise that they are provided by the systems of distribution and exchange of gifts. These are closely connected (Chs. V to VII) and in turn lead us to the study of the storehouse (Ch. VIII) as an implement for preserving and handling taytu. In order to estimate the importance of all this we must see how taytu is used in ceremonial distributions, in trade and in political tribute.

The third line of attack would consist not merely in the study of relations between the various part institutions of agriculture, such as the marriage gift in relation to production, garden magic in relation to garden work. It would proceed to a careful synthesis of the interrelations of aspects into one general assessment of the part played by agriculture as a whole within tribal life. This synthesis, however, transcends the proper task of the field-worker. To lead up to it should be his constant inspiration; he may have his private view about it; but it is not his duty nor yet his prerogative to lay it down in the record of field-work. Exactly as I have abstained from giving my private theory as to the function of the Eastern Papuo-Melanesian institution of *kula*, so here also my views as to the general function of Trobriand agriculture and my theoretical interpretation of the "social value of taytu" (to use the phraseology of my friends Mrs. Winifred Hoernlé and Professor Radcliffe-Brown) can be elicited, I trust, by the comparative sociologist, but are not given explicitly.

I am reaffirming my innocence of any ultimate theoretical assessment of either the institution of *kula* or of agriculture, even after having carefully re-read the last chapter of my *Argonauts of the Western Pacific*, entitled "The Meaning of the Kula". I give there, indeed, a summary of the relations between the component aspects of the *kula*. I analyse the influence of magic on overseas expeditions and the rôle of the individual ambitions and desires integrated in the activity. I also briefly survey a range of collateral facts, culled from other aspects of native life, which throw light on the use of valuables outside the *kula* and on some characteristics of native exchange in general. But I never enlarge there upon the integral function of this institution, though I hope to be able to do so shortly in a theoretical book on primitive warfare and other methods of "heroic enterprise". I shall then attempt to show that, for the Trobrianders at least, the *kula*, as a cultural activity, is to a large extent a surrogate and substitute for head-hunting and war. I shall also develop the view which I have already indicated (in my article *s.v.* "Culture" in the American *Encyclopaedia of Social Sciences*, edited by E. R. Seligman and Alwin Johnson) that in the *kula* the most important economic fact is that the non-utilitarian exchange of valuables provides the driving force and the ceremonial framework for an extremely important system of utilitarian trade. Out of methodological puritanism I have refrained from making any of these points in the record of my field-work on this subject.

The same puritanism has made me stop short of this last stage of theoretical analysis or synthesis in my book on the *Sexual Life of Savages*, though I have thus dealt with some aspects of sexual life in other publications. So in this volume the function of agriculture will have to be formulated by the reader himself from the facts presented to him.

Returning now to method in field-work: I have just been arguing that the first layer of approach, or layer of investigation, consists in the actual observing of isolated facts, and in the full recording of each concrete activity, ceremony or rule of conduct. The second line of approach is the correlating of these institutions. The third line of approach is a synthesis of the various aspects. As I look through the long list of entries in my field notes, I see that, to a large extent, this gradual deepening of my knowledge of the relational aspects of agriculture was a later achievement than the piecemeal study of details. At the same time—and this I have already indicated in Part I—the appreciation of the general value of harvested crops and of their great importance in tribal life was brought home to me by the sheer chaotic welter of observed details from the very outset. At first I was still receiving such information as that reproduced in Section 1 of Chapter IX. I find myself struggling early with the native calendar where a reference to agricultural activities was imposed on me by my native informants. I find entries recording my inspection of the gardens, early plans of garden plots, and of arbours erected on them at harvest; detailed descriptions of the carrying of the taytu, of counting the baskets, and of the display and storing of crops. The term *tokwaybagula*, 'good gardener', was recorded in the first few weeks, giving me an inkling already of the high value attached to efficiency in gardening. Then came my first witnessing of the *vilamalia* rite, studies of technique in planting, ethnographic descriptions concerning seed yams (*yagogu*) and the classification of the various types of yam; and then the long list of magical ceremonies witnessed one after the other, analysed and commented upon. During my first stay in Omarakana I was able, through the good offices of Bagido'u, to obtain an exceptionally full and well documented insight into this aspect of garden work.

The principle of a relationship between magic and work I had very clearly in mind as one of the guiding rules in field observations. My earliest ethnographic publication was on the "Economic Aspect of the Intichiuma Ceremonies", published in a *Fest-schrift* offered to Professor Edward Westermark. Long before I went to the field I was deeply convinced that the relation between religious and magical belief, on the one hand, and economic activity, on the other, would open important lines of approach. The remarkable development of agricultural magic, fishing magic, and magic connected with trade and sailing among the Mailu impressed me strongly in the course of my first survey work among the Motuan tribes near Port Moresby and among the Southern Massim.[1] This relation, needless to say, is perhaps the dominant motive throughout this book. Some of the other mutual dependencies became clearer to me as I worked, notably the extraordinary importance of agriculture in Trobriand political life, achieved through the numerous *urigubu* gifts and the fact that polygamy is one of the principal prerogatives of rank and power. In the field I always found it an invaluable device to map out the facts already obtained, to consider how they were related to each other and to proceed with the investigation of the bigger, more widely integrated type of fact thus arrived at. At times relational phenomena are discovered in the study of concrete documentary data. Thus the principle of *urigubu*—that from each man's garden produce a large quota has to be given to his sister—I discovered by hearing the word *urigubu* used in the classification of crops at harvest. But the meaning of the word became clear to me only when I followed the life history of a taytu tuber from the time it leaves the soil to the time when it comes to rest in the storehouse of the cultivator's sister's husband. The further history from there to some consumer's belly is not less instructive, as the reader of the previous pages, above all of Part I, knows.

3. GAPS AND SIDE-STEPS

But there are dangers in integrating facts, above all in integrating them prematurely. And this brings me to the main theme of this appendix: the statement of the mistakes I have committed, of the traps and blind alleys into which I have been led. Some of them I discovered before leaving the field, though in one or two cases I was able only partly to remedy them. Some of them have emerged out of the comparative treatment and full writing up of my material. Other gaps I can only suspect but not definitely locate.

Returning now to the relation between magic and organised agricultural production, I had made, in my arm-chair work on the Intichiuma Ceremonies and later on magic in general, a discovery which I regarded as of real importance. It was the discovery of a general theoretical principle of sociology and cultural relations: namely that the real function of

[1] Of this I have published only the material collected in Mailu (*Transactions of the Royal Society of South Australia*, 1915). My "survey work" will never be published. It contains scraps of material, some of fairly good quality, I think, some entirely superficial, but nowhere sufficient to give even an approximate picture of the tribal culture.

magic from the sociological point of view consists not merely in giving a public magician the prestige of an individual with supernatural powers, but in placing in his hands the technique of actually controlling work.[1] This discovery led me to direct my attention very largely to what might be called the inaugurative rôle of magical rites; a rôle which fitted beautifully more than three-quarters of Trobriand garden magic. After I had discovered that the cutting of the scrub, the burning, preliminary planting and cleaning, the main planting and so on were thus each introduced by a rite; after I had noted that some of these rites imposed taboos; that in connexion with others the magician directed the work in the sense of publicly announcing the time of its commencement and by supervising it, etc., I constructed a synoptic table somewhat on the pattern of the one in Appendix I. Going over the remaining activities in the gardens, I noted the inauguration of weeding and of the thinning out of roots. Just at the time when growth magic would have obtruded itself on my attention I went away from Omarakana for a few weeks to the western coast. Also I was working at that time on other subjects and, having obtained the full system of spells and a detailed account of the rites, put the gardens on one side. Had I not returned for a third time to New Guinea, my account of garden magic would have been completely faulty because of the absence of growth magic. As a matter of fact, gardening was the one subject which I had fully written up at that time, and I have in my possession a bulky manuscript on the subject in which the account stops short somewhere after the end of the *kamkokola* rite and, after a brief account of what weeding and thinning out means, proceeds to the magic of harvesting.

It was only well after the beginning of my third expedition, that is, during my first visit to Vakuta, early in March 1918, when I was on my way to Dobu, that I discovered the existence of growth magic. M'Bwasisi, the garden magician of the village, whom I had not hynoptised as I had Bagido'u into believing that what I wanted was inaugurative rites, gave me a full set of his ceremonies and explained to me the theory of growth magic. When I returned to Omarakana, in June 1918, Bagido'u on being questioned told me at once that his magic contained spells of growth and in two days I had obtained the full formulae with free translations. By that time this work no longer required the months of painful probing and searching necessary at the beginning of my magical education. As it was, however, I was only able to witness a few of these ceremonies, and the quality of my information on growth magic would unquestionably have been better had I not been influenced by the idea that all magic had an inaugurative function.

[1] Besides the somewhat short article on the "Economic Aspects of the Intichiuma Ceremonies" already quoted, I have written a book in Polish which I wanted to entitle *Primitive Religion as a Force controlling Social Differentiation* (in Polish this phrase can be formulated more succinctly and elegantly). The book was published by the Polish Academy of Science in Cracow during the war and in my absence, and the title was changed into *Primitive Religion and Forms of Social Structure*.

This is a good example of how indispensable it is to check up on material obtained from one informant and collate it with material from other informants and localities. Also, of how indispensable it is to retain fluidity of ideas. The organisation of evidence throughout field-work is indispensable; but premature and rigid organisation may easily become fatal.

Another serious gap in my information refers, as I have already indicated in the text, to taro gardens. Here I was misled by weighty considerations. Taytu is certainly economically more important than taro. The possibility of storing it gives it an importance in the creation of wealth, in exchange, in the ceremonial associated with sociology, which surpasses both that of yams and taro. On the other hand, there are many indications that taro is the crop of more ancient cultivation. The preponderance of taro in magic, the special part which is assigned to it in the gifts given to the spirits at *Milamala*, indicate, I think—even discounting any undue antiquarian or historical bias—that this vegetable was once of greater economic importance.

As the ethnographer ought to keep his eyes open for any relevant indications of evolutionary lag or historical stratification, taro gardening should have been studied as fully and seriously as the gardening of taytu. But not until I came home did I realise that the collation of the two types of gardening and the full discussion of them with some of my expert friends, even with Bagido'u himself, might have shed valuable light on historical or evolutionary questions. I therefore want to state definitely that here is a serious inadequacy in my material. It may be that further enquiry will not reveal very much. It may be, on the other hand, that a few months in the field and as minute a study of the *tapopu* ritual and work as of the *kaymugwa* and *kaymata* would open unexpected perspectives. I still hope that one of the exceptionally intelligent resident magistrates in the Trobriands, or a well qualified missionary, or even a field ethnographer may be able to supplement my remissness.

Another important inadequacy refers to what might be called the quantitative assessment of certain material aspects of gardening. Thus only a very approximate estimate of the extent of the garden lands of a community will be found. The sizes of fields and plots might have been measured even without the help of surveying instruments. Again, what theoretical vistas this would open, it is difficult to say. But were I able to embark once more on field-work I would certainly take much greater care to measure, weigh and count everything that can be legitimately measured, weighed and counted. The weight of a typical basketful of yams would have been easy to estimate. I failed to do this. The number of basketsful produced by an average gardener I have roughly estimated, (see Document II). A far more precise study would not have been difficult. The consumption of taytu per day per head would have been extremely interesting. There is no reason why this should not have been ascertained and I simply have to mark a gap.

My botanical ignorance has been a great handicap to me. Some knowledge of tropical cultivated plants would have been an immense help. I was not able to judge for myself where rational procedure ended and which were the supererogatory activities, whether magical or aesthetic. Thus the whole question of the training of vines, the method of planting taro, taytu and large yams, lacked one important cultural dimension. Above all, I was not quite able to see whether some aspects of the native technique and theory of planting, thinning out and weeding were definitely dictated by scientific principles empirically reached and correctly translated into practice. I do not regard my technological account of gardening as half as good as that of the ceremonial which surrounds it, and that is a terrific indictment of my material. The handicaps imposed on me by my botanical ignorance will be specifically clear to a reader who knows something about tropical botany, and who peruses divisions III and IV in Part V on native theory about the growth and development of plants, their classification and the terminologies of their various parts and aspects.

The difficulties which I still have in putting order and consistency into the Chart of Time-reckoning (Fig. 3) have been indicated; above all, the fact that I did not make absolutely sure in the field and, owing to inconsistencies in my notes, cannot now satisfactorily decide as to the exact place of the garden council, *kayaku*. On the other hand, I would like to say that, barring this and one or two minor points which have been indicated in the course of the narrative, the collation of innumerable scattered entries has produced a fairly adequate chart.

My ignorance of certain technological principles comes out clearly and has been specially indicated in Chapter VIII, where I discuss the *bwayma*. There a lack of competence in one aspect—that is, technology—has not, perhaps, resulted in an inadequacy within its own domain. By dint of hard work I succeeded, I think, in giving a fairly accurate description of the structure of the storehouse. It is rather the relation between the technical product, on the one hand, and native theory of stability, foundations and ventilation, on the other, which has suffered. As a sociologist, I have always had a certain amount of impatience with the purely technological enthusiasms of the museum ethnologist. In a way I do not want to move one inch from my intransigeant position that the study of technology alone and the fetishistic reverence for an object of material culture is scientifically sterile. At the same time, I have come to realise that a knowledge of technology is indispensable as a means of approach to economic and sociological activities and to what might be adequately called native science. A thorough grasp of how natives construct a yam-house would have enabled me to judge why they construct it in that way, and to discuss with them, as between equals, the scientific foundations of their manual systems. It would have also enabled me to assess more rapidly the sociological implications of technological and structural details. Here again the reader of Chapter VIII will find frequent references in the text to certain gaps in my material.

One capital blot on my field-work must be noted; I mean the photographs. Perhaps if you compare my books with other field-work accounts you will not realise how badly mine are documented in their pictorial outfit. The more reason for me to insist on it. I treated photography as a secondary occupation and a somewhat unimportant way of collecting evidence. This was a serious mistake. In writing up my material on gardens I find that the control of my field notes by means of photographs has led me to reformulate my statements on innumerable points. In doing this I have also discovered that in gardening, even more perhaps than in the two previous descriptive volumes, I have committed one or two deadly sins against method of field-work. In particular, I went by the principle of, roughly speaking, picturesqueness and accessibility. Whenever something important was going to happen, I had my camera with me. If the picture looked nice in the camera and fitted well, I snapped it. Thus certain phases of harvesting, i.e. display of taytu in the village and in the gardens, the *kamkokola* ceremonies with their attractive framework of magical structures are well represented. But the first ceremony in the gardens I only witnessed once and then in bad weather and with very little light; also for some reason I had no camera with me. Again one rite of the *vilamalia* I saw performed in pouring rain, and the other at dawn. Thus instead of drawing up a list of ceremonies which must at any price be documented by pictures and then making sure that each of these pictures was taken, I put photography on the same level as the collecting of curios—almost as an accessory relaxation of field-work. And since photography was no relaxation for me, because I have no natural aptitude nor bent towards this sort of thing, I only too often missed even good opportunities.

There is no reason at all why I should not have been able to show you the cutting of the scrub, the thinning out of the tubers, women weeding their plots and, most of all, every single phase of harvesting. If I have seen these acts once, I have seen them hundreds of times. Some of them, notably the cutting of the scrub, were distinctly ungrateful to photography. The men do not detach themselves neatly from the tangled background, and *takaywa* on the mat-glass of a reflex camera looks very much like men loitering about on the outskirts of the jungle. Harvesting, on the other hand, is attractive, is carried out usually in a good light and presents a great many characteristic details of emotional expression and the interest of natives in food, as well as of technology. Day after day I sat and looked on, with—perhaps the most unpardonable of sins—the feeling that to-morrow is also a day. In some acts again, such as the first ceremony in the gardens, or the *vilamalia*, it would have been infinitely better to pose the natives; to invite Bagido'u to come out and reproduce on a fine day the same gesture and pose which he had adopted on a rainy one, or at dawn or sunset. If you know a subject-matter well and can control native actors, posed photographs are almost as good as those taken *in flagranti*. Yet I am sad to say that I have never resorted to this device except when, as in one or two photographs of war magic, I knew that I should never be able to see it performed in earnest. That I have never photographed an actual *kayaku* infuriates me now, though *de facto* a real garden council does not differ at all from any ordinary social gathering. But it certainly possesses a sentimental documentary value and should have been done.

A general source of inadequacies in all my material, whether photographic or linguistic or descriptive, consists in the fact that, like every ethnographer, I was lured by the dramatic, exceptional and sensational. I have indicated how terribly vitiated my linguistic material is by the fact that I failed to record *the* most important types of speech—those embodied in ordinary everyday activities. In photography, my failure to snap a group of men sitting in front of a hut because they just looked like an everyday group of men sitting before a hut is an example of this. It was also a deadly sin against the functional method, the main point of which is that form matters less than function. Twelve people sitting round a mat in front of a house, because they came there by accident and stayed gossiping, have the same "form" as the same twelve people collected on important garden business. As a cultural phenomenon, the two groups are as fundamentally separate as a war canoe from a sago spoon. As the reader who has read Chapter VIII and followed the indictments there expressed will find, I have also neglected much of the everyday, inconspicuous, drab and small-scale in my study of Trobriand life. The only comfort which I may derive is that, in the first place, functional field-work, which after all started to a large extent in the Trobriands, has begun a change in this respect; and secondly that my mistakes may be of use to others.

4. SOME DETAILED STATEMENTS ABOUT ERRORS OF OMISSION AND COMMISSION

Having thus laid down the main sources of inadequacy and of positive mistakes or distortion of perspective, I will list the specific qualifications, doubts, or methodological references which I wanted to make on a number of points in the text, but which, if made there, would have destroyed the unity of the narrative:—

Note 1.—CHART OF TIME-RECKONING (Ch. I, Sec. 1)

The method by which this chart has been arrived at is important in the assessment of its value and its inadequacies. This chart is the result of computing a great number of chronological entries in my field-notes. Most, though unfortunately not all, of my observations are dated exactly. Towards the end of my work this was done consistently. From this computation I was able to arrange events in their chronological sequence and put them into their proper place in every column. Had this work been done in the field, as was the case with several other synoptic instruments of research, I could have checked them with native informants and eliminated all contradictions. As it is, I have to register several points on which my entries contradict one another, and on which I could not arrive at a decisive solution. The most important of these is the *kayaku* (see Note 6). Another difficulty is the shifting position of the *Milamala*. I am certain that the treatment of the subject in Chapter I (Sec. 1) is correct, i.e. that the *Milamala* falls at different stages in different districts; and also that this does not affect the relation between the gardening cycle and the European months. It means, however, that the sequence of native moons changes, since *Milamala* is always the first native moon. It means also that the explanation of the somewhat vague and uncertain nomenclature of the last three moons which I gave in my article on the "Lunar and Seasonal Calendar in the Trobriands", *Journal of the Royal Anthropological Institute*, Vol. LVII, 1927, holds good with regard to two districts, Kiriwina and Vakuta, and to an extent also to Sinaketa. It does not hold good with regard to Kitava where the last three moons of the time-reckoning would fall just on the harvest. Here obviously a systematic enquiry in the field might throw new and interesting light on the subject.

Another difficulty which arose when drafting the chart was the exact relation between the early, the main and the taro gardens (cf. Note 2). It is a matter of regret to me that the habit of charting and preparing synoptic instruments occurred to me only in the course of my field-work instead of having been drilled into me before I started out.

Note 2.—KAYMATA AND KAYMUGWA (Ch. I, Sec. 3)

Here notably I have serious doubts as to whether the early gardens do not sometimes run much more synchronically with the main gardens; also as to how far the divergence in time depends on whether the year is wet or dry, and upon the length of the dancing period. From some of my notes and reminiscences it would appear that the early gardens, which are on a much smaller scale, are always made immediately after the harvest is started on the main gardens, so as to provide the minimum of food necessary towards the hunger month. But I have not investigated this very important subject as fully as it deserves.

In the second place, though I am quite sure that the early gardens are the gardens of the mixed crops as against the later ones, which are the gardens of taytu, I have neither investigated this distinction as fully as it deserves nor yet documented it by enumerating, let us say, the quantity of taro, sugar-cane, pumpkins and peas planted per square division (*gubwatala*) on an early garden, in comparison to the number of taytu vines to be found on the same space. The documentation of the "obvious", "everyday" and easily accessible features is one of those points which I have much neglected in my field-work. It could have been done without any difficulty whatever and might have yielded very important theoretical sidelights. For the neglect of the obvious and everyday, see also Ch. VIII, Introductory Remarks.

Note 3.—Sociology of the Leywota (Ch. I, Sec. 3)

Much fuller documentation could easily have been obtained concerning the sociology of cultivating *leywota*. As regards general information, I was told repeatedly that these plots are usually taken over either by the chief or by one of his next of kin, or by a distinguished *tokwaybagula* (good gardener). A few questions would have elicited who were the cultivators of the *leywota* in, say, Omarakana for several successive years. From a series of such concrete observations the general rule might easily have been documented and exceptions registered.

Note 4.—The Ever-Nascent Myth (Ch. I, Sec. 7)

The outstanding theoretical importance of what might be called the constantly recurring contemporary mythology was not appreciated by myself in the field. It was on the basis of my own field-material and after my return that I formulated the view concerning the nature of myth which I have stated in my book on *Myth in Primitive Psychology*. The theory of myth there developed insists on the fact that "myth, as a statement of primeval reality which still lives in present-day life and as a justification by precedent, supplies a retrospective pattern of moral values, sociological order, and magical belief. It is, therefore, neither a mere narrative nor a form of science, nor a branch of art or history, nor an explanatory tale. It fulfils a function *sui generis* closely connected with the nature of tradition, with the continuity of culture, with the relation between age and youth, and with the human attitude towards the past. The function of myth, briefly, is to strengthen tradition and endow it with a greater value and prestige by tracing it back to a higher, better, more supernatural reality of initial events" (p. 124). Immediately connected with this main principle is the view that every myth carries in its wake a constant string of corroborative stories. To take an example from our own culture: the myth of the first appearance of the Virgin at Lourdes is constantly corroborated by miracles which are wrought every day at the blessed source. Every great mythology of a saint and of his holy deeds in the Roman Catholic Church produces such stories about the miracles connected with the relics, about the influence of the Saint, of his shrine, of his tomb, of his places of worship. Such living, recurrent, regenerated myth "is a constant by-product of living faith which is in need of miracles; of sociological status which demands precedent and example; of moral rule which requires sanction" (p. 125). Now returning to the ever-nascent myths of gardening and the miracles of its magic, I can emphatically state their presence; I cannot, unfortunately, document them. Time after time I have listened at the fireside to the braggings of one garden magician or another; been present at disputes between two villages or two garden magicians; dismissed as irrelevant invaluable historical data supplied me by Bagido'u and others of his colleagues with regard to past achievements of their predecessors. I classed all such phenomena as examples of 'Melanesian bragging' and neglected them. Once more we see how a wrong theoretical attitude in the field has led me either to neglect fundamental and important facts or to set them in an entirely wrong perspective.

Note 4a.—Chief Working with Wives (Ch. I, Sec. 8)

I did not enquire into this question very directly when in the field. I can remember seeing To'uluwa at work with Kadamwasila, Isupwana and Ilaka'isi; but never with the other wives. The position of the wives was by no means equal. In the first place, the chief inherited a number of wives from his predecessors and these remained always very much in the background of his interests and affections. Such wives would usually have grown-up sons, many of whom were resident in the capital, and would help with their mothers' husbandry. The chief had two or three favourite wives, the wives of his own wedding who personally suited him. With these he would garden.

Note 5.—Black Magic (Ch. I, Sec. 8)

Had I been more clearly aware of the existence of black magic in gardening I might have followed it up in connexion with native ideas about the causes of blight, pest, and other mishaps to gardens. I think that it would be possible to state generally that in the Trobriands all mischance and adversity is attributed to the agency of black magic. If this is so, we could say that the Trobriander believes that all natural forces are adequately overcome by intelligent and empirically founded work. Magic, on the other hand, is necessary in order to counteract the effect of mishaps caused by various forms of sorcery or black magic. I have the feeling that this generalisation is correct, but I am not able fully to substantiate it as regards gardening, and this is one of the most important activities where the Trobriander copes with nature both practically and magically.

Comparing the data which will be found in Chapter III (Sec. 2) about the relation between the destructive bush-pigs and the 'magician's bush-pig', comparing what is said there about the dangers of breaking the taboo against sexual intercourse or excretory functions being carried on near gardens; and collating that with my vague information about direct black magic against gardens—the assumption seems very plausible. But obviously here lies a whole domain of Trobriand tribal lore, which I could have explored, which I would explore now were I to return there, but which I missed completely because the problem was not present to my mind.

Here once more I want to drive home the most important principle of method in field-work: that without a theoretical grasp of the problem it is impossible to make relevant observations. An empirical and scientific theory of fact is necessary as a chart for observations and as a check on observations. Without theory there can be no fruitful research. I was fully prepared for studying the problem of the relation of magic to work.

I was quite aware that the facts had to answer the crucial question: is magic a sort of pseudo-scientific technique, does magic at an early stage of development replace science and empirical technique? The answer the facts gave me was clear and definite. Magic is not a technique which ever encroaches on the domain of empirical knowledge and mental achievement. Competent practical work has its own realm based on knowledge, and on that realm magic never trespasses. Magic moves within the realm of uncontrollable forces, chance, accident, damage due to blights and pests, with which human beings are unable to cope. But the further affirmation that all this mischance, disaster and accident is always due to human magical influences, that it is always engineered by human malice—this I am not able fully to substantiate.

Note 6.—KAYAKU (Ch. II, Sec. 3)

On the question of the garden council, I have a number of queries which I am unable satisfactorily to settle from my field notes, and this matter could undoubtedly have been easily solved by very simple enquiry, had I collated all my data in the field, and had I there constructed, as I should have done, a complete chart of time-reckoning and followed it in all its detail. I had a rough seasonal chart; but as a warning to field-workers I would like to say that the time spent on completing such instruments in the field, far from being wasted, is the best investment for research. I find in my notes several entries about proceedings at the *kayaku*, which I either witnessed or which were reported to me. They range over the moons of *Yakoki*, *Kaluwalasi* and *Kaluwasasa*; that is, from some time late in May to early in July. Now if, as is my final assumption embodied in the chart, there is a separate garden council for the early gardens and one for the main gardens, these data are easily accounted for. This assumption, however, stands in slight contradiction to one or two of my earlier entries where I find that only one *kayaku* is made for all gardens. The discrepancy in time then would be due either to local causes: the *kayaku* might take place earlier in those districts which celebrate the *Milamala* earlier, or else it might be shifted according to whether the beginning of gardening is modified by weather conditions. But the discrepancy would still be too great. Another alternative would be that no *kayaku* is made for the early gardens at all, and the disposition of plots is arrived at informally. On balance, I think that the assumptions that one *kayaku* is held for all gardens or that no *kayaku* is held for the early gardens are incorrect, and that there are two garden councils for the early and main gardens respectively; one some time late in May or early June, the other in July.

Note 7.—NUMBER OF PLOTS CULTIVATED (Ch. II, Sec. 3)

The number of plots cultivated by a gardener is an important numerical piece of evidence. My material here is again inadequate. I have very often discussed the matter with natives and have a number of figures indicating that so-and-so makes so-and-so many *baleko*. The figures for the same man differ from one time to another. Such discrepancies in information are not difficult to account for. As in all matters where vanity and ambition play a large part, the man himself, his friends and kinsmen would exaggerate his contribution. Lying in such cases is to the Trobriander a very venial offence. The only certain method would have been to register at harvest exactly the number of plots harvested by each man. This I failed to do—a very serious gap.

Note 8.—VISITS OF THE SPIRITS (Ch. I, Secs. 5–7; Ch. II, Sec. 4; Ch. III, Sec. 1; Ch. V, Sec. 2; Ch. IX, Sec. 2)

Here again, as in most of these notes, I have to insist on the fact that successful research depends upon the synthesis and organisation of evidence done in the field. The greatest source of all the inadequacies and gaps in my own field-work has resulted from the dire methodological fallacy: get as many 'facts' as you can while in the field, and let the construction and organisation of your evidence wait till you write up your material. This fallacy would have been much more pernicious had I not had time partly to cure myself of it during the two intervals between my field-work, which I devoted largely to the organisation of my material. Even so, there remained a great many lacunae in my data, simply because I did not spend time enough in the field collating and synthesising them.

Take, for instance, the problem of the part played by the spirits in general, and ancestral spirits in particular, in native tribal life; here more particularly in agriculture. There is no doubt that had the spirits played a very prominent and direct part this would have obtruded itself on my notice. In other words, the Trobriand belief in spirits and the part they play is vague and shadowy. Yet the fact remains that the vaguer a native belief the more incumbent is it on the field-worker to draw it in precise outline; and by this I do not mean that he should introduce a precision which is not in native belief, but rather that he should outline precisely the character of native vagueness.

Let me here briefly collate the facts and submit to the reader a brief theoretical summary, which I preferred not to give in the text because it goes a little beyond my directly recorded and digested data. On the other hand, it may be useful in gauging the nature of these data. The spirits, then, appear on three distinct occasions and in three distinct rôles:—

1. The visit of the spirits at *Milamala* (compare also for this my article "Baloma—the Spirits of the Dead in the Trobriand Islands", *J.R.A.I.*, 1916. The evidence there given is incomplete as it contains only the results of my first two expeditions to New Guinea. But the main outline of the belief there stated was found correct during my third expedition). On this occasion the spirits of the whole village attend. They return regularly every year to their community and the belief in their presence is clear, concrete, not questioned except by natural sceptics and agnostics, and

expressed in definite institutionalised arrangements. Food is displayed for them and they enjoy its sight. Food is cooked for them and they consume its spiritual substance. A special *kubudoga*, or very high platform, is built for them so that they may sit on it, high above every mortal. Their presence there is associated with the feeling of *malia* (plenty) and on a few days there is a display of valuables, *vaygu'a*, to gladden their eyes. The villagers also dance and carry out other festive activities to honour the spiritual guests, and to commune with them in this manner. The spirits, moreover, usually manifest their presence by material tokens of a character which would probably satisfy most of our contemporaries who believe in rapping tables, ectoplasm, apports, and other tokens of spiritistic mediums. Such manifestations of the presence of spirits are often used by the chief to endorse his wishes; or by the elders to back up traditional behaviour. The real presence of the spirits at *Milamala* enhances the belief which exists throughout the year that if tradition is not properly observed, if custom and law are not obeyed, the ancestor spirits will be displeased and will bring some form of bad luck on the natives. What exactly is the relation between the mischance brought about by the offended spirits and mischance brought about by malicious magic? I cannot say, for again I have not investigated this problem as fully in the field as I should have done. The fact is that I very often heard and noted down casual remarks that such and such mishap—slight drought, blight, pests, attacks of bush-pigs; unsuccessful *kula* or a bodily accident—was due to the wrath of the *baloma* (*pela baloma igiburuwasi*). I also occasionally enquired whether it was really wrath of the *baloma* or the evil intent of magic. But the answer would usually be "I do not know" (*ayseki wala*).

2. The ancestral spirits appear in our account in so far as they figure in the spells; but here they function in a quite different manner: the use of their names is a form of verbal magic, the conjuring up of something by the name of somebody. Also a special class of spirits is invoked: the lineal predecessors in magic of the officiating *towosi*. And here we have to make a further distinction: (*a*) The list of names may appear merely in the formula (M.F. 2, 38, 42, and also M.F. 3, where we only find a reference to ancestral spirits as "old men"). In this case there is no reason to suppose that the spirits are invoked to be present, to take part in the rite and to help the living with their ghostly co-operation. I think that this is an example of verbal magic without any implications of real presence. (*b*) Offerings may be made to the spirits, as in the first inaugural rite or in uttering the spell at one of the harvest rites, or in the *kamkokola* ceremony described in Ch. IX (Sec. 2). Then their presence is much more real and effective. But here again I have not gone deeply enough into the subject to ascertain what they do and whether they are really believed to be there, at least in the same way as they are believed to be present during the *Milamala*. I did not by direct questions and discussion with the natives collate my observations concerning the *Milamala* with my knowledge of the *ula'ula* offerings. Therefore I can only

show the lacunae and state that these are not due to the intrinsic impossibility of answering the question, but merely to my neglect.

3. Least real and effective are such spirits of the deceased as are invoked in one or two spells under the name of the mythological heroes Tudava and Malita (M.F. 10, 22, 29), Iyavata and Vikita (M.F. 1), Tokuwabu (M.F. 19), Yayabwa and Gagabwa (M.F. 5), Botagara'i and Tomgwara'i (M.F. 31) and Seulo and Milaga'u (M.F. 41). Here once more, though I have not probed into the question to my entire satisfaction, I am quite certain that the invocation of such spiritual entities has no ritual consequences or correlates. Rather it is a celebration of the memory of these culture heroes, while, at the same time, by the utterance of the names of such supreme gardeners as Tudava and his companion Malita, fertility is bestowed on the garden.

Note 9.—Rest Days (Ch. III, Sec. 1)

Concerning magical taboos on work, it is clear that these introduce rest days or holidays into the gardening cycle. I think that the natives have no idea about the need of resting or its influence on work, but unfortunately I did not go very deeply into the native point of view by the frontal method of attack.

Note 10.—Black Magic of Bush-pigs (Ch. III, Sec. 2)

The deficiency of my information on this subject and the reasons for it have been fully stated in the text. Here I wish to add that a future enquiry into it would be very useful in answering the problem stated above in Note 5 as to the relation between black and beneficent magic. If I am right in assuming that the natives ascribe most, perhaps all, untoward events which happen in the course of nature to black magic, then the destruction of the crops by bush-pigs would also be ascribed to sorcery. It would always and inevitably be the sorcerer who brings the bush-pigs from Tepila and Lukubwaku to the gardens of the natives, as stated in Motago'i's account, and sorcery would not be just one among many causes. I think my generalisation that "only sorcery lets loose the pests and plagues" is well founded, since it corresponds to native belief in other aspects of reality, above all as regards human health. But I want to repeat here that I have not to my satisfaction scrutinised this belief as regards adverse happenings in gardens, and certainly I have not sufficiently documented it. In this context I should like the reader also to consult the account I have given in another place (*Sexual Life of Savages*, Ch. XI, Sec. 9) of black magic as the cause of truancy and infidelity in domestic pigs, wives and sweethearts.

Note 11.—Taro Magic in Taytu Gardening (Ch. III, Sec. 2)

This subject, that is, the "survival" of taro ritual in taytu magic, is also commented upon in Section 3 of this Appendix. It is related to what is perhaps one of the main deficiencies in my account of Trobriand gardening—the unequal treatment of the agriculture of taro as compared with that of taytu.

Note 12.—The Function of Small Squares (Ch. III, Sec. 3)

Here again I have to insert a methodological caveat to this generalisation. I do not think that there is any part of gardening technique which I have thrashed out more fully in conversation with the natives, very often in the gardens and while the work of *tula* laying was in progress. But were I to return to the Trobriands I would attack the problem not by question and answer, nor even by discussion and argument. The best way to proceed would be to make, so to speak, a diary of the use of the squares. How often is a *baleko* (plot) subdivided into squares as between different gardeners? How often does this happen with regard to several wives of the same man? Again, in the progress of work, how often do the natives actually calculate the number of seed yams per square subdivision? The distinction between such actual observations, which might have been embodied in a document, and mere talk round the problem is considerable, as I was able to appreciate in the course of the working up of my material. Here I want to state emphatically that in my opinion the documentation of this problem could have been easily achieved; and, since it bears on the important distinction between the clearly utilitarian function of the *gubwatala* (square subdivision) and the merely psychological effect of them, it was important to obtain as much material as possible.

Note 13.—Crops on the Kaymata and Kaymugwa Respectively (Ch. III, Sec. 3)

In this context also a much fuller documentation would have been of great value (cf. previous note, and Note 2). A seasonal series of diagrams showing exactly what was planted day after day on representative plots in the *kaymata* and *kaymugwa* would have been most useful.

Note 14.—The Kamkokola Rite as Magic of Planting (Ch. III, Sec. 4)

The deficiency in my information is indicated in the text, where I also give a plausible apology for it. There is no doubt at all that this is not an easy point to settle. But here again, if I had, so to speak, set a definite experiment and observed the gardens day by day, keeping a diary of what was happening, I could have answered by personal and direct observation easily what it is by no means easy to elicit by the question and answer method, which I used to a large extent in this matter. Here once more the ideal method would have been to concentrate on one or two plots and draft their precise history throughout the seasons. I would advise future field-workers seriously to adopt this method of selecting what might be called "experimental garden plots" and following their fate stage by stage.

Note 15.—THE CONSTRUCTION OF FENCE AND KAMKOKOLA (Ch. III, Sec. 4)

In the description of my walk with Bagido'u I reproduce the substance of my field notes. I remember, however, and wish to add here, that in certain cases the *kamkokola* was constructed before the fence had been made. I seem to remember having been told once that in the early mixed gardens, the *kaymugwa*, the fence is made some time before the *kamkokola*; and that, on the other hand, in the main or late gardens, the *kaymata*, the fence is made after the *kamkokola* has been constructed. This gap in my information is connected with my insufficient knowledge and documentation concerning the relation of early to late gardens (see Notes 2 and 13). If the early gardens are used for mixed crops, the planting of which, as we know, precedes the *kamkokola* ceremony; if in the late gardens taytu is planted only after the *kamkokola* is erected, then obviously the late gardens do not need a fence before the *kamkokola* is made, while the early gardens would need such a fence. After one has perceived such correlations, a short conference with a few intelligent informants and some walks in the garden can give a decisive answer to a query. To drive in once more a point of method: the more constructive the handling of one's material in the field, the better will this material become.

Note 16.—OBSERVATIONS ON THE GROWTH OF THE TAYTU (Ch. IV, Sec. 2)

Although in the sentence to which this note is appended there is implied a confession of failure, which is ascribed to my ignorance of botanical processes, I want to add that, were I able to return to the Trobriands, I would in spite of my botanical deficiencies attack the subject far more energetically than I actually did. Here once more it is the documentation which is incomplete. Many of my discussions with natives were held actually in the garden and the natives demonstrated on the actual plants the processes of growth and the correlated magical beliefs. But I relied too much on mere verbal statements, instead of which I should have taken photographs of the various stages of development, made diagrams of them and, once more, concentrated on the study of one or two representative plots. This ought especially to have been done during the technical activities of training, weeding and thinning. Since the problems discussed in this chapter throw an interesting light on the relation between magic and work, the deficiency here noted is not without its theoretical consequences.

Note 17.—GROWTH OF TAYTU AS DUE TO MAGIC (Ch. IV, Sec. 2)

As with black magic (cf. Note 5), I feel that here again a much fuller analysis was necessary. The distinction between the growth of wild plants and of garden crops is constantly made and referred to by the natives; but unfortunately I have not collected a sufficient number of relevant statements. It often happens in field-work that a point which is so constantly brought home to the ethnographer that he takes it for granted will not be sufficiently documented. Towese'i's remarks reproduced in the text are not sufficient by themselves to give the native attitude on this subject.

Note 18.—DISPROPORTION BETWEEN MAGICAL AND TECHNOLOGICAL INFORMATION (Ch. IV, Secs. 2 and 3)

In connexion with what has been said in Note 16, I wish to point out once more that throughout Sections 2 and 3 of this chapter there is a certain lack of balance between the information on magic, which is full and well documented, and that concerning natural processes of growth and the technique of handling. I would make fuller observations on the latter if I could go over the ground again.

Note 19.—PRIVATE GARDEN MAGIC (Ch. IV, Sec. 4)

I am convinced that private garden magic is in reality as "inconspicuous" as asserted in this chapter. I acknowledge there that my information about it is not as complete as it is about official magic. But I must make it clear that I did not attack this problem as directly as I could have done; that is, I did not find out in Omarakana, nor, more important, in other communities where the chief's influence is less pronounced, whether private garden magic is as completely subordinate as it appears. Considering that, as I have stated in the text, private magic would be regarded with disfavour by the garden magician, which means by the chief or headman also, it is natural that private garden magic should be least pronounced in the paramount chief's community. Thus I ought to have studied this subject more especially in the smaller villages.

Note 20.—STORIES ABOUT FAMINE (Ch. V, Sec. 1)

Unfortunately I have noted down only one of these stories—the one reproduced in this chapter. The value of the information would have been much greater if, instead of one text, I had been able to collect a dozen or so. Here the question of time comes in. Towards the end I was able to take down native texts rapidly, but of course it is at best very slow and cumbersome work, especially as every text has to be commented on in the field. A much fuller linguistic documentation would have been feasible had I stayed in the field for another year or so.

Note 21.—THE SOCIOLOGY OF FIRST FRUIT DISPLAY (Ch. V, Sec. 2)

My information on this point is not satisfactory. Here once more a few documents, that is, a diagram or two showing the actual disposition of the heaps displayed in the village, genealogies or relationship diagrams concerning the person displaying the heap, the person to whom it is given, and the relation of these to the deceased, would have been the best methodological device for throwing light on the subject.

Note 22.—THE SYMBOLISM OF TUM (Ch. V, Sec. 3)

In the case of this ceremony I failed to ascertain what the symbolism of pressing down really means to the natives, though it will be seen that in most other rites I have entered into such details with my informant. Such vagaries of observation in the field ought to be indicated by the ethnographer because the absence of knowledge may be due either to neglect or else to the absence of native beliefs or ideas on the given subject.

Note 23.—INFORMATION ABOUT BURITILA'ULO (Ch. V, Sec. 6)

The descriptions given in this section are based on one case actually seen and on a good deal of discussion. I have indicated that the subject was brought to my notice by the actual occurrence of such a transaction in June 1918. Since everyone was much interested in what was happening in the neighbouring villages of Wakayse and Kabwaku and was discussing these events eagerly, I found no difficulty in making my informants talk. But, as in all cases, information must mature; it must be rediscussed at the several stages of observation and after it has been digested by the ethnographer. This was impossible because I left the district soon after the *buritila'ulo* occurred and I left the Trobriands five months later. Hence several points are still obscure to me in my material as it stands. They refer not so much to details, but rather to the main principle of the transaction. The *buritila'ulo* is not a competitive or sportive event in our sense of the term. First of all, there is no umpire, no one to adjudicate or arbitrate between the competitors. Obviously then each village will claim that it has given more. In this connexion the question arises as to whether such a display always did give rise to quarrelling and fighting in the old days when there was no restraint on the natives. In other words, can we say that whenever a *buritila'ulo* was started, fighting between the two villages would follow? Or did they sometimes remain satisfied with the mere bragging, with claims and counter-claims? Also I have not followed in detail the reapportionment of yams after the return gift has been made to the first donors.

Note 24.—TAYTU FOR URIGUBU AND FOR OWN CONSUMPTION (Ch. VI, Sec. 1)

Throughout my account I have adopted the approximate figure of about one-half as the proportion of taytu allotted by a man to his sister's household. Obviously the figure varies. In the case of an old man of high rank who has no sisters and whose distant maternal kinswomen are married to men of much lesser importance than himself, the proportion of the taytu which he gives may be almost negligible. In the case of a strong, energetic and good gardener who has several sisters married, perhaps to people of much higher rank than himself, the proportion which he has to give away may be three-quarters if not seven-eighths. The latter figure might, I think, be true in the case of an unmarried man. Undoubtedly it would have been difficult to give exact documentation on this matter, partly because of the differences here indicated, partly also because the crops used for own consumption are harvested and garnered in a different manner. The *urigubu* is given in bulk and publicly. It is boasted about and numerically recorded. The *taytumwala* is garnered in instalments, very often its amount is understated on purpose. Some of it is consumed as harvested. The boasting is not only about *urigubu* given, but also about that received; for a large *urigubu* brings more honour to the recipient than the crops he has produced for his own consumption. With all this a much fuller documentation was feasible and necessary. To follow the budget of one man or of one household, almost on the principle of Le Play's "family budgets", would have been the best manner to set the experiment; but this would have been impossible in the Trobriands.

Note 25.—Filling a Man's Bwayma after His Wife's Death (Ch. VI, Sec. 1)

To the rule that after the death of a man's wife his yam-house is not filled by his relatives-in-law there is one exception: when a wife dies leaving grown-up sons these will make gardens for their father. This has been indicated in the text. Here I wish to make it clear that this is an exception from the native point of view, because to the Trobriander a son is not a kinsman but simply a relative-in-law of his father. Not that sons would ever be spoken of as relatives-in-law, but their relation to the father is only through his wife. But here my information is not quite full enough to decide the exact technique by which this is done. While living with their father the sons contribute in a two-fold manner to the household. They work the father's plots directly, and thus help to produce his *taytumwala*. They also work with their maternal uncle and thus contribute to their father's household under the title of *urigubu*. A few documents giving what actually happens as regards the residence and work of sons after their mother's death would have cleared up this point better than general and abstract statements.

Note 26.—No Arbours on Gubakayeki (Ch. VI, Sec. 1)

Here obviously is a gap in my material. By walking over any garden at harvest and noting which plots were described as *gubakayeki* and which as *urigubu*, and on which there was an arbour and on which none, the problem would have been definitely solved. Without putting too much trust in my memory I would, however, like to say that I feel certain that I found no arbour which did not contain an *urigubu* heap. This means that arbours are only constructed on *urigubu* plots. It means also that crops from *gubakayeki* plots are garnered as soon as harvested.

Note 27.—The Chief's Urigubu and Tribute (Ch. VI, Sec. 1)

The sociological aspect of the *urigubu* on which my material is least complete is connected with the cases in which the harvest gift is presented by a chief's wife's father or some other fictitious kinsman of hers. A reference to Document II will remind the readers that I have had to make several distinctions there, classifying the *urigubu* gifts from the sociological point of view. I have insisted several times, in narrative and document alike, that such exceptional types of *urigubu* do occur when given to people of rank, above all to the paramount chief. But I have to state here plainly that a much fuller study of the harvest gifts received by people of low rank would have been an invaluable supplement to the information contained in Documents I–III about the chiefs' *urigubu*. Document IV, in which the *urigubu* of commoners is recorded, is incomplete out of all proportion to the importance of the subject.

Note 28.—One-sidedness of Documentary Evidence (Ch. VI, Sec. 3)

In this document I have only recorded the heaps received. Had I made a counterpart to it, showing how many heaps are given away by the same men, its value would have been considerably greater. It would have been a much more complicated business, partly because the donor is apt to boast and exaggerate and partly because, whereas heaps received are displayed all together and on one occasion, heaps given away are handled in different instalments. Nevertheless, only a study of the exports from a man's budget as well as of the imports could have given the full picture.

Note 29.—Estimate of Sociological Information in this Chapter (Ch. VI, Sec. 3)

Looking back on this chapter, I should like to add that, as in every treatment of sociological intricacies, small gaps could be indicated here and there. Those already mentioned in the course of my narrative or in the preceding few notes will indicate the type of query which might have been raised and the type of observations by which it could be answered. The more abstract and complex the subject, the easier it is to push it into such detail that unanswerable problems crop up here and there. In order to do justice to my account, however, I should like to say that, on the whole, the material of this chapter is as good as any material presented by myself. Corresponding to my greater interest in sociological and economic problems, I collected fuller data concerning duties—legal, economic and ceremonial—and with a better understanding, than on technological or botanical facts.

Note 30.—Malia—Mana? (Ch. VII, Sec. 1)

Had I been aware of this interesting etymological query in the field, I could have made a much fuller collection of linguistic texts showing the use of the word *malia* in all its meanings.

Note 31.—Binabina Stones (Ch. VII, Sec. 1)

The function of these stones, the symbolism of pressing, the correlation between the rite here described and that of *tum* at harvest, all these are correlated queries which I would now discuss much more fully with my native friends than I did when in the field. There is obviously a general idea of stability as induced by weight running through all these rites. How far this idea is connected also with the belief that the *vilamalia* magic affects the appetite is difficult for me to decide on the basis of my present information (cf. also above, Note 22).

Note 32.—Theory and Practice of the Vilamalia Magic (Ch. VII, Sec. 4)

I do not think that the discrepancy here noted is due to the quality of my material. As regards the documentation on this subject, I think I have collected as full and well assorted a budget of texts as on any point in my field-work. The only criticism which could be passed with justification on my field material is that it is derived only from two sources—Omarakana and Oburaku—as far, at least, as documentation goes. I would like to repeat that my actual experiences and data are always, and naturally, much richer than the evidence which I can adduce in the form of texts, statements or diagrams. I checked and sampled such important and interesting facts as the "ignorance of nutritive function", embodied in the belief about the efficacy of *vilamalia* magic, in my walks and visits to other villages. It is to a certain extent the difficulty of quality that made me, in selecting the documentation, prefer to adhere to the community from which I had collected the bulk of my evidence.

Note 33.—Filling of Small Bwayma (Ch. VII, Sec. 5)

In connexion with the deficiency indicated in the text, I would like once more to emphasise the fact that unless the ethnographer has the problem clearly in his mind, he not only is unable to register facts and make notes, but even his capacity of observing and of taking mental notes will be limited. The obsession by the ostentatious and grandiose is so universal and pernicious that I do not scruple to indicate it as against myself time after time.

Note 34.—The Consumption of the Crops (Ch. VII, Sec. 5)

The story of what happens to the crops after they have been garnered is not by any means as full as my material concerning their cultivation. In other words, my work is much better on the side of production than it is on the side of consumption or even of exchange. Since the "Functional Method" should lay stress more on the use than on any other aspect of a commodity, there is a slight irony in the deficiency of my material on this point. Unfortunately I had not quite formulated the functional approach when I went into the field. It developed in the course of my field-work and the subsequent elaboration of my material during the last fifteen years or so. The very existence of the "Functional School in Anthropology" was first announced in an article *s.v.* "Anthropology" which I wrote for the three additional volumes, also described as the thirteenth edition, of the *Encyclopaedia Britannica*, 1926. Since then I have given a much fuller statement of the Functional Method in an article *s.v.* "Culture" in the *Encyclopaedia of the Social Sciences*. The data concerning the uses to which the crops are submitted are given in Part I (Introduction), especially in Sections 7 to 9.

Note 35.—The Study of the Uses of Bwayma (Ch. VIII, Introduction)

The critical analysis given in the introductory remarks to this chapter covers most of the deficiencies in my material about the construction of the *bwayma*. The whole chapter is in a way an extensive methodologically annotated gap note; but I am convinced that were I able to return to the field, these deficiencies could easily and rapidly be covered. Here again I would have to study the use, the "consumption" from the economic point of view, rather than the "production". I have not sufficiently penetrated in the field into the interior of the yam-house or assisted at its filling from the inside. Again I have no record of the day-by-day processes of taking out the yams from the *bwayma*. A diary showing what happens, with a representative storehouse taken from each type, would have produced a document containing all the information necessary. On the basis of such knowledge I could have then observed the construction of several yam-houses and found out the correlation between structural elements and the functional characters of a storehouse.

Note 36.—Methodological Problems (Ch. IX, Introduction)

This chapter, like the previous one, includes a direct statement concerning the problem of method and technique in field-work. Most of the gaps and inadequacies are there indicated. The limits within which the information stands as sound and valid are also outlined, especially in Sec. 1. In the last paragraphs of that section emphasis is also laid on the inevitable gaps due to limitations of time. The ideal would, of course, be a series of volumes doing for each community what has here been done for Omarakana. But it is not a feasible proposition. On the other hand, I do not regard it as a shortcoming that throughout my field-work I preferred to carry out intensive research within a restricted area rather than a more superficial survey of an extensive one.

Note 37.—The Value of Sampling (Ch. IX, Sec. 2)

The inferior quality of the information here given, as compared with that of the previous chapters, is obvious and has been stated in the text. I have also drawn attention there to the value of such sampling. The "sampling" here given from Kurokaywa ought to be compared with that contained in Documents VI and VII and in Section 1 of Chapter X.

Taken together these data will enable the reader to assess within what limits the information about Omarakana can be taken as representative.

Note 38.—The Inclusion of Gaps (Ch. X, Sec. 1)

In this chapter also the nature of the information given has made it necessary for me to indicate the gaps throughout the narrative and to give the reasons for them.

Note 39.—Proportion of Taro to Taytu in the South (Ch. X, Sec. 1)

In my field notes I find a precise numerical assessment to the effect that "taytu does not supply the South with more than one-third of its nourishment". This statement, however, I did not document any further in my field notes and, as far as my memory goes, it was merely a rough guess of mine, based on what I had actually observed by visiting the larders and kitchens of Oburaku during my few months' stay there. Numerical assessments of this type are worthless unless fully documented, and material for such documentation I did not collect. I think this would have been a difficult but not impossible task. Whether the value of such information would prove commensurate to the expenditure of time and energy on it is not easy to say. Were I back now in the field I would certainly tackle this type of problem; on a small scale first and then, if it proved amenable to an intelligently set system of observations, I would extend my enquiry. (See also what I have said about the deficiencies in my material on the quantitative side in Section 3 of this Appendix.)

Note 40.—Lack of Time and Opportunity (Ch. X, Sec. 1)

It is hardly necessary to emphasise once more the patent insufficiency of my material in this and the other samples of ceremonies observed at random. When I have seen only one ceremony from a system of magic—and that only once—gaps, oversights and over-emphasis and a general false perspective are inevitable. The value of such samples has been mentioned several times (cf. Note 37). The remedy also is obvious: more time spent and more frequent opportunities for full observation.

Note 41.—The Neglect of Taro Gardens (Ch. X, Sec. 2)

"Roughly speaking"—this phrase and explicit statements in the text indicate clearly the relative weakness of my information on taro gardening. I also give, in the paragraphs which follow, the reasons why I missed, presumably, a great many points in this subject. In a way, as I have indicated already in Section 3 of this Appendix, I am not satisfied with the whole perspective of taro gardens within the context of native agriculture. Compare also Notes 2, 11 and 13 above.

Note 42.—Hearsay Evidence (Ch. X, Sec. 3)

The reader is perhaps aware that my information on the *kaytubutabu*, received at second-hand only, is relatively lively and full of detail. This was due to the quality of my informant, Tokulubakiki. He was, as it were, the superintendent of my staff of informants; the best in a picked lot. At the same time, there were not many subjects on which he could speak with personal authority, since he was not a specialist of outstanding importance in any economic or ceremonial domain. This ritual and a few spells of black magic of a specially abominable type were the only two things of which he could boast. Here, however, he was dealing with a possession of his own: his own hereditary magic which only he could perform in Kiriwina. He worked on this information with greater gusto and interest than on any other subject. He recited to me the spells several times and made sure that I had written them correctly; he brought me the magical substances; he even organised a full-dress performance for my benefit. I have regretted ever since that I failed to photograph this staged ceremony of *kaytubutabu*.

Note 43.—DECAY OF CUSTOM UNDER EUROPEAN INFLUENCE (Ch. X, Sec. 3)

The reader may have noted that on several points I have registered the influence of European culture on native belief and custom. I have mentioned elsewhere European pearling (Part I, Sec. 5); I have spoken about the premature burning of a garden by a European (Ch. III, Sec. 1); I have indicated that some European traders request the services of the native garden magician (Ch. I, Sec. 5); I have reproduced the native view on our Christian garden magic (Ch. I, Sec. 5); and also the revolutionary effect of the game of cricket (Ch. VI, Sec. 3). These were rather amusing touches giving the specific additional colour which the incongruities emerging out of culture contact inevitably produce. But some of them, notably the introduction of pearling, were important economic influences.

The reader who has followed the sociological parts of the book carefully will have noted that the fundamental modification in the constitution of Trobriand society has been produced by the eclipse of the chief's power. The paramount chief and his peers in other districts are no longer the only people or even the main people who wield power and of whom one has to be afraid. There is a resident magistrate who can put you in gaol, fine you, or even—as has happened once or twice—hang you. His law has to be obeyed. There are the missionaries who moralise, pester and shame you into doing this or abstaining from that. There are the traders who exercise a different but not less powerful influence by giving or withholding things which have become almost a necessity. All this has affected the chief's tribute, especially in an indirect manner through the limitation of polygamy (Ch. VI). Warfare also has ceased; the *buritila'ulo* is no more what it was (Ch. V, Sec. 6), and gardening and exchange of produce can be done safely and without armed guards.

Agriculture is also affected by the introduction of European implements, or at least European iron instead of stone (Ch. III, Sec. 5). European crops and fruit trees—though adopted to a very limited extent—still change the finer balance of gardening. Some of them alleviate the pinch of hunger, seasonal and exceptional. The possibility of buying rice or getting it in advance for services and goods—notably pearls to be delivered later—the natives' knowledge that the Government would even feed them in case of extreme famine, affect profoundly their outlook on the relation between *molu* 'famine' and *malia* 'plenty' (Ch. V, Sec. 1).

All these points have been indicated; a good deal of information on contact and change is included in this volume. I was often in my observations driven to note certain phenomena, very largely because my earliest training in exact science made it to me almost physically impossible to neglect the full reality which I had before my eyes. The empirical facts which the ethnographer has before him in the Trobriands nowadays are not natives unaffected by European influences but natives to a considerable extent transformed by these influences. The Trobriander as he was, even two or three generations ago, has become by now a thing of the past, to be reconstructed, not to be observed. And the scientific way to reach even a careful reconstruction is through the observation of what actually exists.

I was thus forced to observe the facts of contact and change, but I want emphatically to state that my attitude, both in theory and practice, on this point was false. The Anthropology in which I was brought up was still mainly interested in the "real savage" as representative of the stone age; in origins; in the history of mankind to be read in the variety of primitive customs still existing. From the purely antiquarian obsession I had freed myself before ever I went into the field. Even in my earliest book, published in England, I criticised the evolutionary and reconstructive writings on the history of marriage, which give "the impression that the Australian tribes were a museum of sociological fossils from various ancient epochs of which the petrified form has been rigidly preserved, but into whose inner nature it is quite hopeless to enquire. The understanding of actual facts is sacrificed (in current anthropological theory) to sterile speculation upon a hypothetical earlier state of things" (*The Family among the Australian Aborigines*, 1913, p. vii of the Foreword).

I also soon discovered that the diffusionist school, whether in Germany, America or Great Britain, is interested in diffusion as it happened four thousand or four hundred years ago, but not as it is happening nowadays. On the other hand, the functional method, or at least that branch of it with which I am associated, was very largely born in the field. There I began to realise that even the reconstruction of all pre-European natives of some fifty or hundred years ago is not the real subject-matter for field-work. The subject matter of field-work is the changing Melanesian or African. He has become already a citizen of the world, is affected by contacts with the world-wide civilisation, and his reality consists in the fact that he lives under the sway of more than one culture.

The principle of studying the changing native as he really is enables us, on the one hand to reconstruct his pre-European culture, not by guess-work or by fortuitously brushing away a piece of calico, a Christian belief, an irksome European taboo, but by studying how these things work, how they clash with his original culture, or else how they have been incorporated into it. On the other hand, the process of the diffusion of culture, as it is going on now under our very eyes, is one of the most important historical events in the development of mankind. To neglect its study is definitely to fail in one of the most important tasks of Anthropology.

I have developed this point of view in one or two articles, notably "Practical Anthropology", *Africa*, January 1929, and "Rationalisation of Anthropology and Administration", *Africa*, October 1930. I am embodying it now in my teaching—but I was not yet under its influence in doing my field-work. This perhaps is the most serious shortcoming of my whole anthropological research in Melanesia.

Note 44.—Scattered Data on Minor Subjects (Ch. X, Secs. 4 and 5)

The information contained in these two last sections is, I feel, by no means satisfactory. It still remains to a large extent disjointed and unorganised. To use a newly coined word dear to the Functional school: it could be a great deal more "contextualised". I obtained lists of plants used and exploited by the natives; very often on my walks I saw people getting down some *wakaykosa*, and I lived through several seasons of fruit gathering and eating. But I do not feel yet that I have either mastered or given a clear picture of native bush-lore.

Note 45.—Assessment of this Chapter (Ch. XII, Intro.)

From the point of view of method, and quality of material, this chapter closely corresponds to Chapter VI. I have very little doubt that in its main sociological outline; in the legal principles of land tenure; in its methodological foundation and its economic working, the information here given is correct. It is based on much richer material, of course, than could be even briefly summarised here. For a great deal of it refers to the social structure of the Trobrianders, to principles of filiation, to the organisation into clans, village communities and political districts. The documents, such as pedigrees, clan table, lists of headmen, one or two samples of village census, I could not adduce here. But my sociological and legal generalisations are based on as satisfactory a body of documentation as I have for any aspect of my material.

The main gaps which I indicate refer to the more detailed and specific information, and even there they refer rather to neglect in recording concrete data than to a confession of having failed to test my conclusions to the full.

Note 46.—Holes of Emergence (Ch. XII, Sec. 1)

Were I to return to the Trobriands I would not rest quite as satisfied with having obtained the general principles of emergence from underground as I was when I left the archipelago. I would try to document my data by much fuller photographic records and of these I have only one or two in my possession (compare Pl. 89 in *Sexual Life of Savages*). I would also draft a number of detailed plans and diagrams. I was not able to discover any ancestral cult connected with the spots of emergence, no ceremonies or offerings at them, no specific taboos. The spots are of course kept free of any defilement, but beyond that I was not able to find any other symptoms of veneration.

Note 47.—Maps of Village Grounds (Ch. XII, Sec. 4)

I regret to say that the map of Omarakana garden lands is the only full record of this type which I drafted in the field. I had great difficulty in drawing it, since all the surveying had to be done in the most rudimentary manner. I attempted to draw a similar map in Oburaku, but found it very much more difficult, partly because the fields there are very broken up by swamps and the sinuosities of the coral ridge. I would like to add, however, that I have checked all the theoretical principles embodied in the map and analysis of Document VIII in Oburaku, as well as in Kurokaywa, Kaybola and Bwoytalu. But it is one thing to test certain legal and topographical principles with the greatest precision, and another to draft a complete map. I spent most of my time in the Trobriands in or near Omarakana, and I knew the territory there as well perhaps as any spot on earth. Towards the end of my stay I could have drawn the map from memory, although that was not the case at the time when I laboriously and painfully constructed it.

Note 48.—Transactions at Kayaku (Ch. XII, Sec. 4)

The principles here laid down—and which I am certain are correct—should have been documented by a number of records of actual transactions; with the name of lessee and owner of the plot, and the quantity of gifts which changed hands. In obtaining the information I discussed the general rules on concrete instances, but did not make notes sufficiently detailed to be available for a document. Nothing would have been easier and more profitable than to collect sufficient material for the presentation of such documents.

Note 49.—Stealing of Produce (Ch. XII, Sec. 5)

Some of the statements here given could be more fully documented. Thus I have collected, for instance, a few formulae of protective magic which, though not among the best recorded spells, yet contain ample documentation for the general statements in the text. They do not, however, belong to this context, since on the one hand they are a part of native sorcery and on the other they refer to legal institutions.

INDEX

Acacia leaves (*vayoulo*), 112, 113
Acacia wood, 312
Adze (*ligogu*), 62, 167, 170, 244, 289; ceremonial use of, 170; charming of, for *tum*, 170
Agriculture. *See* Gardening
Amphletts, the, 68, 73, 74, 75
Animal ancestors, 342
Animals: domestic, 52; wild, 67
Arbours (*kalimomyo*) for stacking *taytu*, 172, 173, 177, 213, 215, 231; division of produce shown in, 173; not on *gubayeki* plots, 194; over *urigubu heaps*, 230
Areca nut leaves, 105
Argonauts of the Western Pacific, cited, 104, 238, 246, 287, 322, 345, 435, 444, 452, 453, 454, 455
Arm-shells, 65
Aromatic herbs. *See* Herbs, and Leaves
Avocado pear (*aguacate*), 315
Axe (*kema*), 62, 132, 153, 167, 244, 422; ceremonial (*beku*) blade, 99, 283; private magic for, 155; steel in general use, 132, 283; charming of, 95, 96, 98, 99, 108, 279, 281, 428
Aykare'i, 386
Ayuvi kakavala rite, 153

Bachelors' house (*bukumatula*), 232
Bagido'u: his credentials, 456; his family, 84, 85, 86, 150; his duties, 78, 84, 93, 94, 99, 101–2; his harangues: before *kayaku*, 67, 94; before *yowota*, 88, 127; taboos kept by, 106; the scrub-burning, 111, 112; the *kamkola* rites, 129; his harangue for *kamkola*, 129; *tum*, 170–171; his illness, 78, 85, 107, 115; estimate of, 85, 94; valuable assistance from, 85, 153; his system of magic, 116, 153, 278; private magic formulae obtained from, 153, 155
Baku (central place of village), 87, 111; crates of yams on, 184; activities occurring at, 231; yam-houses on, 184
Baleko. *See* Garden Plots
Baloma. *See* Spirits
bam, 299
Banana leaves, in magic (*siginubu*), 95, 233, 235, 304, 307, 421; for skirt fibre, 312
Bananas: fruit, 312; magic for, 313; planting of, 181, 312, 313; trees, 58
Basi. *See* Thinning
Basi valu, rite and spell, 225, 237
Baskets: round (*peta*), 177, 179; small oblong (*vataga*), 105, 236, 237, 287, 429; made of coconut leaves, 300
Bastards, 202
Bathing: at water-holes, 177; in the sea, 177
Ba'u sub-clan: status of, 364; village, 342
Beku (ceremonial axe), 283
Bellamy, Dr., *quoted*, 103
Betel-nut (*buwa*), 65, 70, 73, 88, 103, 107, 156, 175, 176, 177, 184, 185, 186, 187, 214; a stimulant, 300; gifts of, 301; chewing of, 315; trees individually owned, 300
Binabina stones, 221, 222, 250, 258, 476
Bird-traps (*sikuna*), 299
Bisalokwa magic system, 277
bisiboda, 254
Bisikola ceremony, 115, 289, 297, 422; as private magic, 153; as banana magic, 313
bisiya'i, 244, 255
Black Magic, 78, 118, 306, 465–6; of Bush-Pigs, 469. *See also Bulubwalata*
Blight, effect of, 52
Boda (medicated leaves), 292
Bokaluva'u, 343
bokavili, 75
Bokuyoba, 111, 393, 394, 407
Boma (sacred grove), 57
Bomaliku. *See* Store-houses
Bomapolu, 395, 408
Bomawise, 394, 407
Bomigawaga, 136
Bomilala, 258
Bomisisunu, 258
Bomyototo, 393, 407, 415

Bopadagu, 343
Borogu (croton), 422
Bo'utukema, 395, 408, 417
Bovagise, 71
Boyeya, 304
Boys, gardening by, 60, 357
Bread-fruit (*kum*), 160, 228, 314; leaves in magic, 236
Brothers: sisters endowed by, 198; duties of elder, 200; of younger, 65; sister's guardian, 202, 204, 206–7; sister taboo, 208, 351
Brudo, M. and Mme., *quoted*, 454
Bubukwa, 220, 243, 244, 250, 251, 254
Bubwaketa, yam, 312
Budaka, 248
Budayuma, 71
Bugwabwaga, 393
Bugwalamwa, 71
Bukubeku, 415–417
Bukumatula (bachelors' house), 232
Bulabula tree and leaves in magic, 235
Bulaviyaka, 232
Bulimaulo. *See* Obukula; *also* Omarakana
Bulubwalata (evil magic), 78. *See also* Sorcery
Bulukaylepa rite (one form of Yowota), 107
Bulukwa gado'i (pig of the fence stake), 118, 298
Bulumaduku, moon of, 149
Burakema rite (one form of Yowota), 107, 108
Burayama: headman of, 367; sub-clan, 86, 342, 346, 347, 364, 433
Burials. *See* Mortuary rites
Buritila'ulo: seriousness of, 74, 176; quarrels leading to, 185; nature of, 213, 473; description of, 181, 182, 183
Busa tree, 92, 304
Bush (*odila*). *See* Scrub
Bush fruits. *See* Fruits, and *Kavaylu'a*
Bush-hen (*kwaroto, mulubida*), 107; soil from nest of (*ge'u*), in magical mixture, 94. 105, 131
Bush-pigs: pig of the garden stake, 117; guard against, 67; spell against, 101, 117, 118, 120, 281; *kakokola* a defence against, 131; the *bulukwa gado'i* (pig of the fence stake), 117, 118; homes of, 117, 118
butia, tree, 312
Butuma, 89
Butura (renown), 82, 231, 331
Buwa. *See* Betel-nut
Buwana, 203, 204, 301
Buyagu, 110
Buyavila Kiriwila, 414
Bwabodila (magical mixture), 285, 286, 287, 288
Bwabwa'u ("black") taytu, 152
Bwadela, 92
Bwaga, 316
Bwaga'u (sorcerer), 415
Bwala, spot of emergence so called, 342
Bwala tapwaroro, 63
Bwanawa, early taytu tubers so-called, 140, 149, 150, 151, 230; distinguished from taytuva'u, 107, 165
Bwasita, 434
Bwaydeda: his magic, 277
Bwayma, 82, 240–272, 474, 477, *and see* Store-house
Bwayma goregore. *See* Store-house
Bwayowa, 307
Bwoysabwoyse, boy named, 60
Bwoytalu: village, 255, 277; emergence holes in, 342

Cannibalism, 162
Canoes (laden with crops); legends as to, 72; in spell, 71, 72, 73; analogy between store-house and, 248–9
Casuarina leaves in magic, 129, 235
Central place. *See Baku*
Chiefs: polygamy among, 56, 79, 80, 192, 360; their monopoly of first-class gardens, 84; ownership of the soil vested in, 361; their store-houses, 229; taboos on, 165; personal power of, 67, 164. *And see* To'uluwa
Chiefs, minor, 192, 210
Chieftainship, foundation of, 217
Children at work, 60, 191, 205
Clans, rank vested in sub-clans of, 64
Clearing of plots. *See Koumwala*
Coconut palms: ownership and cultivation of, 75, 300; individually owned, 300, 308; native attitude to,

300, *et seq.*; sound of, in wind, 308; uses of, 300; leaves in magic, 300.
Coconuts, 71, 73, 107; leaves for roof repairs, 255; for floor mats, 300; for thatching (*yoyu*), 105, 116, 245, 255, 300; in magic, 302; for *bwayma* wall coverings, 300; shoots for torches, 112, 116; cream from, 300; cream of, for anointing, 99; green: collection of, 233, 293; importance of, 300; as food and drink, 88, 103, 300, 309; quality of, in Trobriands, 301; taboo on, 301, 307. *And see* *Kaytubutabu*
Communal labour, 157–8; on the *leywota*, 102, 103, 135, 136; by men only, 134; customs connected with, 59, 80, 122, 157, *et seq.*; by women. *See* Weeding
Competitive challenges, 183
Competitive display. *See* Display
Competitive exchanges. *See Buritila'ulo*
Coral, bits of, in magical mixture, 94, 105
Coral ridge. *See Rayboag*
Cricket, 211, 212, 213
Cries: at work, 103, 134, 135; harvest screams, 178
Crime and Custom, *cited*, 417
Croton tree, and leaves, in magic, 235
Cycas leaves, for tallying, 177

Dadam (a reed), 421
Dadeda, 142, 143
Daga (long pole or ladder), 254, 315
Dagiribu'a, 88
Dakuna, 222
Dala. *See* Sub-Clans
Dayagila, 365
Dayboya, 361, 371
Dayma. *See* Digging-stick
Death, sons' duties in connection with fathers', 206
Deliviyaka, 386
d'Entrecasteaux Isles, 134, 135, 221, 222, 307; legends in, 68, 73, 74
Digging-stick (*dayma*), 62, 132, 133, 145, 152, 153, 167; charming of, 291–2; in Oburaku, 293; private magic for, 155
Digumenu, 68, 71, 72
Dikoyas, village of, 68, 324
Dimkubukubu, or *katakudu* (small stick), 225
Dipapa, 417
Display of crops at harvest, 82, 173, 183, 207, 212
Dobu, 73, 74, 307
Dodige bwala, 231. *And see* Store-house
Dokonikan, mythological grotto of, 177
Domdom, 307
Drought, chief's displeasure expressed in, 111, 163. *And see* Hunger
Du'a'u, 70, 73, 74
Dubwadebula (grotto), 308, 342
Dudubile kwaya'i, 220
Duguvayusi, 88, 90, 432–3
Dumya, 102, 112; taro gardens on, 296, 299
Dwelling-houses, construction of, 232

Exchange, fish and vegetables (*wasi*, *vava*), 94, 162, 163; show *taytu* required for, 231

Fairy-tales (*kukwanebu*), 156–7
Famine, conditions in, 160, 161, 162, 163, 472. *And see* Hunger
Fencing of gardens (*kali*), need for, 120; construction of fence, 61, 123, 126, 357–8, 471; its repair, 76, 77; for *urigubu* heap (*lolewo*), 173, 215
Ferguson, I., legend as to, 70, 73, 135
Fibre (*im*), 312
Field (*kwabila*): named, 55, 87, 88, 348, 372, 373; subdivision of, 88, 89, 331; list of *kwabila* in Vakuta, 420
Field-work in ethnography, requisites in, 452, *et seq.*
Fire by friction, 283
Fish: exchange of *kaulo* with *wasi*, 94; for feasting and *ula'ula*, 95; tabooed, list of, 107; staple diet in Oburaku, 238, 290
Fishing: gardening interrelated with, 52, 94; expeditions, season for, 139, 162; east coast impossible for, in dry season, 162
Food: giving of, a privilege (*ula'ula*), 65; taboos on, 66, 107
Fowls, 52

Fruits of wood and wild, 52, 160, 310, *et seq.*

Gaboyi, 428
Gabu, 110, 116, 120, 121, 129, 136, 289, 312, 313, 422; in *taro* gardens, 297, 298. *See also* Scrub-burning
Gado'i (sticks), 123
Gala kam (insult), 216
Galauwa, 89
Gam (band of coconut leaf), 302, 304, 307; *kivila gam*, 309
Gapulupolu, 386
Garden plots (*baleko*), 90, 121, 150, 284 allotting of, 56, 91, 102, 104, 372, counting of, 88, 89, 90, 91; on the site, 91, 93; number of, per man or boy, 91, 102; inaugural ceremony, 63, 92, 93, 94, 104, 107, *et seq.*
Garden site, 90
Garden wizard, 64, *et seq.*
Gardener, good, term for (*tokwaybagala*), 61; too good, danger of, 83, 175
Gardens: best, chief's possession of, essential, 60; ritual burning of the, 110–11; a walk through the, 57, *et seq.*; clearing of. *See Koumwala*; diversity of, 56, 58; fencing of. *See* Fencing (*kayaku*), 63, 94; mythological background of, 68, *et seq.*; *kamkokola*, ceremony in, 59, 152, 285, *et seq.*; *and see Kamkokola*; layout of, 90; subdivision of, into squares, 58, 91, 121, 123; magic of, 62, *et seq.*, 137, *et seq.*, 470; list of systems of, 419. *See also* Magic; Magical Corner; Main. *See Kaymata*; planting of. *See* Planting; quarrelling over, 93, 103, 104, 175; road through, 90, 91; rubbing the ground of, 93, 101, 102, 108; selection of sites for, 88; situation of, 91; standard plots. *See Leywota*; striking the ground of, 93, 152, 279, 280; visitation of, at harvest, 173; weeding of. *See* Weeding
Gardening: importance of, supreme, 80, 157; magic closely connected with, 55, 59, 60, 62; legal aspect of, 93; sociological aspect of, 59, 157; aesthetics of, 59, 80, 81, 120, 123, 128; four main divisions of, 61; times of (hours), 61; co-ordination of work in, 104; implements used in, 132, 133; similarity between various systems, 273, *et seq.*; distinction between a man's and woman's part in, 79. *See also* Magic
Gardening team: of co-operative workers, 320, 331, 336; composition of, 157, 355, 356, 357, 371; sub-clan transcended by, 356; importance of, 79, 357, 369
Gawa, 71, 73
Gawa'i, 417
Gayasu, plant, 92
Gayewo (pandanus leaf petals), 312
Gayga'i magic system, 277
Gegeku leaves in magic, 236
Geguda, season of, 52
Gelivilavi, moon of, 149, 161, 311
Gere'u, 69, 70, 73
Ge'u (bush-hen's nest), 105
Gibuviyaka rite, 113, 115, 131, 281, 422, 425; in Sinaketa, 298. *See also* Scrub-burning
Gifts: customary law of, 188, *et seq.*; *youlo*, 190; *Takola*, 190; for sexual services (*buwana*, or *sebuwana*), 203, 204, 301; reciprocity in, 199; to magician, 63, 65. *See also* Valuables, *Taytupeta*
Gimwali (barter), 209
Ginger, wild, in magic, 72, 163, 223, 225, 233, 235, 312
Ginuvavaria (mollusc), 155
Gipita, 95
Gipware'i. *See* Lalang grass
Giribwa, 117, 290, 316
Giyauri, 423
Giyokaytapa, 279, 417, 418
Giyotala, 407, 410
Giyulutu magic system, 277
Gogo'a, 393, 395, 408, 410
Gomaya, 293, 294
Gomila, 414
Goodenough, I., 74
Gourds, modes of cooking, 315
Grove, sacred (*boma*, *kapopu*), 57, 278, 281, 286
Growth: magic of, 136, 137, 139, 141,

145, 146, 149, 151, 168, 288, 294, 433; phases in, 299; *megwa geguda* and *megwa matuwo*, 149; seasons of, in native calendar, 149; its inaugural aspect, 149
Gubakayeki, 91, 194, 411, 413, 417, 474; plots for, 91, 411
Gubayladeda, 386
Gubilakuna, 413
Gubwatala (squares), 121, 122
Gugu'a (objects of use), return gift of, 372
Gugula (heap), 172, 214, 387, 392
Gumabudi, 386
Gumanuma (foreigners), 415
Gumasila, 307
Gumguya'u (title of sub-chiefs), 258; lesser chief, 113, 179, 192, 210, 385, 386
Gumigawaya, Tom, 125, 173, 325, 414
Gumilababa, 211, 328, 347, 355, 361, 365
Gumla'i, 434
Gumlu'ebila, 386
Gusaweta, 453
Gutaguta (plant), 421
Guvagava, 427
Guya'u, 414, 417
Gwadila, or *kum* (bread-fruit), 160, 228, 314; leaves in magic, 236

Hancock, Billy, 217, 453
Harvest, the, 61, 159, *et seq.*; allocation of, to matrilineal kindred, 189; gifts of (*kovisi*), 188, *et seq.*; division of produce shown at all stages, 159; duties of, recapitulated, 194–5; family work, 171, 176, 185; festive character of, 82, 138, 158, 213; work and pleasure of, 171, *et seq.*; cries at (*sawili*), 160, 178–9, 214; gleanings of, 176; jealousies regarding, 175, 182, 183, 185; gifts, theory and practice of, 181, 210, *et seq.*; decay of the chief's, 406–11. *See also* Okwala, Tum, Tayoyuwa, Urigubu, Taytumwala, Taytupeta, Sawili, Isunapulo
Herbs, in magic, 225, 307; aromatic, 72; for armlets, 94. *And see* Magical Herbs, Magical Mixture
Hibiscus: flowers, for personal adornment, 99, 312; significance of, 235
Hoernlé, Mrs. W., 455
Hornets' nest (*kabwabu*): bits of, 106; in magical mixture, 94; as headgear, 214
Hunger: famine (*molu*), calamitous nature of, 52, 160, *et seq.*, 163, 238–9; *tubukona molu*, 160; the great *molu*, 161; shame attaching to, 227; coconuts a staple in, 238
Hunting, 52

Ibomala, 417
Ibo'una, 175
Ibutaku, 88, 90, 433
Idaleyaka, 342
Ilabova, 394
Ilagaulo, 415
Ilaka'isi, 394, 410
Ilakasila, 415
Ilapotu, 415
Ilaybisila, moon of, 171, 302, 311
Ilaykumila, 433
Inkuwa'u, 386
Insects and grubs, 138
"Intichiuma Ceremonies, Economic Aspects of," *quoted*, 456–7
Ipikwanada, 106
Isakapu, 140
Isunapulo ceremony, 149, 165, 291; in Oburaku, 237; in *tapopu*, 299; and spell, 149, 165, 167, 292, 295, 424, 428, 429
Isupwana, 407, 408, 414; his storehouse, 393, 395, 415
Italala leywota spell, 421
Iwa, I., 68, 70, 72
Iwaya'i gado'i rite, 422
Iyavata, 101

Kabisitala, 190, 251
Kabisivisi (*bwayma* compartment), 251, 252, 255
Kaboma (garden taboo), 111, 125, 278
Kaboma (wooden dish), 285, 286, 287
Kabulula, 434
Kabululo, 277
Kabutu, 123, 158
Kabwabu. *See* Hornets' nest

Kabwaku (village), 162, 182, 183, 186, 187, 210, 228, 245, 258; magic of, 277; Toliwaga of, 192, 328, 365; cry, 134, 135; component villages of, 346; magician of, 303
Kabwaynaya, 366
Kadala, 386, 417
Kadamwasila, 410
Kadinaka, 408
Kadubulami, 85
Kaduguya, 386
Kadumilagala valu, 225
Kaduwaga, 365
Kaga, 421
Kakaveyola, 393
Kakema (dwarf tree), 220
Kakulumwala (stone-heap), 121, 251, 255
Kala'i, 423
Kalamata, 185, 187
Kalamelu, 185
Kalapisila o valu, kalapisila kuli (stile), 90, 100, 102
Kalava'u, taytuva'u, 171
Kalaviya kalasia, 182
Kalawa (measuring, counting or enumeration), 90, 177, 215, 230
Kali (fence), 123. *See also* Fencing
Kalibudaka, 91
Kaliguvase, 255
Kalikutala, 251
Kalimamata rite, 114, 171, 422; spell, 281, 297; special yam planted at, 281
Kalimomyo. *See* Arbours
Kalogusa, 417
Kalubaku, 408
Kaluluwa leaves in magic, 286, 421
Kalumwaywo, 91, 386
Kaluva'u sub-clan, 86, 342, 343, 346, 347, 367, 414, 431
Kaluwalasi, moon of, 302
Kaluwayala, 283, 312
Kamkokola, 58, 77, 102, 116, 127, 128, 132, 172, 224, 283, 298, 299, 423; (pole) *kwanada* supported by, 124, 127, 130; ritual striking of, 130, 171, 284; rite, 124, 154, 299; only magic, 59, 77, 81, 129, 131; as magic of planting, 470, 471; description of, 131, 132; ceremony, 281, 282; spell chanted over, 283; construction of fence and, 471; ceremony in Momtilakayva system, 285, 288; magical structures called, 59; site of, 102; taboos connected with, 282; its significance, 129, 131, 132; better in the *leywota*, 59; an inaugural planting rite, 127; in Sinaketa, 299
Kammamala spell, 150, 151
Kamtula'i, 414
Kamtuya (supports), 137, 138, 143, 299
Kanibogina (*gwadila* kernels), 311
Kaniku, 172
Kaniyu of Liluta, 213, 215
Kapopu, 126, 286
Kapuwa, 307, 309
Kapwani, 276, 342
Karibudaboda, 372; valuables as, 295
Karigava'u (the newly dead), 293
Karige'i (boundaries), 432
Karisibeba, 413, 414, 415
Kari'ula, 158
Karivisi (decorative triangles), 127, 128, 286
Kariyala (magical portent), 107, 130
Kariyayeli sapi, 145
Kasana'i, 211, 225, 228, 245, 258, 362, 411, 434; magic of, 225, 277
Kasaylola spell, 151
Kasisuwa, 315
Kasivi, 170
Kasiyena yam, 156, 278, 429
Katakubile, 356, 431
Katakudu (stick), 225
Katawabula kavatam, 423
Katekewa (carrying-pole), 179, 286
Katulogusa, 156
Katupwena, 394, 408, 409
Katusakapu, 143
Katutauna rite, 299
Katuva, 255
Katuveyteta, 251
Katuvisa kaydabala, 309
Katuvisa kaykapola, 308
Kaulagu, 286
Kaulasi, 290
Kaulo (vegetable food), 81, 87, 227, 246; exchange of. *See* Exchange; taytu, the staple, 81
Kaututa'u, 113, 413, 415
Kavakaylige, 88, 89, 90, 433
Kavalapu, 244 253, 255

Kavapatu (leaves of covering), 285, 288; herbs for, 285
Kavatam (vine supports), 77, 124, 125, 127, 138, 139, 146, 168, 171; digging-stick planted at foot of, 292; Sinaketan magic with, 294, 315, 428
Kavataria, 94, 211, 277, 328, 361, 365; component villages of, 346
Kavaylu'a (jungle fruits), 160
Kavega'i tree and leaves in magic, 235
Kavilaga, 244
Kavituvatu, 254
Kawatalu, 91
Kayaku, 54, 87, 88, 91, 103, 108, 278, 279, 294, 302, 371, 421, 466. *And see* under Garden Plots: inaugural ceremony
Kayasa, nature of, 177, 179, 211, 212, 213, 214, 215, 216; in Omarakana (1918), 179, 212, 392-7
Kayaulo (totemic tree), 220; leaves of, in magic, 223; wood, 225
Kaybaba (slanting poles), 128, 168, 235, 283, 284, 288; medicated leaves tucked under, 285, 288, 292
Kaybola, 186
Kaybomatu (mussel-shell), 105
Kaybu'a (coral boulders), 106, 421
Kaybudaka, 248, 250
Kaybudi, 138
Kaybwagina, barren tracts in, 290
Kaybwibwi (scented pandanus), 106, 304. *See also* Pandanus
Kaydabala spell, 145, 146; stick: magical insertion of, in Oburaku, 292
Kaydavi (yam on sticks), 184, 185, 187
Kayeki, 167
Kaygaga spell, 100, 101, 102, 108, 279
Kaygogwa'u (wind-rattles), 299
Kaygum, 137, 138
Kaykapola (young coconut leaves in magic), 116. *See also* Torches
Kaykeda (harvest gift), 372
Kaykosa (hooked stick), 311, 314
Kaykubwaya, 423
Kaylagila, 254
Kaylagim, 248, 250
Kayla'i, 63, 362, 371
Kaylavasi sub-clan, 343
Kaylepa (magic wand), 93, 99, 102, 108, 169, 428
Kayle'ula I., 74, 365, 369
Kaylola lola (mooring stake), 249
Kayloulo, 308
Kaylu'ebila system of magic, 84, 101, 276, 278, 349
Kayluvalova stick, 125, 126, 282
Kaymata (main gardens): *Kaymugwa* distinguished from, 58, 87, 121, 194, 413, 463; harvest of, fixing time for *Milamala*, 54, 86, 87; *taro* harvest from, 296; planting in, following *koumwala*, 299; starting of, prior to *Milamala*, 87; variety of crops in, 122; *taro* planting in, 296, *et seq.*
Kaymwila, 422
Kaynubilum, 254, 256
Kayowota (sapling), 101, 102, 108, 280, 428
Kaysalu, 138
Kaysususine, 142
Kaytagem leaves in magic, 280, 286
Kaytapaku, 375
Kaytaulo, 244, 248, 250, 256, 293
Kaytubutabu, 301, 302, 303, 304, 305, 306, 307, 308, 309, 310, 348
Kaytukwa (ornamental staff), 169
Kaytumla bubukwa, 222, 250
Kayvala, 410
Kayvaliluwa, 138
Kekewa'i leaves in magic, 291
Keliviaka ceremony, 285, 288
Kema. *See* Axe
Kiluma, 252, 253, 254, 255
Kinship grouping, 199, 206, 207
Kiriwina, 54, 56, 68, 69, 70, 73, 74, 75, 82, 85, 86, 119, 121, 124, 145, 162, 235, 290, 366; date of spirit visits to, 54, 74; harvest in, 158, 213; the 1918 *urigubu*, 211, 212; bareness of its magic, 282. *And see* Omarakana; To'uluwa
Kisi (polishing implement), 244
Kitava I., 54, 70, 72, 74; legend of its origin, 68; theory of wind-borne disease from and to, 238
Ki'ula'ola, 313
Kivi, 255
Kokola, 253, 255
Kokouyo, 187
Kolova (loud calling of names), 216, 286
Kommatala, 410

Kotila (a plant), 131, 427
Koumwala (clearing), 80, 111, 113, 115, 120, 122, 125, 136, 281, 299, 422, 423; planting immediately following, 123
Kovalawa (lagoon beach), 344
Kovisi (return gift), 192, 223
Koya, the, 135, 245
Koyatabu, 73, 135
Kropan, 69
Kubila (scented plant), 106
Kuboma, 54, 121, 328, 365, 369; ruler of, 277
Kubudoga (raised platform), 84
Kubuna yamada (own food divisions in taro garden), 298
Kubwaya, 423
Kudayuri, myth of, 104 *footnote*
Kudukwaykela, 277
Kukwanebu (fairy-tales), 156
Kula (expeditions), 74, 294, 455; postponement of, for gardening exigencies, 53
Kulubwaga, Headman of Wakayse, 186
Kulumata, 54, 121, 163, 166
Kuluvitu, 276
Kuluwa, 277, 373, 375
Kuluwasasa, moon of (harvest), 295, 296, 449
Kuluwolu, moon of, 302, 311
Kum. *See* Bread-fruit
Kumatala of Kwaybwaga, 213, 216
Kumilabwaga, 316
Kumkumli (baking in earth), 285
Kupwakopula, 277, 278, 285, 346
Kuria (cooking pots), 221
Kurokayva and Kurokaywa: its two gardens and one magician, 355; *towosi* of, 278; villages, 153, 277, 278; system of magic, 154, 278, 281, 282, 283, 289
Kuroroba, 221
Kutugogova (hand-clapping on mouth), 307
Kuvi (large kind of yam), 57, 72, 73, 76, 153, 183, 184, 193; legend of Tudava and, 72; compared with *taytu*, 81, 137; harvesting of, 167, 237; *Kalimamata* spell over, 281
Kuwo'igu, 413, 415
Kwabila. *See* Field
Kwabulo sub-clan, 342
Kwaduya shell, 299
Kwa'iga (vivi kernels), 311
Kwaku sub-clan, 342, 347
Kwanada yam, 114, 124, 135, 171; spell, 281; its habitat, 312
Kwapatu, 374
Kwaroto. *See* Bush-hen
Kwaybwaga, 79, 86, 186, 210, 211, 303, 342, 392; feud of, with other villages, 212, 215, 410; component villages of, 245, 346
Kwaygagabile tree, 304
Kwebila (aromatic herbs), 72
Kweta (bread-fruit seeds), 314
Kwibanena. *See* Yams
Kwita (cuttle-fish), 107
Kwoyavila of Liluta, 212, 213, 407
Kwoylabulami, 421
Kwoynama sub-clan, 85, 362, 364, 385, 415, 417

Laba'i, 84, 276, 365, 367
Labelaba, 415
Labour. *See* Communal Work, Planting, Weeding, etc.
Lagim, 248
Lalang grass (*gipware'i*), 112, 113, 146, 429; for thatch, 245, 254; tuft of, planted in growth magic, 292
Lalogwa (decorative frames), 184
Land tenure, 56, 60, 89, 93, 294, 316, *et seq.*; complications of, 317, 319, 323; practical aspect of, 339; need for scientific organization of evidence referring to, 322; legal aspect examined, 318; various claims in, 331, 335, 348; central claims, 344; consumer's claims, 344; table of claims, 328, *et seq.*; absentee owners, 329, 353, 371; social and constitutional structure determined by, 332; producer's manifold interest in, 331, 333; consumer's interests in, 331, 333; synoptic table, 338, 339; doctrine of First Emergence, 336, 338, 341, 342, 343, 366, 368: its many-sided importance, 369; legal and mythical foundation of, 319, 321, 335, 338; Adelphic line of succession, 345; rank as affecting, 358–9, 360,

362, 363; summary of the system, 339, 376, 381
Lapu (stout poles), 125, 126, 127, 282
Lasawa, 423
Laughlan Is. (Nada or Nadile), 69, 71, 72, 73, 75
Lawa fruit, 311; leaves, 421
Lawaywo, 342
Leaves: in magic, 168, 235; pandanus and banana, 304, 307; *Kaykapola*, 63, 126, 291; edible, modes of cooking, 311; of covering (*Kavapatu*, *yayu*, *youlumwala*), 129, 130. *See also* Magical Mixture
Leo (missionary teacher), 190
Leria (plague), 132
Lévy-Bruhl, Prof., *quoted*, 63
Lewo, tree, and leaves in magic, 223, 225, 235
Leya (wild ginger root) in magic, 223, 225, 312
Leydoga, 408
Leywota "standard plots," 59, 108, 136; situation of, 90, 136; allotting of, 100; magic on, 64, 108, 113; punctual development of, 102, 104; sociology of, 64; clearing on (*koumwala*), 113, 120; *kamkokola* on, 286; the *kamkokola* ceremony, 131; "leaves of covering" rite in, 168, 288
Ligabe, 181
Ligou. *See* Adze
Liku (crate), 184, 185, 186, 187, 231 (log-cabin of store-house), 225, 227, 244, 248, 252, 256; (large beams), 245, 246, 249, 250, 251, 254
Likula bwayma, 189, 190
Lileykoya plant and leaves, 170, 421, 423, 429
Liluta, 79, 177, 210, 211, 212, 215, 342, 392
Lily (native), *moroba'u*, 301
Lime-pots, 315
Lisiga (chief's dwelling), 88, 232
Lobu'a, 346, 410
Lokwa'i tree, 311
Lolewo, 173, 215
Lomilawayla, 88, 433
Luba, 369, 431
Lubagewo, 386
Lubalabisa, 158
Lu'ebila, 276, 349, 367
Lukuba clan, 343
Lukubwaku (bush-pig's home), 117, 118
Lukulabuta clan: myth of its rank, 312, 343; rank of its sub-clans, 362
Lukwasisiga clan, 85, 86, 346, 385, 386, 414, 417
Lumlum (symbolic refuse heap), 113, 298
Lunar calendar, 53, 54, 55; different in different districts, 54
Lupilakum, 150
Lusançay Is., 73, 175, 307
Luya, 346

Madawa, 68
Magic (*megwa*): association of, with work, 55, 60, 61, 62, 93; communal, 153; list of systems of garden, 419; analysis and chart of organising function of garden magic, 435–51; importance of, 62, 223, 310; a social force, 310; Inaugural ceremonies (*yowota*), 93, 107: two forms of, 94, 95, 315; methods of charming, 287; payments for, 62, 153, 303: for private magic, 153; private, 78, 152, *et seq.*, 155, 472; chief claims to superiority in, 153; rain and drought attributed to (*tourikuna*), 83; rubbing the ground, 101, 281; systems of, peculiar to each village, 275; almost identical, 275, 276, 277; differences, 276; *kaytubutabu*, 301–6; an organising force, 336; *towosi*, 63, 64, 83; power of, and the efficiency of work, 75; performed officially by garden magician, 77. *See also* Magical formulae, magical corner, magician (garden), growth
Magic of the Garden, 62, *et seq.*
Magical corner, 108, 114: in garden, *kamkokola* erected at, 59, 123, 124; in Oburaku, 292
Magical Formulae, 95, 96, 114, 131, 143, 144, 145, 233–4, 236, 278, 279, 280, 289, 309; *vatuvi*, 96–8, 111, 249; *kaygaga*, 100; *kayowota*, 101; *yowota*, 101; *kaylepa*, 102; *pelaka'ukwa*, 114; *kalimamata*: chanted loud, 114; over "leaves of covering," 129–30;

vakalova, 131; *vaguri* (or *tavisi sobula*), 142–3; *katusakapu sobula*, 143; *kari-yayeli sapi*, 145; *kaydabala*, 146; *kaylavala dabana taytu*, 147; *kasayboda*, 148; *vapuri*, 149–50, 249; *kammamala*, 150; *kasaylola*, or *talola silisilata*, 151; private, of the tula-*yagogu*, 153, 154; over the digging stick, 155; *isunapulo*, 165; *vakam kuvi, vakam uri*, 167; *okwala*, 169; *tum bubukwa*, or *kaytumla bubukwa*, 221; *kayaulo* and *leya*, 223–4; *pelaka'ukwa*, 281; *Kalimamata*, 281; weeding spell, 284, 288; *Kaytubutabu*, 305–6; to palms to fruit and mature, 308, 309; banana spell, 313; spoken over a boundary stick, 156; loud chanting of, 96, 105, 277; spells well known, but recitation restricted, 96

Magical herbs: naming of, 220, 221, 225, 285, 427; chanting over, 287, 307. *See also* Herbs, and Magical mixture

Magical mixture, 94, 105, 112, 127, 129, 131, 152, 281, 421, 427; preparation of, 94, 95, 236–7, 304; for "leaves of covering," 129, 130

Magical prism. *See Kamkokola*

Magician, garden (*towosi*): his office, 84, 85, 87, 357; his rank, 60, 64; his work, 61, 66, 77, 139; the inaugural ceremony, 63, 110, 111, 152; the series of rites, 219; gifts to, 65, 287; his helpers, 64, 65, 66; his harangue before *kayaku*, 59; *kayaku* before his house, 88; the scrub-burning, 110, 111; practitioner in private magic, 78, 152, *et seq.*; work organised by, 66, 152; payment to, 63; his position, 64; his influence, 291; as 'master of the soil,' 328–9. *See also* Bagido'u, magic, taboos

Mailu, 453

Makita (aromatic creeper), 304, 307

Malasi clan, 86, 368; rank of its sub-clans, 84, 342

Malay apple, 126, 177, 236

Malelo'i, 414

Mango, 228, 312, 314

Mangrove, 92; poles cut from, 312

Marita, 69, 70, 73

Marriage, law and contract of, 199, *et seq.*; constitution of the contract, 199; family founded on, 203; patri-local, 203, 205, 206, 207; status conferred by: on the man, 201, on the woman, 201; nature of recipro-city in, 199, 291; economic aspect of, 204; *urigubu* a main element of stability in, 205, 207; cross-cousin, 205, 206, 207, 232, 349, 353, 354, 362; (exogamous) as effecting land tenure, 336

Marshall Bennet Is., 68

Matala, 140

Matrilineal society, 171, 189, 190, 199, 200, 201, 336, 341, 360

Mats: used in magic, 95, 96, 99, 111, 113, 116, 127, 129, 170, 225, 274, 280, 292, 304, 305; of coconut, 245, 250, 251, 254; on *byayma* floor, 257; of coconut leaves, 300

M'Bwasisi, 458; garden magician of Vakuta, 70

Measurement, standard of, 177, 184, 185

Megalabwalita, 386

Megwa. *See* Magic, also Growth

Menoni: fruit, 160, 311; leaves in magic, 236, 291

Milamala (palolo worm), 54; moon of, 53, 54, 110; festivities, times of, 86, 104, 110, 115, 136, 302, 348; offerings made at, 296; food offerings at, 287

Millipede in magic spell, 147

Mimosa in private magic, 284

Misima, 70, 73

Mitakata (assistant to Bagido'u), 85, 111, 113, 127, 417; of Gumilalaba, 176, 361

Mitayuwo, 90, 413

Modigiya, 304

Modulabu, garden magician, 108

Mokakana fruit, 311

Mokatuboda, 104

Mokilavala, 407

Mokolu fruit, 311

Moliasi, chief, 182, 186, 187

Moligilagi, 277, 342

Mollusc, boring (*ginuvavarya*), 155

Molu. *See* Hunger

Molubabeba, 111, 113, 163, 413, 414

Mom'la ceremony, 152, 294
Momola (the eastern sea-shore), 125, 235, 343-4
Momtilakayva magic system, 277, 278, 283, 285, 288, 445
Momyaypu (mummy apple), 314
Mona (taro pudding), 155. *See also* Taro
Monobogwo, 408
Monori, 386
Monumadoga, 408
Moon sequences. *See* Lunar calendar
Moraywaya, 386
Moresby, Port, 453, 457
Moroba'u, 301
Mortuary rites, 53, 166, 175, 206; exhibition of valuables, 287
Mosagula Doga, 347
Motago'i of Sinaketa, 67, 110, 117, 294, 371; his reliability, 117, 118, 297
Motaniku, 413
Motuan tribes, 457
M'tabalu, 362
M'tawa, 79, 210, 211, 212, 392, 410
Mukava'u, 386
Mukumku, 413
Mulobwayma sub-clan, 342
Mulubida (small bush-hen), 107
Mumwalu yam, 312
Muruwa, or Muyuwa, 73
Mussel-shell (*kaniku, kaybomatu*), 105, 172
Mwadoya, 426, 429
Mwagwaya, 407
Mwakenuva, 163
Mwamwala, 255
Mwauri sub-clan, 342, 364
Mwaydayli, 415
Mwaydola, 386
Mweyoyu (man of Wakayse), 182
Myth in Primitive Psychology, quoted, 95, 312, 344, 368, 464
Mythology, 55, 59, 60, 64, 75; influence of, on land tenure, 338

Naboyuma, 395
Nabwasuwa, 415
Nada, or Nadile (Laughlan Islands), 69, 71, 72, 73, 75
Nakoya, 150
Namtawa, 394, 407, 415
Namwama Guya'u, eldest son of To'uluwa, 394, 407, 417; tension following his expulsion, 361, 417; his private magic, 153
Nasibowa'i, 153, 156, 278, 279, 280; his private magic, 282, 283, 289
Natives of the Mailu, The, cited, 323
Natu fruit, 311
Navavile (*tovilamalia* of Oburaku), 67, 235, 236, 238, 291, 347
Nene'i, 423
Nigada, act of, 372
Noku fruit, 160, 161, 164, 312; leaves in Omwala rite, 168, 169
Noriu fruit, 304, 312
Normanby I., 73
Numakala, Chief, 163, 245, 393, 415
Nunuri leaves in magic, 280, 283
Nuts, 311

Obowada, 277, 313; magician of, 303
Obrebski, Dr., *cited*, 117, *footnote*
Obukula (hole of emergence, Bulimaulo), 342, 343, 346, 365, 414
Oburaku, 233, 237, 238, 291, 355; tale of fight near, 162; *vilamalia* in, 233; magic in, for hunger times, 237; influence of garden magician in, 291
Obwabi'u, 88, 433
Obweria, 108; magician of, 277, 303; one gardening team of, 355
Odaybaybona, 427
Odila. *See* Scrub
Ogayasu, 92
Okayaulo, 290, 316
Okaybu'a, 92
Okaykoda, 346; colony of, 373
Okopukopu, 54, 313; sub-clan, 342
Okwala rite, 167, 168, 169, 170, 424, 429; in Sinaketa, 294, 295
Olivilevi, 54, 211, 228, 365, 375, 431; Kiriwinian terminology, 196; magic system of, 276
Omarakana, 84, 127; rank of, 63, 88; gardens of, 90, 110, 111; its garden magic considered best, 84, 85, 86, 101, 119; origin of the system, 277; its situation, 87, 90, 92; *kayasa* harvest in (1918), 180; *urigubu* of 1918, 211, 212; paramount chief of, 78; description of map of, 431-4;

Bulimaulo, hole of emergence in, 414; water hole in grove of (Bulimaulo), 342, 343, 346; big storehouse in, 249; harvest gift after the Kayasa, in 1918, 392–7
Omlamwaluwa, 276, 365, 411
Omwaydogu, 92
Opikwakula, 90, 433
Orayayse, 417
Ornaments, 312. *See* Shell ornaments
Osapola, 85, 407, 417
Osayuyu, 365
Ovaboda, 88, 89, 90, 433
Ovavavile, 278, 281
Oyweyowa, 365

Paku (medicated leaves), 293
Palms: extensive planting of, 300, *et seq.*; names given to, 308. *And see* Betel, and Coconut
Palolo worm, 54
Pandanus, 126; leaves as decorative streamers, 228; for wall partition, 245; its many uses, 312; streamers in wind-rattles, 299; leaves and petals in magic, 99, 106, 304, 307; fruit (*vadila*), 312
Papaia fruit, 314
Pawpaw (*carica papaia*), 161; leaves in magic, 236
Payments, *vakapula*, 176, 177, 284, 303
Pearl shell, used in magic, 165, 167
Peas, 111, 193, 315
Pelaka'ukwa rite, 113, 117, 281, 292, 422; in Sinaketa, 294, 297, 298
Peraka, 422
Peta (baskets), 179, 285
Pigs, devices for keeping off, 76. *And see* Bush-pigs
Pineapple, 314
Pipi fruit, 312
Piribomatu, 407, 410
Pisila. *See Kalapisila*
Planting (*sopu*), 53; preliminary, 110, 115, 125, 133; main: its rites, 132, 136; method of, 132, *et seq.*
Plenty (*malia*), 52, 54, 59, 81, 160, 211, 220; magic of, 226
Plot. *See* Garden plots
Pokala (system of land transference), 192, 326, 338, 345, 349, 369
Polygamy of chiefs, 56, 79, 191, 232, 360
Po'u, 244, 248, 250, 251, 252, 254
Procreation, Trobriand theory of, 199
Pruning (*basi*). *See* Thinning
Pulitala, 361
Pulukuwalu (small fights), 182, 183
Pumpkins, 193
Pupwaka'u ("white" *taytu*), 152
Purayasi, 85, 163
Puri, 138, 149
Puwaya, 177, 216
Pwakova. *See* Weeding
Pwana (hole of emergence), 342
Pwata'i, 184, 409
Pwawa, 278
Pwaypwaya, 76

Radcliffe-Brown, Prof., *quoted*, 452, 455
Rain: taboo regarding, 107; magic of, 78, 104, 163, 343, 362
Rattan, 249
Rayboag (coastal ridge), nature of, 57, 76, 94, 126, 285; holes in, 76, 315; *kuvi* on, 315; magical ingredients collected from, 94, 105, 106; yams grown there, 291, 315
Reciprocal principle, 181–3, *et seq.*, 199, 206, 290. *And see* Exchange
Rest Days, 469
Reuta'ula, 421, 422, 429
Road entries to villages, 344

Sagali: preparations for, 246, 285, 286, 287, 302; contributions to (*dodige bwala*), 193; show *bwayma* opened for, 87, 231
Sago: leaves used in the south, 245; for thatching, 245
Sakapu (hole of emergence), 88, 90, 143, 342, 346, 433
Saka'u (bird), 220
Sakaya, 150
Samugwa, 386
Sanaroa, 307
Sapi. *See* Weeding
Sapona (ground between stony heaps), 299
Sasana (*lawa* kernels), 311
Sasari plant, 166
Sasoka tree, 106

Sasova leaves, 427, 428
Sawili (harvest screams), 160, 178–9, 214
Sayboda spell, 147
Sayda leaves in magic, 236; nuts, 311
Saysuya leaves in magic, 236
Scarcity. *See* Hunger
Scare-crows, 58, 77, 156
Scrub (*odila*), 76, 89, 181, 286, 304, 310, 312, 343
Scrub-burning (*gabu*), 53, 61, 104, 110, 111, 170, 281, 422; double function of, 110; *gibuviyaka* rite, 112, 113, 281; scrub-cleaning (*koumwala*), 120; scrub-cutting (*takaywa*), 53, 61, 93; quarrels at, 93, 103; starting of, 89; on the *leywota*, 102–3; *vakavayla'u*, 111, 116
Seasons, native calendar of, 54
Seburada, 395, 408, 409, 410
Sebuwana, 203
Setagava (a weed), 220
Seuse'u, 421, 422
Sexual freedom, 202, 304
Sexual Life of Savages, cited, 119, 202, 203, 205, 219, 232, 238, 293, 322, 342, 345, 413, 417, 434, 444, 456, 469, 482
Shark skin and tooth, tools of, 72, 244
Shell ornaments, 64, 228, 290, 291
Sibugibogi (ritual present to magician), 65
Sickness and skin-disease, 163, 164
Siginubu, 304, 307, 313
Sikuna (bird-traps), 299
Silakutuva (taro pudding), 296
Silakwa system of magic, 277
Silasila, 427
Silisilata, 140, 146
Simdarise Wawa of M'tawa, 407, 417
Simsim I., 74
Simsimwaya. *See* Sweet Potatoes
Sina (black bird), 107
Sinaketa and Sinaketans, 117, 258, 291, 346; diet of natives in, 290; garden magician in, 92, 291; magic in, 291, 294
Siribwobwa'u spell (*kaydabala*), 145
Sisiye'i (bracken), 281
Sister: special claim by, 189; brother's responsibility for, 198, 202. *See also* Brothers
Sobula, 140
Soil, classification of, 76, 89
Sokwaypa, description of, 256
Songs (*Wosi*), 134
Sopu, term limited to *taytu* planting, 124–5, 132, 299. *See also* Planting
Sorcery: witchcraft (*bulubwalata*), 78; death attributed to, 18, 119; fear of, 243; attempts to compass death by, 181; garden misfortunes attributed to, 78, 119; famine ascribed to, 119
Sousula (gifts to magicians), 65, 421
Spirits: of the departed, return, 53, 59, 287; ancestral (*baloma*); house of, in gardens, 77; its construction, 115; offerings of food to, 63, 65, 66, 95, 109, 166, 278, 296; at *yowota*, 93, 171, 348; first fruits to, 166; *sagali*, an offering to, 287; feasting and expulsion of, 110; invocation of, 96; Momtilakayva magic regarding, 278–9; visits of the, 467–8
Spitting of ginger root, 72, 225, 235
Spondylus shells, 290
Stick, hooked, for fruit-picking, 311
Sticks (*gado'i*), 123, 311
Stile (*pisila, kalapisila*), 90, 99
Stone heaps, 121, 133, 135, 342
Stone quarries, 68–9
Stones: used in *pelaka'ukwa* rite, 117; in black magic, 118; sacred, on the *baku*, 235; foundation of store-houses (*kaylagila*), 247, 248, 254. *See also* Binabina
Store-house (*bwayma*): kinds of, 255, 256; taboos on, 258; situation of, 218; the function of, 228, *et seq.*, small boys', 232; structure of smaller, 255, *et seq.*, floor, the least carefully constructed part of, 250; ventilation of, 241, 242, 251; magical consecration of, 219; function of the, 288, *et seq.*, *bomaliku*, 257; domestic: their importance, 242; compared with show ones, 82; situation, 242; privacy, 232; types of, 219, 228, 232; boys', 257; platforms of, 226, 232, 241, 246, 247, 256, 258; summary of facts regarding, 240, *et seq.*; structural and economic characteristics of: Linguistic terminology, 257, *et seq.*; show: construction of:

proportions, 228, 229, 242, 243, 245, 247, 259; only *vilamalia* magic at, 217, 247; no platforms to, 231, 247; characteristic of, 231; diagrams of, 263–272; materials used for, 243, 244; tools required, 244; platforms used in constructing, 247; three main parts, 246; foundations, 241, 243, 247, 248; log-cabin, 228, 248, 249, *et seq.*; construction of its compartments, 240, 243, 245; roof, 242, 244, 246, 252, *et seq.*; thatching, 254, repairing, 211, 245, 246; decoration of, 228; painting and carving, 228, 244, 245; decoration of *baku* by, 229, 258; filling of (*dodige bwayma*), 82, 87, 180 184, 188, 189, 193, 195, 196, 210, 219, 222, 229, 241, 246, 251, 258, 410; persons engaged in, 196, 246; method of, 222; time required, 247; ceremonial at, 152, 193, 195, 210, 228; storing of *taytu*: the *urigubu*, 189; the *taytumwala*, 230; no additions to, save in coastal villages, 231; tubers never returned to, 231; in the *yalumugwa*, 385–387; situation of, 218, *et seq.*; "untying of," 189; summary of facts regarding, 240, 241

String and Rope, 312

Sub-clans, 84, 85, 86; rank vested in, 332; as 'masters of the soil,' 329; displacement of, by those of higher rank, 369; rights of, 74; *dala*, 385; structure of, 345, 346; status and power of the head of, 363; several to one village, 342, 346

Sugar-cane, 57, 58, 88, 110, 126, 177, 184, 187, 193, 293, 315

Suloga, 68

Sulumwoya (aromatic herbs), 72, 312; leaves in magic, 237

Sunini (penetration), 93

Supi, 423

Suvayalu, 407

Suviyagila, 277, 346

Swamps (*dumya*, *podidiweta*), 161; in the South, 290, 291; products of, 291, 312

Sweet potato (*simsimwaya*), 58, 161, 180, 181; fodder for pigs, 181

Tabalu, 84, 328, 366, 368, 431

Tabalu chief, 84, 328

Tabalu sub-clan, 85, 276, 293, 294, 336–337, 342, 347, 365, 368

Taboos, 106, 119, 135, 166, 428; on chiefs, 106; on cooking, 232; brother-sister, 351; on *bwayma*, 225; imposed by garden magician, 62, 63, 66, 77, 164; at *kamkokola* erection, 123; on magician as to special foods, 66, 106–7, 165; against *bwanawa*, 258; hardship of, in *molu*, 164; ceremonial lifting of, 164, 167, 171; by magician in *mom'la*, 294; on palms, 301, 304, 305; on coconuts, 301, 302; on magician, 167, 276, 307; *kaboma*, 111. *And see* Coconuts, *Kaytubutabu*

Tabubula (kind of tribute), 192

Tagwala, 327

Takaywa, 93, 134, 286, 422. *And see* Scrub-cutting

Takola (repayment gift), 190, 372

Takwalela, 372

Talala rite, 101, 422, 428

Talola silisilata spell, 151

Tamgogula, 157, 158

Tamna, 140

Tamwana (litany of spell), 255, 306

Tapopu (taro gardens), 58, 295, *et seq.*; in the South, 295–6

Taro: magic in taytu gardening, 469–470; gardens of (*tapopu*), 295, *et seq.*; soils for, 296; no special magic for, 297, 298, 299; harvesting of, 296, 297, 299; perishable nature of, 296; pudding of (*mona*), 99, 155, 156, 177; cultivation of, in the South, 296; perishability of, 81, 296; pudding and porridge of (*silakutuwa* and *uripopula*), 296; contrasted with yams and taytu, 296

Tasasali rite, 294, 424

Tata'i tageguda, 424

Tata'i tamatuwo, 424

Ta'ukuwakula (scream), 103. *And see* Cries

Ta'ula, 158

Ta'uya, *Tauyo* (shell trumpet). *See* Trumpet

Tavakanisi yena, 424

Tavile'i, 157

Tavisi sobula, 142
Tawa'i, 407
Tayoyuwa, 53, 165, 167, 171, 230, 233
Taytu: the staple crop in the North, 137, 291; various forms of, 173; *lupilakum*, 150; planting of, 132, 134, 140, 141 (*and see* Planting); stages of its growth, 138, 140, 149; foliage of, 137; spells for foliage, 146; spells for closing up (canopy), 147, 148; training the vines, 137; harvest of (*tayoyuwa*), 165; cleaning tubers of, 172; inferior (*unasu*), 173, 193, 230; seed (*yagogu*), 173, 194; storing of, 160; observations on growth of, 471. *See also* Store-house, filling of
Taytukulu (displayed taytu), 173
Taytumwala (own grown for own consumption), 193, 194, 208, 229, 251, 258, 413; difficulties of computing, 230
Taytumwaydona, 196, 208, 295
Taytupeta (gifts), 189, 223, 413, 414, 415; Vakutan custom with, 295. *See also* gifts
Taytuwala (*urigubu* crops alone so-called), 171, 194, 411
Tepila (home of the pigs), 117, 118
Teta, 251, 252, 255
Tewara, 307
Teyava (village), 170, 242, 250, 253; magic of, 294; Garden magic in, 427–429
Thinning out of tubers (the *basi*), 61, 62, 107, 138, 151, 152, 164, 165, 230, 429; inaugural rite for, 152, 288
Tilakayva, 210, 216, 245, 277, 278, 346
Tilataula, 162, 182, 277, 369; ruler of, 328
Tilaykiki (yell), 95, 102
Time: reckoning of, by crops, 52, 53; sequences of years, 64
Tobacco, payment in, 85, 176
Tobiyumi, 386
Toboyowa, 408
Tobuguya'u, 386
Tobutusa'u, 413
Tobwabodila (acolytes), 288
Togabutuma, 387
Tokabitam (master carver), 244, 246
Tokaraywaga, 386, 417
Tokavataria, 386
Tokay, 385
Tokolibeba, 413
Tokubukwabuya (bastards), 202
Tokulubakiki, 163, 215, 306, 413, 414, 415; his banana magic, 303, 313
Tokunasa'i of Kaytagava, 213, 215
Tokuwabu, 150
Tokwamnapolu, 407
Tokway, 375
Tokwaybagula (the perfect gardener), 61, 62, 67, 83, 86, 91, 126
Toli: right to use of, in title, 344, 364; *pokala* acquisition of, 345
Toliboma, 344
Tolikwabila, 346
Tolipwaypwaya, 328, 344, 346, 350
Tolitaytu, 214
Tolivalu, 344, 350
Toliwaga (sub-chief), 192
Toliyavata, moon of, 136, 149, 160; early taro crops harvested in, 296
Tomako'u, 417
Tomnavidala, 413
Topinata'u, 386
Torches: charming of, 111, 112, 116, 170; charmed, 113, 281, 298
To'udawada, 92
To'ulatile, 202
To'uluwa: Paramount Chief, 84, 111, 175, 176, 214, 367, 393, 414, 415, 417, 418; at the *kayaku*, 87, 88; at the *yowola*, 93; his wives, 191; his *urigubu*, 217; harvest gifts to, in 1918, 392, *et seq.*; rebuilding of store-house, 211
Tourapata, 408
Tourikuna magic (of paramount chief), 83
Tovakakita, 176, 417, 418
Tovilamalia (*towosi* so named in *Vilamalia*), 67, 77, 88, 219, 223, 233, 235, 237
Toviyama'i, 407
Towese'i, assistant to Bagido'u, 85, 111, 127
Towosi, 67, 77, 88, 281, 283, 285. *And see Tovilamalia*
Toyoyuwa, 165
Trade winds, native belief as to, 54

Trumpet shell, 179, 214, 215, 233, 235, 307; spell, 235
Tubowada, 277
Tubukona molu, 160
Tubuloma, 88, 89, 433
Tubwalasi, 423
Tubwoysewaga, 394, 407
Tudava, 70, 71, 72, 73, 74; legend of, 343; myth of, 68, 69, 74
Tuginitu, 386
Tukwalapi, 386
Tukwa'ukwa, 277, 302, 342, 366, 413, 429
Tula (boundary of sticks), 102, 121, 122, 124, 128, 153, 154, 423; interconnection of, with *karivisi* and *kamkola*, 131
Tukulumwala (boundary), 133
Tum ceremony, 168, 170, 171, 424, 429, 473; in Sinaketa, 295
Tum bubukwa spell, 226, 250
Tuma, 115
Tutuya, 253
Tu'utauna, 307
Tuvata'u leaves in magic, 237
Tuwaga (stone heaps), 121, 133, 135

Ubwana, 304
Ubwara, leaves in magic, 106; tubers, 106, 107
Ulapa yagogu, 423
Ula'ula (food offering to spirits), 65, 279, 424
Ula'ula walaka, 429
Ulumdala (gleaned taytu), 173, 180, 193, 194, 230
Unasu. *See* Taytu: inferior
Unu'unu (hair), 172
Urigubu: (harvest gifts), 91, 172, 176, 191, 195, 196; strict and loose usages, 193, 196; extension of its meaning, 189, 193, 474, 475; connotation in the south, 196; harvest gifts to kinsfolk, 189; chief's tribute, 191, 192, 210; certain garden plots, 91; principles of, 210 *et seq.*; distinction of crops so called, 194; best *taytu* allotted to, 189, 194; givers of, 189; festivities and proceedings, 159, 160; transporting of crops to sister's husband, 91, 189, 190, 295; *vilakuria* the first gift of, 199; principle of, the core of social and economic life, 189, 190, 199, 208, 209; gifts in return for, 190; explanation of, 196, 208; significance of, 209; central feature of marriage contract, 193, 290, 328; importance of, 209; chiefs' compared with commoners', 191; typical examples of, in 1918, 211; harvest gifts to kinsfolk, 196; representative examples of, 415–418; explanation of, 207, *et seq.*; land-tenure aspect of, 353, 354, 359, 369; various categories of, 393–397; after the *kayasa* in Omarakana, in 1918, 392–397
Ulilaguva, 247, 254
Urinagula (fire-place stones), 237
Uripopula (taro porridge), 296
Usigola, 302
Utokakana, moon of, 171, 302
Utukwaki (*youmwegina* seeds), 311

Vaboda kadumalaga magic, 237
Vaboda kaukweda magic, 237
Vadila (pandanus fruit), 312
Vaguri sobula spell, 141, 142, 423
Vakolova rite, 131
Vakam taytu rite, 171
Vakapula vakapwasi (cooked food payment), 176
Vakavayla'u. *See* Scrub-burning
Vakuta, 54, 117, 190, 211; *rayboag* in, 57, 74; legend in, 70, 74, 108; date of spirit visits to, 54; list of *kwabila* in, 420; Kiriwinian marriages in, 291; magic of, 294, 295, 420–426
Vakwari, 423
Valuables, gifts of, 65, 72, 149, 189, 190, 246; return gift of, 199; exchange of, for seed yams, 163
Vapopula digadaga, 109
Vapuri, 149, 249, 424
Vapwanini yotata rite, 299, 424
Vasakapu soluba, 143, 423
Vataga (oblong baskets). *See* Baskets
Vataulo, 255
Vatuvi spell, 96, 111, 113, 127, 130, 152, 155, 170
Vava. *See* Exchange
Vayaulo (acacia), use for wood of, 312; leaves, 112

Vaygua, 295, 372, 373
Vayla'u (theft of food), 232, 374
Vayoulo, 112
Vewoulo, 193, 231, 372
Veyola, 199, 386, 387, 417
Vikita, 101
Vilakuria (first instalment of *urigubu*), 199. *And see Urigubu*
Vilamalia: importance of, 82, 83, 219, 226, 227, 237; the rite, 223, 233, 476; object and function of, 226, *et seq.*; carry-over idea connected with, 239; spoken over *urigubu* crops (in north), 194; at building of a store-house, 217; parts of *bwayma* mentioned in spell, 224; in Oburaku, 233, 291; in times of hunger and sickness, 163, 237; mentioned, 83, 116
Vilaylima, 313, 365, 408, 409; a community of rank, 410
Village: central place. *See Baku*
Villages: system of magic for each, 62; sub-clans in, 329, 331, 332, 346, 347, 355
Vilomugwa (low-rank vassals), 86
Vinavina (chants), 136. *See also Sawili*
Vise'u, 85
Vivi nuts, 311
Viyoulo, 415

Wabu (marks on corpse), 175
Wabusa, 92
Wagwam, 347
Wakaya (large banana), 106, 107; leaves, 106, 170, 312
Wakayluva, 277, 432
Wakayse, 186, 187; *buritalaulo* in, 182–3; magic of, 277
Wakaysi, 210
Wallabies, 67; bone of, for carving, 244; guard against, 120
Wamea, 307
Wamwara, 71
Wanuma, 69
Waribu, 90, 433
Wasi. *See* Exchange
Water-holes, 163; bathing in, 99, 177; rights as to, 432; special (holes of emergence), 342
Wawela, 290, 316
Waya (inlet), 342
Waywo. *See* Mango
Weather. *See* Drought, and Rain
Weeding (*pwakova, sapi*), 53, 144, 145; inaugural ceremony, 149, 428; spell for, 288; accounted women's work, 61, 79, 123, 138, 144, 145; communal, by women: the *yausa*, 122, 138, 144, 158; Sinaketan rites of, 294
Weyana, 415
Weyka (village grove), 286, 304
Wind-rattles (*kaygogwa'u*), 299
Witchcraft. *See* Sorcery
Wokubila leaves in magic, 280, 286
Wokunu, 91
Women: pre-marital sexual freedom of, 202, 304; economic position of, 353. *See also* Sister, Weeding, Marriage
Woodlark Is. (Murwa or Muyuwa); legends in, 68–9, 70, 73, 74, 453
Woydagobu, 433

Yabuna, 386
Yagogu, 134, 140, 153, 154, 173, 193, 238, 284, 423. *And see* Yams
Yakala, 103
Yakoki, moon of (August), 87, 302, 311, 314
Yakosi, moon of, 110, 314
Yalaka, 355
Yalumugwa, 386; *urigubu* in, 179, 210, 418; store-houses of, 222, 249, 385; magic systems of, 277; component villages of, 346
Yam-house. *See* Store-house
Yams: making of supports for, 137; native names for varieties of, 76, 153; *Kwanada*, 171; Kwibanena, 183. *See also Kuvi* and *Taytu*
Yaulo (gifts), 231
Yausa, custom of, 144
Yavatakulu, moon of, 136, 149, 311, 314
Yavatam (moon), 149, 311
Yaya'i, 179
Yayu (casuarina tree), 129, 304, 427
Yells, 125. *See also* Cries, and Harvest
Yeye'i, 137
Yoba (driving away of the spirits), 110
Yobilabala (horizontals in lattice), 254
Yobukwa'u, 85
Yobunatolu, 424

Yogaru, 175
Yogwabu, 414, 431
Yokunukwanada, 106, 427
Yokwa'oma, 106
Yola'ula, 427
Yolukula, 423
Yonakui tree, 304
Yosewo (uncut jungle), 100, 117
Youla'ula, 106, 421
Youlo, 190
Youlumwala, 129
Youmwegina fruit, 311
Yo'uribwala, 423
Yourawotu, 176, 210, 277, 278, 284, 286, 346, 418
Yovisi, 386, 418
Yowana, 85, 362, 371
Yowota, 422; two forms of, 93, 94, 102, 107; insertion of *si gado'i baloma* at, in Oburaku, 292; not on the *tapopu* in Sinaketa, 298
Yoyu (coconut leaves), 105. *And see* Coconut
Yuvata'u, 387